Harald Ibach

# Physics of Surfaces and Interfaces

Harald Ibach

# Physics of Surfaces and Interfaces

With 350 Figures

 Springer

Professor Dr. Harald Ibach
Forschungszentrum Jülich GmbH
Institut für Bio- und Nanosysteme (IBN3)
Wilhelm-Johnen-Straße
52425 Jülich
Germany
e-mail: h.ibach@fz-juelich.de

ISBN 978-3-642-07107-2        e-ISBN 978-3-540-34710-1

Springer is a part of Springer Science+Business Media

springer.com

© Springer-Verlag Berlin Heidelberg 2010

Cover Design: eStudioCalamarS.L., F. Steinen-Broo, Pau/Girona, Spanien

# Preface

Writing a textbook is an undertaking that requires strong motivation, strong enough to carry out almost two years of solid work in this case. My motivation arose from three sources. The first was the ever-increasing pressure of our German administration on research institutions and individuals to divert time and attention from the pursuit of research into achieving politically determined five-year plans and milestones. The challenge of writing a textbook helped me to maintain my integrity as a scientist and served as an escape.

A second source of motivation lay in my attempt to understand transport processes at the solid/electrolyte interface within the framework of concepts developed for solid surfaces in vacuum. These concepts provide logical connections between the properties of single atoms and large ensembles of atoms by describing the physics on an ever-coarser mesh. The transfer to the solid/electrolyte interface proved nontrivial, the greatest obstacle being that terms such as *surface tension* denote different quantities in surface physics and electrochemistry. Furthermore, I came to realize that not infrequently identical quantities and concepts carry different names in the two disciplines. I felt challenged by the task of bringing the two worlds together. Thus a distinct feature of this volume is that, wherever appropriate, it treats surfaces in vacuum and in an electrolyte side-by-side.

The final motivation unfolded during the course of the work itself. After 40 years of research, I found it relaxing and intellectually rewarding to sit back, think thoroughly about the basics and cast those thoughts into the form of a tutorial text.

In keeping with my own likings, this volume covers everything from experimental methods and technical tricks of the trade to what, at times, are rather sophisticated theoretical considerations. Thus, while some parts make for easy reading, others may require a more in-depth study, depending on the reader. I have tried to be as tutorial as possible even in the theoretical parts and have sacrificed rigorousness for clarity by introducing illustrative shortcuts.

The experimental examples, for convenience, are drawn largely from the store of knowledge available in our group in Jülich. Compiling these entailed some nostalgia as well as the satisfaction of preserving expertise that has been acquired over three decades of research.

I pondered long and hard about the order of the presentation. The necessarily linear arrangement of the material in a textbook is intrinsically unsuitable for describing a field in which everything seems to be connected to everything else. I finally settled for a fairly conventional sequence. To draw attention to relationships between different topics the linear style of presentation is supplemented by cross-references to earlier and later sections.

Despite the length of the text and the many topics covered, it is alarming to note what had to be left out: the important and fashionable field of adhesion and friction; catalytic and electrochemical reactions at surfaces; liquid interfaces; much about solid/solid interfaces; alloy, polymer, oxide and other insulator surfaces; and the new world of switchable organic molecules at solid surfaces, to name just a few of a seemingly endless list.

This volume could not have been written without the help of many colleagues. Above all, I would like to thank Margret Giesen for introducing me to the field of surface transport and growth, both at the solid/vacuum and the solid/electrolyte interface. This book would not exist without the inspiration I received from the beautiful experiments of hers and her group and the almost daily discussions with her. I should also be grateful for the patience she exercised as my wife during the two years I spent writing this book.

Jorge Müller went through the ordeal of scrutinizing the text for misprints, the equations for errors, and the text for misconceptions or misleading phrases. I also express my appreciation for the many enlightening discussions of physics during the long years of our collaboration.

I greatly enjoyed the hospitality of my colleagues at the University of California Irvine during my sabbatical in Spring 2005 where four chapters of this volume were written. On that occasion I also enjoyed many discussions with Douglas L. Mills on thin film magnetism and magnetic excitation, the fruits of which went into the chapter on magnetism. In addition, the chapter on surface vibrations benefited immensely from our earlier collaboration on that topic.

Of the many other colleagues who helped me to understand the physics of interfaces, I would like to single out Ted L. Einstein and Wolfgang Schmickler. Ted Einstein initiated me in the statistical thermodynamics of surfaces. Several parts of this volume draw directly on experience acquired during our collaboration. Wolfgang Schmickler wrote the only textbook on electrochemistry that I was ever able to understand. The thermodynamics of the solid/electrolyte interface as outlined in chapter 4 of this volume evolved from our collaboration on this topic.

With Georgi Staikov I had fruitful discussions on nucleation theory and various aspects of electrochemical phase formation which helped to formulate the chapter on nucleation and growth. Guillermo Beltramo contributed helpful discussions as well as several graphs on electrochemistry. Hans-Peter Oepen and Michaela Hartmann read and commented the chapters on magnetism and electronic properties. Rudolf David contributed to the section on He-scattering. Claudia Steufmehl made some sophisticated drawings. In drawing the structures of surface, I made good use of the NIST database 42 [1.1] and the various features of the package.

Last but not least I thank the many nameless students who attended my lectures on surface physics over the years. Their attentive listening and the awkward questions it led to were indispensable for formulating the concepts described in this book. Finally, I beg forgiveness from my colleagues in Jülich for having been a negligent institute director lately.

Jülich, May 2006                                    *Harald   Ibach*

# Contents

# 1. Structure of Surfaces

Surface Physics and Chemistry flourished long before anything was known about the atomic structure of surfaces. Chemical, optical, electrical and even magnetic properties were investigated systematically, sometimes in great detail and not without lasting success. The concept of an ideally terminated bulk structure with its assumed physical properties frequently served as a base for the rationalization of the experimental results. Examples are the postulation of specific electric properties that would arise from the broken bonds at surfaces of semiconductors and the high chemical activity that might be associated with defects on the surface. Quantitative understanding on an atomic level could not be achieved however without knowledge the crystallographic structure of surfaces. Vice versa, a tutorial presentation of our present understanding of the physics of surfaces and interfaces requires the fundament of facts, concepts and the nomenclature that has evolved from the analysis of surface structures. The first chapter of this treatise is therefore devoted to the structure of clean and adsorbate covered surfaces, the important defects at surfaces and the structural elements of the solid/electrolyte interface.

As for Solid State Physics in general, the quantitative understanding on an atomic level greatly benefits from the periodic structure of crystalline matter since the periodicity reduces the electronic and nuclear degrees of freedom from $10^{23}$ per $cm^3$ to the degrees of freedom in a single unit cell. However, at surfaces the reduction in the degrees of freedom by periodicity is less, as the three-dimensional symmetry is broken. Near surfaces, material properties may differ from the bulk in several monolayers below the surface. The surface unit cell of periodicity therefore necessarily contains more atoms than the corresponding unit cell of the bulk structure. Not infrequently, the unit cell of a real surface is substantially larger than the surface unit cell of a terminated bulk, which increases the number of atoms in the surface unit cell further. For example, the surface cell of the clean (111) surface of silicon contains 49 atoms in one atom layer and the restructuring involves 4-5 atom layers! Solving a bulk structure with that many atoms per unit cell is not an easy, but nowadays tractable problem, but structure analysis at surfaces has to be performed in the presence of the entire bulk below the surface. It is still one of the greatest successes of surface science that after decades of research and literally thousands of papers the structure of the Si(111) surface was eventually solved.

Substantial advances in surface crystallography are owed to the experimental and theoretical achievements in *Low Energy Electron Diffraction* (LEED) and *Surface X-Ray Diffraction* (SXRD). *Scanning Tunneling Microscopy* (STM) and other scanning microprobes contributed by providing qualitative images of surfaces, which reduced the number of possibilities for surface structure models. Presently, the structures of more than 1000 surface systems are documented, and the number keeps growing [1.1].

# 1.1 Surface Crystallography

## 1.1.1 Diffraction at Surfaces

The first section of this volume is devoted to the essential elements of surface crystallography: Laue-equations, Ewald-construction, and symmetry elements.

Elastic scattering of X-rays or particle waves from infinitely extended three-dimensional periodic structures undergoes destructive interference, which leaves scattered intensity only in particular directions. The conditions under which diffracted intensity can be observed are described by the three Laue-equations, which can be expressed in terms of a single vector equation

$$k - k_0 = G \,, \tag{1.1}$$

in which $k$ and $k_0$ are the wave vector of the scattered and incident wave, respectively, and $G$ is an arbitrary vector of the reciprocal space. At the surface, the bulk periodicity is truncated and the three Laue-equations reduce to two equations concerning the components of the incident and scattered wave vectors parallel to the surface.

$$k_\parallel - k_{0\parallel} = G_\parallel \tag{1.2}$$

$G_\parallel$ is a vector of the reciprocal lattice of the two-dimensional unit cell at the surface. Diffracted beams are therefore indexed by two Miller-indices $(h,k)$. The reduction to two Laue-equations has the consequence that scattering from a surface lattice leads to diffracted beams for <u>all</u> incident $k_0$, unlike for bulk scattering where diffracted beams occur only for particular wave vectors of the incident beam. As for the bulk, the Laue-condition is best illustrated with the Ewald-construction. Figure 1.1 shows the Ewald-construction as it is typical for LEED: A beam of low energy electrons (energy $E_0$ between 20 and 500 eV, corresponding to a wave vector $k_0 = 5.12 \, \mathrm{nm}^{-1} \sqrt{E_0 / \mathrm{eV}}$ ) with normal incidence is diffracted from the surface lattice. Depending on the energy, the {01}, {11}, {02}... beams are observed in the backscattering direction, providing direct information on the surface reciprocal lattice.

Early experiments used a Faraday cup for probing the diffracted beams [1.2]. More convenient is the experimental set-up introduced by Lander et al. [1.3], which is displayed in Fig. 1.2. The equipment was primarily designed for a qualitative quick overview on the diffraction pattern.

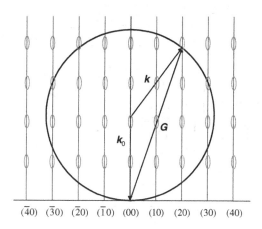

$(\bar{4}0)$  $(\bar{3}0)$  $(\bar{2}0)$  $(\bar{1}0)$  $(00)$  $(10)$  $(20)$  $(30)$  $(40)$

**Fig. 1.1.** Ewald-construction for surface scattering. The magnitude and orientation of $k_0$ (normal incidence) is representative of a LEED-experiment. Diffracted beams occur if the wave vector of the scattered electron ends on one of the vertical rods (*crystal truncation rods*) representing the reciprocal lattice of the surface. Diffracted electrons are therefore observed for all energies of the incident beam: The scattering intensity is particular large if the third Laue-condition concerning the perpendicular component of the scattering vector (indicated by ellipsoids) is approximately met.

Later the same equipment has been used also for the quantitative analysis of diffracted intensities by monitoring the spots on the screen with the help of a video camera and specially developed image processing software (Video-LEED).

Like all other experiments using low energy electrons, LEED gains its surface sensitivity from the relative large cross section for inelastic scattering. The prime source of inelastic scattering is the interaction with collective excitations of the valence electrons electron, the plasmons (Sect. 2.2.2, 8.1). The mean free path of electrons in the relevant range is of the order of 1 nm. All elastically backscattered electrons therefore stem from the first few monolayers of the crystal. This is the reason that intensity is observed even for energies for which the third Laue equation for the vertical component of the scattering vector $K = k_0 - k$ is not fulfilled. The few monolayers, from which the diffraction originates, however, suffice to impose a weak Laue-condition on the vertical component of the scattering vector $K$. In Fig. 1.1 this weak Laue condition is indicated by the ellipsoids. Figure 1.3 displays the measured diffracted intensity of the (10) beam from a Cu(100) surface [1.4] together with the position of the expected intensity maxima according to the third Laue-condition. The experimental intensity curve indeed displays pronounced maxima, but only very roughly where they are expected from single scattering (*kinematic scattering*) theory. Surely, the complexity of the various features in the intensity curve cannot be explained based on single scattering

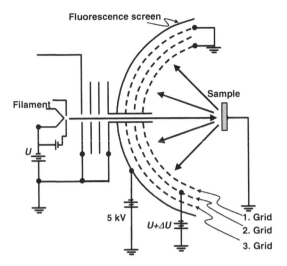

**Fig. 1.2.** Instrument for low electron energy diffraction. Diffracted electrons are observed on a fluorescent screen. The grids serve for various purposes. Grid 1 establishes a field free region around the sample, grid 2 repels inelastically scattered electrons so that they cannot reach the screen, grid 3 prevents the punch-through of the high voltage applied to the screen to the field at grid 2.

events. Multiple elastic scattering of the electron has to be taken into account (*dynamic scattering theory*). The difficulty to describe multiple elastic scattering of electrons theoretically has been a major impediment in the development of surface crystallography. As Fig. 1.3 demonstrates [1.4-6], theory is now able to describe the observed intensities quite well. A quantitative structure analysis is performed by proposing a model for the structure and by comparing experimental and theoretical LEED intensities as a function of the atom position parameters (*trial and error* method). Comparison of theory and experiment is quantified in the Pendry R-factor $R_p$ which is defined on the basis of the logarithmic derivative of the intensities $I$ with respect to the electron energy $E_0$.

$$R_p = \frac{\sum (Y_{\text{theory}} - Y_{\text{exp}})^2}{\sum (Y_{\text{theory}} + Y_{\text{exp}})^2}$$

$$Y = \frac{I_{\log}}{1 + (I_{\log} V_{0i})^2}$$

(1.3)

with

$$I_{\log} = \frac{\partial I}{I \, \partial E_0}$$

(1.4)

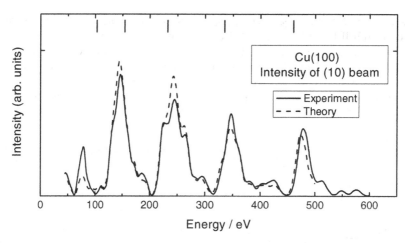

**Fig. 1.3.** Intensity of the (10) beam diffracted from a Cu(100) surface vs. beam energy for normal incidence. Experiment and theory are plotted as solid and dashed curves, respectively. The positions of the maxima according to the simple single scattering theory are indicated as vertical bars.

Here, $V_{0i}$ is the imaginary part of the inner potential (approximately the width of the intensity peaks on the energy scale) and the sum is over all energies and diffracted beams. The agreement between theory and experiment in Fig. 1.3 corresponds to an $R_p$-factor of 0.08. In general, $R_p$-factors below 0.20 are considered as good.

Compared to LEED, X-ray scattering has the definite advantage that X-rays are scattered only once. The scattering amplitude is therefore the Fourier-transform of the scattering density [1.7] and intensities are easily calculated for any given structure. Schemes for direct structure determination via the Patterson function can be employed. Surface sensitivity is achieved by working under condition of grazing incidence. Since the photon energy is well above all electronic excitations the complex refraction index $\tilde{n}$ for X-rays is described by the dielectric properties of the free electron gas in the high frequency limit. The real part of $\tilde{n}$ is therefore smaller than one. Total reflection of the X-ray beam occurs at grazing incidence if the angle between the beam and the surface plane $\alpha_i$ is smaller than a critical angle $\alpha_c$. Typical values for $\alpha_c$ are between 0.2° and 0.6° for an X-ray wavelength of 0.15 nm [1.8]. Under condition of total reflection the X-ray intensity inside the solid drops exponentially with a decay length $\Lambda$ of about 10 nm. All diffraction information therefore concerns no more than about 50 atom layers. Information of just the surface layer is contained in diffracted beams of a surface superlattice. The intensity of such beams is sufficiently large for detection and stands out from the diffuse background. The technique is called **Grazing Incidence X-Ray Diffraction** (GIXRD). Figure 1.4 shows the *structure factor* (the modulus of the scattering amplitude as due to the structure) as a function of the perpendicular component of the scattering vector [1.9] (a) for a bare Cu(110) surface and (b) for a Cu(110) surface covered with oxygen. The parallel component of the scattering vector is chosen to fulfill the (01) surface diffraction condition. The full line is calculated using the structural parameters, which gave the best fit to <u>all</u> measured structure factors (about 150). Note that comparison between experiment and theory is made for the intensity outside the L=1 peak that results from the third Laue condition. The intensity in that peak contains mostly information about the structure of the bulk inside the decay length $\Lambda$.

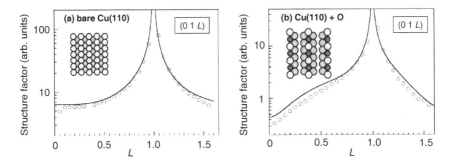

**Fig. 1.4.** Structure factor along the (01) *crystal truncation rod* as a function of the vertical component of the scattering vector $L$ expressed in units of the reciprocal lattice vector **(a)** for a bare Cu(110) surface and **(b)** for a Cu(110) surface covered with oxygen [1.9]. The insets display a top view on the first two layers of surface atoms (see also Sect. 3.4.3). The structure with oxygen is the so-called added row structure where every second row is formed by a chain of oxygen atoms (dark circles) and Cu-atoms. Experimental data and theory for the optimized geometry data are shown as circles and solid lines, respectively.

The applicability of single scattering theory also provides the possibility to use the elastic diffuse X-ray intensity for an analysis of non-periodic features on surfaces, such as defects or strain fields associated with domains of adsorbates [1.9, 10]. Furthermore, vacuum is not required, which makes X-ray scattering a technique suitable also for studies on the solid/liquid interface [1.11] if the liquid layer is thin enough.

The question which of the two methods LEED or SXRD is the method of choice depends on circumstances. In principle, both methods can provide equally precise atom positions for a large number of atoms per unit cell. The scattering cross section for X-rays scales with the square of the atom number Z. Light elements contribute little to X-ray scattering and data are not sensitive to the position of light element. LEED does not suffer from that to the same extent. Because of the larger momentum transfer in the direction of the surface normal, LEED has a better sensitivity to the vertical atom coordinates, while SXRD is more sensitive to the lateral position. X-ray scattering experiments require extremely flat surfaces because of the grazing incidence condition while LEED is more forgiving with respect to sample quality. At present, most of the surface structure determinations are based on the quantitative analysis of LEED-intensities. However, the balance may tip as improved synchrotron sources become more available.

### 1.1.2 Surface Superlattices

#### *Notation*

The positions of surface atoms differ from the bulk because of the broken symmetry and the broken bonds. The modifications are referred to as *relaxations* if the surface unit cell remains that of the truncated bulk. If the surface unit cell is different, then the corresponding changes in the structures are addressed as *reconstructions*. The lattice of an adsorbed phase with a unit cell larger than the surface cell of the truncated bulk is called a *superlattice*, the associated structure a *superstructure*. Adsorbate superstructures frequently go along with a reconstruction of the substrate. The nomenclature therefore is not unambiguous.

Base vectors of the unit cell of superstructures and surface reconstructions are expressed in terms of the base vectors of the unit cell of the truncated bulk. With $s_1$ and $s_2$ as vectors spanning the surface unit cell of a truncated bulk lattice, the lattice vectors of the actual unit cell on the surface, $a_1$ and $a_2$, are described by the matrix $t$

$$\begin{pmatrix} \mathbf{a}_1 \\ \mathbf{a}_2 \end{pmatrix} = \begin{pmatrix} t_{11} & t_{12} \\ t_{21} & t_{22} \end{pmatrix} \begin{pmatrix} \mathbf{s}_1 \\ \mathbf{s}_2 \end{pmatrix} \qquad (1.5)$$

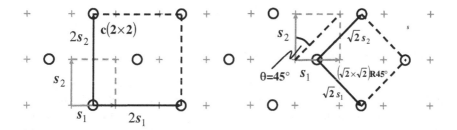

**Fig. 1.5.** Illustration of the notation of the c(2×2) unit cell of the surface lattice and its alternative notation as $\sqrt{2}\times\sqrt{2}$ R45°.

In most cases, this unambiguous notation introduced by E. A. Wood in 1964 [1.12] is unnecessarily complicated and inconvenient. If the surface lattice vectors are just multiples of the lattice vectors $s_1$ and $s_2$ unit cells are denoted as (2×1), (2×2), (3×1), etc. Centered and primitive unit cells are denoted by adding a "c" and a "p" to the notation, e.g. p(2×2) and c(2×2). This type of notation is not always unique: The c(2×2) lattice on a (100) surface of a cubic crystal can also be noted as $\sqrt{2} \times \sqrt{2}$ R45° in which the R45° stand for a rotation by 45° (Fig. 1.5). The unambiguous matrix notation $\begin{pmatrix} 1 & 1 \\ 1 & -1 \end{pmatrix}$ is rarely used in that case, as it is more difficult to quote.

A few common adsorbate superlattices are displayed in Fig. 1.6 together with their notation.

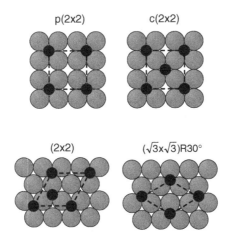

**Fig. 1.6.** Typical adsorbate superlattices on surfaces together with their trivial notation. Substrate and adsorbate atoms are displayed as black and grey, respectively.

### Diffraction pattern of superlattices

The existence of a superlattice on a surface is most easily discovered in a diffraction experiment because the larger unit cell produces extra, fractional order spots in the diffraction pattern between the normal $(hk)$ spots of the truncated bulk lattice. The determination of the base vectors of the unit cell frequently requires the consideration of domains. For example, the diffraction pattern of a (1×2) unit cell on a (111) or (100) surface of a cubic material has half order spots in terms of the Miller-indices of the substrate at $(h \pm 1/2)$, $(h \pm 3/2)$, etc.. The equivalent second (2×1) domain, which is rotated by 90°, has spots at $(1/2\ k)$, $(3/2\ k)$, etc. (Fig. 1.7). The pattern is distinct from the pattern of a (2×2) lattice since the latter would produce reflexes also at $(\pm 1/2, \pm 1/2)$, $(\pm 3/2, \pm 3/2)$, etc., which are absent in the diffraction pattern of the (1×2), (2×1) superlattice (Fig. 1.7).

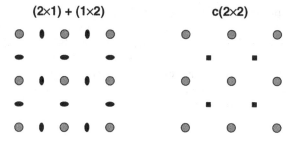

**Fig.1.7.** Diffraction pattern of two domains of a (1×2) superlattice and a c(2×2) superlattice on a (100) surface of a cubic material.

Centered unit cells and unit cell containing glide planes can be identified because they give rise to systematic extinctions. The extinctions of reflexes (h, k) are calculated from the surface structure factor $S_{hk}$.

$$S_{hk} = \sum_{\alpha} \exp\left[-i\pi(h'u_\alpha + k'v_\alpha)\right] \tag{1.6}$$

Here, $h'$ and $k'$ are the Miller-indices of the superlattice, $u_\alpha$ and $v_\alpha$ are the components of the vector $r_\alpha$ pointing to the atom $\alpha$ in the unit cells in terms of the base vectors $a_1$ and $a_2$.

$$r_\alpha = u_\alpha a_1 + v_\alpha a_2 \tag{1.7}$$

We consider the c(2×2) superlattice as an example. Because of the (2×2) lattice, the Miller-indices of the superlattice $h'$ and $k'$ in terms of the Miller-indices of the substrate lattice $h$ and $k$ are $h' = 2h$ and $k' = 2k$. The components $u_\alpha$ and $v_\alpha$ are $u_1 = v_1 = 0$ and $u_2 = v_2 = 1/2$. The structure factor is therefore

$$S_{hk} = 1 + (-1)^{2(h+k)} = \begin{cases} = 0 & \text{if} \quad 2(h+k) \quad \text{uneven} \\ = 2 & \text{if} \quad 2(h+k) \quad \text{even} \end{cases}. \tag{1.8}$$

The c(2×2) structure is therefore identified by characteristic extinctions in the half-order spot of the (2×2) lattice. In particular, these extinctions occur for all half-order spots along the $\langle h\,0 \rangle$ and $\langle 0\,k \rangle$-directions (Fig. 1.7).

### Point group symmetry of sites

A very important element of the surface structure is the symmetry of various sites on surfaces, important, because the local symmetry of an atom or molecular complex determines the classification of the eigenvalues of the electronic quantum states as well as the selection rules in spectroscopy. The fact that the surface plane is never a mirror plane reduces the number of possible point groups on surface to those, which have rotation axes and mirror planes perpendicular to the surface. These point groups are $C_s$, $C_{2v}$, $C_{3v}$, $C_{4v}$, $C_{6v}$, $C_3$, $C_4$, $C_6$. Figure 1.8 illustrates the most important point groups $C_s$, $C_{2v}$, $C_{3v}$, $C_{4v}$ together with the point groups $C_3$ and $C_4$. For the purpose of analyzing and classifying spectroscopic data, it is useful to have the character tables of the point groups at hand. Characters tables for $C_s$, $C_{2v}$, $C_{3v}$, $C_{4v}$, and $C_{6v}$ are listed in Table 1.1.

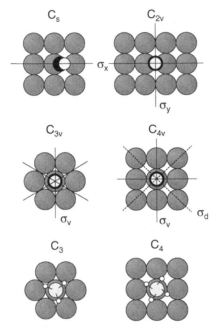

**Fig. 1.8.** The point groups $C_s$, $C_{2v}$, $C_{3v}$, $C_{4v}$, $C_3$, and $C_4$. The upper four point groups are illustrated with a diatomic molecule like CO or NO (black and gray circle). The species representing $C_3$ and $C_4$ are hypothetical. The point groups $C_s$, $C_{2v}$, $C_{3v}$, and $C_{4v}$ are frequently encountered with molecules like CO, NO, or $NH_3$.

**Table 1.1.** Character tables of surface point groups. The upper left corner notes the point group. The first column are the irreducible representations, the following columns are the characters of the classes of the group. The last column describes to which irreducible representation the translations along the $x$, $y$ and $z$-axes and the rotations around these axes belong. This is important since the translations and rotations of a molecule turn to vibrations when the molecule is adsorbed. (see Sect. 7.2.2).

| $C_s$ | I | $\sigma_{xz}$ | |
|---|---|---|---|
| $A'$ | +1 | +1 | $z, x, R_y$ |
| $A''$ | +1 | -1 | $y, R_x, R_z$ |

| $C_{2v}$ | I | $C_2$ | $\sigma_{xz}$ | $\sigma_{yz}$ | |
|---|---|---|---|---|---|
| $A_1$ | +1 | +1 | +1 | +1 | $z$ |
| $A_2$ | +1 | +1 | -1 | -1 | $R_z$ |
| $B_1$ | +1 | -1 | +1 | -1 | $x, R_y$ |
| $B_2$ | +1 | -1 | -1 | +1 | $y, R_x$ |

| $C_{3v}$ | I | $C_3$ | $\sigma$ | |
|---|---|---|---|---|
| $A_1$ | +1 | +1 | +1 | $z$ |
| $A_2$ | +1 | +1 | -1 | $R_z$ |
| $E$ | +2 | -1 | 0 | $x, y, R_x, R_y$ |

| $C_{4v}$ | I | $C_4$ | $C_4^2$ | $\sigma_v$ | $\sigma_d$ | |
|---|---|---|---|---|---|---|
| $A_1$ | +1 | +1 | +1 | +1 | +1 | $z$ |
| $A_2$ | +1 | +1 | +1 | -1 | -1 | $R_z$ |
| $B_1$ | +1 | -1 | +1 | +1 | -1 | |
| $B_2$ | +1 | -1 | +1 | -1 | +1 | |
| $E$ | +2 | 0 | -2 | 0 | 0 | $x, y, R_x, R_y$ |

| $C_{6v}$ | I | $C_6$ | $C_6^2$ | $C_6^3$ | $\sigma_v$ | $\sigma_d$ | |
|---|---|---|---|---|---|---|---|
| $A_1$ | +1 | +1 | +1 | +1 | +1 | +1 | $z$ |
| $A_2$ | +1 | +1 | +1 | +1 | -1 | -1 | $R_z$ |
| $B_1$ | +1 | -1 | +1 | -1 | +1 | -1 | |
| $B_2$ | +1 | -1 | +1 | -1 | -1 | +1 | |
| $E_1$ | +2 | +1 | -1 | -2 | 0 | 0 | $x, y, R_x, R_y$ |
| $E_2$ | +2 | -1 | -1 | +2 | 0 | 0 | |

| $C_2$ | I | $C_2$ | |
|---|---|---|---|
| $A$ | +1 | +1 | $z, R_z$ |
| $B$ | +1 | -1 | $x, y, R_x, R_y$ |

| $C_3$ | I | $C_3$ | |
|---|---|---|---|
| $A$ | +1 | +1 | $z, R_z$ |
| $E$ | +1 | -1 | $x, y, R_x, R_y$ |

| $C_4$ | I | $C_4$ | $C_4^2$ | |
|---|---|---|---|---|
| $A$ | +1 | +1 | +1 | $z, R_z$ |
| $B$ | +1 | -1 | +1 | |
| $E$ | +2 | 0 | -2 | $x, y, R_x, R_y$ |

## Space groups

Space groups combine translations with point symmetry operations. In three dimensions, the combination of the 14 Bravais-lattices with the 32-crystallographic point groups yields the 230 crystallographic space groups. In two dimensions, only 17 space groups exist. Three important ones are illustrated in Fig. 1.9.

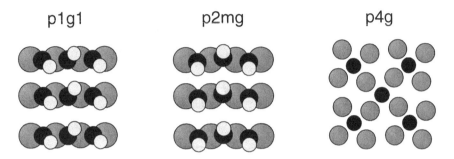

**p1g1**      **p2mg**      **p4g**

**Fig. 1.9**. Illustration of common space groups at surfaces. All structures contain a combination of translation and mirror symmetry, a glide plane. The p2mg structure contains an additional mirror plane perpendicular to the glide plane.

## 1.2 Surface Structures

Many materials, notably metals, have a surface lattice, which corresponds to the bulk crystallographic (*hkl*) plane. Merely the atomic distances vertical to the surface plane are changed to a larger or lesser degree, depending on the material, the surface orientation, and the type of bonding. The surfaces of some 5d-transition metals, however, reconstruct to form large, sometimes even incommensurate surface cells. Reconstructions are also typical for covalently bonded semiconductors. This section presents the surface structures of common materials [1.1].

### 1.2.1 Face Centered Cubic (fcc) Structure

Many metal elements crystallize in the face-centered cubic (fcc) structure. Among them are the coinage metals copper (Cu), silver (Ag), gold (Au), as well as the catalytic important metals nickel (Ni), rhodium (Rh), palladium (Pd), iridium (Ir) and platinum (Pt). Surfaces of these metals have been studied intensively since the early days of Surface Science. We therefore begin the presentation of surface structures with the low index surfaces of fcc-crystals. Following the convention in crystallography, we denote a set of equivalent faces by braced indices, e.g. {100}, and particular faces like (100), (010), or (001) by indices in parenthesis. The three most densely packed, and therefore the most stable {111}, {100}, and {100} surfaces of unreconstructed fcc-crystals are depicted in Fig. 2.1. The packing density is the highest for the {111} surfaces, followed by the {100} and {110} surfaces. The coordination numbers of surface atoms are 9, 8 and 7 for the {111} , {100} and {110} surfaces, hence the number of broken bonds are 3, 4 and 5 per surface atom.

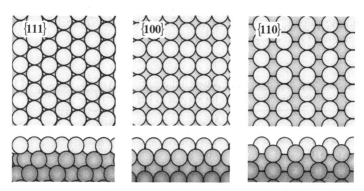

**Fig. 1.10.** {111}, {100}, and {100} surfaces of fcc-crystals; bottom row displays side views.

The open structure of the {110} surface has the peculiar feature that atoms in the surface layer have nearest neighbor bonds not only to the next, but also to the third layer. Vice versa, the second layer atoms have one broken bond oriented perpendicular to the surface plane. Hence, this surface has 5 broken bonds per surface atom, but 6 broken bonds per surface unit cell. In a nearest neighbor model, each broken bond on a surface corresponds to 1/12 of the cohesive energy $E_c$. Accordingly, the surface energies per surface unit cell are $E_c/4$, $E_c/3$, and $E_c/2$, for the {111}, {100} and {110} surface, respectively. Since the atom packing density also decreases in that sequence, the three surfaces differ by a lesser amount in their surface energy per area. In units of $E_c/a_0^2$, in which $a_0$ is the lattice constant, the energies of the {111}, {100} and {110} surfaces are 0.577, 0.666 and 0.707, respectively. The actual differences between the surface energies are even smaller because of next nearest neighbor and many-body contributions to the surface energy.

The surface layer of the {111} surface has a six-fold rotation axis and three non-trivial mirror planes. Together with the second layer underneath the symmetry reduces to a three-fold rotation axis. The highest symmetry of a molecular species site on that surface is therefore $C_{3v}$. However, if the adsorbate species has a six-fold rotation axis and interacts only with the first layer atoms the effective point group symmetry is $C_{6v}$. The {100} surfaces have four-fold symmetry and two non-trivial mirror planes. The highest symmetry of an adsorbate is thus $C_{4v}$. The {110} surface has a two-fold axis and two mirror planes. The highest point group symmetry is $C_{2v}$.

Unreconstructed surfaces as depicted in Fig. 1.10 are found on $\alpha$-cobalt ($\alpha$-Co), Ni, Cu, Rh, Pd, and Ag. Atoms in the surface layer assume a position as in the bulk save for a possible relaxation of the vertical distance between the surface layer and the layer underneath. This relaxation is very small (1-2%) for the {100} and {111} surfaces, hardly outside the error of the best structure determinations. The relaxation is larger for the {110} surfaces, and even the distance between the second and the third layer differs notably from the bulk. Table 1.2 lists mean relaxations on the {110} surfaces for a few materials.

**Table 1.2.** Relaxation of the distance between the surface layer and the second layer $\Delta d_{12}$ and the second and the third layer $\Delta d_{23}$ for several {110} surfaces.

| Material | $\Delta d_{12}$ | $\Delta d_{23}$ |
|---|---|---|
| Cu{110} | −9% | +3% |
| Ag{110} | −8% | 0% |
| Ni{110} | −9% | +3.5% |
| Pd{110} | −5% | +1% |
| Rh{110} | −7% | +2% |

Most surfaces of the 5d-transition metals Ir, Pt and Au reconstruct. The nature of the reconstruction is such that the surface plane of the reconstructed surface contains more atoms per area than an unreconstructed surface. The guiding principle therefore appears to be to compensate the loss of coordination caused by the "broken bonds" at the surface by increasing the effective coordination in the surface plane. The 5d-metals resort to this method because the stiffness of the 5d-orbitals induced by the two nodes prevents compensation by relaxation of the interatomic distances. Relativistic effects on the late 5d-metals and the associated contraction of the s-shell may also play a role. Some authors have attributed the propensity to reconstruct to the large tensile stress on the 5d-metal surfaces [1.13]. However, later experimental [1.14] and theoretical [1.15] investigations concerning the reconstructions on {100} and {110} surfaces did not confirm this view. When attempting to understand the reconstruction phenomenon on bare metal surfaces one should keep in mind that the energy gain in the reconstruction is very small. Investigations on the gold/electrolyte interface show that the difference in the free energy for the reconstructed and unreconstructed Au(100) surface is 0.05 N/m [1.16] which amounts to less than 4% of the surface energy [1.17].

The reconstructed {100} surfaces of Ir, Pt and Au all involve a nearly hexagonal packing of atoms in the surface layer. For Iridium this leads to a (5×1)

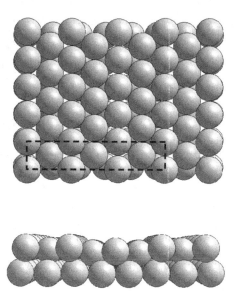

**Fig. 1.11**. Top and side view of the (5×1) reconstructed Ir(100) surface in which the surface layer consists of a buckled quasi-hexagonal overlayer of atoms. The buckling depends on the lateral position of the surface atom with respect to the second layer atoms and amounts to 0.48 Å at the maximum [1.18]. The dashed rectangle indicates the unit cell. The {100} surfaces of platinum and gold feature the same quasi-hexagonal arrangement of atoms in the first layer, but the surface layer is more densely packed and incommensurate with the substrate.

**fcc-sites**          **hcp-sites**          **fcc-sites**

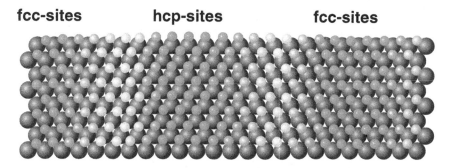

**Fig. 1.12**. Reconstruction on the Au(111) surface by an uniaxial compression of the surface layer. The position of the surface atoms with respect to the second layer change from fcc-sites, to bridge sites, to hcp-sites, to bridge-sites and back to fcc-sites. The height corrugation induced thereby is easily seen in an STM-image (Fig. 1.13).

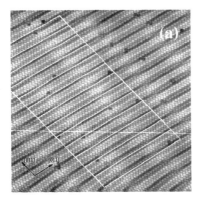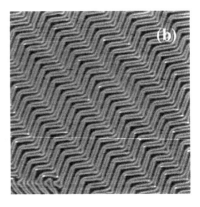

**Fig. 1.13.** (a) STM-image of a reconstructed Pt(100) surface [1.19]. On Pt(100) as well as on Au(100), the surface layer is slightly rotated with respect to the substrate causing an incommensurate structure. (b) STM-image of reconstructed Au(111) [1.20]. The height corrugation of the primary reconstruction (Fig. 1.12) is seen as white stripes. The superimposed secondary "Herringbone"-reconstruction reduces the elastic strain energy in the substrate [1.21].

reconstruction (Fig. 1.11) so that the density of atoms in the surface layer is 6/5 of the unreconstructed surface. Even higher atom densities (~125%) are realized with the quasi-hexagonal but incommensurate overlayers on Pt(100) and Au(100) [1.22-24].

Of the {111} surfaces of 5d-metals, only the Au(111) reconstructs. The reconstruction involves an uniaxial compression of the surface layer along a$\langle 110 \rangle$-direction by about 4.5% to a (1×22) unit cell (Fig. 1.12).

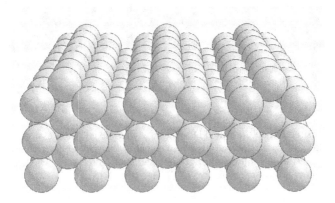

**Fig. 1.14.** The (1×2) reconstruction on {110} surfaces of Ir, Pt and Au.

Superimposed on the (1×22) reconstruction is a secondary reconstruction, the "Herringbone" reconstruction which helps to reduce the elastic energy in the substrate [1.21] (see also Sect. 3.4.3). The reconstructions on the {110} surfaces of Ir, Pt and Au are of a different nature: By removing every second row of atoms, (111) microfacets are formed (Fig. 1.14) [1.25-27]. The reconstruction involves a multi-layer reconstruction consisting of a buckling in the third and fifth layer and a row pairing in the second and fourth layer.

## 1.2.2 Body Cubic Centered (bcc) Structure

Typical metals with bcc-structure are tungsten (W), molybdenum (Mo), niobium (Nb), and iron (Fe). Spurred by the interest in their use as thermionic electron emitters, surfaces of tungsten have drawn the attention of researchers since the early years of the 20th century. Studies included measurements of the work function of various crystal faces and the influence of adsorbates, in particular of alkali atoms, on the work function. Later on, tungsten surfaces were considered as a model for surface phenomena in general, partly for that history, partly because the metallurgy of single crystal preparation was well developed for tungsten, and last not least, because tungsten surfaces are comparatively easy to prepare clean in ultra-high vacuum vessels made from glass (Sect. 2.2.3).

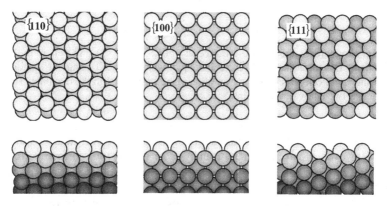

**Fig. 1.15.** Top and side view of the {110}, {100}, and {111} surfaces of a bulk terminated bcc-structure. The very open {111} surface is formed by three layers of atoms that are missing some of their nearest neighbor bonds.

The bulk-terminated surfaces of bcc-crystals are displayed in Fig. 1.15. The atom density on the most densely packed {110} surface amounts to 91.8% of a hexagonal close packed surface. The atoms form a compressed hexagon with each atom surrounded by four atoms in nearest neighbor distance, and two atoms in the 15.5% larger second nearest neighbor distance. The {100} surfaces possess 65.1%

of the density of a hexagonal close packed face, which amounts to 70.9% of the density of {110} surfaces. The {111} surfaces have a very open structure. The atom density is down to 41.1% of a close packed surface, or 44.7% of {110} surfaces. Atoms in three layers are missing nearest neighbors. Since the distance to the second nearest neighbors is merely slightly larger than to the first neighbors, an estimate of the surface energies based on the coordination numbers is not meaningful.

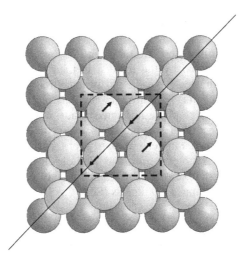

**Fig. 1.16.** Structure of the W(100) surface at 150K. The (2×2) unit cell is indicated by dashed lines. The surface atoms in the cell are pair wise displaced along one diagonal (solid line) of the (2×2) cell which produces a glide plane orthogonal to the diagonal. The reconstruction has two equivalent domains.

For a long time it was believed that neither surface of the bcc-metals would reconstruct. However, a careful structure analysis of the W(100) surface performed at 150 K [1.28, 29] revealed a (2×2) reconstruction of the space group pmg (Fig. 1.16). The reconstruction of the W(100) surface escaped detection for experimental reasons. Firstly, the majority of surface studies were performed at room temperature and above where the reconstruction is disordered and the surface therefore appears as being unreconstructed with a high Debye-Waller factor. Secondly and probably more importantly, adsorbed hydrogen produces a c(2×2) structure (Sect. 1.2.4). The pmg diffraction-pattern is easily mistaken for a c(2×2) pattern since it has the same extinctions along the h, k-axes as the pmg. The additional reflexes along the four diagonal ($|h| = |k|$) directions exist for both structures, for the pmg-structure because of the two equivalent domains. (The structure as drawn in Fig. 1.16 would have extra reflexes for the h = k direction, not for the h = −k direction). Hydrogen adsorbs dissociatively with a high sticking coefficient on the W(100) at room temperature and below. Hydrogen is also the prime residual gas in stainless steel vacuum chambers (Sect. 2.2.1). As tungsten surfaces are prepared by high temperature oxidation and annealing and some time is required to cool the crystal down to 150 K, hydrogen adsorption is hard to avoid unless special precautions are taken. Hence, even when researchers found a low temperature (2×2) might have attributed it to a hydrogen induced reconstruction. Presently, also the clean Mo(100) surface is believed to reconstruct at low temperatures, but no structure analysis is available at this time.

### 1.2.3 Diamond, Zincblende and Wurtzite

The group IV-elements carbon, silicon and germanium crystallize in the diamond structure in which each atom is surrounded by a tetrahedron of neighboring atoms, providing optimum overlap of the $sp^3$-type covalent bonds. The diamond structure can be viewed as two fcc-structures displaced along the cubic space diagonal by a vector $(^1/_4, ^1/_4, ^1/_4)a_0$ with $a_0$ the lattice constant (Fig. 1.17a). The structure has its name from the diamond phase of crystalline carbon although diamond is not the most stable phase of carbon, which is graphite. The III-V and II-VI compounds are likewise primarily covalently bonded in a tetrahedral configuration. The III-V compounds and some of the II-VI compounds crystallize in the diamond structure with each of the two atoms of the compound occupying one of the fcc-substructures. The structure is then named zincblende, after the mineral name of the II-VI compound ZnS. A ZnS-crystal has four polar axes oriented along the tetrahedral bonds. A dipole moment can arise if the tetrahedral symmetry is distorted, e.g. by shear stresses. Crystals with ZnS structure therefore have merely non-diagonal components of the dielectric tensor.

**(a)**                                    **(b)**

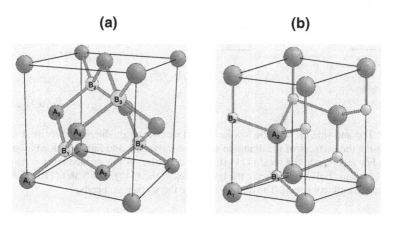

**Fig. 1.17**. Structure of **(a)** zincblende and **(b)** wurtzite. The zinblende structure reduces to the diamond structure if A- and B- atoms are identical. The zincblende structure has eight the wurtzite structure 4 atoms in the unit cell.

Most of the II-VI compounds crystallize in the hexagonal wurtzite structure. In wurtzite, the local configuration is as in zincblende (Fig. 1.17b). The arrangement of the tetrahedrons in space differs, however. When build with ideal tetrahedrons, wurtzite has a c/a ratio of $\sqrt{8/3} = 1.633$. However, the symmetry of the structure is compatible with the tetrahedrons being distorted along the c-axis. Since the c-axis is a polar axis, wurtzite crystals are pyroelectric (pyroelectricity denotes a variation of a permanent polarization with temperature), and possess one non-zero diagonal and off-diagonal elements of the piezoelectric tensor.

Because of the covalent nature of the bonding (with some ionic character in the III-V and II-VI-compounds) the termination of the bulk structure at the surface means broken bonds, also called *dangling bonds*. To minimize the energy associated with the dangling bonds nearly all surfaces of the group IV elements and of the III-V and II-VI compounds reconstruct in one or another way. In order to be able to describe and understand nature of the various reconstructions involved it is necessary to know the reference frame of the low index bulk terminated structures. We therefore depict the surfaces as they arise from the truncated bulk structures of zincblende and wurtzite, before entering the discussion concerning reconstructions.

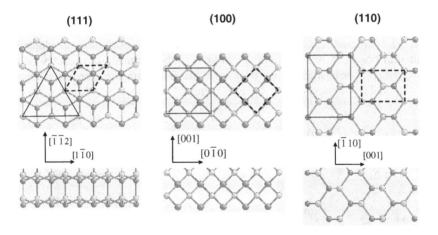

**Fig. 1.18**. Top and side view of the low index surfaces of the zincblende structure. Pictures also represent the surfaces of the diamond structure if the dark and lightly shaded atoms are identical. For zincblende the ideal (111) surfaces are polar, as the surface layer consists of one type of atoms. Full lines indicate the boundaries of the (111), (100) and (110) planes as drawn into the bulk cubic cell. The dashed lines are the surface unit cells.

Figure 1.18 shows top and side views of the {111}, {100} and {110} surfaces of the zincblende structure, as they arise from the truncated bulk structure. The surface layer of a {111} surface consists of only one type of atoms and has therefore a polar character. The (111) and ($\bar{1}\,\bar{1}\,\bar{1}$) surfaces are not identical. On {111} and

{110}, surfaces atoms have one dangling bond, on {100} surfaces each surface atom has two. The illustrations in Fig. 1.18 represent the surfaces of the diamond structure when dark and light atoms represent the same element.

Figure 1.19 shows top and side views of two surfaces of wurtzite. As for the {111} surfaces of zincblende, the surface layer of the {0001} surfaces consist of atoms of one type; the surfaces are therefore polar. Because of the arrangement of the tetrahedrons, wurtzite appears as rather open when viewed along the c-axis, compared to the zincblende and diamond structure.

**(0001)**                    **(11$\bar{1}$0)**

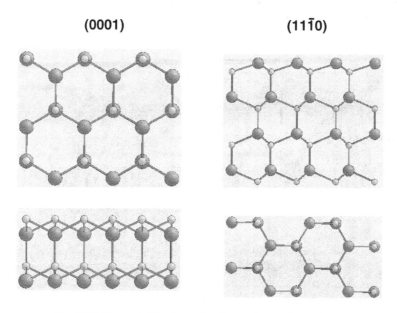

**Fig. 1.19**. Top and side view of surfaces of wurtzite surfaces.

### The Si(111) surface

Some of the early work in surface science is associated with the Si(111) surface prepared by cleaving a silicon crystal along the (111) plane in ultra-high vacuum. Low energy electron diffraction (LEED) revealed that the surface is reconstructed to a (2×1) unit cell [1.30] and transforms to a surface with a (7×7) unit cell upon annealing. For a long time, research concentrated on the (2×1) surface for various reasons. Firstly, cleaved surfaces are easily prepared and one could rest assured that the surface was clean (Sect. 2.2.3), whereas it was debated for a long time whether the (7×7) reconstruction was really a property of the clean surface. Secondly, strong Fermi-level pinning was found on the (2×1) surface [1.31], providing evidence for a high density of surface states. The high density of states was directly associated with the dangling bonds on the silicon surface. Furthermore, the (2×1) surface displayed interesting spectroscopic features, both with respect to surface vibrations [1.32] and optical properties [1.33, 34]. Much of this early work remained speculative with respect to the interpretation of the experimental results, because the surface structure was unknown. When surface structure analysis became feasible and the structure of the (2×1) surface was solved around the mid 80-ties of the last century, interest in the (2×1) surface had already declined in favor of the (7×7) reconstructed surface. The structure of the (2×1) surface as determined by Sakama et al. is shown in Fig. 1.20 [1.35]. The structure is characterized by chains of surface atoms, which are π-bonded by the electrons in the dangling bonds. The hybridization reduces to $sp^2$, so that the surface atoms form a more planar structure.

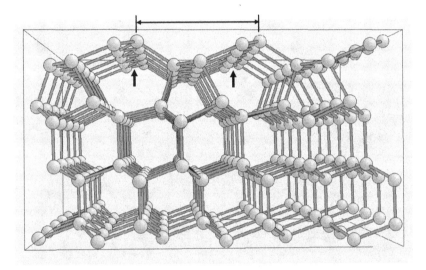

**Fig. 1.20.** Perspective side view on the reconstructed Si(111)(2×1) surface. Electrons in dangling bond establish a π-bonding between surface atoms, so that they arrange in chains (marked by arrows) along a $< 1\bar{1}0 >$ - direction. The size of the unit cell along a $< \bar{1}\bar{1}2 >$ - direction is thereby doubled (double headed arrow).

Unlike the Si(111) (2×1) surface, the Si(111)(7×7) surface is an equilibrium phase. The structure involves a rearrangement of the position of many atoms as well as additional atoms. A migration of atoms from an atom source, e.g. steps, is therefore necessary to build the (7×7) structure which explains that the structure is not formed directly after cleaving the crystal at room temperature. The complexity of the structure has challenged researchers for a long time. Hundreds of papers were published proposing and considering possible elements of the structure without getting a grasp on the full complexity of the problem. Even advanced techniques of LEED-intensity analysis could not solve the puzzle, as a successful structure analysis by LEED requires a trial structure fairly close to the final one. At last, scanning tunneling microscopy provided decisive clues that narrowed the number of possibilities for the structure. Figure 1.21 shows an STM-image of the Si(111)(7×7) surface. Within the unit cell, STM finds 12 bright spots. If these bright spots are identified with atoms, this means the structure features twelve atoms in a particular elevated position. These atoms were later identified as extra silicon atoms (*adatoms*) sitting on three dangling bonds of silicon atoms, thereby reducing the number of dangling bonds by a factor of three. The STM-image shows further one deep, wide hole per unit cell.

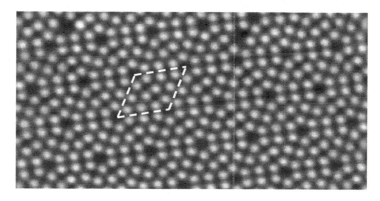

**Fig. 1.21**. STM-image of a Si(111)(7×7) surface. Dashed lines mark the unit cell. The image shows twelve bright spots and one deep and wide hole per unit cell. The bright spots correspond to silicon adatoms bonding to three dangling surface bonds.

With these clues, a further one stemming from medium energy ions scattering stating that the structure should involve a stacking fault, and his own Patterson analysis of high energy electron diffraction data, Takanayagi et al. were able to propose the currently accepted model [1.36]. The model has been termed the *D*imer-*A*datom-*S*tacking fault (DAS) model after its key structural elements. The structure is shown in Fig. 1.22. The atom coordinates are taken from the LEED-structure determination of Tong et al. [1.37]. We begin the discussion of the various structural elements with the stacking fault. The top view on the two uppermost Si-double layers in the right and left side of the rhombic unit cell differs. In the right half, the structure is as in bulk silicon, in the left half the arrangement of the first two double layers is as in wurtzite. Hence, this section is faulted with reference to the silicon structure. At the domain boundary between the faulted and non-faulted area six silicon atoms pair up to form three dimers (textured arrows in Fig. 1.22). The adatoms are best seen in a side view. The side view in Fig. 1.22 displays the three sheets of atoms that lie between the dotted lines drawn in the top panel. This section of the unit cell has four adatoms between the large holes at the apices of the rhombic unit cell. The positions of these adatoms correspond to the white spots along a line connecting the two apices of the unit cell in the STM-image Fig. 1.21. Four more adatoms exist on either side, above and below the dotted lines in Fig. 1.22. While the adatoms reduce the number of dangling bonds per unit cell, some of the silicon surface atoms retain their original dangling bonds. Four of them are shown in the side view. These surface atoms are called *rest atoms*.

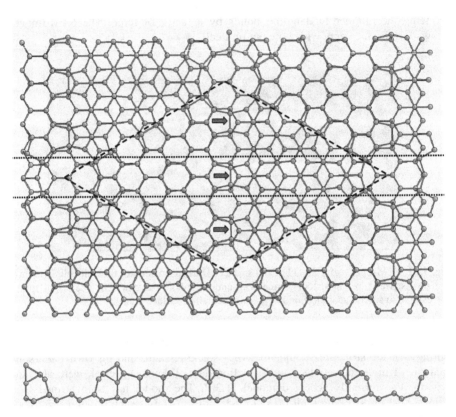

**Fig.1.22.** Structure of the Si(111)(7×7) surface according to the **D**imer-**A**datom-**S**tacking fault model (DAS). The lower panel displays a side view of the atoms residing between the dotted lines shown in the top panel. Four adatoms sitting on a triplet of Si-atoms mark positions of height maxima that are the salient feature in STM-images (Fig. 1.21). The (7×7) unit cell (dashed lines) comprises 12 adatoms. The top view shows clearly that the arrangement of tetrahedrons in the first two double layers is as in wurtzite on the left hand side of the rhombic unit cell. Hence, the stacking of layers is faulted with respect to the silicon structure. Dimerization of surface Si-atoms occurs along the domain boundary between the faulted and non-faulted section of the (7×7) unit cell (textured arrows).

### The Ge(111) surface

The cleaved Ge(111) surface exhibits the same (2×1) reconstruction as the Si(111) surface. The (2×1) reconstructed surface transforms irreversibly into a stable structure around 200 °C [1.30]. Again, minimizing the number of dangling bonds is the driving force for the reconstruction. For germanium the resulting equilibrium structure is, however, much simpler since the reconstruction involves merely an ordered array of adatoms on the otherwise unreconstructed, though distorted Ge(111) surface. Figure 1.23 shows the c(2×8) structure that is obtained after annealing the surface [1.38].

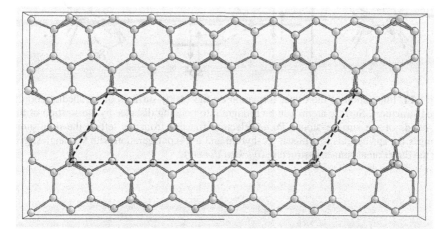

**Fig. 1.23**. Perspective view on the topmost layers of the reconstructed Ge(111)c(2×8) surface. Each unit cell (dashed lines) contains three adatoms. The adatoms cause a distortion of the germanium structure, clearly visible in the panel. The distortions extend several layers deep into the bulk.

### The Si(100) and Ge(100) surface

Atoms on the {100} surfaces of the tetrahedral coordinated crystals would have two dangling bonds each if the surface existed as a truncated bulk (Fig. 1.18). Clearly, such a surface must be even less stable than the ideal (111) surface, and reconstructions, which reduce the number of dangling bonds, are expected. The geometry of a {100} surface permits a way to saturate 50% of the dangling bonds by pairing the surface atoms (Fig. 1.24). The moderate energy needed to distort the bond angles of the $sp^3$-bonded surface atoms is overcompensated by the gain in energy due to the formation of dimers. The symmetric dimer has still two dangling bonds, i. e. half-filled electron states of the same energy. Breaking the symmetry lifts the degeneracy of the electrons states, which allows for the filling of the lower energy state with two electrons whereby the electronic energy is reduced (Fig. 1.24). This general principle of energy reduction by symmetry breaking is called *Jahn-Teller effect*. In this particular case, the Jahn-Teller effect involves a partial transfer of one electron to one of the two atoms in the dimer. Electrons of that atom then form a $p^3$-configuration with 90° natural bond angles. The electrons of the donating atom form a planar $sp^2$-hybrid. The state of lowest

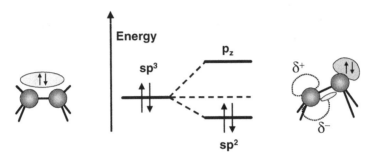

**Fig. 1.24**. Illustration of the dimer formation on the {100} surfaces of tetrahedral coordinated structures. Surface atoms can be brought into bonding distance by a distortion of the $sp^3$-bonds of the surface atoms. Partial electron transfer from the left to the right atom changes the $sp^3$-hybrids to a planar $sp^2$-hybrid and a $p^3$-configuration with 90°-angles, giving the dimer an asymmetric structure (*buckled dimers*).

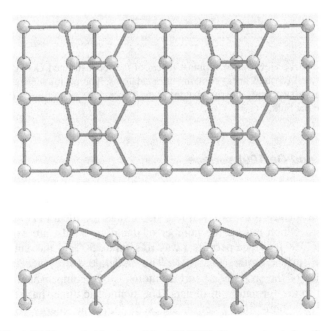

**Fig. 1.25**. Top and side view of the Si(100)(2×1) reconstructed surface.

energy is therefore an asymmetric dimer (*buckled dimer*). Note that the buckled dimer has no half-filled electron orbitals and therefore no remaining dangling bonds.

The electrons in the filled and empty states form bands of surface states [1.39] that are energetically located in the band gap between the valence and the conduction band (Sect. 8.2.4). The total density of these states is two states per surface atom (Sect. 3.2.2). Many different ordered structures can be realized with the asymmetric dimers as building blocks. The simplest structure is with all dimers tilted in the same direction. The resulting reconstruction is a (2×1) structure which exists in two domains. An example is shown in Fig. 1.25 with the Si(100)(2×1) reconstructed surface. The structure analysis was performed using LEED at 120K [1.40]. A simple structure with an equal number of dimers of either orientation is the c(4×2) reconstruction which can exist on Si(100) as well as on Ge(100). Figure 1.26 shows top and side view of Ge(100)c(4×2). The unit cell (dashed rectangle) contains two asymmetric dimers of either type. The energies of the various arrangements of the asymmetric dimers differ only because of elastic interactions between different dimers. These interactions are comparatively weak. Entropy plays therefore an important role in the free energies of various surface

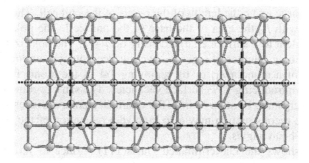

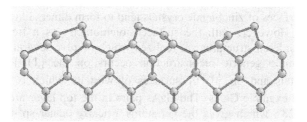

**Fig. 1.26.** Top and side view of the Ge(100)c(4×2) reconstructed surface. For clarity, the side view displays merely three planes of atoms along the dotted line. The unit cell (dashed rectangle) contains two asymmetric dimers of each type.

configurations. Phase transitions between the ordered structures occur as a result. Another consequence of the weak interactions between dimers is that at room temperature dimers flip back and forth between the two asymmetric states. STM-images average over the two configurations so that the dimers appear to be symmetric in such images.

### Surfaces of zincblende and wurtzite

The truncated bulk {111} and {100} surfaces of zincblende and the {0001} surfaces of wurtzite are polar, that is the outermost surface layer consists of one of the two types of atoms only (Fig. 1.18). Because of the ionicity of the bonds in zincblende and wurtzite, the outermost layer would bear an uncompensated surface charge. A zincblende crystal terminated by a (111) surface on the one end and a $(\bar{1}\,\bar{1}\,\bar{1})$-surface on the other, or a wurtzite crystal terminated by (0001) and $(000\bar{1})$-surfaces would bear a net permanent polarization giving rise to electric fields in the adjacent vacuum. If, furthermore, the surfaces were planar extended, there would be no countercharge to terminate the electric field, which means that the field energy adds an infinite amount of energy to the surface energy (see also Sect. 4.2.1). For the wurtzite structure the permanent polarization $P$ is easily calculated with the help of Fig. 1.17. The dipole moment $p$ per unit cell is $qc/4$ when $q$ is the ionic charge and $c$ the length of the c-axis of the unit cell. The polarization $P$ is $P = p/cF_b$, with $F_b$ the area of the base of the unit cell. The polarization $P$ is therefore equivalent to a surface charge density of $q/4F_b$. The polarization is compensated by placing counter charges on the surfaces, which amount to the ion charges of 1/4 of a monolayer, or by removing 1/4 of the atoms in the surface layer, that is by introducing 25% surface vacancies. The same argument applies to the zincblende crystals. In other words, the nominally polar surfaces are prone to reconstruct. The reconstruction may also involve a relaxation of the bond lengths and a change of bond angles.

Figure 1.27 shows the Ga-terminated GaAs(111) surface as an example. As suggested by the considerations above every fourth Ga-atom is missing. Furthermore, the first double layer of Ga- and As-atoms is relaxed to a nearly planar $sp^2$-type configuration.

The {100} surfaces of zincblende crystals tend to form dimers like the diamond type structures. However, with the surface stoichiometry as a free parameter, many complex ordered structures are realized which involve several atom layers. A relatively simple generic reconstruction occurs on the {110} surfaces of zincblende crystals and on $\{10\bar{1}0\}$ surfaces of wurtzite which is displayed in Fig. 1.28 for the example GaAs. The GaAs pairs in the top layer are tilted by an angle of about 28°, which gives the Ga-atoms a nearly planar $sp^2$-type bonding and the As-atoms a $p^3$-type configuration. Both electronic configurations are natural for neutral Ga- and As-atoms with their three and five valence electrons, respectively.

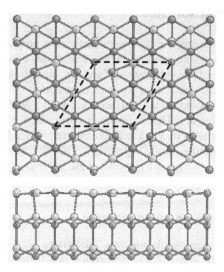

**Fig. 1.27.** Top and side view of the GaAs(111)(2×2) surface. Ga-atoms are displayed in light grey. One quarter of the Ga-surface atoms is missing, i. e. one Ga-atom per unit cell (dashed line). Furthermore, the first double layer of Ga- and As-atoms is relaxed to an almost planar $sp^2$-type configuration (side view).

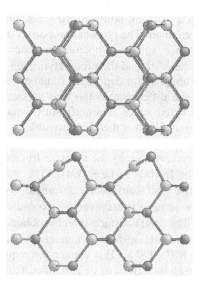

**Fig. 1.28.** Top and side view of the GaAs(110) surface. Light grey shaded balls represent Ga-atoms. The tilt in the surface bonds by about 28° is caused by the different hybridization of the electrons of the surface atoms. Ga-atoms assume a $sp^2$- and the As-atoms a $p^3$-configuration.

## 1.2.4 Surfaces with Adsorbates

The large diversity in the structures of bare surfaces is surpassed by the diversity of structures of adsorbates covered surfaces. Adsorbates can have three different effects on the structure of the substrate surface. By saturating the dangling bonds, adsorbates may eliminate the reason for a reconstruction of the bare surface and thereby restore the unreconstructed surface. Secondly, adsorbates may cause a restructuring of the substrate. This happens in particular for strongly bonded adsorbates. Thirdly, adsorbates may leave the substrate surface largely unaltered and merely form ordered commensurate or incommensurate superlattices on top of the substrate. In this section, we briefly consider these general aspects. Specificities of individual adsorbates are presented in Sect. 6.4.

### *Lifting of reconstructions by adsorbates*

Except for the case of very weak bonding, adsorbates alter the reconstruction of bare surfaces. Frequently the effect of adsorption is a lifting of the reconstruction and a return to the unreconstructed substrate. For covalently bonded substrates, this is the case if the adsorbate bonding involves merely the dangling bonds of the truncated bulk structure. A well studied and illustrative example is the hydrogen covered Si(111) surface with one hydrogen atom bonding to each surface atom. Such a surface can be prepared in air by wet chemistry, inserted into a vacuum chamber and investigated at length under vacuum conditions without becoming contaminated. In addition to being non-reconstructed, surfaces thus prepared are rather flat and stable even in air. The physical properties of this ideal surface serve as a benchmark for intrinsic surface properties in general. Electronic as well as phonon properties have been studied therefore. Hydrogen terminated Si(111)(1×1) surfaces also serve as a template for deposition of nanostructures.

For metal surfaces, the difference in the free surface energies of the reconstructed and unreconstructed states is significantly smaller than for covalently bonded materials. Occasionally, it is therefore possible to lift the reconstruction by adsorption and subsequent removal of the adsorbate by gentle heating or by a catalytic reaction and preserve thereby the unreconstructed surface as a metastable state. An example is the Ir(100) surface [1.41]. The metastable Ir(100)(1×1) is obtained by exposing the (5×1)-surface for 2 min to $O_2$ at 475 K, followed by annealing to 750 K. The oxygen is removed by exposure to $5×10^{-7}$ mbar $H_2$ at a temperature of 530 K. The (1×1) surface thereby obtained persists at room temperature. The metastable (1×1) surface is converted back into the stable (5×1) surface by annealing to 800 - 900 K. This conversion requires the incorporation of 20% additional Ir-atoms, which have to be generated from kink site at steps. The generation of adatoms from kink sites requires energy which explains that the (1×1) surface is metastable. The measured activation energy for the conversion of 0.9 eV [1.41] appears to be a reasonable number for the formation energy of surface adatoms from kink sites.

The (2×1) missing row reconstructions of the 5d-metal (Fig. 1.14) can also be removed by adsorption processes, e. g., on the Pt{110} surface by adsorption of CO [1.42]. The considerable relaxation times associated with the transport of Pt-atoms over large distances which is required to lift the reconstruction gives rise to oscillatory catalytic reactions under steady state conditions and fascinating spatio-temporal patterns [1.43, 44].

### *Restructuring of substrates by adsorbates*

In the early days of surface crystallography, substrate surfaces were considered rigid templates, merely providing specific sites on which adsorbate lattices would unfold. It was not before methods of surface analysis had progressed and the coordinates of many atoms per unit cell could be determined with accuracy that the restructuring of substrate lattices under the influence of adsorbates was revealed. The hydrogen covered W(100) surface is an interesting example. Hydrogen prefers adsorption in bridge sites, however with the tungsten atoms closer than they would be on a (100) surface with a truncated bulk structure. The fact that on the clean surface the tungsten atoms are laterally displaced from their bulk positions (Fig. 1.17) shows that the atoms on the W(100) surface have some degree of flexibility with respect to sideward motion. With half a monolayer of hydrogen atoms the tungsten atoms pair under upon adsorption of hydrogen in bridge sites, and these pairs order into a c(2×2) pattern (Fig. 1.29). Upon adsorption of two hydrogen monolayers, all possible bridge sites become occupied with hydrogen atoms and the tungsten atoms return to their bulk positions with respect to lateral displacements.

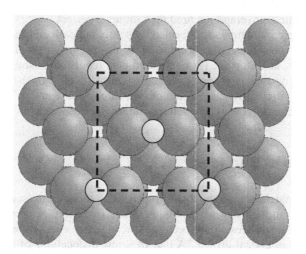

**Fig. 1.29**. The W(100)c(2×2) structure with half a monolayer of hydrogen atoms. Hydrogen adsorbs in bridge sites with the tungsten atoms paired.

While the restructuring of the W(100) surface upon adsorption of hydrogen is revealed only by quantitative structure analysis, the massive restructuring of the Ni(100) surface upon adsorption of nitrogen and carbon becomes apparent by the systematic extinctions in the diffraction pattern due to the glide planes involved in the restructuring. The resulting p4g-structure was already shown in Fig. 1.9.

Adsorption of alkali atoms likewise habitually induces major reconstructions and can give rise to very complicated ordered structures with alkali atoms in the outermost surface layer as well as in layers buried in deeper layers. The resulting structures are better described as ordered surface alloys than as adsorption structures (Sect. 6.4.6).

### Adsorbate lattices on rigid substrates

In cases when the substrates structure does not change upon adsorption, save for a minor rearrangement of the atom coordinates, the structure and periodicity of the adsorbate lattices is determined by the interplay of the site specificity of the adsorption energy and the interaction between the adsorbate atoms. The distinction between these two energetic contributions is somewhat artificial as the interaction potential also depends also on the adsorption site. The interaction between adsorbates can be purely repulsive, but more typically consists of a combination of attractive forces at longer distances, and repulsive forces at shorter distances. In many cases, the adsorption energy has a pronounced maximum for one particular type of adsorption site, so that only this site is taken at any coverage. An example is the adsorption of oxygen and sulfur on transition metals where the oxygen and sulfur atoms assume the site of highest coordination, the fourfold hollow site on a {100} surface and the threefold hollow site on the {111} surfaces. The resulting ordered lattices (Fig. 1.6), e.g. the p(2×2) lattice, are usually observed at coverages below the nominal coverage required for forming a perfect lattice. This means that islands of ordered structures are formed, which is indicative of attractive interactions. On the other hand, the occupation of nearest neighbor sites (Fig. 1.6) is excluded. Thus, at least some form of a hard-core repulsion must exist.

Rare gases on metals represent systems with a small, though not vanishingly small, site specificity. The lateral interactions are of the van-der-Waals type. Ordered commensurate as well as incommensurate structures exist in that case.

# 1.3 Defects at Surfaces

Ever since researchers thought about properties of surfaces, defects have played a prominent role in their considerations. In 1925, Taylor [1.45] proposed that catalytic reactions at surfaces would occur at special active sites. Defects also play an important role in crystal growth. A pair of screw dislocations of opposite sign of the Burgers-vector, a Frank-Read source, promotes nucleationless growth at the step sites on surfaces. The relevance of steps and defects at steps in epitaxial growth was discussed in the seminal work of Burton, Cabrera and Franck [1.46]. In this early work, defects were discussed mostly from the standpoint of theory since hardly a technique was available whereby defects could be made accessible to experimental investigation. The only possibility – decoration of steps and point defects by large Z-elements (mostly Au) and imaging the decorated defects in an electron microscope – produced images of "dead" defects: After decoration, steps cannot change position with time, nor are steps catalytically or otherwise active. The invention of the scanning tunneling microscope (STM) and the subsequent development of other scanning microprobes changed that situation completely. Not only that line and point defects have become visible objects, one can even track their motion as they migrate across the surface, in the course of thermal fluctuations, catalytic reactions, epitaxial growth, or abrasion. A remaining, yet essential limitation is the large discrepancy in the time scale of the scanning probes and the time scale of defect migration.

## 1.3.1 Line Defects

Line defects on surfaces are steps, boundaries between different domains of adsorbate structures, and dislocations, but also non-structural defects as the boundaries between magnetic or ferroelectric domains. Although these line defects are of a very different nature, they also have certain things in common. For example, work is required to create the defect and the work depends on the orientation. In a coarse-grained description, the static and dynamic properties of all different line defects are therefore treated by the same statistical theory.

### *Steps*

The easiest access to steps of defined orientation, conceptually as well as technically is via surfaces that are inclined with respect to a low index surface by a small angle. These surfaces are called *vicinal surfaces* (Fig. 1.30). If all steps are one atom layer high, which is the generic form of bare vicinal surfaces after preparation in ultra-high vacuum, the mean number of steps per length is unambiguously determined by the angle of inclination $\theta$. Figure 1.30 shows a schematic view of a vicinal surface, together with the most important point defects. For particular azimuthal directions, the steps are oriented along a direction of dense atom packing. They are nominally free of kinks. In the following, we consider vicinal orientations on cubic materials. We begin with the vicinals of the {100} surfaces. These have Miller-indices of the type {1 1 n}. The surface consist of terraces, each $n/2$ atom diameters wide, separated by monatomic steps along a $\langle 110 \rangle$ direction. The $\langle 110 \rangle$ direction is the direction of nearest neighbors on {100} surfaces. Steps along this direction are therefore (ideally) free of kinks. Figure 1.31 shows a ball model of the (1 1 9) surface as an example. The illustration shows that the steps form {111} microfacets. A nomenclature for stepped surfaces, that is more descriptive than the Miller indices denotes vicinal surfaces by the type of microfacets and the number complete of atom rows on the terraces: the (1 1 9)-surface, e.g., is described as 4(100)×(111) in this convention. The latter notation immediately conveys the atomic picture. The Miller-index notation, on the other hand, is unique, and the

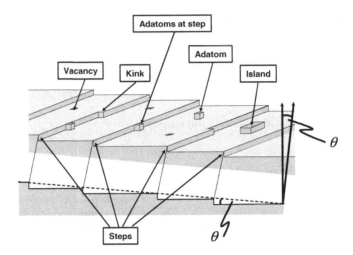

**Fig. 1.30.** Schematic illustration of a vicinal surface with one-atom layer high steps, kinks in steps, adatoms on terraces and at steps, vacancies and islands formed by a group of adatoms.

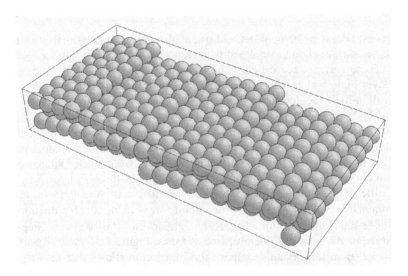

**Fig. 1.31.** Ball model of the (1 1 9) surface of an fcc-crystal. The surface consist of 4.5 atom wide terraces, separated by monatomic step along a ⟨110⟩ direction.

rotation angle $\theta$ with respect to the low index surface can be calculated exactly from the scalar product of the normal vectors. For the (111) vicinals one has

$$\cos\theta = \frac{(0,0,1)\cdot(1,1,n)}{|0,0,1||(1,1,n)|} = \frac{1}{\sqrt{1+2/n^2}}. \tag{1.9}$$

For small angles $\theta$

$$\theta \approx \sqrt{2}/n \tag{1.10}$$

is a useful approximation. The error is smaller than 4% for $n > 4$.

Two different types of close packed steps exist on the vicinals of the {111} surfaces of fcc-crystals. Depending on the direction of inclination with respect to the $\langle 111 \rangle$ direction, steps can display either {100} microfacets (A-steps) or {111} microfacets (B-steps). Figure 1.32 and 1.33 show the (7 7 9) and (9 9 7) surfaces as examples. The notation for general {111} vicinals with A-steps is $\{n\,n\,n+2\}$. The width of the terrace is $n+2/3$ atomic rows, each having a width of

$$a_\perp = \frac{1}{2}\sqrt{3}a_\| = \frac{1}{4}\sqrt{6}a_0 \tag{1.11}$$

in which $a_\|$ is the atom diameter or the atomic length unit parallel to the step direction and $a_0$ is the lattice constant. The angle of inclination with respect to the (111) surface is

$$\cos\theta = \frac{(1,1,1)\cdot(n,n,n+2)}{|(1,1,1)||(n,n,n+2)|} = \frac{n+2/3}{\sqrt{n^2+4n/3+4/3}} \tag{1.12}$$

A good approximation for the angle $\theta$ is

$$\theta \approx 2\sqrt{3}/3(n+1) \tag{1.13}$$

The error is smaller than 5% for $n > 4$.

The notation for general {111} vicinals with B-steps is $\{n+1\,n+1\,n-1\}$. The terrace width is $(n+1/3)a_\perp$. The angle of inclination with respect to the (111) surface is

$$\cos\theta = \frac{(1,1,1)\cdot(n+1,n+1,n-1)}{|(1,1,1)||(n+1,n+1,n-1)|} = \frac{n+1/3}{\sqrt{n^2+2n/3+1/3}} \tag{1.14}$$

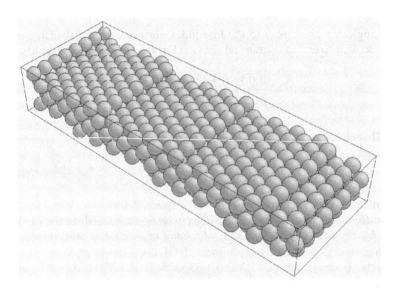

**Fig. 1.32**. Ball model of the (7 7 9) surface with steps showing {100} microfacets (A-steps).

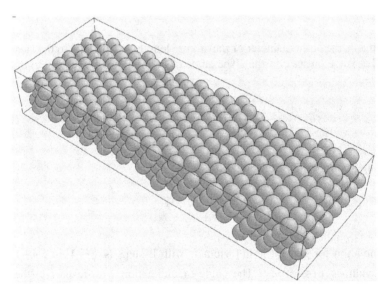

**Fig. 1.33**. Ball model of the (9 9 7) surface with steps showing {111} microfacets (B-steps). If one more atom row on the terrace is included in the consideration then the local structure at a B-step is as on the (110) surface (Fig. 1.10).

Here $\theta$ is approximated by

$$\theta \approx \sqrt{2}/3n \qquad (1.15)$$

The error is below 5% for n > 6.

The {100} surfaces of the fcc- and bcc-structure have fourfold rotation symmetry. On the {100} surfaces of the likewise cubic diamond structure the symmetry is reduced to a two-fold rotation axis. Vicinal surfaces of the (1 1 n) type have therefore two types of terraces, one with the dangling bonds oriented parallel [$1\bar{1}0$] the other parallel to [$\bar{1}10$] (Fig. 1.34). Consequently, two types of monolayer high steps exist, one with the bonds on the upper terrace perpendicular to the step ($S_a$-*steps*), the other with the bonds parallel to the step ($S_b$-*steps*). We note that vicinal Si(100) surfaces with a miscut angle around 4° form double steps so that only a single reconstruction domain exists with the dimer rows perpendicular to the step orientation [1.47].

An even larger variety of steps exists on zincblende surfaces since the top layer consists of different elements on adjacent terraces. This makes the $(1\ 1\ n)$- and the $(1\ 1\ \bar{n})$-surfaces nonequivalent. $S_a$- and $S_b$-step atoms can be either of the Zn- or the S-type atoms. The pairing row reconstructions on Ge(100) and Si(100) leads to rows which are parallel to the step on the upper terrace of an $S_a$-step and perpendicular for an $S_b$-down step (Fig. 1.35 [1.48]). The two types of steps have different equilibrium morphology because of the different energies associated with the formation of kinks on the two types of steps [1.49].

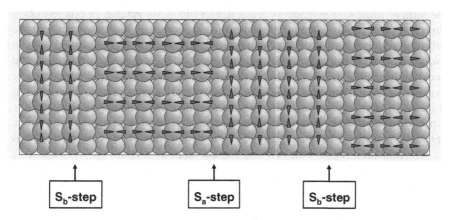

**Fig. 1.34.** Ball model of the unreconstructed (1 1 17) surface of the diamond structure with one atom layer high steps. Steps descend from left to right. The dangling bonds are drawn as triangles. They are rotated by 90° with respect to each other on adjacent terraces. The paired surface atoms in the reconstructed phase are like wise mutual orthogonal.

**Fig. 1.35**. STM image of a Si(100) surface with steps. The $S_b$-steps are rougher than the $S_a$-steps since the kink energy is lower on $S_b$-steps (image from [1.48], original reference B. Swartzentruber et al.) [1.49] .

### Domain walls

Ordered periodic structures exist in patches of finite size, called *domains*. A simple form of finite size domains is illustrated in Fig. 1.36. Patches of ordered (2×2) superstructures are displaced with respect to each other by one base vector of the substrate lattice. The line defects between the different domains are called *domain walls*. Depending on the density of atoms in the wall, one distinguishes *light* and *heavy* walls. Structural domains may also differ in the type of superstructure. For example, ordered domains of a (2×2) and c(2×2) superstructure may coexist (Fig. 1.37). Domain patterns of this type are typical for systems which realize one type of superlattice at one particular coverage and another at a higher coverage ($\Theta = 0.25$ and $\Theta = 0.5$ in the example) when the coverage is between the two limits. The transition region between one domain and the next in Figs. 1.36 and 1.37 is abrupt, in other words, the thickness of the walls is only an atomic distance. This is typical for adsorbate systems for which the corrugation of the adsorbate/substrate potential is large compared to the adsorbate/adsorbate interaction, so that all adsorbate atoms reside in the same defined sites. If the corrugation of the adsorbate/substrate potential is small compared to the lateral interaction, then the thickness of the domain wall increases. An example are the walls between the domains of fcc- and hcp-site occupancy on the reconstructed Au(111) surface (Fig. 1.12/1.13) which are several atoms wide. If the corrugation of the adsorbate/substrate potential is very small compared to the lateral interaction potential, the structure eventually becomes incommensurate. The structure of Au(111) is on the borderline between the latter two situations.

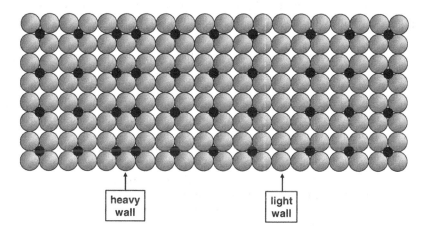

**Fig. 1.36**. Schematic illustration of light and heavy domain walls in an adsorbate lattice. Atoms in the domains are displaced only in one direction. More realistic is a displacement also along the vertical direction, permitting the formation of heavy domain walls in which the atoms are less close.

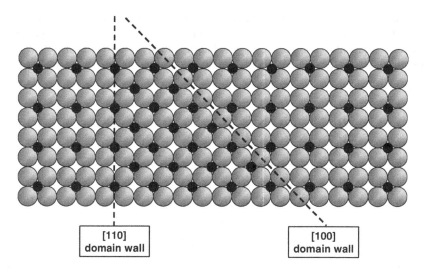

**Fig. 1.37**. Domains of (2×2) and c(2×2) superstructures on a (001) fcc/bcc-surface with domain walls oriented along the [110] and [100] direction.

The problem of finding the equilibrium position of atoms in and near a domain wall can be mapped onto a simple one-dimensional mathematical model, the Frenkel-Kontorova-model (see e.g. [1.50, 51]). The model considers the equilibrium states (and in an extension the dynamics) of a linear chain of atoms coupled by springs in a sinusoidal potential. The Hamiltonian is of the form

$$H = \sum_n \frac{1}{2} k (x_{n+1} - x_n - a)^2 + U\left[1 - \cos(2\pi x_n / b)\right] \tag{1.16}$$

Here, $a$ and b are the natural lattice constant of the chain (without the potential) and of the potential, respectively; $k$ and $U$ are the spring constant and the potential depth; the atom positions are denoted by $x_n$. Despite its simplicity, the model is very rich in properties (in particular if a kinetic energy term is added [1.51]) and describes essential features of domains, domain walls and transitions between commensurate and incommensurate phases. Here, we discuss merely the static solutions of the model that are relevant for domain walls. By differentiation with respect to the atom position $x_n$ one obtains the condition for equilibrium as

$$k(2x_n - x_{n+1} - x_{n-1}) + U(2\pi/b)\sin(2\pi x_n/b) = 0. \tag{1.17}$$

We consider the case where the number of atoms is half the number of minima of the potential, that is a coverage of $\Theta = 0.5$ and assume that the springs are not loaded if the atom distance equals $2b$ ($a = 2b$). Then the equilibrium condition is fulfilled when the atoms reside in every second potential minimum. We now insert one extra atom into the system and assume that the spring constant is very soft, $kb^2 \ll U$. The inserted extra atom then assumes a position close to the potential minimum adjacent to another atom, which is thereby pushed out of the position of minimum potential (Fig. 1.38a). For symmetry reasons, the displacements $u$ of the

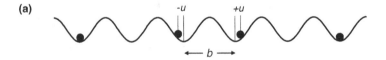

**(a)**

**(b)**

**Fig. 1.38.** Illustration for two limiting solutions of the Frenkel-Kontorova-model. (a) The heavy wall solution for very soft springs ($kb^2 \ll U$, (b) the incommensurate solution for very hard springs $kb^2 \gg U$.

two atoms have the same magnitude, but are of opposite sign. The change in energy with respect to the energy where all atoms sit in the potential minimum is

$$\Delta E = ku^2 + \frac{1}{2}k(b-2u)^2 + 2U\left[1-\cos(2\pi u/b)\right] \qquad (1.18)$$

Expanding $\Delta E$ for small displacements $u$ one obtains

$$\Delta E = \left[3k + \frac{4U\pi^2}{b^2}\right]u^2 - 2kbu + \frac{1}{2}kb^2 \qquad (1.19)$$

The displacements $u$ which lead to a minimum in $\Delta E$ are

$$u_{min} = br^2/(1+3r^2), \qquad (1.20)$$

with

$$r = \sqrt{\frac{kb^2}{4\pi^2 U}} \qquad (1.21)$$

The displacement $u$ vanishes when the spring constant approaches zero, as they should. The energy associated with the two displaced atoms, the domain wall energy is

$$E_w = \frac{kb^2(1-2r^2)}{2(1+3r)}. \qquad (1.22)$$

The solution has acceptable accuracy as long as r < 0.5.

The other extreme case, infinitely stiff spring constants, keeps the atom distance at $a$, and the adsorbate lattice becomes incommensurate with the substrate lattice (Fig. 1.38b). However, an analytical solution exists for an interesting intermediate case for which the displacements of the atoms $u_n$ differ from zero over a wider range of n, but deviate little from one atom to the next. In that case, the difference $u_{n+1} + u_{n-1} - 2u_n$ can be replaced by the second derivative of $u(n)$ with respect to n, $n$ now considered as a continuous variable. Inserting $x_n = 2b+u_n$ into 1.17 one obtains the sine-Gordon equation

$$\frac{d^2 u(n)}{dn^2} = \frac{2\pi U}{kb}\sin\left(\frac{2\pi u(n)}{b}\right) \qquad (1.23)$$

This non-linear differential equation is solved by a trick. One multiplies with $du/dn$ to obtain

$$\frac{du}{dn}\left\{\frac{d^2u}{dn^2} - \frac{2\pi U}{kb}\sin\left(\frac{2\pi u}{b}\right)\right\}$$

$$= \frac{d}{dn}\left\{\frac{1}{2}\left(\frac{du}{dn}\right)^2 + \frac{U}{k}\cos\left(\frac{2\pi u}{b}\right)\right\} = 0 \tag{1.24}$$

Partial integration yields

$$\frac{1}{2}\left(\frac{du}{dn}\right)^2 + \frac{U}{k}\cos\left(\frac{2\pi u}{b}\right) = K \tag{1.25}$$

We are looking for a particular solution which satisfies the boundary condition that $u(n) \equiv 0$ for $n \to -\infty$ and $u(n) \equiv b$ for $n \to +\infty$. Hence $K = U/k$. Further calculation yields

$$\left(\frac{du}{dn}\right)^2 = \frac{2U}{k}\left\{1 - \cos\left(\frac{2\pi u}{b}\right)\right\} = \frac{4U}{k}\sin^2(\pi u/b) \tag{1.26}$$

and

$$\int \frac{du}{\sin(\pi u/b)} = \sqrt{\frac{4U}{k}}\, n \tag{1.27}$$

After integration and solving for $u(n)$ one obtains

$$u(n) = -(2b/\pi)\tan^{-1}(\exp(n/r)) \tag{1.28}$$

Figure 1.39 shows solutions for $r = 1$ and $r = 5$. The width of the domain wall is about $4r$. The domain wall energy in this case is

$$E_\mathrm{w} = 2kb^2/(\pi^2 r) \tag{1.29}$$

The solution has acceptable accuracy for $r > 1$.

Because of the periodicity of the potential, the domain wall has the same analytical form if the center is shifted to any arbitrary value of $n$. Localized nonharmonic excitations like this are generally called *solitons*. More than one soliton and a periodic arrangement of solitons are further possible solutions of the sine-Gordon equation (1.23).

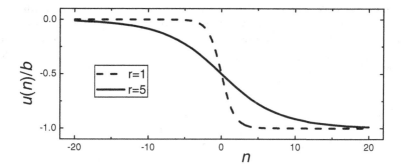

**Fig. 1.39.** Relative displacement $u(n)/b$ of the atom positions according to the continuum solution (1.28) of the Frenkel-Kontorova model. The width of the domain wall increases proportional to $r$.

### Dislocations

Domain walls in adsorbate lattices (Figs. 1.36, 1.37) may be considered as special forms of dislocations, which have their core at the interface between the adsorbate and layer and the first substrate layer. The direction of the dislocation line coincides with the direction of the domain wall. A ball model of the solution of the Frenkel-Kontorova-model discussed in the previous section is shown in Fig. 1.40.

The Burgers-vector of the dislocation as constructed by a closed lattice path (*Burger's circle*) is perpendicular to the dislocation line (and parallel to the interface). The dislocation is therefore an edge dislocation. Interface dislocations are ubiquitous in the heteroepitaxial systems.

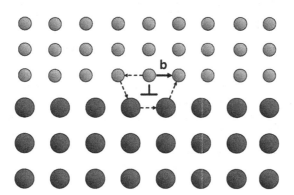

**Fig. 1.40.** Ball model of a dislocation at the interface between a substrate (large dark balls) and an epitaxial pseudomorphic layer (small light balls). The Burgers vector $b$ is parallel to the interface and perpendicular to the direction of the dislocation line.

Dislocations cannot end inside bulk material. The dislocation depicted in Fig. 1.40 could extend (perpendicular to the plan of drawing) until it ends at the boundary of the crystal, which is unlikely because of large lateral extension of the film compared to its thickness. Alternatively, the dislocation may form a closed loop at the interface, or it could bend upwards, go through the deposited film and end at the surface. In the latter case, one speaks of *threading dislocations*. As threading dislocations exist inside the film, they may have detrimental effects on its electronic properties. One therefore attempts to keep the dislocations at the interface, if possible.

On the reconstructed Au(111) surface, the surface layer atoms alternate between fcc- and hcp-sites (Fig. 1.12). In the language of dislocation theory, this type of dislocation is called a Shockley *partial dislocation*, "partial" because the glide vector amount to a fraction of a lattice unit. Two of these partials make one full displacement from one fcc-site to the next (Fig. 1.12). The atom positions of the Au(111) system may be described by a Frenkel-Kontorova type of model with an alternate sequence of two different potential valleys [1.14]. Similar cases that have been studied extensively are the monolayer deposits of Ag on Pt(111) [1.52] and Cu on Ru(0001) [1.53].

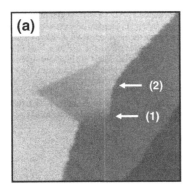 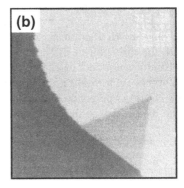

**Fig. 1.41.** STM images (50 nm × 50 nm) of dislocations at Ag(111) surfaces. **(a)** A step originates at the point where the core of a screw dislocation meets the surface. The step height increases gradually to the height of one atom layer. When this height is reached, the step appears fuzzy due to rapid kink motion along the step. The sharp lines forming a 60° angle are due to Shockley partial dislocations, i.e. due to stacking faults in the $(11\bar{1})$ and $(1\bar{1}1)$ planes inside the bulk. **(b)** STM image displaying the full base triangle of a tetrahedron with stacking faults in the $(11\bar{1})$, the $(1\bar{1}1)$, and the $(\bar{1}11)$ plane. The step height from the lower terrace onto the triangular plane is 2/3 of step height of a monolayer, the step height from the triangle to the next terrace amounts to 1/3 of a monolayer.

Surface line defects are also produced when a dislocation line of a bulk screw dislocation or of a dislocation with some screw character emerges at the surface. A step originates at the point where the dislocation core meets the surface. Such steps are frequently observed in STM images of metal surfaces, in particular when the samples have experienced a longer history of sputtering and annealing cycles, a procedure prone to generate dislocations in ductile metal crystals. An early report on dislocations concerns the Ag(111) surface [1.54]. Figure 1.41 shows STM images of an Ag(111) surface with several features caused by dislocations. In Fig. 1.41a a step emerges on the surface (arrow (1)) which increases in height until the height of a monolayer is reached at (2). From thereon the step appears rough due to the rapid motion of kinks along the step edge on Ag-surfaces at room temperature (Sect. 1.4.2). The sharp lines forming a 60° angle are due to Shockley partial dislocations. These consist of stacking faults in the $(1\,1\,\bar{1})$ and $(1\,\bar{1}\,1)$ planes inside the bulk. In Fig. 1.41b a full base triangle of a tetrahedron next to a normal step is visible. The tetrahedron is terminated by $(1\,1\,\bar{1})$-, $(1\,\bar{1}\,1)$-, and $(\bar{1}\,1\,1)$-planes in the bulk of the material along which a plane of atoms reside in hcp-sites. These stacking faults give rise to non-integer steps (in units of a monolayer). The step height from the lower terrace to the plane of the triangle is 2/3 of a monolayer, the step height from the triangle to the next layer 1/3 of a monolayer. Since the non-integer steps represent the protrusion of a bulk defect, no kinks exist on these steps. The edges of the triangle appear therefore sharp in the STM-image.

Bulk defects, which appear as protrusions on the surface are frequently observed on lattice-mismatched, epitaxial films. Thin films of Cu grown on the Ni(100) surface are an example [1.55]. The compressive strain in the Cu-

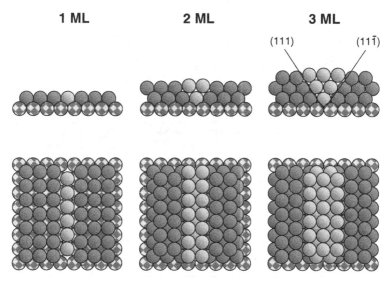

**Fig. 1.42.** Epitaxial growth of copper on Ni(100). Strain in the first monolayer is relieved by displacing a row of Cu-atoms (light balls) with respect to the other Cu-atoms. An additional row of Cu-atoms is inserted with each further layer, whereby internal {111} facets are formed. The faulted areas are higher at the surface and are therefore visible in STM-images as rectangular shaped area of slightly larger height (after Müller et al. [1.55]).

films due to the 2.6% misfit of the lattice constant is relieved by the introduction of stacking faults into the first Cu-layer, and a gradual build-up of internal {111} facets. The process is illustrated in a ball model in Fig. 1.42.

## 1.3.2 Point Defects

### *Kinks*

The most important point defect on surfaces is the kink. Figure 1.43 shows kinks in a step running along the direction of dense atom packing. Atoms in this particular site have the coordination 6. In German, this site is called a *Halbkristallage* (literal translation: *half crystal position*). A Halbkristallage is a very special site indeed: it takes exactly the cohesive energy per atom to move one atom from this site into the vacuum. The reason is that the kink reproduces itself when an atom is removed. By taking one atom after another, firstly an entire row of atoms is removed, until one comes to the end of the surface, then one might start on the next step. Every once in a while, one removes an atom of higher or lower coordination in the process, but for an infinite crystal the number of such atoms is vanishingly small compared to atoms at kink sites. The atoms in kink sites representing a Halbkristallage therefore determine the vapor pressure of solids.

Kinks in steps exist as thermal excitations. Their equilibrium concentration depends on the energy[1] $\varepsilon_k$ required to generate a kink (Fig. 1.44). For a densely

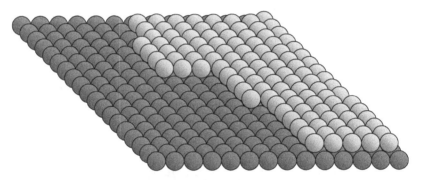

**Fig. 1.43.** Ball model of a (100) surface with a step along the [011] direction, which has kinks: two single atom kinks of opposite sign and a kink, which has a length that corresponds to three atom diameters.

packed step, this amounts to half the energy required to break one bond, alternatively, the energy involved in the reduction of the coordination number of a step atom by one. If the energy were a linear function of the coordination, then the kink energy would be one 12th of the cohesive energy. For copper that would amount to 290 meV, much larger than the experimental values ($\varepsilon_{k\{100\}}$= 129 meV and $\varepsilon_{k\{111\}}$= 117 meV [1.56]).

---

[1] More precisely, it is the work required to generate a kink (See Sect. 3). For a surface in vacuum, the work is the change in the Helmholtz-free energy, which is approximately equals the energy if the temperature is low.

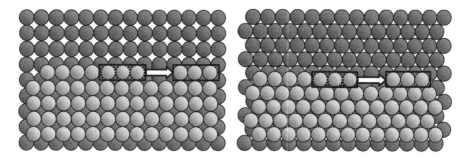

**Fig. 1.44.** Two kinks are generated by shifting a row of step atoms sideways. On the {100} and {111} surfaces of an fcc-material step atoms at densely packed steps have the coordination 7 and a kink atom has coordination 5. Two atoms are therefore brought from coordination state 7 to 6 in the process depicted above.

A better estimate for defect energies is obtained by scaling the binding energies of atoms with a fractional exponent of the coordination $\alpha$

$$E(C) = E_{coh} (C / C_{bulk})^{\alpha}. \tag{1.30}$$

Figure 1.45 shows the binding energy of an atom as a function of the coordination number relative to the cohesive energy for Cu, Ag, and Au according to the *effective medium theory* (EMT) in lowest level of approximation [1.57]. The theory is equivalent to the *embedded atom model* (EAM) [1.58].

The good fit of the "data"-points in Fig. 1.45 to the solid curve ($\alpha = 0.3$) should not be overrated; it is a consequence of the basic structure of the model, and not necessarily realistic in detail. By applying (1.30) with $\alpha = 0.3$ to Cu (cohesive energy 3.49 eV) one obtains for the kink energy $\varepsilon_{k\{100\}} = \varepsilon_{k\{111\}} = 134$ meV, which is close to the experimental numbers.

Since the kink energy is merely a small fraction of the cohesive energy, a considerable concentration of kinks exists at steps even at moderate temperatures. The concentration of kinks of either sign is

$$P_k = 2\exp(-\varepsilon_k / k_B T) \tag{1.31}$$

in which $k_B$ is the Boltzmann constant and $T$ the temperature. Eq. (1.31) holds if $k_B T$ is a fraction of $\varepsilon_k$ so that $P_k \ll 1$. To stay with the example copper the kink concentrations are $P_k(111) = 0.022$ and $P_k(100) = 0.014$ at 300 K.

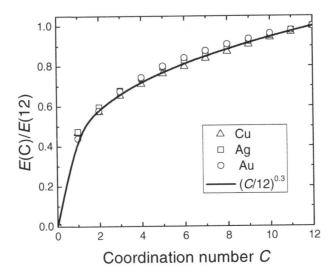

**Fig. 1.45.** Binding energy of atoms vs. coordination number for Cu, Ag, and Au, according to effective medium theory (EMT) in its lowest order approximation [1.57]. The solid line is a fit according to (1.30) with an exponent $\alpha = 0.3$.

In Fig. 1.35 a vicinal Si(100) was displayed with two different types of steps, the $S_a$-steps parallel to the dimer rows on the upper terrace, and the $S_b$-step perpendicular to the dimer rows on the upper terrace. Both steps contain kinks corresponding to an equilibrium concentration at about 600 °C [1.49]. The kink concentration on the $S_b$-steps is significantly higher than on the $S_a$-step, calling for a lower kink energy on the $S_b$-steps. In addition to kinks of one dimer unit there are many kinks which have a length of several dimer units, in particular on the rough $S_b$-step. Swartzentruber et al. have fitted the observed concentrations of kinks of various lengths to a model, which assumes that the kink energy is proportional to its length plus a corner energy.

$$\varepsilon(n) = \varepsilon_{corner} + \varepsilon_k n \qquad (1.32)$$

with $n$ the length of the kink in dimer units. The corner energy $\varepsilon_{corner}$ was determined to 80 meV, and the kink energy per length $\varepsilon_k = 28$ meV and 90 meV, for the $S_b$- and $S_a$-steps, respectively. The kink energies on Si(100) are much smaller than for Cu, in particular in relation to the cohesive energy (3.5 eV/atom for Cu, 4.6 eV/atom for Si). This is because no nearest neighbor bond breaking is required to generate kinks on the Si(100) surface due to the reconstruction.

In addition to kinks in thermal equilibrium, *forced kinks* exist on surfaces if steps run along a direction off the direction of dense atom packing. One possible reason is pinning of steps by impurities. Another typical case is a step as a boundary of a two-dimensional island. The step has then has segments of all directions and the concentration of forced varies along the perimeter of the island. A systematic way to produce steps with a uniform orientation in a particular direction is to cut the surface as a vicinal surface. Figure 1.46 shows a ball model and an STM-image of a Cu(5 8 90) surface as an example [1.59]. By virtue of these forced kinks, surfaces acquire a chiral character, which can be exploited to achieve enantioselectivity in catalytic reactions [1.60, 61].

**(a)**                                             **(b)**

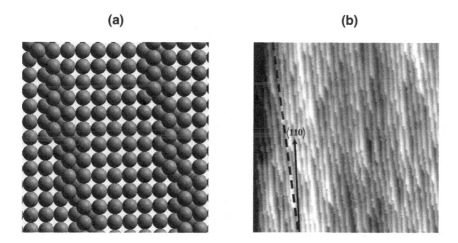

**Fig. 1.46**. Ball model of a (5 8 90) surface of fcc **(a)** and a STM image of the Cu(5 8 90) surface **(b)**. The real surface has a concentration of forced kinks according to the angle of cutting. Locally, the kink concentration fluctuates. Note that the surface has a chiral character (courtesy of Margret Giesen, [1.59]).

We note further that the product of the equilibrium concentrations of the kinks of the majority and minority type is $P_k^+ P_k^- = \exp(-2\varepsilon_k / k_B T)$. The equation is the same as for the product of electrons and holes in semiconductors (eq. 3.14): in both cases, a thermal activation process creates an equal number of species of either type (positive and negative kinks, electrons and holes). For kinks, the *law of mass action* has little practical value as hardly any kink of the minority type is found in STM-images (Fig. 1.46).

## Adatoms and vacancies

Atoms in kink sites (Halbkristallagen) determine the vapor pressure since the kink reproduces itself in the process of evaporation. By the same argument, the "vapor pressure" of a dilute two-dimensional lattice gas, in other words the concentration of atoms on a terrace is determined by the energy $E_A$ required to bring an atom from a kink site onto the terrace. Expressed in units of available sites, the equilibrium concentration of adatoms on terraces $\Theta_{adatom}$ is

$$\Theta_{adatom} = \exp(-E_A/k_B T). \tag{1.33}$$

The same equation hold for the equilibrium concentration of vacancies when the energy required taking an atom out of the surface and moving it to a kink site is inserted in (1.33). Note that the equilibrium concentration is independent of the position of the adatom or vacancy on the terrace, as long as the energy does not change as a function of position. The equilibrium concentration of adatoms and vacancies is very low at room temperature: On Cu-surfaces $10^{-12}$ ($E_A = 0.7$ eV) is a typical number. Despite their low concentration, adatoms and vacancies mediate noticeable mass transport processes on Cu-surfaces at room temperature because of their rapid movement. For the same reason, adatoms and vacancies cannot be observed in STM-images unless the surface temperature is very low (Sect. 1.4.2). Then, however, only atoms deposited at low temperature or vacancies generated at low temperature, e.g. via sputtering can be seen. The equilibrium concentration escapes observation. Thus, there is no direct experimental access to the equilibrium concentration and the energy of formation for these defects, in contrast to kinks in steps. Vacancies in surfaces can be made visible by decoration experiments. Deposition of a small amount on Mn or In leads to a decoration of vacancies as Mn or In atoms move into the empty sites and appear as an immobile protrusion of the surface there. Since vacancies are generated at steps, the decorated, immobilized vacancies form a spotted band along both sides of steps [1.62, 63]. These experiments reveal something about the kinetics of vacancy generation and diffusion, but not about their equilibrium concentration.

## Cluster, Islands, Mounds

A two- or three-dimensional compound of a few identical or different atoms on surfaces is called a *cluster*. The term cluster is used in particular when the unit forms a chemically different species, such as a $C_{60}$-cluster. Sometimes, in particular in theory, the term is also employed for units of a very few atoms (2-10) of the same element or compound as the substrate. The term *island* is used for an ensemble of many atoms (> 100) on a surface. If the island is only a monolayer high, then it is called a two-dimensional (2D) island. Three-dimensional islands are sometimes called *stacks, mounds* or *hillocks* depending on their shape.

Cluster, islands, and mounds on surfaces escape the scheme of classification employed here, insofar as their physical properties have 0-3 dimensional aspects.

Islands may trap an electron wave function, in particular that of a surface state (Sect. 8.3.3), thereby bearing the property of a quantum dot, hence a 0D property. The outer boundary, a step is a 1D feature. Larger islands have an extended, in case of heteroepitaxial islands possibly strained surface, giving them a particular 2D aspect. Finally, large multilayer islands represent 3D solid matter in a particular state.

## 1.4 Observation of Defects

Scanning probe microscopies, in particular the STM, have become the most important tool to observe and investigate defects on surfaces. Their great advantage is obvious: the nature and shape of the defects is identified from the images, their atomic structure can be seen directly, and their motion on the surface can be tracked. Nevertheless, diffraction techniques stand their ground. At elevated temperature, often even at room temperature, defects migrate across the surface far too fast to be imaged by the comparatively slow scanning microprobes. Diffraction techniques can still provide information on average properties, such as the mean concentration of defects or the mean shape of islands. For these reasons, diffraction, primarily electron diffraction, is still employed in surface defect analysis (for a review see [1.64]).

### 1.4.1 Diffraction Techniques

#### *Stepped surfaces*

A classical diffraction experiment concerns vicinal surfaces with regular step arrays. Such a surface has two periodicities, one is the atomic periodicity of the flat terraces, and the other one is the periodic array of steps and terraces. Diffraction therefore requires constructive interference with respect to both periodicities: a LEED spot arising from the surface unit cell can only appear as a (single) spot if one has constructive interference also with respect to diffraction from different terraces. This amounts to a third Laue condition concerning the wave vector components perpendicular to the terraces.

$$k_\perp - k_{0\perp} = G_\perp = 2\pi\, n/h \tag{1.34}$$

Here, n is an integer, $h$ is the vertical distance between terraces and $k_{0\perp}$ and $k_\perp$ are the components of the wave vector of the incident and the diffracted electron perpendicular to the terraces (not to the macroscopic surface!). If one has destructive interference ($n = 1/2$, $3/2...$) then the intensity is zero at the position of the beam diffracted from the flat surface. The diffracted intensity then appears in two spots, symmetric on both sides of the original spot. Figure 1.47 displays the Ewald-construction for diffraction from a vicinal surface with periodic steps. The spot intensities can be calculated in kinematic approximation in a little model. The model considers only s-wave scattering from the surface atoms. The surface atoms are assumed to be arranged in rows oriented parallel to the step direction. We are interested in the case in which the scattering vector $\Delta K = k - k_0$ is perpendicular to the steps.

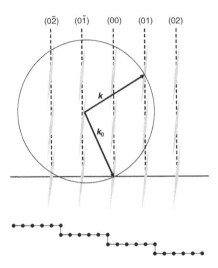

**Fig. 1.47.** Ewald-construction for diffraction from a vicinal surface with periodic steps. Diffracted intensity occurs when the scattering vector $\Delta K = k - k_0$ ends in the shaded areas. Diffraction spots alternate between singlet and doublets.

The position of the atom rows on the surface $r_{n,m}$ are denoted by the vector

$$r_{n,m} = n a e_x + m h e_z \tag{1.35}$$

with $e_x$ and $e_z$ the unit vectors along and perpendicular to the surface of a terrace, respectively, and $a$ the distance between the atom rows. The scattered intensity is

$$I \propto \left| \sum_{n,m} e^{i \Delta K \cdot r_{n,m}} \right|^2 \tag{1.36}$$

We assume a regular array of $M$ terraces each possessing $N$ atom rows and obtain.

$$I \propto \left| \sum_{n,m} e^{i \Delta K_x an} e^{i \Delta K_z hm} \right|^2 = \left| \sum_{n=1}^{N} e^{i \Delta K_x an} \right|^2 \left| \sum_{m=1}^{M} e^{i(\Delta K_x Na + \Delta K_z hm)} \right|^2$$

$$= \frac{\sin^2 (\Delta K_x aN/2)}{\sin^2 (\Delta K_x a/2)} \frac{\sin^2 \left[ (\Delta K_x aN + \Delta K_z h)M/2 \right]}{\sin^2 \left[ (\Delta K_x aN + \Delta K_z h)/2 \right]}$$

(1.37)

The second term in (1.37) is the stringent interference condition describing the spot size if the number of interfering terraces $M$ is large compared to the number of atoms on an individual terrace $N$. The smoother interference function of the $N$ atoms modulates the intensity of the spots. Figure 1.48 illustrates the two terms for $N = 5$ and $M = 50$, for the case of destructive and constructive interference between terraces, respectively. For constructive interference between terraces, that is for a particular energy, a diffracted beam appears as a single sharp diffraction peak. As the energy is lowered or increased the relative position of the sharp peaks (short-dashed lines in Fig. 1.48) and the broad peak (long-dashed line) shift with respect to each other. The product function then has its peaks shifted either towards the left or to the right of $\theta = 0$. Eventually a second peaks appears, and gains intensity. Under condition of destructive interference between terraces the two peaks have the same intensity and appear symmetric around $\theta = 0$. The width of the splitting is a measure of the terrace width: The magnitude of the splitting in units of the distance between diffracted spots is the inverse of the width of the

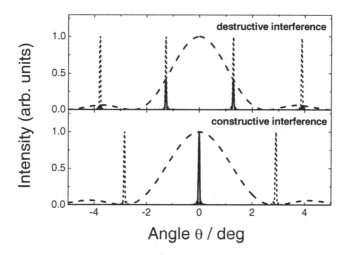

**Fig. 1.48.** Diffracted (00) intensity for perpendicular incidence from a vicinal surface with 50 terraces possessing 5 atoms each for destructive (*top*) and constructive (*bottom*) interference between the terraces (solid line). The long-dashed line is the scattered intensity from a single terrace (first factor in eq. (1.37)). The short-dashed line represents the interference between the terraces (second factor in eq. (1.37)).

terraces measured in number of atom rows. We have assumed that the terrace widths are all the same. Then, the sharpness of the spots depends only on the number of terraces contributing coherently to the scattering. In the calculated example 5×50 = 250 atoms contributed coherently. In reality, one has a distribution of terrace widths. The broader the terrace width distribution, the broader are the spots. The terrace width distribution can therefore be determined by measuring the intensity of diffracted spots with high accuracy. Due to the finite angular and energy resolution, conventional LEED-display systems (Fig. 1.2) do not suffice for that purpose. **S**pot **P**rofile **A**nalysis **L**ow **E**nergy **E**lectron **D**iffraction (SPALEED) requires special equipment, which is commercially available.

### Diffraction from other defects

Each type of defect gives rise to a particular diffraction pattern according to the structure of the defect and its embedding into the surface matrix [1.64]. This specific diffraction pattern can be used to identify the nature of the preponderant surface defect and to analyze its shape, size and concentration. However, with the availability of scanning microprobes, diffraction patterns have lost importance in studies of surface defects. Presently, diffraction techniques are mainly employed in cases where the defects exist merely on a very short time scale, move rapidly about on the surface in the temperature range of interest, or exist only under experimental conditions where scanning microprobe cannot be brought to bear. We therefore focus our discussion on one situation of this type and the typical defects, which are encountered there, and consider a surface that is subject to exposure by a constant flux of atoms or molecules. In order to ensure the epitaxial growth of a smooth surface the substrate temperature is typically chosen such that adatoms and other defects have a high mobility. With the exception of a very short period of nucleation, steady state concentration of the deposited atoms is small since they are quickly captured by the already existing nuclei (Sect. 11.1.1). The deposited single atoms have therefore no effect on the scattering. The preponderant defects during steady state growth are three-dimensional crystallites or monolayer high islands of deposited material. We consider the case of *layer-by-layer growth*: Then, the surface alternates between two limiting states, a flat surface and a surface covered with half a monolayer of islands of a particular size. The island size is determined by the nucleation density on the flat surface (Sect. 11.1.1). In the latter state, the surface has therefore a high concentration of steps that form the perimeter of the islands. In contrast to vicinal surfaces, these steps have all orientations. Constructive and destructive interference between scattered beams therefore concern all directions. Depending on the energy, the lattice diffraction spots would therefore alternate between the sharp spots and rings. As the islands generated in a nucleation process have a distribution of sizes, the diffraction spots alternate between sharp spots for constructive interference, and blurred spots for destructive interference between terraces. The mean size of the spots when they are blurred, measured in terms of the distance to the next reciprocal lattice rod, is the inverse of the mean island size, measured in terms of corresponding surface lattice. The shape of the blurred spot reflects the shape of the islands: square shaped-islands produce square-shaped spots. If the islands have rectangular shape, the spot are likewise rectangular with their long edge in the direction of the short edge of the islands in real space.

Since the time scale of the interaction of the electron with the solid is of the order of $10^{-16}$ s, which is much shorter than the time scale of any configuration changes on the surface, the scattering pattern reflects an average over snapshot images of the surface structure under the particular growth conditions.

### 1.4.2 Scanning Microprobes

The first instrument that enabled the imaging of surfaces with atomic resolution was the *Field Ion Microscope* (FIM) developed by E. W. Müller [1.65]. Though this technique is applicable only to the low index surfaces of a limited number of materials, many results of persistent importance on the diffusion of single atoms and of clusters consisting of a few atoms were obtained by this technique [1.66]. The major drawbacks of the method are the restrictions concerning sample material and crystal faces and the smallness of the surface areas that can be investigated. The *Scanning Tunneling Microscope* (STM) invented by Binnig and Rohrer in 1982 [1.67] does not suffer from those shortcomings. Over the years, STM has been developed to become the instrument for the investigation of surface morphologies on all length scales as well as of atom-size surface defects. It is no overstatement to talk about a Surface Science before and after the appearance of STM. Before the advent of STM, the focus was on the periodic, crystallographic structure of surfaces. The STM has opened our eyes to the richness of large scale morphological (mostly non periodic) features on surfaces and also to the fact that there is a rapid migration of atoms on surfaces at room temperature or above, even on those surfaces which appear as totally calm and rigid in their diffraction pattern. STM and other microprobes are not confined to vacuum environment. The *Electrochemical STM* has become a standard tool for atomic scale studies of processes on surfaces in contact with an electrolyte. STM is even more dominant in that field than in vacuum surface science because most classical surface probes that employ electron-, atom-, or ion-beams are not applicable in an electrolyte environment.

In the more recent years, various derivatives of the STM such as the *Spin Polarized Scanning Tunneling Microscope* (SPSTM) have debuted. The *Atomic Force Microscope* (AFM) originally suffering from a lower, non-atomic resolution has now a lateral resolution comparable to the STM, and has found widespread application in the investigation of insulator and soft matter surfaces. Nevertheless, for the surfaces addressed in this volume, STM remains to be the most important microprobe. We briefly describe the experimental technique in the following, as well as some particular features, which have to do with the time scale of the STM imaging process in relation to the persistence time scale of surface phenomena, and consider tip surface interactions.

## *The scanning tunneling microscope*

The basis of the STM is the quantum mechanical tunneling process and its extreme sensitivity on the width of the tunnel gap. A sharp metallic tip (ideally terminated by a single atom) is immersed into the evanescent electron wave functions outside the solid surface. For small tunneling voltages, the tunneling current is proportional to the density of those electrons in the tail that have their energy at the Fermi-level. As the electron density decreases exponentially with the distance, the tunnel current also decreases exponentially with the distance. In the standard mode of operation, the tunnel current is held constant by a feedback loop, so that the tip follows the contour of constant density of electrons at the Fermi-level. An alternative mode of operation is the constant height modus. The latter is employed in particular for fast scans (video frequency) over small and mildly corrugated surface areas.

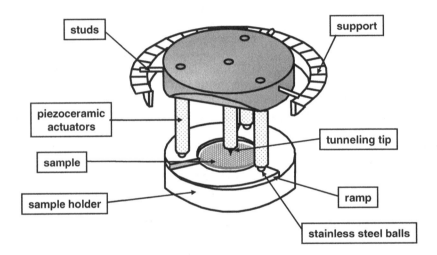

**Fig. 1.49**. The STM of Frohn, Wolf, Besocke and Teske [1.68].

The lay-out of tunnel microscopes has changed considerable since the first design of Binnig and Rohrer [1.69]. Here, we review the design of Frohn et al. [1.68] (frequently also referred to as the Besocke-microscope) which has the advantage of a very small temperature drift due to its built-in compensation of thermal expansions. The microscope is shown in Fig. 1.49. The central piezoelectric actuator carrying the tunnel tip is fixed to a tripod of three further piezoelectric actuators of the same type each resting on a 120° ramp. Oxidized stainless balls are mounted on the latter three actuators. The central position of the actuator carrying the tip and the fact that all four actuators are alike ensures an almost perfect thermal compensation, so that the tip has very little thermal drift with respect to the surface, neither laterally nor vertically. Scanning along the *xy*-directions is performed either by bending the actuators of the tripod or by bending the central actuator. In order to operate the STM in the latter mode the central actuator must feature two pairs of metal plates for bending (preferably, in the upper part close to where the actuator is fixed to the tripod) in addition to the tube metal plating for the vertical motion of the tip. The STM is positioned onto the sample holder by lowering the entire four-actuator unit by the support ring. Because of the tripod arrangement, the system is mechanically quite stable and insensitive to vibrations. Atomic resolution is obtained with a minimum of vibration insulation, e.g. by placing the sample on a stack of metal plates with Viton™ dampers in between. Better insulation, in particular against sound, is obtained by placing the microscope on a base plate that is suspended by soft, damped springs.

The microscope works also in an inverted form: the four-actuator unit is mounted on a base plate with the tip pointing upwards while the ramp/sample unit rests on the three stainless steel balls of the outer three actuators. This latter version recommends itself in cases where in-situ samples changes via a transfer system are required. Controlled heating of the sample to temperatures above room temperature is also possible in that case. One has the additional technical advantage that the STM head with its large number of electric leads need not be moved about. The version shown in Fig. 1.49 is more suitable if cooling of the sample below room temperature is desired.

Despite its undisputable advantages, the Besocke-microscope has also the disadvantage that the coarse approach is delicate and requires a good match of sample and tip adjustment within a margin of a few tenth of a millimeter. Commercial microscopes for standard applications tend to trade off stability and small thermal drift for easier handling by employing xyz-linear motors for the tip manipulation and a sample mount that permits an easy exchange of samples.

### The time structure in scanning probe images

The time required for a single STM image ranges between 50 ms and a few minutes. The shortest times per frame are obtained by using the constant height mode with the feedback loop shut off. A video-frequency repetition rate of STM-frames is thereby achieved. However, even a video STM is a slow instrument the time scale of atom diffusion, unless the temperature is correspondingly low. Imaging single atoms requires therefore cooling of the sample, in many cases down to the liquid helium range. The situation is different when the atoms are part of a periodic structure. Then the lateral interactions keep the atoms in place even at room temperature and above. Atomically resolved images of periodic adsorbate structures are therefore readily obtained.

Similar considerations apply to images of other morphological features. In all cases, the effect of the relative time scales of the imaging process and of possible structural changes on the surface have to be taken into account. A good example for the importance of the time scale is the *fuzzy* appearance of monatomic steps (Fig. 1.41 and 1.46). The STM tip finds the step at a different position in each scan line (see also Sect. 10.5).

### Tip surface interactions

The original use of the STM was to investigate surface structures. In the course of these investigations, researchers frequently found that the tip had an effect on the image. Atoms appeared to be dragged along with the tip and occasionally entire steps were drawn from one position to another [1.70]. These effects frequently depend on tunneling voltage and current, unfortunately not always [1.71]. The very presence of a tip above an atom can have an effect on the activation energy for diffusion. Papers have been published which called the entire use of STM for quantitative analysis of surface diffusion studies into question, but this is throwing the child with the bathtub. In many cases, the tip has no effect on the observed processes and there are simple test to prove that this is so. One simply has to make sure that no quantitative aspects of the data depend on the time the tip spend over the investigated area and also not on the number of times the STM tip had visited the area in question. Such test are performed by comparing data obtained with different scan speeds, or even better, with data that are obtained by introducing a variable pause in the scan process after each scan line, and compare that data with those obtained in continuous scanning.

There is also a good side to tip/surface interaction: the tip can be used to manipulate atoms on surface, move them about and place them into patterns in a controlled way [1.72, 73]. In these days, it seems that every university or research institute likes to see its logo made with atoms using STM-tip manipulation. Atoms can be moved by using the tip as a push rod. One can also employ attractive forces between the tip and an adatom or adsorbed molecule [1.73]. Alternatively, an adsorbed molecule can be attached to the tip by applying a suitable potential, and then dropped at the desired place on the surface. Aside from making logos one can do other, scientifically more interesting things with atom manipulation. Studying and inducing chemical reaction on a single molecule [1.74, 75] is one, building special quantum structures [1.76, 77] another (see Sect. 8.3.3).

## 1.5 The Structure of the Solid/Electrolyte Interface

This section is brief for two reasons. One reason is that experimental results on the crystallographic structure of the solid/electrolyte interface are scarce. Secondly, some structures of solid surfaces in vacuum as discussed in the preceding sections persist in an electrolyte environment. For example, the reconstructions of the Au(100), (110) and (111) surfaces exist also when the surface is in contact with an electrolyte. However, one feature is unique to the interaction with the electrolyte and the associated charging of the surface. That is the phase transition to an unreconstructed state of the surface at positive potentials of the gold sample. The potential induced phase transition is a consequence of the interfacial thermodynamics, which is discussed in detail in Section 4.2.3. The definition of an electrode potential and the standard experimental setup in electrochemical work is

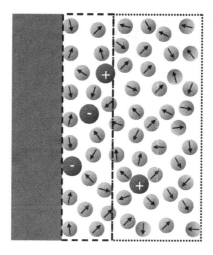

**Fig. 1.50.** Model for an electrolyte near a solid surface. See text for discussion.

described in Section 2.3. Basic properties of the electrolyte in the vicinity of the solid are described in Section 3.2.3. Here, we focus on a general introduction to structural properties at the solid/electrolyte interface.

It is convenient to divide the interfacial region conceptually into three zones (Fig. 1.50), the solid, a layer of strong electrolyte/solid interaction called *Stern layer* (circumscribed by dashed lines), and the electrolyte in the vicinity of the solid but far enough to maintain its character of a bulk liquid electrolyte (circumscribed by dotted lines). We assume the electrolyte to be an aqueous one. Within the electrolyte, ions are dressed by solvation shells of polarized water molecules, which screen the fields originating from the ions (Fig. 1.50). The ions with their solvation shell carry a net charge, which moves in an electric field. A space charge layer can build up near the solid surface, with a characteristic decay length depending on the ion concentration in the electrolyte. The electrical properties of this zone can be treated in a continuum approximation (Sect. 3.2.3).

The Stern layer consists (i) of water molecules, (ii) of ions, which have kept their solvation shell and (iii) ions in a chemisorbed state. Electrochemists call these adsorbed species *specifically adsorbed ions* to distinguish between species that adsorb directly on the solid surfaces and those ions that keep their solvation shell. The term is somewhat unfortunate as it carries the connotation that the ions retain their ionic character at the surface, which is not at all the case (see Sect. 3.1.3). It is a well-known result of Solid State theory (although not always appreciated in full consequence) that the electrons of the solid screen the ions almost completely leaving only a modest polarization, which gives rise to a modification of the work function. Examples for "specifically adsorbed ions" are Cl, Br, or $SO_4$, to name a few. When these species are adsorbed on the surface, they form structures and superlattices, in principle as known from vacuum physics. For a limited number of systems a structure analysis with a limited scope has been performed using GIXRD (see e.g. [1.78]). The surface concentration and thereby the structure is controlled by the surface charge on the solid. This aspect is considered later in Sect. 6.2.5.

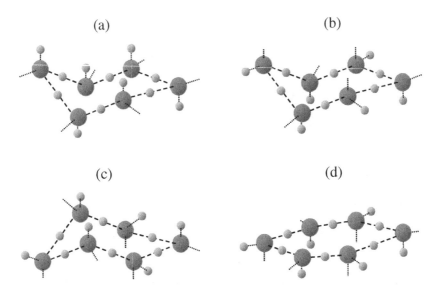

**Fig. 1.51.** (a) The bi-layer structure of water in densely packed adsorbate layers as proposed in 1980 [1.79] from vibration spectroscopy. In order to form a 2D-network three of the hydrogen atoms pointing upwards or downwards have to be committed to bond oxygen atoms to neighboring rings. In 2002 Meng et al. [1.80] showed that the structures shown in (b) and (c) with hydrogen and oxygen lone pair bonding to the surface have nearly the same energy. According to Ogasawara et al. [1.81] the oxygen atoms should be nearly coplanar and bond to the surface alternatively by the oxygen-lone pair orbital and by hydrogen bonding (d).

The bonding of water molecules has been investigated theoretically and experimentally for metal surfaces. The bonding of single water molecules is discussed in Sect. 6.4.4. More relevant to the solid/electrolyte interface is the structure of water that forms at higher coverage. From vibration spectroscopy is was concluded that water molecules form hexagonal rings of bi-layers and that the water molecules bond to the Pt-surface in two ways, via the oxygen lone pair bonds and via hydrogen bonding with the hydrogen atoms pointing downwards [1.79] (Fig. 1.51a). In order to form a defect free 2D-network three of the hydrogen atoms pointing upwards or downwards have to be committed to bond the oxygen atoms of neighboring rings. The lateral position of the oxygen atoms in the network is as in graphite [1.82, 83]. In 2002 Meng et al. [1.80] showed that the structures displayed in Fig. 1.51b and 1.51c with hydrogen and oxygen lone pair bonding to the surface have nearly the same energy on Pt(111). In the same year, Ogasawara et al. proposed [1.81] that the oxygen atoms should be nearly coplanar with a buckling of merely 0.25Å (Fig. 1.51d). According to their model, the oxygen atoms bond to the surface alternatively by the oxygen-lone pair orbital and by hydrogen bonding. There is consensus that the oxygen atoms are placed in the a-top position on Pt(111). For the case of Ru{0001} a structure involving a partial dissociation of water was proposed [1.84, 85]. However, the structure was not confirmed for other surfaces. It is therefore probably less relevant for solid/water interfaces at room temperature.

The theoretical and experimental studies cited above concern adsorbed water layers on surfaces in vacuum that are stable only at temperatures below 150K because of the small binding energy. There is also consensus that the energy differences between the various proposed structures are small. As neither theoretical calculation involves entropy, which should be relevant as the binding energy of water is small, theory cannot convincingly converge on one particular structure model, not even for one particular surface. With regard to the solid/water interface at room temperature, possible entropic contributions to the free energy and the volatility of the hydrogen bond make it likely that all structure elements considered above are simultaneously present at the solid/water interface with a rapid flipping back and forth between various configurations. At room temperature, water at the interface is therefore in a state that is between a solid and a liquid. The models for the ice-layer predict the oxygen atoms to be at distances between 1.31Å and 1.98Å from the boundary of the Pt-atoms which is consistent with the pronounced peak in the mean density of the oxygen atoms in a water layer as a function of the distance from a solid surface [1.86]. Hydrogen bonding by water molecules is presumably also important in the ordered structures formed by "specifically adsorbed ions" on surfaces. This concerns in particular the oxygen containing species $ClO_3$, $SO_4$, and structures investigated by STM and infrared spectroscopy have been interpreted that way.

# 2. Basic Techniques

## 2.1 Ex-situ Preparation

### 2.1.1 The Making of Crystals

Surface Science is built on single crystals. The starting point for experimental work on surfaces is therefore the preparation of oriented surfaces on single crystals. Either that surface is the object of study or it serves as a template on which the material to be investigated is grown epitaxially. Depending on the material, single crystals are commercially available or are grown by research institutions for specific purposes. Rarely does a surface scientist grow crystals for her/himself. The interest in the surfaces properties of a material usually encompasses the bulk properties of the material as well, and should so, because the understanding of surface properties requires a good knowledge of bulk properties. For some materials, semiconductors in particular, the commercial interest has created a market for semi-finished products in the form of large area wafers of a particular surface orientation. The technology of crystal growth and wafer preparation is most advanced for silicon. Wafers with 12-inch diameter represent the current industrial standard. Wafers are also available for III-V compounds. Because of the stringent requirements of industry, the surfaces of these wafers are extremely flat, oriented with a high precision, and free of contaminants. The wafers are typically coated by a protective coating, which is easily removed after insertion into an *ultra-high vacuum* (UHV) chamber. For silicon, the coating is a thermally grown $SiO_2$-layer. The oxide layer is removed by heating to about 1000 °C in UHV and a well-ordered surface is the result. Creating a clean surface that way is not only easy, but also superior to any conceivable homemade preparation method, because of the enormous industrial R&D-efforts that went into silicon wafer technology.

The situation is different for metals surfaces. Here, the typical starting point is a single crystal from some commercial source. For some materials however, small single crystal beads are grown easily from high-purity wires. This is by far the cheapest method of crystal growth, and even superior the conventional methods regarding the purity and absence of dislocations. The method is therefore described in the following.

Figure 2.1 displays the equipment schematically. The end of a pending metal wire of 0.5-1 mm diameter is molten by a propane/air or a hydrogen/air flame to form a droplet of 2-5 mm diameter. A stream of argon coming from the side serves two purposes; one is to surround the metal droplet with inert gas, the other to push the flame gently to the side. By varying the argon flow and/or the gas supply to the burner the temperature of the droplet can be controlled accurately. Upon cooling, the solidification begins at the wire. With lowering temperature, the solid/liquid phase boundary moves away from the top (Fig. 2.2) until the entire drop is solidified. The result is not necessary a single crystal. However, by processing the crystal repeatedly between melting and solidification, and by beginning the solidification process at the wire end very slowly one eventually may grow the crystal from a single nucleus, and thereby grow a single crystal.

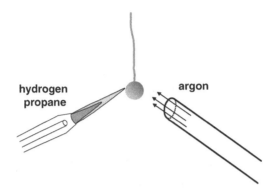

**Fig. 2.1**. Growing single crystal beads from a wire.

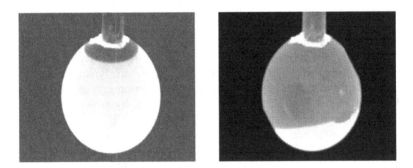

**Fig. 2.2**. Growing a bead crystal from a molten droplet. The pictures show a platinum bead. The lower light part is the melt. Upon cooling, the solidification begins at the wire (left) and eventually the entire droplet is solidified. The clearly visible {111} facets indicate the growth of a single crystal. The bright ring-like structure at the top is some dirt that accumulates in the transition area between the wire and the bead crystal in the course of repeated melting and solidification (courtesy of Udo Linke).

By repeated melting and solidification, one may further drive impurities contained in the original material upwards to the wire end of the bead and accumulate the impurities there: a poor man's way of refining by zone melting! Single crystals of high purity are obtained for Au, Pt, Ir, Rh and Pd. Propane gas is used for Au; hydrogen for the other materials. Palladium dissolves large concentrations of hydrogen. Repeated zone melting avoids an accumulation of hydrogen, however. Bead crystals can also be grown from other metals, e.g. Ag and Cu, however only in vacuum. The wire is then heated at the end by electron bombardment. Single crystals of alloys, e.g. Pt/Rh alloys, can likewise be grown.

The final single crystal beads displays {111} facets (Fig. 2.2) as the {111} facets do not undergo a roughening transition below the melting point (Sect. 4.3.1).

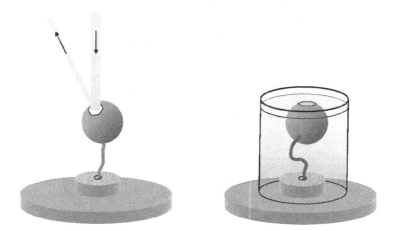

**Fig. 2.3**. Bead crystals can be oriented to within 1°-2° by bending the wire such that the reflection of a laser beam from the {111} facets is in the desired direction. Subsequently the entire sample is embedded into meta-acrylate or an epoxy-resin.

## 2.1.2 Preparing Single Crystal Surfaces

In order to prepare surfaces of the desired orientation from bulk single crystal material the crystal has to be first oriented and then cut along an oriented plane, and finally polished to a shiny surface.

### Bead crystals

An approximate orientation is easily arranged for bead crystals with the help of the {111} facets: The bead crystal is mounted by fixing its wire to the head of the goniometer that is used later for x-ray orientation and polishing (see below). If the bead crystal is illuminated by laser beam that has a diameter larger than the bead crystal, the {111} facets exposed to the beam produce defined reflected beams (Fig. 2.3). A rough orientation to the desired direction is then obtained by simple bending the wire by hand. Once the rough orientation is achieved, the wire and the bead crystal is fixed by embedding it into meta-acrylate (or into an epoxy resin).

The embedded bead crystal is mounted on a polishing jig as shown in Fig. 2.4. The jig is designed for fine adjustment of the orientation within a range of 6° for the polar angle and 360° for the azimuth.

**Fig. 2.4**. A polishing jig that allows for adjustment of the tilt angle in an arbitrary azimuthal direction (courtesy of Udo Linke). A bead crystal is mounted for illustrative purposes. The bead crystal is embedded into meta-acrylate before polishing.

The embedded crystal together with the resin is abraded using SiC abrasive paper to expose the bead. The polishing jig is then fixed to an X-ray equipment to check and further adjust the orientation of the crystal according to the Laue pattern, or the diffraction pattern obtained by characteristic X-ray emission lines. After this final orientation, the polishing procedures can begin. These typically comprise the following steps:

1. Wet grinding with SiC- paper, grain size 400, 800, 1200.
2. Cleansing to remove SiC particles.
3. Wet polishing on silk cloth with diamond paste, grain size 6μ
4. Wet polishing on nylon cloth with diamond paste, grain size 3μ
5. Wet polishing on velvet cloth with diamond paste, grain size 1μ
6. Wet fine-polishing on velvet cloth with $Al_2O_3$ paste, grain size 0.03μ
7. Cleansing and drying with alcohol.
8. Final check on orientation using X-ray methods (e.g. rocking curves).

9.  Removing the resin.
10. The cleaned and dried crystal is placed into a quartz furnace and annealed in an $H_2$/Ar or an $O_2$/Ar atmosphere (depending on the material) to heal-out surface defects and to leach-out common bulk impurities like sulfur and carbon.
11. The sample is probed in a scanning electron microscope using *Selected Area Channeling Patterns* (SAPS)

An example of the final product is displayed in Fig. 2.5.

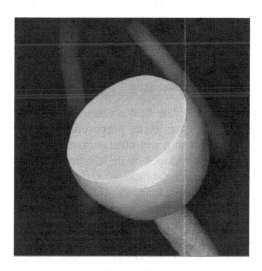

**Fig. 2.5**. The ready-to-use bead Pt-crystal exposing a (111) surface. Note the other {111} facets on the side in trigonal symmetric arrangement (courtesy of Udo Linke).

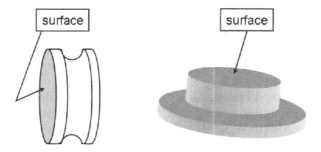

**Fig. 2.6**. Two common forms of ready-to-mount metal crystals. The left form is for use with a tungsten or molybdenum wire loop. The hut-crystal is fixed to the sample holder by clamping the crystal on the rim of the hut. The thereby induced plastic deformations stay away from the surface, at least for some time.

*Samples cut from bulk single crystals*

A few additional steps are required if metal samples are to be prepared from a bulk single crystal. These crystals come from the manufacturer in the form of oriented rods with diameters between 5-20 mm. The first preparation step is a rough orientation using the Laue pattern. A disk is then cut off the rod using a spark erosion wire saw or an acid wire saw for ductile metals. Brittle materials are cut using a peripheral saw or an annular saw with diamond or SiC fortified blades. The use of spark erosion has the advantage that the sample form can be chosen more freely. After cutting, the damaged surface layer is removed by etching. The crystal is annealed as described in step 10 above. After that, the crystal is mounted on the polishing jig using dental wax, oriented, and polished following the procedures described before. Figure 2.6 shows two commonly used forms of metal crystals.

*Preparation of Si-wafers by wet chemistry*

Single crystal surfaces of silicon are best obtained directly from wafers as they are provided by industry for commercial purposes. No further polishing is required, but a final treatment using wet chemistry may be recommended. Such preparation procedures were developed by the electronic industry for the purpose of wafer cleaning prior to device fabrication. Originally, the main goal was to achieve a surface free of metal contaminants. Nowadays, wafer quality assurance concerns the control of many impurities such as O, C, F and H, as well as the control of the roughness of surfaces. The silicon preparation methods split into two categories as they generate either hydrophilic or hydrophobic surfaces. Hydrophilic surfaces are terminated by polarized bonds, such as Si-O or Si-OH, which bind water molecules via hydrogen bonding. Hydrophilic surfaces are used as intermediates to prepare well-defined $Si/SiO_2$ interfaces. Hydrophobic surfaces are covered mostly by Si-H, i.e. by non-polarized bonds.

Among the methods that employ hydrophilic surfaces, the most widely used method is the so-called *"RCA cleaning"*, which involves a set of different cleaning steps specifically designed to remove a particular contaminant or a particular class of contaminants. All steps contain hydrogen peroxide as oxidizing agent, and other chemicals to eliminate a selected contaminant. Three basic steps are:

1.  $H_2O_2$ + $H_2SO_4$, 10 minutes at 80°C, (Removes the residues of a photo resist or other organic contaminants, and forms a 1 nm layer of $SiO_2$).
2.  $H_2O_2$ + 1 $NH_4OH$ + 5 $H_2O$, 10 minutes at 80°C (Removes organic contaminants and some metals and forms a 1nm layer of $SiO_2$).
3.  $H_2O_2$ + 1 HCl + 5 $H_2O$, 10 minutes at 80°C (Removes heavy metals and forms a 1nm layer of $SiO_2$).

In each step, a 1nm layer of $SiO_2$ is formed which serves to protect the surface against reactive contaminants from air such as unsaturated hydrocarbons. The

1nm $SiO_2$ must be removed before the next cleaning step. This is done by a hydrophobic etch procedure which is customarily a HF-dip (hydrofluoric acid) followed by rinsing with ultra-pure, de-ionized water.

It is known for a long time that HF etches $SiO_2$ very effectively. The polarized Si-O bonds are broken by the polarized HF molecule. The $F^-$ -ion binds to the positively charged Si-atom and the proton is captured by the O-atom. The reaction mechanism can be described as

$$(\equiv Si-O-Si\equiv) + 2HF \Rightarrow 2(\equiv Si-F) + 2H_2O$$

It had been suggested initially that the resulting chemical stability of the silicon surface is due to a passivation of silicon by fluorine. This hypothesis was supported by the fact that the Si-F bond strength ($\approx 6.0$ eV) is much greater than that of Si-H ($\approx 3.5$ eV). Nevertheless, fluorine has been observed to be a minority species on the etched surface. The surface is terminated mostly by hydrogen, but hydrocarbons and oxygen have also been found [2.1-4]. Several investigations showed the latter to be contaminants so that the surface is genuinely H-terminated. This striking fact has been explained on the basis of reaction kinetics by Ubara et al. [2.2]. They postulate that fluorine terminated silicon complexes are unstable in HF solution due to strong polarization of the Si-Si back bonds and are removed from the Si-surface by releasing $SiF_4$ into the solution, leaving a H-terminated surface behind.

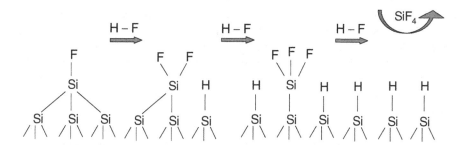

Quantum chemical calculations of the activation barriers for these types of reactions [2.5] show that a reaction such as $\equiv SiH + HF \rightarrow \equiv SiF + H_2$, though exothermic, has an activation barrier, which is significantly higher than that of the Si-Si bond cleavage reaction. As the hydrogen termination results from kinetics it is clear that some fluorine remains on the surface. The amount depends on the concentration of the hydrofluoric acid and varies between 6% and 12 % for HF concentrations between 24% and 48% [2.6].

The HF-dip not only produces unsatisfactory results concerning the H-termination but also a considerable surface roughness. Smoother surfaces are obtained when HF-solutions buffered at higher pH-values are employed. The following recipe produces very smooth, H-terminated Si(111) surfaces. Si(100) surfaces produced that way are rougher and the H-termination is less perfect. All agents must be *"electronic grade"*. The water must be de-ionized and free of hydrocarbons (total hydrocarbon concentration, TOC < 3 ppb).

1.  Oxidize wafer thermally ($900\,°C$, under $O_2$ flow, oxide thickness 100 nm).
2.  Rinse sample using trichloroethylene, acetone and alcohol.
3.  Place teflon beaker and tweezers in a glass beaker and clean to remove the residual organic substances, using $H_2O:H_2O_2:NH_4OH$ 4:1:1 at $80\pm2\,°C$ for 10 minutes.
4.  Repeat step 3 with the sample inside the glass beaker.
5.  Pre-etch the thermal oxide in $H_2O:H_2O_2:HCl$ 1:1:4, 5 minutes at $80\,°C$.
6.  Rinse thoroughly in water.
7.  Remove remaining oxide using buffered HF (pH = 5.0), 5 minutes in the teflon beaker.
8.  Rinse.
9.  Oxidize the sample softly: $H_2O:H_2O_2:HCl$ 4:1:1, 10 minutes at $80\,°C$.
10. Rinse.
11. Remove oxide layer built in the preceding step and H-terminate surface using 4 % $NH_4F$ (pH = 7.8) for 6.5 minutes in the Teflon beaker.
12. Rinse thoroughly.

It is very important that samples have no contact with air after step 5, or at least after step 7. Pulling the sample through a liquid-air interface surely contaminates the surface, since hydrophilic organic contaminants decorate the surface of the liquids. Thus, solutions are removed by diluting with water and the sample is brought into the active agents while being covered with water. This is far more important than keeping the exact concentrations.

The sample is now H-terminated and largely inert. It contaminates only very slowly in air, mainly through an uptake of unsaturated hydrocarbons. Storing the sample in hydrocarbon-free argon or nitrogen helps to keep the sample clean for a longer time if desired. The sample can be inserted into a vacuum-chamber via an air lock pumped by an oil-free pumping system. It has been recommended that the initial pumping down from atmospheric pressure should be performed slowly via a leak valve to avoid turbulences, as these turbulences could bring organic compounds onto the surface. The importance of this procedure presumably depends on the cleanliness of the air lock and the pumping system used.

# 2.2 Surfaces in Ultrahigh-Vacuum

## 2.2.1 UHV-Technology

### Basic issues

Solid surfaces are investigated in *Ultra-High Vacuum* (UHV) environment in order to minimize the interaction with foreign materials. The number of molecules that arrive on a surface per time on an area $A$ equals the number of molecules contained in cylinder of a height $\langle |v_x| \rangle$ with the base area $A$. The mean velocity $\langle |v_x| \rangle$ of gas molecules with a velocity direction towards the surface is

$$\langle |v_x| \rangle = \sqrt{k_B T / 2\pi m} \qquad (2.1)$$

Here, $m$ is the mass of the molecule, $k_B$ the Boltzmann constant and $T$ the temperature. With the state equation of the ideal gas of pressure $p$, $pV = Nk_B T$ the number of molecules in the said cylinder is

$$\frac{N}{V} \langle |v_x| \rangle A = p A / \sqrt{2\pi m k_B T} . \qquad (2.2)$$

The flux $F$ of molecules impinging on the surface per area and time is therefore

$$F = p / \sqrt{2\pi m k_B T} \qquad (2.3)$$

A standard time span for surface studies is about $10^4$ s. If one wishes to have less than $10^{13}$ molecules per $cm^2$ on the surface (corresponding roughly to 1/100 of a monolayer) a pressure of $5\times10^{-10}$ Pa ($Nm^{-2}$) ($\cong 5\times10^{-12}$ mbar) is required. A pressure as low as this is in the range described as *Extremely High Vacuum* (XHV). As most of the residual gas molecules are harmless to the surface, a pressure of about $10^{-8}$ Pa suffices for most experiments. The density of molecules in a "vacuum" of that pressure is $10^6$ $cm^{-3}$ and the free mean path of the molecules is about $10^6$ m. A typical UHV-vessel may have a volume of 100 l and an inner surface area of 1 $m^2$. The total number of molecules in the gas volume is then $10^{11}$. Because of the large mean free path, these molecules never meet each another in the gas phase; rather they traverse the vessel from wall to wall. Hence, the walls of the vessel determine the properties of the gas phase in the UHV-regime. The walls also host the vast majority of the molecules inside the vessel in the form of an adsorbate phase. Let us assume for the purpose of illustration that the walls are covered by a monolayer of molecules. The total number of molecules on the walls is then about $10^{21}$, which exceeds the number of molecules in the volume of the vessel by ten orders of magnitude! If all these atoms would desorb and be in the gas-phase, the pressure would increase by ten orders of magnitude, i.e. to 100 Pa. Even though the coverage of the walls may be much less than a monolayer, it is clear from the example

that the walls control the pressure in a UHV-vessel. This has important consequences: if a part of the surface of the vacuum vessel is heated, molecules desorb
from there in enormous quantities, giving rise to a pressure burst. Maintaining
UHV-conditions during heating some parts of the vessel or the sample in its sam-
ple-holder requires a careful out-gassing of those parts by heating during the initial
pump-down. The de-gassing process as such cannot take place once the pressure is
already in, or near UHV-range. Staying with the example above and assuming a
pumping speed of 100 l/s, it would take $10^{10}$ seconds to remove the monolayer of
gas from the walls if pumping were performed at $10^{-8}$ Pa, but it takes merely one
second at 100 Pa. The natural coverage of air exposed surfaces with gases, in particular water, is removed and UHV-pressures are eventually achieved only by
baking the entire vacuum chamber while pumping the vessel at moderately low
pressures of $10^{-2}$-$10^{-4}$ Pa.

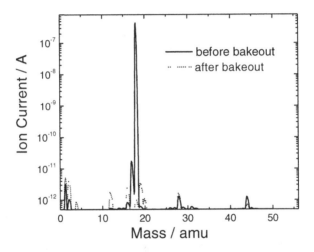

**Fig. 2.7.** Mass spectrum of the residual gas in a UHV-vessel (logarithmic scale) before and
after bake-out (solid line and dashed line, respectively). Before bake-out, the mass spectrum
is dominated by water (mass 18 plus some 17). After bake-out the spectrum contains hydrogen (mass 1 and 2) methane (mass 16), CO (mass 28), and $CO_2$ (mass 44) as well as some
cracked hydrocarbons as evidenced from the peaks to the left and right of mass 28.

Typical bake-out temperatures are 150-250 °C. Construction materials must withstand these temperatures and must have a low enough vapor pressure. Because of
its zinc content, brass is not a suitable UHV-material, for example. Other unsuitable materials comprise rubber, plastics including Teflon (because of its fluorine
content), solders containing Cd and porous materials which tend to release gases
forever. Classic construction metals for UHV are stainless steel for most parts of
the vacuum chamber, copper for gaskets and as a good heat conducting metal,
tungsten for filaments, molybdenum and tantal for parts to be heated to high temperatures, and aluminum/magnesium alloys for those parts of the construction for

which low weight is required. Suitable insulating materials are sintered, impervious ceramics made of $Al_2O_3$, sapphire for good heat conduction combined with electrical insulation, and machinable ceramics such as Macor™. A good flexible material for electrical insulation is Kapton™, which withstands heating in vacuum up to 250 °C.

The standard construction material for UHV-vessels is stainless steel. In its natural state, stainless steel contains large quantities of dissolved hydrogen. This hydrogen outgases only very slowly during normal bake-out procedures so that the walls of the vessel keep their hydrogen inventory over a lifetime of many years. This dissolved hydrogen therefore determines the residual pressure of a well-baked system. Figure 2.7 displays mass spectra of the residual gas in a vacuum chamber before and after bake-out. Before bake-out, the spectrum is completely dominated by the water peak at mass 18 and its cracking pattern. After bake-out, hydrogen is the prevailing residual gas. The atomic hydrogen originates from hydrogen molecules dissociated at the hot filaments of the mass spectrometer and the Bayard-Alpert pressure gauge. Further components in the residual gas are CO, $CO_2$ and hydrocarbons. A peak at mass 32 ($O_2$) would be a sure indication of a leak. A UHV-vessel never contains oxygen unless there is a leak because oxygen is reduced to water by atomic hydrogen.

### Pumps

Pumps fall into two categories, roughening pumps to evacuate the vacuum vessel down to a pressure of $10^0$-$10^{-2}$ Pa and the pumps for the UHV-regime. Roughening pumps are oil lubricated *rotary pumps* or oil-free pumps that use bellows and valves. In the early days of UHV-technology, oil or mercury diffusion pumps were employed for the high-vacuum regime. These are now completely replaced by turbo-molecular pumps or ion getter pumps. *Turbo-molecular pumps* (called turbo-pumps in the following) are mechanical pumps. Their active part is a fan of rotating blades (Fig. 2.8). The pumping action results from an asymmetry of the molecular flow between the left and the right side of the fan. In order to calculate the flux of molecules from left to right and from right to left we need to consider the velocity distribution with respect to the reference frame of the rotating blades. The spaces between the blades form long channels so that only the molecules in the direction $\theta_b$ (Fig. 2.8) can pass. In the rotating reference frame, the thermal velocities of those molecules are shifted by $v_b\cos\theta_b$ (Fig. 2.8), with $v_b$ the circumferential speed of the blades.

$$v \Rightarrow v - v_b \cos\theta \qquad (2.4)$$

The current density $j$ of molecules possessing a particular velocity is given by the product of the velocity and the density $\rho$. The current densities from left to right and right to left are therefore

$$j_\rightarrow = \rho_1 \int\limits_{v_b \cos\theta_b}^{\infty} (v - v_b \cos\theta_b) F(v) dv$$

$$j_\leftarrow = \rho_r \int\limits_{-v_b \cos\theta_b}^{\infty} (v + v_b \cos\theta_b) F(v) dv$$

(2.5)

in which $\rho_1$ and $\rho_r$ are the densities left and right of the rotor, respectively.

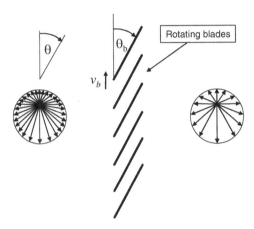

**Fig. 2.8.** The figure illustrates the pumping effect of a molecular pump. The blades move with a circumferential speed $v_b$ (upwards in the figure). The arrows illustrate the thermal velocity of gas molecules relative to the moving blades.

The lower boundary for the integral is where the velocity towards the channel between the blades is zero. $F(v)$ is the Maxwell velocity distribution of particles moving in a particular direction

$$F(v) = \left(\frac{m}{2\pi k_B T}\right)^{1/2} \exp(-mv^2 / 2k_B T).$$

(2.6)

In steady state condition $j_\rightarrow = j_\leftarrow$ and the ratio of the particle densities and therefore the ratio of the pressures, the *compression ratio* is then

$$K = \frac{\rho_1}{\rho_r} = \frac{p_1}{p_r} = \frac{\int\limits_{-v_b \cos\theta_b}^{\infty} (v + v_b \cos\theta_b) F(v) dv}{\int\limits_{v_b \cos\theta_b}^{\infty} (v - v_b \cos\theta_b) F(v) dv}$$

(2.7)

After some algebra one obtains

$$K = \frac{2 - e^{-\kappa} + \sqrt{\pi\kappa}\left(1 + \dfrac{2}{\sqrt{\pi}}\displaystyle\int_0^{\sqrt{\kappa}} e^{-y^2}\,dy\right)}{e^{-\kappa} - \sqrt{\pi\kappa}\left(1 - \dfrac{2}{\sqrt{\pi}}\displaystyle\int_0^{\sqrt{\kappa}} e^{-y^2}\,dy\right)} \quad \text{with} \quad \kappa = \frac{mv_b^2\cos^2\theta}{2k_BT}. \tag{2.8}$$

The coefficient $\kappa$ is essentially the square of the ratio of the circumferential speed of the blades and the mean velocity of the gas molecules (2.1). For small $\kappa$ the Gaussian integrals can be approximated by $2\sqrt{\kappa/\pi}$, and $K$ is then approximated by

$$K = 1 + (6 + \pi + 6\sqrt{\pi})\kappa + 2\sqrt{\pi\kappa} \tag{2.9}$$

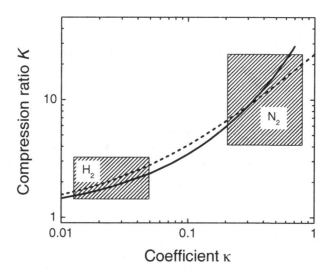

**Fig. 2.9.** Compression ratio $K$ as a function of the coefficient $\kappa$. The dashed curve is the approximation (2.9). The shaded areas mark the ranges obtained by a commercial pump (Leybold 340M).

Figure 2.9 shows the compression ratio as a function of $\kappa$. For a significant pumping effect, the circumferential speed $v_b$ must be of the order of the mean thermal velocity of gas molecules (120 m/s for $N_2$, 450 m/s for $H_2$). Materials of sufficient mechanical strength for the rotors and suitable ball bearings for operation became available only in the last decades. The shaded areas in Fig. 2.9 mark the ranges of compression ratios realized by the rotor of a commercial turbo-pump (*Leybold 340M*) which features magnetic bearings of the **System FZJ™** developed in the Research Center Jülich. The upper and lower limits correspond to the circumferential speeds of 350 m/s and 175 m/s at the outer and inner diameters of the rotor blades. To bridge the pressure gap between UHV and the vacuum provided by the roughening pump the turbo-pump posses a stack of 11 bladed wheels. The total compression ratio for nitrogen is more than sufficient. The overall compression ratio for hydrogen of about $2.04^{11} \cong 2500$ is less impressive. The turbo-pump with magnetic bearings is free of lubricant. Backed by an oil free roughening pump a completely oil free pumping system can be realized. Since the final pressure of oil-free roughening pumps is about 10 Pa a molecular drag pump (working on the principle of an archimedian screw) is placed between the turbo-pump and the roughening pump.

*Ion getter pumps* come as diodes or triodes. Figure 2.10 shows a triode consisting of a collector, an anode and a cathode made from titanium. Titanium combines good mechanical properties with a high reactivity to most residual gases, except the noble gases. A freshly prepared titanium films absorbs all but the noble gases readily and can therefore be employed as a *getter*. A voltage of about 5-7 kV is applied between anode and cathode. The high voltage in combination with the shape of the cathode causes high local electric fields at the cathode surface, and electrons are field-emitted from there. The magnetic field forces these electrons into an orbital path, which enhances the probability for collisions with the residual gas molecules. The electrons ionize the molecules and the positive ions are accelerated towards the cathode. Their path is nearly straight because of the larger mass. Due to the geometric arrangement and the shape of the cathode, the ions strike the cathode at grazing incidence. Titanium material from the cathode is thereby sputtered mostly in forward direction and deposited on the collector. The amount of sputter-deposited titanium is proportional to the number of molecules hitting the cathode, i.e. proportional to the pressure. This ensures that the supply of titanium atoms freshly deposited on the collector matches the number of adsorbed atoms from the residual gas. The steady stream of titanium atoms onto the collector buries the adsorbed residual gas molecules or atoms inside the growing bulk material. In that way, not only reactive gases are effectively gettered but also non-reactive gases in particular noble gases which normally would reside only briefly on the collector surface and be released into back into the gas phase. A triode ionization pump is therefore capable of pumping also noble gases. Unlike the turbo-pumps, ion getter pumps do not remove the gas from the vessel. Rather the gas accumulates in the titanium matrix. Ion getter pumps therefore tend to have a memory effect. If the high voltage is switched off, e.g. to stop the pumping during a gas inlet, the pump releases less tightly bound gases like methane and ethane. In order to minimize memory effects the majority of the gas inventory of a UHV-vessel released during bake-out should be removed from the vessel by a turbo-pump, so that the ion pump is used only in the high vacuum regime. The ion getter pump is also regenerated to some extent by baking the pump elements at 300-400 °C during bake-out while pumping with a turbo-pump. If used only in this way, ion getter pumps practically last forever. A characteristic defect that can occur after some time is an unstable operation in the high-vacuum regime with pressure bursts from time to time. This is due to needles and flake of material that grow on the cathode. These needles and flakes can give rise to sudden events of electron field emission, which release so much heat locally that titanium material with its gas inventory is evaporated. The pressure bursts can bring the pressure temporarily from $10^{-8}$ Pa to $10^{-4}$ Pa. A remedy is to raise the operating voltage cautiously to about 20 kV, which burns off the flakes and whiskers. After that, a complete bake-out cycle is required.

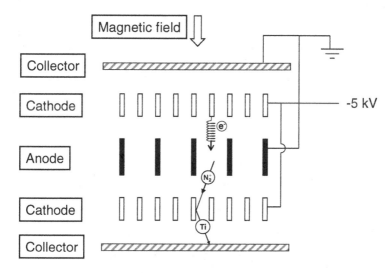

**Fig. 2.10.** Schematic drawing of a triode ion getter pump.

Pumping by an ion-pump is frequently supported by an additional getter film, which is deposited from a heated titanium wire in regular intervals. The intervals are matched to the pressure. The large area of that deposited films provides a high pumping speed for gases that are gettered by titanium. The titanium film is particular effective if it is deposited on a wall cooled by liquid nitrogen. The pumping speed of a perfectly absorbing film is calculated from (2.1) by noting that $A\langle v_x \rangle$ is the volume, which is pumped per time by an absorbing area $A$. For nitrogen, a $100 \text{ cm}^2$ area has a pumping speed of 1200 l/s. A typical vacuum chamber equipped with titanium getter pump that covers the entire bottom of the chamber with the titanium film is emptied about a hundred times per second. Unfortunately, the gas phase is replenished from the permanently out-gassing walls. The considerations above apply also to the layout of the roughening pumps and the dimensions of pump lines. In the regime of molecular flow, a long tube has a conductance that corresponds to its cross-sectional area multiplied by the ratio of the diameter $d$ and the length $L$. The conductance is therefore approximately

$$C = 120 \frac{m}{s} \frac{d^3}{L} \tag{2.10}$$

The characteristic time constant for evacuation of a chamber having a volume $V$ is is $\tau = V/C$. It then takes $\tau \ln 10 = 2.3 \tau$ for every decade of pressure reduction. To pump a volume of $120\,l$ in a characteristic time of $100\,s$ through a tube of $1\,m$ length requires a tube diameter of $2.1\,cm$.

### Vacuum gauges

Measuring the pressure in different pressure regimes requires different gauges. Classical mechanical manometers work down to $100\,Pa$. The range of mechanical pressure measurements is extended to about $10^{-2}\,Pa$ by the **Baratrons** in which the pressure induced deflection of a membrane is measured as a change of capacitance.

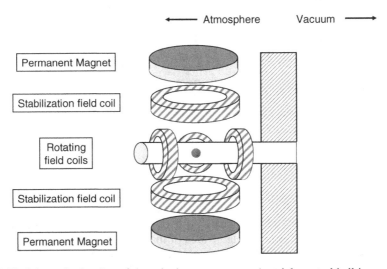

**Fig. 2.11.** Schematic drawing of the spinning rotor gauge. A stainless steel ball is magnetically suspended in vacuum. The friction with the residual gas molecules is measured.

A simple instrument for the range between atmospheric pressure and $10^{-1}\,Pa$ is the *Pirani*-manometer in which the loss of heat of a fine wire due to convection and, in the low-pressure regime, due to heat conduction is measured.

A very accurate instrument for pressures between atmospheric pressure and $10^{-4}\,Pa$ is the **spinning rotor gauge** (Fig. 2.11). The instrument measures the friction of a magnetically suspended stainless steel sphere by the residual gas. The sphere is magnetically suspended in the equilibrium position between two permanent magnets. The equilibrium is stable with respect to lateral displacements of the sphere, but instable with respect to a vertical movement. An additional electro-magnetic field that is controlled by the instantaneous position of the sphere in a feedback loop stabilizes the vertical position. The magnetic suspension is practi-cally frictionless, save for the gas friction. To measure that friction, the sphere is first driven to a certain rotation speed (400-800 Hz), then the drive is shut off and the slow decay in the rotation speed is accurately measured by monitoring the electromagnetic signals induced in the driving coils by the rotating magnetic mo-ment of the sphere. At higher pressures, the power consumption of the rotating magnetic field required to keep the rotation speed constant is measured. The in-strument is completely inert and works in corrosive and hot environments. It represents also the certified international transfer standard for vacuum measure-ment. The determination of the pressure via the gas friction requires some data accumulation time. The time is longer for lower pressures. For every day use, the friction rotor gauge is therefore unpractical. Furthermore, the friction gauge does not cover the UHV-range.

The standard instrument in a wide range of pressures from $10^{-2}-10^{-9}$ Pa is the *ionization gauge* after Bayard-Alpert (Fig. 2.12). Electrons emitted from a hot-filament cathode are accelerated by the positive voltage on the anode, which con-sists of a cylindrical wire mesh. A thin wire is placed in the center of the anode as the ion collector. The potential on the ion collector is negative with respect to the cathode by about 25 V. Gas molecules ionized by the accelerated electrons inside

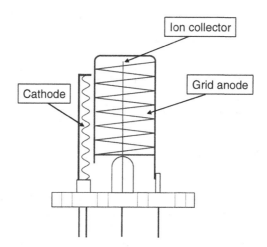

**Fig. 2.12.** Bayard-Alpert ionization vacuum gauge on a metal flange.

inside the anode cylinder travel to the central collector. The current on the collec-tor is therefore proportional the concentration of gas atoms and therefore a measure of the total pressure. Since molecules have a different ionization probabil-ity, the pressure reading depends on the type of gas. The ultimate pressure limit is given by the *X-ray limit*: electrons arriving at the anode cause the emission of soft X-rays from there. The X-rays in turn cause the photoemission of electrons from the surface of the ion-collector. Since this current has the same sign as the current of the collected ions, it fakes a pressure. The photocurrent is proportional to the surface area of the central wire, which is therefore kept as thin as possible. The X-ray limit is typically about $10^{-9}$ Pa.

### Residual gas analysis

Mass spectroscopy of the gas in an UHV-chamber is required
- to define sources of malfunction of the vacuum system,
- to control the exposure of surfaces with specified gases,
- and to monitor desorption from surfaces.

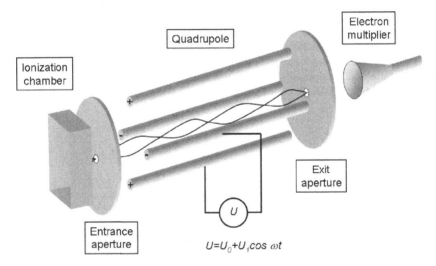

**Fig. 2.13.** Sketch of a quadrupole mass spectrometer. See text for explanation.

The most commonly used type of mass spectrometer is the quadrupole mass spectrometer. This spectrometer consists of an ionization chamber, a system of four metallic rods between two apertures, and a detector for the ions (Fig. 2.13). The quadrupole is a mass-selective filter, which works on the principle of dynamic stabilization. Ions produced in the ionization chamber via electron impact ionization are accelerated to a potential $eU_0$ by the positive bias on the two apertures and the quadrupole. An additional oscillating voltage is applied to the four rods such that opposite rods have the same potential. Ions traveling along the center path between the rods are in equilibrium with respect to the oscillating potential, in stable equilibrium with respect to one pair of rods and in an unstable equilibrium with respect to the other. Ions embarked on trajectories that pass the entrance aperture at a small angle experience a potential that is proportional to the square of the deviation from the center path, and oscillating in its sign. Ion trajectories oscillating around the center path (Fig. 2.13) occur if the phase of the ion path is matched to the phase of the oscillating voltage such that the potential repels the ions towards the center path when the ion is further away from the center and vice versa. Without solving the mathematics in detail one can therefore write down the focusing condition

$$\tau = \frac{L}{v} = n\,2\pi / \omega \tag{2.11}$$

in which $\tau$ is the time during which the ion traverses the distance $L$ between the two apertures, $v$ is the ion velocity, $\omega$ is the frequency of the ac-voltage and $n$ is an integer number. Since the velocity of ions possessing the mass $m$ is $v = \sqrt{2ZeU/m}$ with $Ze$ the ion charge, the focusing condition (2.11) selects a mass, which is proportional to $\omega^{-2}$.

## 2.2.2 Surface Analysis

### Monitoring surface structure during preparation

Sample preparation in UHV requires in-situ control of the surface structure and chemical composition. Experimental tools for an in-depth study of surface structure and morphology were already discussed in chapter 1. Low energy electron diffraction (LEED) is also used for qualitative checks on the surface order. For monitoring the surface structure during preparation procedures, in particular during epitaxial growth *Reflection High Energy Electron Diffraction* (RHEED) is more suitable since the equipment is not in the way of evaporation sources and other tools. The experimental arrangement is sketched in Fig. 2.14a. Electrons with energies in the range between 10-100 keV strike the surface at grazing incidence, and the diffraction pattern is observed in reflection. Because of the higher energy, the diffraction spots are closer together compared to LEED. Sharp diffraction spots occur either if the surface is very flat or if the surface is covered with small three-dimensional crystallites. More often, however the diffraction pattern consists of vertical streaks. Upon surface disorder, the reciprocal lattice rods assume some fuzziness, and this fuzziness elongates the diffraction spot along the vertical direction (Fig. 2.14b). The experimental set-up provides means to characterize the growth-mode. If electron energy and the angle of the incident beam are chosen so that a particular beam, e.g. the (00) beam experiences destructive interference from the beam reflected by consecutive monolayers then layer-by-layer growth causes oscillations in the diffracted intensity (cf. Sect. 11.1). Whenever a layer is complete, the intensity is maximal, and minimal for half a completed monolayer because of the destructive interference. The effect is best seen for medium electron energies (500-1000 eV) because of the higher surface sensitivity of these electrons.

By counting the oscillations, one has an accurate measure of the deposited number of monolayers that can be used to calibrate the evaporator for a particular ingot.

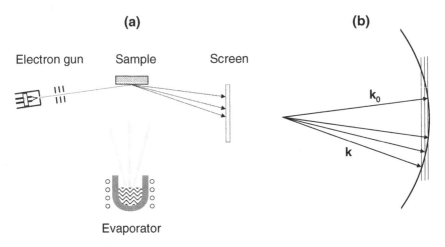

**Fig. 2.14.** (a) Experimental set-up for *Reflection High Energy Electron Diffraction* (RHEED) during epitaxial growth. (b) The Ewald-construction for high-energy electrons in reflection geometry.

### Monitoring surface cleanness and composition

The most common methods for the analysis of the element composition of surfaces are based on the emission of core electrons either by photoemission or via the Auger-process (Fig. 2.15). In both cases, the kinetic energy of the emitted electrons is characteristic of the elements. The spectroscopy of the kinetic energies therefore provides information on the sample composition. The spectroscopy based on the photoemission process is either called *X-ray Photoemission Spectroscopy* (XPS) or, as named somewhat misleadingly by its inventor K. Siegbahn, *Electron Spectroscopy for Chemical Analysis* (ESCA). The spectroscopy based on the Auger-process is termed *Auger Electron Spectroscopy* (AES). The kinetic energy of the photo-emitted electron is determined by energy conservation

$$E_{kin} = h\nu + E(N) - E(N-1) \tag{2.12}$$

Where $E(N)$ and $E(N-1)$ denote the energies of the system with $N$ and $N-1$ electrons, respectively. To a rough approximation, the energy difference can be expressed in terms of the electron energy levels of the $N$-electron state so that in case of photoemission from the K-shell one obtains

$$E_{kin} = h\nu - \left(E_{vac} - E_K\right) \tag{2.13}$$

The Auger-process is the radiationless filling of a core hole after ionization by an electron from an upper shell or the valence band, with the energy transferred to a second electron of the upper shell (Fig. 2.15). The final state in the Auger-process is a two hole-state. The energy of a two-hole system differs from a single hole-system. To a crude approximation the kinetic energy of the Auger-electron is

$$E_{kin} = E_{f1} + E_{f2} - E_I .\qquad(2.14)$$

in which $E_{f1}$ and $E_{f2}$ are the energies with the electron holes in the final state and $E_I$ is the energy of the state that is ionized initially. It is customary to denote the Auger-electron by the notations of the three electron shells involved in the process, KLL, LMM, etc. The valence band is denoted as V. Since electrons of medium kinetic energy are most suitable for surface analysis (see below), Auger-transitions between the higher electron states, including the valence band are employed. For the second row elements from Li to Ne, including the typical surfaces contaminant C, these are the KVV-Auger electrons. For the third row elements (Na – Ne) these are the LVV transitions. For the 3d-transition elements, the 3s, 3p-shells are several tens eV below the 3d/4s valence band. The LMM, LMV, and LVV Auger-electrons form a characteristic triplet.

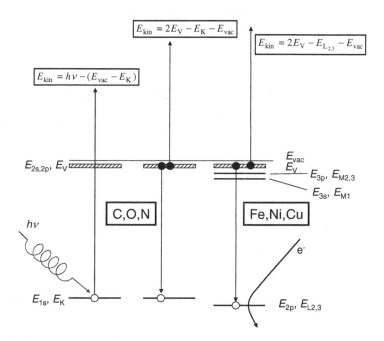

**Fig. 2.15.** Illustration of the Photoemission and Auger-process. Shown are K-shell photo-emission (*left*) and a KVV-Auger-emission (*center*) for second row elements, and a LVV-Auger-emission for 3d-transition metals (*right*).

Neglecting the many-body nature of the excitation, the intensity of an Auger-line is described by the matrix element

$$I_{Auger} \propto \left| \left\langle \Psi_1^{(i)} \Psi_2^{(i)} \,|\, 1/r_{1,2} \,|\, \Psi_1^{(f)} \Psi_2^{(f)} \right\rangle \right|^2 \qquad (2.15)$$

in which $\Psi_1^{(i)}$, $\Psi_2^{(i)}$, $\Psi_1^{(f)}$, and $\Psi_2^{(f)}$ are the wave function of the two electron in the initial and final state, respectively and $1/r_{1,2}$ is the Coulomb interaction between the two electrons. The matrix element vanishes if $\Psi_1^{(f)}$ is an s-state and the initially occupied state is a p-state, vice versa. The intensity of the Auger-lines also reflects the number of participating electrons is a state. For example, the LVV Auger-line for the 3d-transition elements is roughly proportional to the number of electrons in the valence band. The LVV-line is the strongest line in the L-triplet for Cu and becomes very weak in the case of Sc.

Photoemission-spectroscopy and Auger-spectroscopy gain their surface sensitivity from the fact that electrons have a high cross section for inelastic scattering. The information depth for surface spectroscopies is given by the mean distance, which the electron can travel in the material without loosing a significant amount of energy. Energy losses due to phonons do not count in this regard, as their energy is too small! Because of the inelastic scattering events the flux of electrons of certain energy decays exponentially as

$$I = I_0 \exp(-x/\lambda) \qquad (2.16)$$

where $\lambda$ is the mean free path. The mean free path is inversely proportional to the imaginary part of the electron self-energy [2.7, 8] which may be calculated in the *Random Phase Approximation* (RPA) from the dielectric response function $1/\varepsilon(\omega, q)$ with $\omega$ the frequency and $q$ the wave vector. The inverse of the mean free path $\lambda$, the *stopping power* for electrons with the energy $E$ and wave vector $k$ is given by [2.9]

$$\lambda^{-1} = \frac{\hbar}{\pi a_B E} \int \frac{1}{q[\partial \varepsilon / \partial \omega]_{\omega_q}} dq \int_0^{\omega_{max}} \omega_q \, d\omega_q \, \mathrm{Im} \frac{-1}{\varepsilon(\omega, q)} \Theta(E - E_{min} - \hbar\omega) \quad (2.17)$$

$E_{min}$ is the Fermi energy or the conduction band edge in case of an insulator and $a_B$ is the Bohr radius. The Heavyside $\Theta$-function takes care of energy conservation. The upper frequency limit $\omega_{max}$ corresponds to the maximum energy loss for a given $q$-vector that occurs when the $q$-vector is oriented opposite to the $k$-vector of the electron. In that case, energy conservation requires that

$$\omega_{max} = \frac{\hbar}{2m} \left[ k^2 - (k - q)^2 \right]. \qquad (2.18)$$

If damping is disregarded the response function has a $q$-independent pole at the plasmon frequency $\omega = \omega_p$ and can be represented by

$$\text{Im}\frac{-1}{\varepsilon(\omega,q)} = \frac{\pi}{2}\omega_p\delta(\omega-\omega_p)\,. \tag{2.19}$$

This simple expression is a good representation of the response function for all materials as long as the electron energy is large and if the plasmon frequency is calculated from the density of valence electrons $n$ as

$$\omega_p^2 = \frac{ne^2}{m\varepsilon_0}\,, \tag{2.20}$$

with $\varepsilon_0$ the absolute dielectric permeability. The result for $\lambda^{-1}$ is

$$\lambda^{-1} = \frac{\hbar\omega_p}{2a_B E}\ln\left(\frac{q_{\max}}{q_{\min}}\right) \tag{2.21}$$

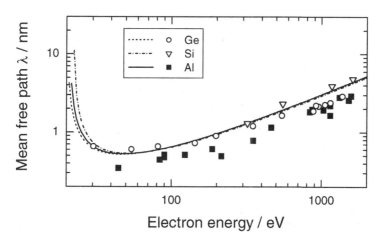

**Fig. 2.16.** Mean free path of electrons with respect to plasmon excitations according to (2.18) for Ge, Si and Al (dashed, dash-dotted and solid line, respectively) together with selected data points (circles, triangles and squares, respectively) [2.10-12].

The maximum wave vector can be taken as the wave vector where the plasmon ceases to be a well-defined excitation, which is the case when the phase velocity of the plasmon $\omega_p/q$ is equal to the Fermi-velocity $v_F$. The minimal wave vector

$q_{min}$ is obtained, if $q$ is oriented anti-parallel to the wave vector of the electron $k$ in which case (2.18) holds. After insertion of $q_{max}$ and $q_{min}$ (2.21) becomes

$$\lambda^{-1} = \frac{\hbar\omega_p}{2a_B E} \ln\left( \frac{\hbar\omega_p}{2\sqrt{E_F E} \, (1 - \sqrt{1 - \hbar\omega_p / E})} \right) \tag{2.22}$$

The result for $\lambda$ is plotted in Fig. 2.16 for the elements Ge, Si and Al together with experimental data. The model describes the general energy dependence quite well, underestimates however the stopping power as it neglects surface plasmon excitations and core excitations (relevant for $E \gg \hbar\omega_p$) as well as electron hole pair excitations (relevant for $E < \hbar\omega_p$). Auger and photoelectrons which involve the upper shells and have therefore energies between 100 and 1000 eV are particular suitable for surface elemental analysis as their mean free path is small.

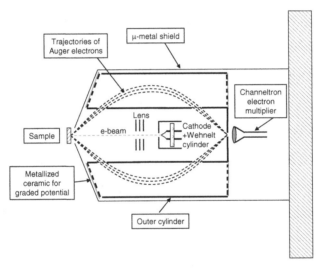

**Fig. 2.17.** Cylindrical mirror electron Auger-spectrometer with an integrated electron gun for electron impact ionization.

For the purpose of surface elemental analysis, Auger-spectroscopy has gained more acceptance than XPS for two reasons. One is that Auger spectrometers can be built rather compact as add-on devices. The other is that Auger-spectroscopy in combination with a highly focused electron beam has spatial resolution. Figure 2.17 shows the standard form of a cylindrical mirror analyzer with a coaxial electron gun for primary ionization. The beam energy for ionization ranges between 3 and 10 keV. The *Cylindrical Mirror Analyzer* (CMA) consists of two coaxial metallic cylinders. The inner cylinder has entrance and exit pupils, which define the angular aperture of about ±6°. The magnitude of this angle is chosen to match the third order angular aberration of the cylindrical mirror analyzer to achieve a good compromise between resolution and transmission. To minimize perturbations on the potential the openings in the inner cylinder are covered by a high-transparency wire mesh. The inner cylinder is on ground potential, as the sample. The outer cylinder is negatively biased. For a given bias, electrons of a particular kinetic energy are focused on the exit aperture. The entrance aperture is provided by the diameter of the focused electron beam used for ionization. To avoid perturbations from the fringe fields at the ends of the two cylinders two partially metalized ceramic plates provide roughly the same $\ln r$ -dependence of the potential as in the interior of the cylinder analyzer. The use of a separate grazing incidence electron gun instead of the integrated gun enhances the surface sensitivity.

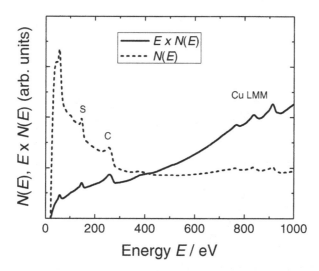

**Fig. 2.18.** Auger-spectrum obtained from a cylindrical mirror analyzer of a Cu(110) surface before cleaning [2.13]. Electron beam energy for primary ionization is 3 keV. The dashed line is the spectrum after correcting for the analyzer transmission.

Since the Auger-analyzer scans through the energies by changing the deflection voltage, the energy width of transmitted electrons is proportional to the pass energy. The intensity of the transmitted signal is therefore also proportional to the pass energy. The current at the detector is therefore proportional to the product of the pass energy $E$ and the intensity of electrons emitted from the sample $N(E)$.

Figure 2.18 displays a spectrum obtained from a contaminated Cu(110) surface. The characteristic Auger-lines of Cu and of the contaminants S and C ride on top of a large background of secondary electrons. The dashed line is the response corrected for the increasing transmitted bandwidth. One sees that the intensity of secondary electrons increases for very low energies. The cut-off at about 20 eV is caused by a loss of analyzer transmission due to stray electric and magnetic fields.

Auger-spectra as displayed in Fig. 2.18 are used for quantitative analysis after subtracting a suitable function to describe the background. For qualitative analysis, differentiated spectra are more convenient. Figure 2.19 displays differentiated spectra of a Cu(110) surface before and after cleaning. On the contaminated surface, the intensities of the high energy Cu-lines are reduced to about 45%. Considering the mean free path of about 1.3 nm and the emission angle of 45°, one can estimate the thickness of the contamination layer to be of the order of 1 nm. The low energy Cu-peak is not suitable for this type of estimate since (i) the mean free path within the dirt layer is uncertain at low energies, and (ii) the Cu-Auger peak involves surface valence band states of Cu, which change due to the chemical bond with carbon and sulfur atoms in the contamination layer.

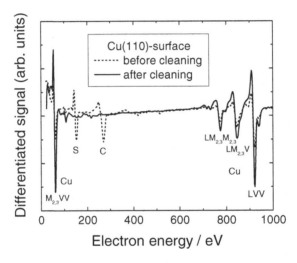

**Fig. 2.19.** Differentiated spectra of a Cu(110) surface before (dashed line) and after cleaning (solid line) [2.13]. The Cu-signal is smaller on the dirty surface as the contaminant layer reduces the thickness of the Cu-layer that contributes to the spectrum.

## 2.2.3 Sample Preparation in UHV

### *Preparation by removing a protective layer*

Si-surfaces prepared by wet chemistry have a protective layer of Si-H bonds, so that the surface stays clean in air for some time (Sect. 2.1.2). Si-surfaces may also be covered by a protective oxide layer, produced either by wet chemistry or by thermal oxidation in a quartz-furnace. Those samples can be introduced into the vacuum system via an air lock and transferred to the sample holder. Once mounted on the sample holder and after reestablishing UHV-condition the protective oxide layer can be removed by heating the crystal. This preparation method produces well-ordered and clean surfaces. For thick oxide layers a temperature above 900°C is required. Heating is performed either by electron bombardment heating or by passing the heating current through the sample. In the case of silicon, the procedure requires the application of high voltages and low heating currents in the beginning and of high currents at low voltages at high temperatures because of the temperature dependent resistance of the semiconductor silicon.

Occasionally, surfaces or thin film systems are prepared in a separate UHV-chamber, coated by protective layer and the transferred to the analysis chamber. The suitable type of coating depends on the material to be investigated. High-temperature metals are best coated by a layer of gold. Metals that cannot be heated to high enough temperatures to flash-off the gold may be coated with iodine or bromine. An established coating for GaAs is *arsenic capping*.

### Preparation by mechanical means

The traditional and safe method to prepare clean surfaces in UHV is cleaving. The method has lost importance in recent years, but still has its place in the preparation of some materials.

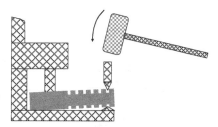

**Fig.2.20.** Crystals are cleaved by driving a wedge into prepared notches that mark the intended cleaving plane.

To facilitate the cleaving in UHV the crystals are cut into a particular shape with little notches on both ends of the intended cleavage plane (Fig. 2.20). Cleaving is performed by driving wedges into the notches to the point of contact, followed by a sudden blow of a hammer. If large single crystals are available, longer bars with several pairs of notches can be prepared for multiple cleavage of one sample. While surfaces prepared by cleaving are clean and maintain the stoichiometry of the bulk crystal, the surfaces may contain an unspecified number of steps of various heights. Furthermore, crystals cleave only along particular crystallographic planes. Ionic crystals cleave along the neutral planes that contain an equal number of cations and anions (e.g. the {100} planes for alkali-halides, the {110} planes for III-V compounds). The II-VI compounds cleave likewise well along the neutral {110} planes, but also along the {0001} planes. Si and Ge cleave along the {111} planes, though not very well. Layered crystals such as intercalates and graphite can also be cleaved by fixing and removing an adhesive tape.

Another way of preparing clean surfaces with bulk stoichiometry, albeit with defects, is by grinding. The technique has been employed in the analysis of the electronic surface structure of perovskites.

### Preparation by sputtering and annealing

Sputtering with noble gas ions is most frequently employed method to remove contaminants from the surface. The method is therefore given a more detailed consideration. Noble gas ions used for sputtering are generated in an ion gun (Fig. 2.21) by electron impact ionization, accelerated to the desired energy and focused onto the sample. The noble gas supply is provided either by feeding the gas into the ion gun housing via a leak valve or by backfilling the entire chamber with the gas. For cleaning, the broad focus of the ion gun depicted in Fig. 2.21 (a few mm diameter, depending on ion energy and sample distance) suffices. For a very homogeneous sputter rate across the sample the beam of a fine-focus ion gun is swept over the sample area.

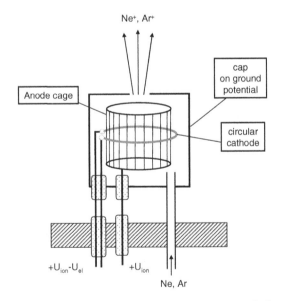

**Fig. 2.21**. A simple ion gun featuring an annular cathode, a cage-anode for the electrons and an outer cap with a hole. Electron emitted from the cathode are accelerated towards the cage-anode, which is at $U_{el} = +150$ V with respect to cathode potential and at $+U_{ion}$ with respect to the grounded cap. The electrons travel back and forth in the cage until they hit a noble gas atom and finally disappear in the cage-anode. The positively charged ions are accelerated towards the cap and a moderately well focused beam of ions leaves the orifice.

Three different processes contribute to the removal of contaminants (Fig. 2.22) [2.14, 15]: The direct knock-off process by the collision of the ion with the contaminant, the knock-off by a reflected ion, and a process in which the contaminant atom receives enough energy for desorption from the outwards flux of sputtered substrate atoms. The cross sections for the three processes are of the same order of magnitude. The cross section of first two decay slowly with energy, the cross section of the last increases with energy and surpasses the cross section of the first two mechanisms at about 1.5 keV [2.14]. Significant sputtering of substrate atoms normally is an undesirable side effect since it disorders the substrate and intermixes the contaminants with the substrate. Ion energies of 500 to 1 keV are optimal for cleaning the substrate of surface contaminants. The cross sections for sputtering adsorbates are of the order of $10^{-15}$ cm$^{-2}$ for 500 eV argon and neon ions. The use of neon has the advantage that a liquid nitrogen cooled getter pump can be operated during sputtering which keeps the background pressure of reactive gases in the chamber low (neon does not adsorb at 77 K, argon does!).

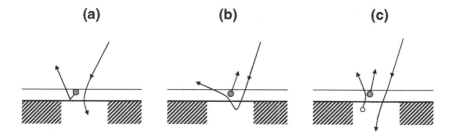

**Fig. 2.22.** Schematic illustration of the three different sputter processes that contribute to the removal of surface contaminants: **(a)** Direct energy knock-off collisions, **(b)** knock-off after recoil of the ion from the substrate, and **(c)** energy transfer from sputtered substrate particles (after Taglauer [2.14])

The sputter rate for contaminants on the surface is proportional to the cross section $\sigma$ and the ion current $j$

$$\frac{dN}{dt} = -N\sigma j \tag{2.23}$$

so that the number of contaminant particles decays exponentially in time

$$N = N_0\, e^{-t/\tau}, \tag{2.24}$$

with a time constant

$$\tau = (\sigma\, j)^{-1}. \tag{2.25}$$

The time constant $\tau$ is of the order of a minute for typical ion currents of $5\mu A/cm^2$. One might therefore expect to have the surface free of contaminants after a few minutes of sputtering. Unfortunately this is not so! Inevitably, contaminant atoms are pushed into the substrate matrix during sputtering and are removed only by sputtering off a few monolayers of substrate material, in the course of which a fraction of contaminant atoms get pushed deeper into the substrate, and so forth. This slows down the cleaning process considerably. The typical contaminant on metal surfaces, carbon and sulfur, can be brought back to the surface by mild annealing of the sample. Sequences of alternate sputtering and annealing are therefore recommended to remove these surface contaminants effectively. Sulfur and carbon contamination arises not only from surface processes. These atoms are also contained in the bulk single crystals with a concentration of the order of ppm. As the surface disorders during sputtering, the crystals have to be annealed to higher temperatures (e.g. up to 0.9 of the melting temperature) in order to restore surface order after sputter cleaning. During this annealing procedure, the bulk impurities can diffuse towards the surface, and segregate there if the free enthalpy in the adsorption sites is lower than in the bulk. In a simple model for segregation the surface equilibrium concentration $\Theta$ is given by

$$\frac{\Theta}{1-\Theta} = c\,e^{-(H_{surf}-H_{bulk})/k_BT} . \tag{2.26}$$

Here, $c$ is the bulk concentration, and $H_{surf}$ and $H_{bulk}$ are the enthalpies per contaminant atom on the surface and the bulk, respectively. In the high temperature limit, the surface coverage is smaller or at most of the order of the bulk concentration. At lower temperatures and if $H_{surf} \ll H_{bulk}$ (i.e. the surface sites are energetically preferred) the surface equilibrium coverage can approach one, even for low bulk concentrations. This is the typical situation for the common metal contaminants C and S (for a detailed discussion see Sect. 5.4.3). Consequently, the sputter-annealing process has to be repeated over and over, until the entire crystal is leached completely. This may take as many as 100 cycles, depending on the required state of cleanliness. Detrimental to effective leaching of the crystal is that the surface concentration of segregated contaminants at a particular temperature, once it is smaller than one, decreases with the decreasing bulk concentration. One therefore frequently abstains from complete leaching and terminates the cleaning procedures with a final long sputter session, followed by a mild annealing to a temperature, which is sufficient for re-crystallization, but too low to let the bulk impurities diffuse to the surface. The surface cleaning procedures can be speeded up by using thinner crystals or by heating the crystals in a hydrogen atmosphere for a longer time prior to mounting in the UHV-chamber.

## Preparation by in-situ chemical reactions

Because of the considerable deficiencies of the sputter-annealing process, alternative methods of cleaning should be considered, if possible. One of them is cleaning by chemical reactions in vacuum. Surfaces of the refractory metals, tungsten, molybdenum and niobium, e.g., are cleaned by high temperature oxidation. This process burns off surface carbon deposits as well as the dissolved carbon, which tends to segregate to the surface even more readily in an oxygen environment since the reaction product CO is a state of lower chemical potential. Oxygen also reacts with the substrate material to form oxides. For the refractory metals, these oxides have a higher sublimation pressure than the substrate material. They are therefore removed by heating without evaporation of substrate material. The required temperatures are listed in Table 2.1. The comparatively high temperatures are achieved by electron bombardment. Since only refractory metals withstand such temperatures, the sample holder, at least the parts that become hot, must be manufactured from the same materials. The temperatures can be measured either by using a pyrometer or by spot welding a W/Re thermocouple to the sample.

**Table 2.1.** Recipes for cleaning by oxidation and reduction for W, Mo, Nb and Re.

| Material | Oxidation | Reduction |
|----------|-----------|-----------|
| W | 24 h, $1 \times 10^{-4}$ Pa, 1000 K | 2500 K |
| Mo | $5 \times 10^{-6}$ Pa, 1400 K | 2000 K |
| Nb | $1 \times 10^{-4}$ Pa, 1000 K | 2h at 2300 K |
| Re | 90 s, $1 \times 10^{-4}$ Pa, 1600 K | 2100 K |

In order to maintain UHV-conditions during sample heating, the sample holder must be degassed during and after chamber bake-out by heating the sample to the same temperatures as during the cleaning procedure. Successful degassing requires that the electrons emitted from the cathode during electron bombardment only strike the intended target. Otherwise, electrons hitting the chamber walls or the sample manipulator would cause *electron stimulated desorption*, with negative consequences for the vacuum. A simple recipe to avoid electron stimulated desorption is to bias the cathode positively with respect to ground by about 50-100 V. In that way even electrons that are elastically scattered from the target cannot arrive on the chamber wall or at those part of the sample mount, which are not on high positive voltage. A positive bias is automatically obtained by using a power supply for the cathode heating current that floats with respect to ground and by connecting the cathode to ground by a resistor. The resistance $R$ is calculated as 50 V/$I_{emission}$ with $I_{emission}$ the cathode emission current required to heat the sample to the desired temperature. Depending on the sample size, temperature and the heat loss via conduction to the sample holder, up 100 W electric power may be required.

Cleaning by oxidation has also been used for cleaning Pt-surfaces. Oxygen burns off carbon very effectively. After prolonged heating in an oxygen atmosphere, Auger-spectra of Pt-surfaces tend to display an oxygen peak that cannot be removed by heating to temperatures where adsorbed oxygen would desorb. This stable surface oxide is due to Si or Ca contaminations in the bulk, which segregate to the surface under the influence of the chemical potential of oxygen. Since Si and Ca are difficult to detect by Auger spectroscopy in the presence of the many Pt-associated peaks, the oxygen signal was mistaken for a form of stable platinum oxide for some time. $SiO_2$ and CaO do not desorb. They are best removed by sputtering. Hence, a combination of oxygen treatment and sputtering appears to be the optimum cleaning procedure for platinum.

Another method of chemical cleaning that works rather effectively without being too aggressive is cleaning with atomic hydrogen. The method can be used to remove most common contaminants from the surface: carbon, nitrogen, sulfur, chlorine, fluorine, etc.. Silicon and germanium surface are etched by atomic hydrogen by the production of $SiH_4$ and $GeH_4$. On II-V and II-VI compounds, the stoichiometry is affected. Atomic hydrogen is produced by dissociation of molecular hydrogen on a hot tungsten filament that is placed near the sample. A better method is to employ a beam of hydrogen atoms. A beam of up to 100% atomic hydrogen is obtained by passing molecular hydrogen through a hot tungsten capillary. Figure 2.23 shows a calibrated source of atomic hydrogen (after K. Tschersich [2.16, 17]).

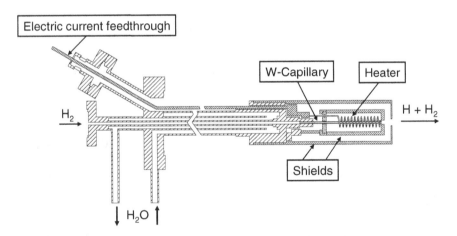

**Fig. 2.23.** Source for atomic hydrogen. Molecular hydrogen is dissociated at the hot walls of a resistively heated tungsten capillary.

Cleaning by atomic hydrogen is a standard method in semiconductor industry. It is less common in research labs, mostly because it is only lately that good sources of atomic hydrogen became commercially available. It is not advisable to use ineffective means of making atomic hydrogen from molecular hydrogen and compensate the ineffectiveness by introducing massive amounts of molecular hydrogen into the chamber: The hot oil in the turbo-pump and the rotary-pump may be crack-reacted to hydrocarbon products that can be disastrous to the vacuum system!

### Preparation by epitaxial growth

For many years, surfaces were considered clean when "the Auger spectra showed no traces of contaminants". The sensitivity of Auger-spectroscopy is not very high, however. The detection limit is 1% of a monolayer, in favorable cases 0.1%. That still means that on a linear scale every tenth to thirtieth atom is a foreign atom. While this contamination level may suffice for some studies of the electronic properties and for structure analysis it would be intolerable for studies on the vibrational properties, on the morphological features of surfaces after annealing, on catalytic surface reactions, or on the growth modes in epitaxial growth, to name a few issues of current research interests. Contamination levels, as e.g. seen in STM-images are frequently of the order of $10^{-6}$. These low contamination levels cannot always be achieved by the classical techniques mentioned before. An alternative at least for element crystals is to evaporate the material and grow several monolayers of the material onto the pre-cleaned crystal surface to bury the remaining contaminant atoms. To obtain a smooth and contamination free surface the substrate temperature need be chosen high enough to facilitate interlayer transport (Sect. 11.1.4), but low enough to avoid segregation of the buried impurities. Naturally, the evaporation source has to be out-gassed carefully, and sufficiently high purity ingots have to be used to achieve good results.

# 2.3 Surfaces in an Electrochemical Cell

## 2.3.1 The Three-Electrode Arrangement

When a piece of material is immersed into an electrolyte, the ions of the electrolyte react with the surface and transfer their charge to the solid. If the solid is electrically isolated otherwise, the electric potential of the solid changes until it reaches an equilibrium value that is characteristic of the electrode/electrolyte combination. The ion/surface reaction can be controlled by applying a potential from an external source. That is why electrochemistry of the solid/electrolyte interface is the science of solids in an electrolyte under potential control. The potential is defined with respect to some reference redox system (see Sect. 3.1.3). Thus, the experimental set-up involves always a *reference electrode*, which is in equilibrium with the electrolyte. Equilibrium means that no current should flow between electrolyte and the reference. The potential of the *working electrode* (the electrode of interest) with respect to the reference electrode must therefore be measured without drawing a significant current on the latter. The current load on the working electrode is picked up by a third electrode, the *counter electrode.* Hence, electrochemical experiments require three electrodes.

We note that in electrochemistry the term *potential* is used synonymically for *voltage* and is therefore measured in units of volts, rather than electron volts. We stay in keeping with that custom throughout this entire volume. Furthermore, electrochemists denote the potential by the letter $E$. To avoid confusion with the energy we use the symbol $\phi$.

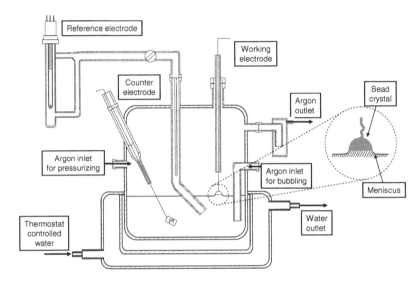

**Fig. 2.25**. Electrochemical cell with reference electrode, see text for discussion.

Figure 2.25 shows a glass vessel used for standard electrochemical measurements. The counter electrode is a sheet of platinum. The reference electrode is electrically connected to the electrolyte in the vessel via the *Luggin-capillary.* The capillary should end near the working electrode so that the measured potential is little affected by potential drop between the working electrode and the counter electrode. The Luggin-capillary is filled by introducing slightly pressurized argon into the inlet on the left side. This drives the electrolyte into the capillary once the valve is opened. Contact with the possibly different electrolyte in the reference cell is thereby established. To remove the dissolved oxygen, argon is bubbled through the cell prior to immersion of the working electrode. The working electrode is immersed into the electrolyte under *potential control*, which means that the potential of the working electrode is fixed with respect to the reference, regardless of the current between the working electrode and the counter electrode. Figure 2.25 shows a bead crystal as the working electrode. The crystal is retracted a little after immersion so that the electrolyte forms a meniscus with the perimeter. This ensures that only the prepared and oriented surface (see expanded view) is in contact with the electrolyte.

## 2.3.2 Voltammograms

The basic experiment in electrochemistry is the *voltammogram*: the potential of the working electrode with respect to the reference is swept at a constant rate up to a maximum and then backwards to the starting potential to complete a full cycle, and the current is measured during the entire cycle. This basic experiment is the starting point for practically all investigations on the solid/electrolyte interface because to the learned scientist the voltammogram immediately reveals the status of the surface in terms of order and cleanness. We discuss this with the voltammogram of a Pt(111) surface as an example (Fig. 2.26). The vertical axis in Fig. 2.26 is the area specific current, the horizontal axis the potential with respect to the *Reversible Hydrogen Electrode* (RHE). This electrode consists of a sheet of high surface area platinum (*black platinum*) with molecular hydrogen bubbling through the electrolyte. The electrolyte at the reference is the same as at the working electrode. No continuous electrochemical reaction takes place in the entire potential range shown in Fig. 2.26. Hence, the current were zero if the potential were kept constant. The finite current arises from loading and unloading the interfacial capacitance (Sect. 3.2.3), from the charging and discharging of ions adsorbing and desorbing from the surface at a particular potential, and from phase transitions that may take place on the surface. The current density $j$ arising from the capacitance is

$$j = C\frac{\mathrm{d}\phi}{\mathrm{d}t}, \tag{2.22}$$

in which $\mathrm{d}\phi/\mathrm{d}t$ is the sweep rate of the potential and $C$ is the area specific interfacial capacitance. The capacitance is about 50 $\mu F/cm^2$ (Sect. 3.2.3). The sweep rate in Fig. 2.26 was 50 mV/s. The capacitive current is therefore about 2.5 $\mu A/cm^2$. Only the current between the peaks in the right half of the figure is therefore a capacitive current. Starting from the left, the nearly constant current is due to desorption of hydrogen from an adsorbed layer of hydrogen on the surface. Hydrogen desorbs as a solvated, positively charged proton, hence the current. Electrochemists call the adsorbed layer of hydrogen *underpotential deposited* (upd-layer, Sect. 6.2.4). Underpotential, because the bulk phase of hydrogen, the $H_2$-gas, develops at a more negative potential. The small peaks in the otherwise monotonous current are due to desorption from A- and B-steps on the surface (Figs. 1.32 and 1.33). The magnitude of these peaks is therefore a good indicator

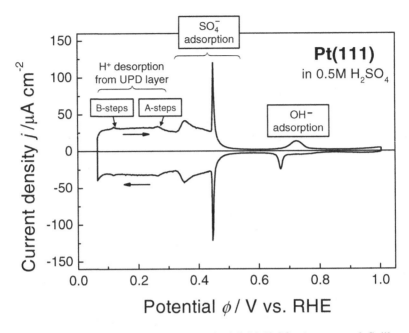

**Fig. 2.26.** Voltammogram of a Pt(111) crystal in 0.5 M $H_2SO_4$ (courtesy of Guillermo Beltramo). The potential is with reference to the ***Reversible Hydrogen Electrode*** (RHE). The sweep rate is 50 mV/s. See text for further discussion.

of the amount of disorder on the surface. The fact that these peaks exist is indicative of the cleanness of the sample and the electrolyte: contaminants would tend to sit in step-sites and block these sites for hydrogen. The assignment of these peaks to step-sites originates from comparative studies on Pt(110) and Pt(100) surfaces. For example, the voltammogram of Pt(110) exhibits shows a strong peak at 0.12 V RHE. Since the local atomic structure of a B-step together with two atom rows on the lower terrace is that of a (110) surface (Fig. 1.33) the small peak at 0.12 V RHE in Fig. 2.26 is assigned to hydrogen adsorption at B-steps. An interesting question is why desorption from the step sites gives rise to a sharp peak as opposed to desorption from the terraces. It is shown in Sect. 6.2.5 and 6.3.2 that the sharpness of a desorption peak is related to the lateral interactions between adsorbates: sharp peaks occur when there are no interactions between adsorbates or attractive interactions, broad peaks or even featureless currents occur for repulsive interactions. In the latter case the adsorption isotherm extends over a large potential range and correspondingly desorption extends over a broad range (see also Fig. 4.7). The broad hump at 0.35V RHE is the initial adsorption of $SO_4^-$-ions. Further uptake causes a phase transition into an ordered sulfate adlayer, which manifests itself by the sharp spike at 0.45V RHE. The sharpness of the peak is an indicator of the domain size, hence of the order and cleanness of the surface. The final peak is caused by the insertion of $OH^-$-ions into the sulfate layer. The peaks have their counterparts in the negative sweep. Except for the OH-peak, all peaks in the negative sweep are at the same potential as in the positive sweep and have the same shape. That means that adsorption/desorption processes were in equilibrium with the electrolyte at the corresponding potential.

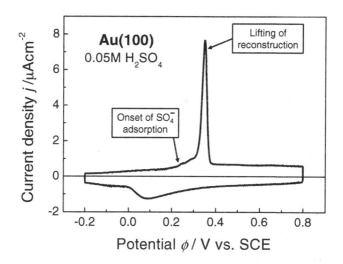

**Fig. 2.27.** Voltammogram of Au(100) in 0.05 M $H_2SO_4$ (courtesy of Margret Giesen). The potential is with reference to the *Saturated Calomel Electrode* (SCE). Sweep rate is 10 mV/s. See text for further discussion.

The voltammograms of gold surfaces are less rich in features, nevertheless interesting. Figure 2.27 displays the voltammogram of an Au(100) surface in 0.05 M $H_2SO_4$. Now, the potential is with reference to the *Saturated Calomel Electrode* (SCE) (0V RHE corresponds approximately to 0.24V SCE). The sweep rate is 10 mV/s. The surface was immersed into the electrolyte at −0.2 V SCE. At this potential the Au(100) surface is reconstructed (Fig. 2.28a, see also Sect. 1.2.1). The reconstruction is lifted when the potential is raised to positive values. Since the *potential of zero charge* (pzc) of the unreconstructed surface is lower by about 0.3 V, the interface capacitor is suddenly charged upon lifting the reconstruction, which causes the peak in the current (for details see Sects. 3.2.3, 4.2.3 and Fig. 4.7). As for the Pt-surface, sulfate ions adsorb for positive potentials, however, the sulfate uptake extends over a wide potential range of 0.6V (Fig. 4.8, 6.19) and no peak arises from that. Instead, the gradual $SO_4^-$-adsorption gives rise to the higher current. The beginning of the $SO_4^-$-adsorption is seen as the onset of a higher current in Fig. 2.27. When the reconstruction is lifted, the surplus atoms

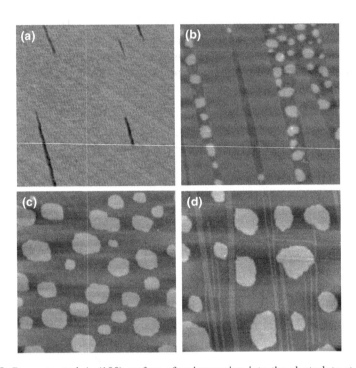

**Fig. 2.28.** Reconstructed Au(100) surface after immersion into the electrolyte at negative potentials **(a)**. The reconstruction is lifted around 0.35 V SCE **(b)**. The 25% surplus atoms (Sect. 1.2.1, Fig. 1.11) form adatom islands on the surface **(c)**. The surface reconstructs again after sweeping back to negative potentials, however the process is kinetically hindered. The reconstruction begins in streaks. Atoms at the island edges jump into the next layer underneath: the island is "eaten up by the streaks" **(d)**. (Courtesy of Margret Giesen).

form islands on the surface (Fig. 2.28b). When the potential is swept backwards, the islands do not instantaneously dissolve into the first layer. Rather, the reconstruction is established along linear streaks that "eat" into the islands (Fig. 2.28c) causing the odd island shapes. Since the process of re-establishing the reconstruction takes time, the reconstruction peak in the negative sweep is broad and shifted towards negative potentials. Its shape depends on the sweep rate.

Voltammetry is sometimes seen as the equivalent to thermal desorption spectroscopy of surfaces in vacuum (Sect. 6.3). In thermal desorption spectroscopy, the species desorbing from the surface are observed while the temperature is raised at a constant rate. Upon a closer look the differences between voltammetry and desorption spectroscopy are however larger than the similarities. On the technical side, voltammograms differ because one runs a complete cycle of the potential with adsorption and desorption. Unlike the desorption in vacuum the desorbing species have to be transported away by diffusion which has a considerable influence on the kinetics and makes the quantitative interpretation of voltammograms more complicated. As we have seen, the most significant features in voltammograms are those associated with phase transitions that occur at a particular potential for which is no equivalent in thermal desorption spectroscopy. Adsorption and desorption on the other hand mostly cause structureless currents. Finally, voltammograms show continuous reactions outside the realm of the ideally polarizable electrode.

### 2.3.3 Preparation of Single Crystal Electrodes

There may be still one or the other surface scientist coming from the UHV-background who would be inclined to denounce work on electrolyte surfaces as "dirty", or dismiss electrolyte surfaces as being less well defined than surfaces prepared in UHV. Latest with the advent of single crystal electrochemistry and the electrochemical STM there is no justification for such arrogance. Experiments on the solid/electrolyte interface can be just as clean and well defined as any UHV-experiment. Merely, the methods by which that result is achieved differ: They largely involve chemical rather than physical preparation techniques. There is definite lack of methods to determine the state of cleanness of a surface in an electrolyte. No equivalent to Auger- or photoemission spectroscopy to reveal impurities exists. With a well-equipped UHV-system at hand a physicists can teach himself how to prepare a surface by trial and error. The preparation of electrochemical experiments and of electrochemical surfaces requires training and experience. This holds even more as the interpretation of the standard check on the experimental conditions, the voltammogram, requires expertise and the comparison to a reference voltammogram of an undisputedly clean and well-ordered surface. In the absence of other techniques, it has taken quite a while until researchers could agree as to how a voltammogram of a clean and well-ordered surface of a specific material in a specific solution should look. Differences in voltammograms frequently concern very subtle features: the height and width of a narrow peak (e.g. the peak caused by the disorder-order phase transition in Fig. 2.26), the magnitude of defect-associated peaks (steps in Fig. 2.26), or the overall slope of the voltammogram ("hanging" voltammogram). Moreover, even if reference voltammograms are available, a deviation from the reference may be for more than one reason. For example, the absence of the small step-associated hydrogen desorption peaks in Fig. 2.26 could be due to a very low concentration of steps on the surface, which would be good. However, it also could (and may more likely) be due to contaminations blocking the step sites, contaminations that came from the crystal preparation procedure, from the electrolyte or from the walls of the glass vessels because of insufficient cleaning. To determine the cause for a failure, or rather to have the right thoughts about a failure, requires experience, skill and considerable training. These capabilities are not acquired in a do-it-yourself training program. Rather the novice should enter a "school" and learn the trade there.

There is another important reason why self-training is not the thing to do, in particular not for a person with a background in physics, that is safety! Cleaning procedures for the glassware require handling of extremely hazardous agents. Strongly oxidizing acids such as Carot's acid (also called Piranja acids since organic material such as ones finger is dissolved in seconds!) or a mixture of concentrated sulfuric and nitric acid are used in liter-quantities and boiling hot. Dealing properly with those hazardous materials requires special laboratory equipment, protective clothing and, above all, training in safe and disciplined working conduct, which is not necessarily one of the virtues of a physicist. It is therefore that we abstain from providing detailed recipes for cleaning procedures of the glassware and the preparation of electrolytes in this volume. In the following, we concentrate on the pre-treatment of a few specific single crystal materials.

Platinum, gold and silver single crystal surfaces are prepared by *flame annealing*. The crystals are gently heated by a Bunsen burner fed by hydrogen and air. Platinum is heated yellow-hot ($\cong 1300\,°C$), silver and gold to dark red-hot ($\cong 700\,°C$). Bead crystals are particularly well suited for flame annealing. The crystals are held with tweezers by their wire end, which minimizes stress on the crystal during the process. The crystals are cooled down slowly in an argon/hydrogen atmosphere (with 5% hydrogen). Once the temperature is below 100 °C, the crystals are immersed into the electrolyte under potential control. The silver crystal requires chemical treatment before flame annealing in $H_2O_2$/cyanide and $H_2O_2$ solutions.

Copper surfaces are in-situ electro-polished in 50-66% orthophosphoric acid by applying a potential of about 2 V between the copper electrode and a platinum sheet for 15-60 s.

# 3. Basic Concepts

The section reviews some basic concepts of Solid State Physics, Chemistry and Electrochemistry in as much as they come to bear in this volume. Elastic properties of crystalline solids are treated more extensively since they have become rather important for surfaces and thin film systems lately. As an application, the elastic interactions between defects and strain-induced self-assembly are considered.

## 3.1 Electronic States and the Chemical Bonding in Solids

### 3.1.1 Metals

The simplest model for the quantum states of electrons in a metal is the particle-in-a-box model, the box being represented by a potential wall. The model neglects all explicit electron-electron interactions. Each electron is considered as being completely independent of each other and described by a single particle wave function. According to the Pauli-principle, the single electron states (the energetically degenerate spin-up and spin-down states are counted as separate states!) are occupied by one electron up to a maximum energy, the Fermi-level. As simple as the model is, it accounts for many essential properties of the electronic structure of metals. In the context of this volume, the model also serves to introduce certain notations. It is therefore briefly sketched in the following. For details, the reader is referred to standard textbooks on solid-state physics.

The electron states are particular simple if one assumes the potential box to possess infinitely high walls.

$$V(x, y, z) = \begin{cases} 0 & for \ 0 \leq x, y, z \leq L \\ \infty & elsewhere \end{cases} \qquad (3.1)$$

The Schrödinger equation inside the box and the boundary condition that the wave function be zero where the potential is infinite is fulfilled by the ansatz

$$\psi(x, y, z) = \left(\frac{2}{L}\right)^{3/2} \sin k_x x \ \sin k_y y \ \sin k_z z \qquad (3.2)$$

with

$$k_i = \frac{\pi}{L} n_i \quad n_i = 1,2,3... \tag{3.3}$$

The energy $E$ is

$$E(k) = \frac{\hbar^2 k^2}{2m} = \frac{\hbar^2}{2m}\left(k_x^2 + k_y^2 + k_z^2\right) \tag{3.4}$$

in which $m$ is the electron mass. The eigenstates labeled with the first four $k_x$-values are displayed in Fig. 3.1. For a bulk solid, these low-energy states are unimportant, since there are only a few of them and they remain occupied under all circumstances. The low-energy states do play a significant role in quantum well structures where only a small number of theses states exist below the Fermi-level (Sect. 8.3). For a box with finite potential walls ($V = V_0$ outside) the wave function develops exponential tails which extend into the region outside the box with a decay length inversely proportional to the square root of $V - \hbar^2 k_x^2 / 2m$ .

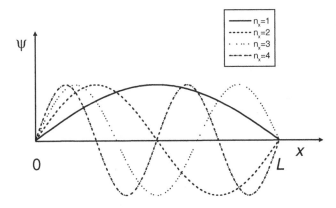

**Fig. 3.1.** The wave function $\psi$ for particles in a box for $n_x = 1,2,3,4$; $n_y = n_z = 1$. The larger the wave vector, the more rapidly the electron density $|\psi|^2$ of the corresponding electron drops to zero at the boundary of the box. For $k = k_F$ the characteristic decay length is about $\pi/2k_F$.

Following the Pauli-principle, the single electron states are filled up to an energy $E_F$, the Fermi-energy that is

$$E_F = \frac{\hbar^2 k_F^2}{2m} = \frac{\hbar^2}{2m}(3\pi^2 n)^{3/2} \tag{3.5}$$

with $n$ the electron density and $k_F = (3\pi^2 n)^{3/2}$ the wave vector of the electron with the energy $E_F$. By counting the number of states in an energy window $dE$ one can easily derive the density of electron states $D(E)$ per energy and volume for the particle-in-a-box-model,

$$D(E) = \frac{(2m)^{3/2}}{2\pi^2 \hbar^3} \sqrt{E} .$$

(3.6)

This expression is a very good approximation to the actual density of states of s-electrons in metals. For metals with d-states in the regime of valence electrons e.g. the 3-5d transition metals, a contribution due to d-states is to be added. Because of the more localized character of d-electrons and the larger number of d-states per atom, the density of d-states is confined to a narrower energy range and the density exceeds that of s-electrons by far.

According to the Pauli-principle, the occupation of an electron state can be either zero or one. This principle also regulates the occupation of electron states at a finite temperature T. The probability for a state to be occupied $f(E,T)$ is

$$f(E,T) = \frac{1}{e^{(E-\mu)/k_B T} + 1}$$

(3.7)

with $\mu$ the chemical potential of electrons and $k_B$ the Boltzmann constant. At $T = 0K$, $f(E,T)$ is 1 for $E < \mu$ and 0 for $E > \mu$. The chemical potential is therefore equal to the Fermi-energy $E_F$ at $T = 0$ K.

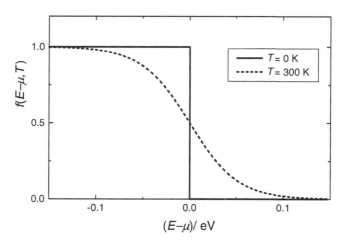

**Fig. 3.2.** Fermi-function $f(E,T)$ for E in the vicinity of $\mu$ for $T = 0$ K and $T = 300$ K. The Fermi-function is symmetric to both sides of $E-\mu = 0$ with a width of $2k_B T$ to either side.

For $T > 0$ K the Fermi function is a smooth symmetric function around $(E-\mu) = 0$ with a width of about $2k_BT$ to each side (Fig. 3.2). The concentration of electrons $n$

$$n = \int_0^\infty f(E,T)D(E)\mathrm{d}E \qquad (3.8)$$

must remain constant with temperature. Since in genral the densities of states above and below the Fermi-level are not equal, and since the Fermi-function is a symmetric function, the chemical potential $\mu$ changes with temperature. The shift with temperature is proportional to the negative slope of the density of states at the Fermi-level, hence downwards for a $D(E)$ as in (3.6). The shift can become quite large for transition metals, which causes the larger thermo-electric power of such metals.

The Fermi-function has its name because it describes the occupation statistics of fermions. The derivation of the Fermi-function makes only use of the possible occupation numbers and not of the spin being ½! Fermi-statistics is therefore also the appropriate statistics for the occupation probability of all other particles that have either zero or one as the possible occupation numbers. Statistical problem of that occur also in other areas of physics, e.g. in the case of a non-interacting lattice gas (Sect. 5.4.1). The relation between the probability of a site being occupied, in other words the fractional coverage, and the chemical potential of the particles is the same as for fermions. This concept immediately leads to the Langmuir Isotherm (Sect. 6.2.1).

## 3.1.2 Semiconductors

Metals are characterized by the fact that the Fermi-level falls inside a band of electron states. Electrons at the Fermi-level can therefore pick up energy in infinitesimal small amounts, and thus can gain kinetic energy in an electric field, which is the origin of the high electric and thermal conductivity of metals. Semiconductors and insulators possess a filled band of valence electrons, which is separated from a band of unoccupied states, the "conduction band" by an energy gap. This band structure is a consequence of the bonding and the electron configuration of the atoms forming the solid. Consider silicon as an example (Fig. 3.3): If one places the Si-atoms at the atom positions of the diamond structure (Fig. 1.17), however at distance r much larger than the equilibrium distance, then the electron orbitals initially remain as they are in the atomic $3s^2 3p^2$-configuration. As the distance r is reduced the electron overlap, form bands and the electronic configuration changes to $sp^3$ hybrids, in order to maximize electron density in the regions of low potential energy in the bonding regions between the atoms. All four valence electrons per atom stay in the lowest band, the valence band, which is fully occupied, while the upper band is completely empty. This immediately raises the question where does one have to place the Fermi-level, respectively the chemical potential at finite temperature. It seems that one might place $E_F$ arbitrarily anywhere between the occupied and the unoccupied band. However, this is not so! To determine the position of the chemical potential at finite temperature and hence the Fermi-level by considering the limit $T \to 0$ one needs to consider the density of the valence and conduction band in more detail. Around the minimum of the conduction band, the electron energy can be expanded into a Taylor series with respect to the wave vector $k$. The lowest term is necessarily the $k^2$-term. The electron energy can thus be described by

$$E(k) = E_c + \frac{\hbar^2}{2}\left(\frac{k_x^2}{m_x} + \frac{k_y^2}{m_y} + \frac{k_z^2}{m_z}\right)\ldots \tag{3.9}$$

in which $m_x$, $m_y$, and $m_z$ denote effective masses of the electron in the three directions and $E_c$ is the edge of the conduction band. Near the conduction band minimum, the density of states has the same square-root dependence on the energy as for a metal (Fig. 3.4). The same argument can be made for the density of states at the upper edge of the valence band.

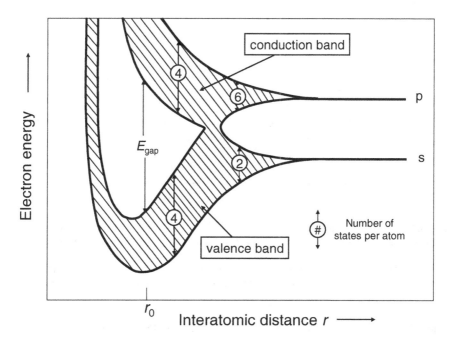

**Fig. 3.3.** Energy bands of electrons in solids formed by the group IV elements C, Si, and Ge as a function of the interatomic distance. Note that the equilibrium separation $r_0$ is not at the minimum of the electron energy because the Coulomb repulsion between the ion cores needs to be balanced.

Because of the rapid decay of the occupation probability with energy, only the densities of states near the band edges are important (Fig. 3.4). The concentration of electrons n in the conduction band is

$$n = \int_{E_c}^{\infty} D_c(E)\, f(E,T)\, dE \cong N_{eff}^c\, e^{-\frac{E_c-\mu}{k_B T}}$$
(3.10)

with the effective density of states

$$N_{eff}^c = \left(\frac{2\pi m_c^* k_B T}{h^2}\right)^{3/2}.$$
(3.11)

Here $m_c^*$ is a mean effective mass of electrons in the conduction band. In (3.10) the Fermi-function is approximated by a Boltzmann distribution, which is a good approximation as long as the chemical potential is several $k_B T$ below the conduction band.

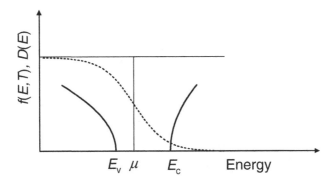

**Fig. 3.4.** Density of states in the conduction and valence band of semiconductors (solid line) and the Fermi-function (dashed line). The chemical potential adjusts itself so that the total number of electrons in the conduction band equals the number of unoccupied states (holes) in the valence band. The chemical potential of electron has therefore a defined value even though there are no states at the energy that corresponds to the chemical potential.

The concentration of unoccupied states, of "electron holes" is correspondingly

$$p = \int_{-\infty}^{E_v} D_c(E)\left[1 - f(E,T)\right] dE \cong N_{eff}^v\, e^{+\frac{E_v-\mu}{k_B T}}$$
(3.12)

with

$$N_{\text{eff}}^{\text{v}} = \left( \frac{2\pi m_{\text{v}}^* k_B T}{h^2} \right)^{3/2} \tag{3.13}$$

the effective density of states at the top of the valence band. The product of the concentration of electrons $n$ and holes $p$ is

$$np = N_{\text{eff}}^{\text{c}} N_{\text{eff}}^{\text{v}} e^{-\frac{E_g}{k_B T}} \tag{3.14}$$

in which $E_g = E_c - E_v$ is the energy gap between the conduction and valence band edge. Eq. (3.14) can be considered as the law of mass action for electrons and holes.

As electrons in the conduction band stem from the valence band the concentration of electrons $n$ equals the concentration of holes $p$. By equating (3.11) with (3.12) and by solving for $\mu$ one obtains the position of the chemical potential as a function of temperature.

$$\mu = \frac{E_c - E_v}{2} + \frac{k_B T}{2} \ln \frac{N_{\text{eff}}^{\text{v}}}{N_{\text{eff}}^{\text{c}}} \tag{3.15}$$

Hence, $\mu$ lies in the middle between the conduction and the valence band edge at $T = 0$ K, and it stays there if $N_{\text{eff}}^{\text{c}} = N_{\text{eff}}^{\text{v}}$. The chemical potential of electrons has therefore a defined value at all temperatures even though there are no states at the energy $\mu$. This statement may appear trivial at this point, but is less trivial when extended to insulators or ionic conductors of solid or liquid phase. As long as the system has states for electrons to be occupied, one has a defined position of the chemical potential of electrons. For two systems in equilibrium, the chemical potential determines the position of the energy levels with respect to each other. If the density of states is low around the chemical potential, the position of the chemical potential changes rapidly when the system is charged. For semiconductors e.g., the chemical potential can vary between the conduction band and the valence band, depending on concentration and sign of additional charges brought into the system by doping or by contact with a metal or another semiconductor. Local variations of the chemical potential are even more facile in insulators. In metals, on the other hand, the chemical potential of electrons hardly changes upon charging.

### 3.1.3 From covalent bonding to ions in solids and liquids

Compounds made from group III- and V-elements and from II- and VI-elements are likewise bonded by sp$^3$-hybrids. The structure of III-V compounds is the diamond structure, albeit with the two atoms in the primitive unit cell occupied by one atom each, whereby the Zincblende structure results (Fig. 1.17). An alternative structure in which the tetrahedral coordination is retained is the wurtzite structure that is predominantly realized for II-VI-compounds. Since the two atoms in the unit cell are different, the bonding has a partial ionic character. It is important to state that the ionicity is really just a partial one. Considering the compound ZnO as an example, which crystallizes in the wurtzite structure, merely a fraction of an electron charge is transferred from zinc to oxygen in forming the bond, independent on how one defines atomic charges in a solid in detail. It is therefore misleading to say that the Zn atom is in the $Zn^{2+}$ oxidation state in ZnO, or, to give another example, that Fe is in both the $Fe^{2+}$ and $Fe^{3+}$-oxidation states in the compound magnetite, $Fe_3O_4$. Only in a few extreme cases of a purely ionic bonding as in the alkali-halides, electronic charge is transferred completely from one atom to another. Assigning spectroscopic energy levels as $X^{n+}$, $X^{m+}$ levels to characterize different electronic structure of an atom X in a different local environment, as it is done sometimes, is therefore a misconception. The situation is even more confusing in chemistry where oxidation and reduction were traditionally conceived as removal and addition of one or more electrons from an atom, processes that actually do not occur. One might wonder how this concept of oxidation/reduction came about, as it evidently has no foundation in the physics of the solid state. It is presumably because chemistry was primarily the science of reactions, and reactions may be accompanied by a charge transfer. This holds in particular for corrosion and galvanic deposition. A zinc atom, e.g., when dissolved in an acidic aqueous solution becomes an ion complex, and the process is accompanied by leaving behind two electrons on the solid zinc electrode.

In the previous section we have learned that electron have a defined a chemical potential even when no electronic states exist at the energy which corresponds to the chemical potential. A chemical potential and hence a Fermi-level exists therefore also in solids with partial or even purely ionic bonding which may be complete insulators with regard to electron transport. If a Fermi-level can be defined there, it exists also in liquids with ionic but no electronic conduction, i. e. in electrolyte solutions. For a semiconductor we have found an easy way to relate the position of the Fermi-level to the electronic states of the system. It is less trivial to establish such a relation with respect to the energy scale of the multitude of redox reactions. The reference point to all redox reactions is the protonization of hydrogen in water.

$$\frac{1}{2}H_2 \Leftrightarrow H^+ + e^- \tag{3.16}$$

Instead of writing $H^+$ sometimes $H_3O^+$ is used in textbooks to indicate that $H^+$ does not exist by itself in water. According to more recent work, this again is a poor representation of the actual state, $H_5O_2^+$ being a better one [3.1]. The reaction (3.16) is a half-reaction, which cannot occur in water as such because water has no state to accommodate the free electron. A supplementing second half of the reaction that takes care of the electron is

$$H_2O + e^- \Leftrightarrow \frac{1}{2}H_2 + OH^- . \tag{3.17}$$

The sum of (3.16) and (3.17) is the water dissociation reaction

$$H_2O \Leftrightarrow H^+ + OH^- , \tag{3.18}$$

which requires a free enthalpy $\Delta G$ (practically equal to the free energy $\Delta F$ at atmospheric pressure) of 0.83 eV per atom. If one assigns by definition the energy level $\Delta G = 0$ to the half reaction (3.16) then $\Delta G = 0.83$ eV is to be assigned to the half reaction (3.17). One way to raise that energy in the appropriate form, namely as a free energy, is to place the electron on a metal with a negative voltage. With respect to the definition of the zero on the redox energy scale the voltage on the metal would have to be −0.83 V. By comparing equilibrium reaction energies of various reactions the complete scale of "Standard Potentials" (potentials in the sense of voltages, not energies) evolves. This energy scale can furthermore be matched to an absolute energy scale by considering certain reaction cycles (see e.g. [3.2]). Reference point of that absolute scale is the vacuum energy of an electron, which is the energy of an electron just out side the solid or liquid (to be specifically defined in the next section). For the water dissociation, we then have the following energy scales.

**Table 3.1.** Energy scales and standard potential for the water dissociation reaction.

| Reaction | Chemical energy $\Delta G$ / eV | Absolute energy scale / eV | Standard Potential $U$ / V |
|---|---|---|---|
| $\frac{1}{2}H_2 \Leftrightarrow H^+ + e^-$ | 0 | −3.67±0.2 | 0 |
| $H_2O + e^- \Leftrightarrow \frac{1}{2}H_2 + OH^-$ | +0.83 | −4.5±0.2 | −0.83 |
| $\frac{1}{2}O_2 + 2H^+ \Leftrightarrow H_2O$ | +1.24 | −4.91±0.2 | −1.24 |

The equilibrium concentration of the ions $H^+$ and $OH^-$ are calculated from the formation enthalpy as

$$\left[H^+\right]\left[OH^-\right] = \left[H_2O\right] e^{-\frac{\Delta G}{k_B T}} \tag{3.19}$$

With $[H^+] = [OH^-]$ one obtains for the relative concentration of protons in water at 298 K almost exactly $1.0 \times 10^{-7}$ which is nice as it provides a simple pH scale, given by the negative logarithm of the $H^+$-concentration: pH = 7.0 denotes neutral water. Equation (3.19) is identical to (3.14) relating the concentration of electrons and holes in an intrinsic semiconductor. The analogy goes even further. By doping the semiconductor with electron donors more electrons are created, the product of the concentration of electrons and holes remains the same. Similarly, when a base NaOH is added to water, more $OH^-$ ions reduce the concentration of $H^+$-ions so that the balance (3.19) is maintained, and the pH shifts to larger numbers. A semiconductor doped with donors corresponds to a base, when doped with electron acceptors it corresponds to an acid. This correspondence can be exploited to define an electron chemical potential in an electrolyte. One may interpret the dissociation of $H_2O$ into the ions $H^+$ and $OH^-$ as a transfer of an electron from one state to another with an activation energy of 0.83 eV. This defines the position of the Fermi-level as

$$\mu = E_{\frac{1}{2}H_2 \Leftrightarrow H^+ + e^-} - pH\,k_B T \ln 10 . \qquad (3.20)$$

The standard potential scale is therefore a function of the pH-value.

## 3.2 Charge Distribution at Surfaces and Interfaces

### 3.2.1 Metal Surfaces in the Jellium Approximation

The particle-in-a-box model describes the metal electrons in terms of single particle wave functions, which are standing waves. For a box with infinitely high potential barrier to the vacuum, all electron wave functions vanish at the surface, independent of the electron energy. The charge density is therefore zero at the surface (Fig. 3.1). Towards the interior, the charge density rises to the bulk value within a screening length which is of the order of a quarter of the shortest possible wave length, the Fermi-wavelength $\lambda_F = 2\pi/k_F$. The bulk charge density is approached in an oscillatory way because of the sharp cut-off at the Fermi-wave length. If the potential well is of finite depth, then the charge density is not zero at the surface. Rather it "spills out" into the vacuum with an exponential decay. The decay length $\Lambda$ is determined by the work function $\Phi$, the difference between the energy of an electron in the vacuum $E_{vac}$ and the Fermi-energy $E_F$.

$$\Lambda = \frac{2\hbar}{\sqrt{2m\Phi}} \qquad (3.21)$$

While the particle-in-a-box model is reasonably realistic concerning the electron charge distribution near the surface, it cannot predict the work function $\Phi$ or, alternatively the depth of the potential well. To this end, one has to move at least to

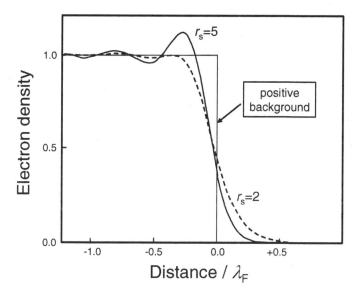

**Fig. 3.5.** Electron density at the surface of a metal in the jellium model [3.5]. The density is plotted for two different bulk electron densities represented by $r_s$, which is the radius of a sphere in units of the Bohr radius $a_0$ describing the volume associated with one electron. Values of $r_s$ range between 2 and 6. Low values correspond to high electron densities. The oscillations in the charge density are known as Friedel-oscillations.

the next level of sophistication and include electron-electron interactions. The density functional theory [3.3, 4] applied to the "jellium-model" [3.5, 6] is the simplest approach. The electron density near a surface according to Lang and Kohn [3.5] is displayed in Fig. 3.5 for two different electron densities in the bulk. As common in the theory of metals, the electron density is described as the radius $r_s$ in units of the Bohr radius $a_0$ of the sphere possessing the volume of one electron. Values of $r_s$ range between 2 and 6, which approximately correspond to the electron densities in aluminum and cesium, respectively. Figure 3.5 shows nicely the spill out of electrons beyond the boundary of the positively charged background of the ions (assumed uniform). Towards the interior, the density approaches the bulk value within about a quarter of the Fermi-wavelength $\lambda_F$. Since the spill-out of the electron charge decays to zero exponentially, the self-consistent potential for an electron leaving the metal approaches the vacuum level within half a Fermi wavelength, according to the jellium model. In reality, the vacuum level is approached more slowly because of the classical image force, which is not accounted for in the quantum mechanical jellium model. An electron in the vicinity of a metal surface experiences a force from its image charge inside the metal, which causes the image potential

$$\Phi_{\text{image}} = -\frac{e^2}{16\pi\varepsilon_0 d} = -0.36\,\text{eV}\,(d/\text{nm})^{-1}. \tag{3.22}$$

Here, $\varepsilon_0$ is the vacuum permittivity. Quantum mechanically, the image force arises from a virtual excitation of surface plasmons. The relatively slow increase of the potential to the vacuum level has interesting practical and conceptual consequences. On the practical side, the effective work of a metal electrode can be reduced by applying a high electric field, which leads to field-assisted thermionic emission. Conceptually, the existence of the image potential and the gradual, yet largely unknown transition into the quantum mechanical regime leads to the difficulty that the potential for an electron a few tenths of a nm away from the surface is not well defined.

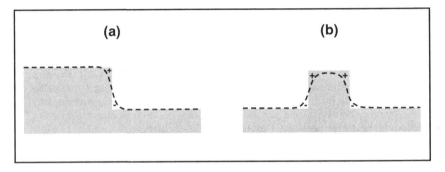

**Fig. 3.6.** Illustration to the Smoluchowski-effect. The density of electrons cannot follow the sharp contour of the jellium-edge at a step site (**a**) or at a single adatom (**b**). The result is a dipole moment pointing with the positive end away from the surface leading to a reduction of the work function. As well as the model describes qualitatively the dipole moment associated with rough edges, it is an artifact of the jellium model and not the real cause for the dipole moment (see Fig. 3.7).

The work function of a metal depends on the crystallographic structure of the surface. In general, the more open the structure the lower is the work function. Furthermore, a rough morphology leads to a reduction of the work function. This reduction is known as the Smoluchowski-effect. Within the framework of the jellium model, the Smoluchowski-effect has the same origin as the smooth contour of the electron density in response to the sharp contour of the potential (Fig. 3.5), the kinetic energy. The finiteness of the kinetic energy imposes a shortest screening length on the electrons. Because of the finite screening length, electrons cannot perfectly screen a sharp structural contour on a surface such as given by a step (Fig. 3.6). Steps should therefore have a dipole moment $p_z$, with the positive end pointing away from the surface, leading to a reduction of the work function for a given concentration $n_s$ of dipoles on the surface (see also section 4.3.5).

$$\Delta\Phi = -e\, p_z n_s / \varepsilon_0 \qquad (3.23)$$

As well as the model describes qualitatively the dipole moment associated with rough surface features, it is an artifact of the jellium model and obscures the real cause for the dipole moments which are associated with surface roughness. On a real surface, the electrons "see" no sharp edges associated with steps or adatoms, the charge contours provided by the ion cores of the solid are already smooth. The real cause for the positive dipole moment, e.g. of an adatom has to do with the formation of bonds with the substrate surface atoms. Bond formation requires the occupation of empty states. The occupation of states right above the Fermi-level cost the least energy. The substrate provides more such states than the single adatom, simply because it contains more atoms [3.7]. Hence, the valence charge density flows towards the surface, thereby creating a net positive dipole (if all atoms involved are of the same type). The effect is illustrated in Fig. 3.7. There, the difference in the charge density caused by the bonding of an adatom at the surface of a large cluster of gold atoms representing a (100) surface is plotted in a plane normal to the surface. Dashed and solid contour lines represent a reduction and enhancement of electron charge, respectively. One sees that in the outer region above the adatom, the electron concentration is reduced, and the charge is moved towards the surface on both sides of the adatom, giving rise to a dipole moment.

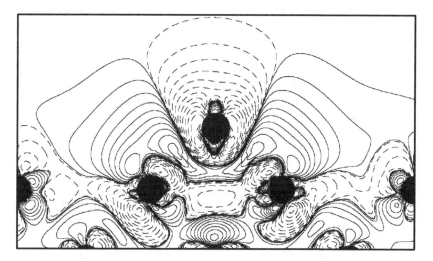

**Fig. 3.7.** Contour lines of the difference in the electron charge density caused by the bonding of an adatom on the Au(100) surface. The plot is a cross section in the (011)-plane through the adatom. Dashed and solid lines correspond to a reduction and increase of the charge density, respectively [3.7].

## 3.2.2 Space Charge Layers at Semiconductor Interfaces

Because of the (orders of magnitude) lower densities of charge carriers, screening in semiconductors involves much larger length scales than in metals and can therefore be treated semi-classically. This section is devoted to the problem of screening of a homogeneous surface by charge carriers, electrons or holes in the bulk of the semiconductor. Surface charges and their corresponding countercharges in the semiconductor bulk are frequently an intrinsic property of semiconductor surfaces because of the existence of surface states and a mismatch of the neutrality position of the chemical potential in the surface states and the bulk. We consider the Si(111) surface as an example.

The unreconstructed Si(111) surface would have one dangling bonds on each surface atom, i.e. each Si-surface atom has a nonbonding orbital which can accommodate two electrons, is, however, occupied only by a single electron. Hence, the surface states form a band of metallic character, albeit of very low conductivity. The conductivity is low since the bands are flat in $k$-space because of the poor overlap between the surface atoms. The effective mass is therefore rather high. The (7×7) reconstruction reduces the number of dangling bonds by roughly a factor of two without changing the metallic character of the surface state band. The center of the corresponding metallic band is in the lower half of the band gap of bulk silicon. The neutrality position of the chemical potential for this band of surface states is therefore also in the lower half of the conduction band. The surface states become negatively charged when the chemical potential rises above the natural level given by the occupation of the participating orbitals, and positively charged when it falls below.

Similar arguments can be brought forward for the Si(100) surface. For the unreconstructed Si(100) surface each surface atom would have two half-filled dangling bond orbitals. One of the orbitals engages in the formation of dimer bonds between adjacent surface atoms (Sect. 1.2.3) leaving one single occupied dangling orbital per Si-surface atom. For symmetric dimers, the half-filled orbitals would represent again a metallic surface state band. As discussed in Sect. 1.2.3, the symmetric, metallic state is instable with respect to a Jahn-Teller distortion into asymmetric dimers. The metallic band then splits into a filled band of the dangling bonds of the "up-atoms" and the empty band of the "down-atoms" of the dimer. The surface state band has therefore semiconducting properties, again with a low conductivity because of the flat bands with high effective masses. The surface neutrality level lies between those two bands in the lower half of the band gap (for details on the surface band structure see Sect. 8.2.3).

Because of the high density of surface states, a small shift of the chemical potential away from the neutrality level causes a large surface charge. Inside the semiconductor bulk, the chemical potential is determined by the neutrality condition for the bulk. The chemical potential is near the center of the band gap for undoped material, near the conduction and valence bands for n- and p-doped material, respectively. Equilibrium between surface and bulk requires that the chemical potentials in the surface $\mu_s$ and in the bulk $\mu_b$ must be at the same energy. This is realized by bending the bulk band structure near the surface (Fig. 3.8) and by moving the chemical potential in the band of surface states. Since the interface must remain neutral as a whole and the density of surface states is large, the shift within the surface state band in negligible, the chemical potential is "pinned" by the surface states. The first experiment that demonstrated the pinning (denoted as *Fermi-level pinning* in the semiconductor literature) due to surface states was performed in 1962 by Allen and Gobeli on cleaved Si(111) surfaces [3.8].

**(a)**                                  **(b)**

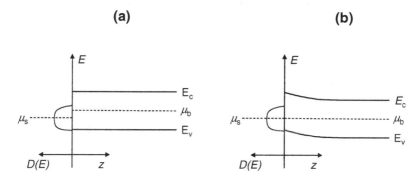

**Fig. 3.8.** A high density of surface states causes a bending of the band structure of the semiconductor near the surface to match the chemical potential in the surface state $\mu_s$ to the chemical potential in the bulk $\mu_b$.

The bending of the band structure corresponds to variation of the electric potential with the coordinate $z$. As the typical length scale for the band bending is much larger than the lattice constant, the dependence of the potential on the position $z$ can be calculated in the continuum approximation. In the following, we denote the electric potential (in the sense of a voltage) as $\phi(z)$, so that the energy levels of the band structure vary as $-e\phi(z)$, with $e$ the charge of an electron. The variation of the potential $\phi(z)$ is then obtained from a self-consistent solution of the Poisson-equation and the charge density $\rho(z)$ as given by the occupation of the various energy levels according to Fermi-statistics

$$\frac{d^2\phi(z)}{dz^2} = -\frac{\rho(z)}{\varepsilon\varepsilon_0} . \tag{3.24}$$

Here, $\varepsilon$ is the relative dielectric permittivity of the material. The charge density has contributions from the concentration of holes $p(z)$, of electrons $n(z)$, of ionized donors $N_D^+$, and of ionized acceptors $N_A^-$.

$$\rho(z) = e\Big(p(z) + N_D^+(z) - n(z) - N_A^-(z)\Big) \tag{3.25}$$

A qualitative insight into the magnitude of the band bending and its dependence on the doping is provided by the Schottky-model. The model assumes a complete pinning of the chemical potential at the surface and one type of doping, n- or p-type, in the bulk. Here, we consider the case of p-doping. The band bending is then as qualitatively depicted in Fig. 3.9a. In the bulk, the charge density $\rho(z)$ is zero and the chemical potential lies between the energy level of the acceptors and the valence band, provided the temperature is not too high and the p-type conduction is dominated by the doping. This is the typical situation for doped silicon at room temperature. Because of the position of the surface states, the bands must bend downwards near the surface and the chemical potential moves closer to the center of the band gap. Thereby, the acceptors become ionized and thus negatively charged in the band-bending region.

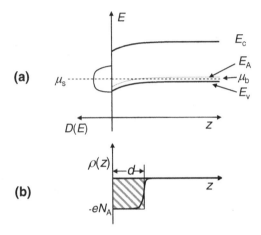

**Fig. 3.9.** (a) Band bending in the case of p-doping when the band of surface states is centered in the lower half of the band gap. (b) The charge density $\rho(z)$ can be approximated by constant value $-eN_A$ in the range between $z = 0$ and $z = d$ (hatched area).

The maximum charge density is given by the concentration of acceptors $N_A$. Because of the exponential dependence of the occupation probability on the position of the energy levels the transition between charge density $\rho(z) = 0$ and $\rho(z) = -eN_A$ is confined to a narrow range (Fig. 3.9b). The charge density $\rho(z)$ can therefore be replaced by a constant $-eN_A$ between $z = 0$ and $z = d$. The solution of the Poisson-equation is then simply obtained by elementary integration as

$$\phi(z) = \phi(\infty) - \frac{eN_A}{2\varepsilon\varepsilon_0}(z-d)^2 \quad z \leq d \tag{3.26}$$

The total band bending is therefore

$$\phi_s = \phi(\infty) - \phi(0) = \frac{eN_A}{2\varepsilon\varepsilon_0} d^2 \tag{3.27}$$

Under the assumption of a complete pinning of the chemical potential at the surface $\phi_s$ is given simply by the difference between the neutrality levels in the surface and the bulk,

$$\phi_s = (\mu_s - \mu_b)/e . \tag{3.28}$$

The thickness of the space charge layer d in terms of that difference is

$$d = \left( \frac{2\varepsilon\varepsilon_0(\mu_s - \mu_b)}{e^2 N_A} \right)^{1/2} . \tag{3.29}$$

Hence, the thickness of the space charge layer is inversely proportional to the square root of the acceptor concentration. For a typical doping of $10^{17}$ cm$^{-3}$ and with $\varepsilon = 12$ for Si, one obtains $d \cong 100$ nm which, in hindsight, justifies the use of the continuum approximation.

Technically more important than the space charge layers on clean surfaces are those in pn-junctions and at the metal semiconductor interface. The very first solid state electronic device, a piece of mineral PbS with a spring-loaded tip of copper bronze which served a rectifier for radio frequencies in the first half of the last century, was based on the electric properties of the metal/semiconductor interface. A typical potential diagram of a metal/semiconductor interface, this time for an n-type semiconductor is displayed in Fig. 3.10. Equilibrium between the metal and the semiconductor is assumed (no current flow). The positions of the chemical potentials in the metal and the semiconductor must match. In order to achieve that, band bending must occur. The magnitude of the band bending is given by

$$\Phi_s \equiv e\phi_s = \Phi - \Delta - \chi + \mu_s - E_c \tag{3.30}$$

in which $\Phi$ is the work function of the metal, $\chi$ the electron affinity of the semiconductor $\mu_s$ the position of the chemical potential in the bulk of the semiconductor and $E_c$ the conduction band edge. The quantity $\Delta$ represents a potential drop due to a dipole layer within the metal/semiconductor interface. This dipole layer is of microscopic origin and due to the atomic structure of the interface.

For the metal/semiconductor interface, the chemical potentials inside the metal and the semiconductor can be shifted with respect to each other by applying an extra voltage $U$, which results in a current flow. The current is a non-linear function of the potential. The current rises exponentially if the applied voltage is positive on the metal (for n-doping), since the band bending is reduced thereby, leading to a higher conductivity of the semiconductor in the space charge region.

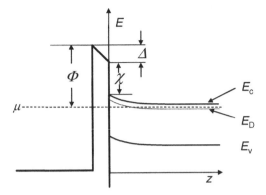

**Fig. 3.10.** Potential diagram of a metal/semiconductor interface with an n-doped semiconductor. The band bending is determined by the work function of the metal $\Phi$, the potential drop $\Delta$ across the microscopic interface, the electron affinity of the semiconductor $\chi$, and the position of the chemical potential in the semiconductor relative to the conduction band edge $E_c$. The thicknesses are not drawn to scale: the microscopic interface is a few tenths of a nm thick, while the thickness of the space charge layer is of the order of 100 nm, depending on the concentration of dopants.

For reverse bias, the band bending increases and the space charge layer becomes semi-insulating. Current can only flow because of thermal generation of holes in the semiconductor within the recombination distance from the interface [3.9]. The metal/semiconductor interface therefore acts as a diode. In addition to the rectifying property, also the behavior with respect to ac-currents is of interest. For ac-current loads, the metal/semiconductor acts as a capacitor. One charge sits on the metal or in interfacial states; the counter charge is in the space charge layer. The capacitance is entirely determined by the thickness of the space charge. The electrical properties of the metal/semiconductor interface with respect to ac-current are determined by the differential interfacial capacity, which is the derivative of the charge on the capacitor with respect to the applied potential difference between the metal and the semiconductor. The capacitance per area is then

$$C = \frac{\partial \sigma}{\partial U} = \frac{\partial}{\partial U}(eN_{\mathrm{D}}d) = eN_{\mathrm{D}} \frac{\partial}{\partial U}\left(\frac{2\varepsilon\varepsilon_0(\phi_{\mathrm{s}} + U)}{eN_{\mathrm{D}}}\right)^{1/2} = \frac{\varepsilon\varepsilon_0}{d}, \qquad (3.31)$$

as if the capacitor were built from two metal plates placed at a distance given by the thickness of the space charge layer with the semiconductor material as a dielectric. The thickness d itself depends on the applied potential U. In the range of reverse bias, the metal/semiconductor interface (as well as any pn-junction) can be employed as a tunable capacitance, which is a standard application in electronic devices.

### 3.2.3 Charge at the Solid/Electrolyte Interface

The objective of this subsection is to calculate the total capacitance of the solid/electrolyte interface. The charge distribution is similar to the metal/semiconductor interface. As for the latter, there is a thin region of the interface with a thickness of one or two atoms/molecules consisting of adsorbed water molecules, solvated ions that are weakly bonded to the surface and, depending on the chemistry and the applied electrode potential, specifically adsorbed ions (Fig. 1.50). Relatively little is known about the atomic structure of this "*Stern-layer*" (see Sect. 1.5). Concerning the electrical properties, macroscopic measurements as well as general principles require a drop or rise of the macroscopic potential across the Stern-layer, in particular if ions are "specifically" adsorbed at the surface (Sect. 1.5). Water molecules are bonded to the surface essentially with their dipole moment parallel to the surface since otherwise the potential drop across the interface (3.23) would become huge because of the large dipole moment of the water molecule. First principle theoretical calculations confirm this orientation (Sect. 6.4.4). However, an externally applied electric field across the interface can reorient the water molecules slightly, giving rise to a polarizability of the interface. Further contributions to the polarizability stem from the solid surface. They have been calculated within the jellium model [3.2]. In a macroscopic measurement, this polarizability as well as the reorientation of the water molecules appear as a capacitance, known as the *Helmholtz*-capacitance $C_H$. The Helmholtz capacitance has a broad maximum around pzc (Fig. 3.12). The decay of the Helmholtz-capacitance on both sides of pzc is caused by a saturation of the dielectric properties due to the extremely high electric fields in the Stern-layer.

Parallel to the displacement current loading or unloading the interfacial capacitance, ohmic exchange currents may exist, by which ions moving from or to the surface are unloaded or loaded with electrons originating from the substrate (or holes in case of semiconductors), thereby causing an electrochemical reaction. Here, we are interested in the physical properties concerning the dynamics and electronics of the solid/electrolyte interface and disregard electrochemical reactions. We therefore focus on solid/electrolyte interfaces in a certain potential window for which no charge transfer, and hence no reactions occur. In this potential window, which can be as large as about one volt, the interface is ideally polarizable, in other words the interface has the electrical property of a pure capacitance. The Helmholtz capacitance is electrically in series with, the capacitance of the adjacent liquid electrolyte layer. This latter capacitance can be calculated macroscopically, similar as for space charge layers in semiconductors. The model was developed by Gouy and Chapman [3.10, 11].

For simplicity we assume the electrolyte to consist of positive and negative ions of the same charge number $Z$. The concentration of positively and negatively charged ions $n^+(z)$ and $n^-(z)$ obeys Boltzmann-statistics

$$n^+(z) = n_0 \exp\left\{-\frac{Ze\phi(z)}{k_B T}\right\}, \quad n^-(z) = n_0 \exp\left\{+\frac{Ze\phi(z)}{k_B T}\right\} \tag{3.32}$$

The zero of the potential $\phi(z)$ is in the interior of the electrolyte, $z \to \infty$, and $n_0$ is the concentration of ions of either type in the neutral electrolyte. The potential $\phi(z)$ is to be calculated self-consistently from (3.32) and the Poisson-equation

$$\frac{d^2\phi(z)}{dz^2} = -\frac{\rho(z)}{\varepsilon\varepsilon_0} \tag{3.33}$$

in which $\varepsilon$ the relative dielectric permittivity (about 80 for water) and $\rho(z)$ the charge density

$$\rho(z) = Ze\big(n^+(z) - n^-(z)\big) \tag{3.34}$$

The solution for $\phi(z)$ is given by the Poisson-Boltzmann equation

$$\frac{d^2\phi(z)}{dz^2} = -\frac{Zen_0}{\varepsilon\varepsilon_0}\left\{\exp\left(\frac{Ze\phi(z)}{k_BT}\right) - \exp\left(-\frac{Ze\phi(z)}{k_BT}\right)\right\} \tag{3.35}$$

For small potentials, the terms on the right hand side can be expanded and one obtains the linearized Poisson-Boltzmann equation

$$\frac{d^2\phi(z)}{dz^2} = -\frac{2Z^2e^2n_0}{\varepsilon\varepsilon_0k_BT}\phi(z), \tag{3.36}$$

which has the solution

$$\phi(z) = \phi_G \exp\{-\kappa z\} \tag{3.37}$$

with

$$\kappa^2 = \frac{2Z^2e^2n_0}{\varepsilon\varepsilon_0k_BT} \tag{3.38}$$

and $\phi_G$ the potential at the interface of the Stern-layer and the liquid electrolyte. The inverse of $\kappa$ is the Debye-length $d_{Debye}$. Table 3.2 shows the Debye-length and the Gouy-Chapman capacitance for typical concentrations. The Debye-length at least for dilute electrolytes is large enough to justify the continuum approximation in the calculation of the potential.

Figure 3.11 displays the potential as a function of the distance from the surface. The potential drop within the Stern-layer is shown as a linear dependence with the distance, which assigns a macroscopic and constant dielectric permittivity to the layer.

**Table 3.2.** Debye-length and Gouy-Chapman capacitance at the potential of zero charge for typical ion concentrations.

| $n_0$ / mol$^{-1}$ | $10^{-4}$ | $10^{-3}$ | $10^{-2}$ | $10^{-1}$ |
|---|---|---|---|---|
| $d_{Debye}$ / nm | 30.4 | 9.6 | 3.04 | 0.96 |
| $C$ / $\mu$F cm$^{-2}$ | 2.33 | 7.36 | 23.3 | 73.6 |

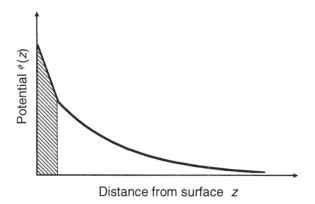

**Fig. 3.11.** Potential at a metal/electrolyte interface. The shaded area represents the Stern-layer of tightly bonded water molecules, solvated ions, and specifically bonded ions. The thickness of the Stern-layer is about 0.3 nm, its dielectric constant less than that of free water since the water molecules in that layer cannot rotate freely as in the bulk of the water. Beyond the boundary of the Stern-layer, the potential decays exponentially. The decay length depends on the concentration of ions in the electrolyte solution.

For potentials $\phi > 2k_BT/e$ the linear solution (3.36) is to be replaced by the general solution of the Poisson-Boltzmann equation (3.35) which is

$$\phi(z) = \frac{4k_BT}{Ze}\arctan\left\{e^{-\kappa z}\tanh\left(\frac{Ze\phi_G}{4k_BT}\right)\right\}. \tag{3.39}$$

The charge density at the interface $\sigma_G$ is given by the derivative of the potential with respect to $z$ at $(z = 0)$

$$\sigma_G = \varepsilon\varepsilon_0\frac{\partial\phi}{\partial z}\bigg|_{z=0} = \frac{2\varepsilon\varepsilon_0\kappa k_BT}{Ze}\sinh\frac{Ze\phi_G}{2k_BT}. \tag{3.40}$$

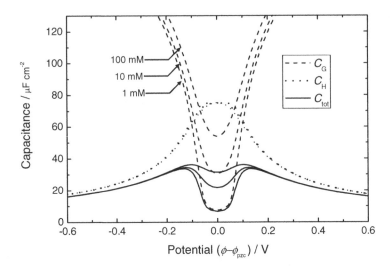

**Fig. 3.12.** Gouy-Chapman capacitance $C_G$, the Helmholtz capacitance $C_H$ and the total capacitance $C_{tot}$ are plotted as dashed, dotted and full lines for electrolyte concentrations of 1 mM, 10 mM, and 100 mM. The total capacitance has a minimum at the potential of zero charge $\phi_{pzc}$. The minimum becomes less pronounced for higher concentrations and is rather shallow for small concentrations.

The differential capacitance of the Gouy-Chapman layer is therefore

$$C_G = \frac{\partial \sigma_G}{\partial \phi_G} = \varepsilon \varepsilon_0 \kappa \cosh \frac{Ze\phi_G}{2k_B T} . \qquad (3.41)$$

The Gouy-Chapman capacitance has a minimum at the potential where the charge is zero. As the Gouy-Chapman capacitance lies in series with the capacitance of the Stern-layer the total capacitance (the only experimentally accessible quantity) is

$$C^{-1} = C_G^{-1} + C_H^{-1} \qquad (3.42)$$

For dilute electrolytes, the Gouy-Chapman capacitance is much smaller than the capacitance of the Stern-layer. Then, the total capacitance is mainly determined by the smaller Gouy-Chapman capacitance and has therefore a pronounced minimum at the *potential of zero charge* (pzc) $\phi_{pzc}$.

The calculation of the functional dependence of the total capacitance as a function of the electrode potential requires a self-consistent solution of expressions for $C_G$ and $C_H$ as both values depend on the potential drop across each capacitor in series. A self-consistent solution is calculated best by expressing $C_G$ and

$C_H$ in terms of the charge density and by observing that the total charge density in the electrolyte is identical to the charge density at the electrode (with opposite sign). In Fig. 3.12 the Gouy-Chapman capacitance, the Helmholtz-capacitance, and the total capacitance are plotted for three electrolyte concentrations: 1 mM, 10 mM, and 100 mM. For all three concentrations, the capacitance displays a minimum at the potential of zero charge $\phi_{pzc}$. Measurements of the capacitance at moderate electrolyte concentrations therefore serve to identify the pzc.

## 3.3 Elasticity Theory

### 3.3.1 Strain, Stress and Elasticity

The elasticity theory of crystalline solid has gained importance in recent years. The elastic energy plays a decisive role in the growth modes and the stability of thin film systems, the interactions between defects and in the self-assembly of periodic nanostructures. Furthermore, strain significantly affects electronic and magnetic properties of thin films. This section briefly reviews some basic elements of elasticity theory and considers homogeneously strained thin film systems.

The state of strain in a solid is described by the dependence of a displacement vector $u$ on the position denoted by $r$. The second rank tensor of infinitesimal strain $\varepsilon_{ij}$ is defined by

$$\varepsilon_{ij} = \frac{1}{2}\left\{\frac{\partial u_i}{\partial x_i} + \frac{\partial u_j}{\partial x_i}\right\}. \tag{3.43}$$

By definition, the tensor is symmetric. The antisymmetric tensor, with a minus sign between the derivatives of displacements $u$, represents a pure rotation of the solid. As a matter of convenience, the components of the tensor are usually expressed in terms of particular cartesian coordinates which are chosen to agree as much as possible with the crystallographic axes. The diagonal elements of the tensor $\varepsilon_{ij}$ are strain components associated with a change in volume (Fig. 3.13). The magnitude of the (infinitesimal) change in the volume is given by the trace of the deformation tensor.

$$\frac{\Delta V}{V} = \sum_i \varepsilon_{ii} = \mathrm{Tr}\,\varepsilon \tag{3.44}$$

The non-diagonal elements $\varepsilon_{ij}$ describe the deformation of a volume element in i-direction as one moves along the j-direction and therefore correspond to a shear distortion (Fig. 3.13).

A solid resists deformations; hence, deformations generate forces. For a homogeneous material, the forces in response to a strain or shear are proportional to the area that is affected by the deformation. One therefore relates the forces to the areas. For a definition of these area related forces, the "stresses", one considers a section through the crystal perpendicular to the $x_j$-axis and removes, in thought, the material on the right hand side of the intersection. The forces per area in the direction $i$ that are necessary to keep the crystal in balance are the components of the stress tensor $\tau_{ij}$ (Fig. 3.13). The stress tensor is symmetric just as the strain tensor: an antisymmetric part of the stress tensor would represent a torque, and in equilibrium all torques must vanish inside a solid.

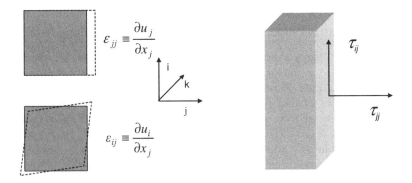

**Fig. 3.13.** Illustration for the definition of the strain and stress tensor components.

Stresses and strains are related by *Hook's Law*. In its most general form Hook's Law reads

$$\tau_{k\ell} = \sum_{ij} c_{k\ell ij}\varepsilon_{ij} ,\qquad(3.45)$$

in which $c_{k\ell ij}$ are the components of the forth rank tensor of the elastic modules. Because of the symmetry of the stress and strain tensors $\tau_{k\ell}$ and $\varepsilon_{ij}$ one has the relations $c_{k\ell ij} = c_{\ell k ij} = c_{k\ell ji}$. The number of independent components of the elastic tensor is further reduced by the requirement that the elastic energy be a unique function of the state of strain. The energy density $U_{elast}$ is

$$U_{elast} = \sum_{k\ell} \int \tau_{k\ell} d\varepsilon_{k\ell} = \frac{1}{2}\sum_{ijk\ell} c_{k\ell ij}\varepsilon_{ij}\varepsilon_{k\ell} .\qquad(3.46)$$

This equation yields the same result independent of the chosen indices for the axes if $c_{k\ell ij} = c_{ijk\ell}$. With these symmetry relations, the number of independent components of the elastic tensor reduces to 21, which permits a shorthand notation by the Voigt indices. The assignment follows the scheme

$$11 \rightarrow 1 \quad 23 \rightarrow 4$$
$$22 \rightarrow 2 \quad 13 \rightarrow 5$$
$$33 \rightarrow 3 \quad 12 \rightarrow 6.$$

Components of the stress and strain tensors can also be denoted using Voigt's notation. In order to ensure that all non-diagonal elements of the strain and stress tensor in the energy density (3.46) are properly accounted for (i.e. $\varepsilon_{ij}$ and $\varepsilon_{ji}$) a complete transition to Voigt's notation would require the introduction of redefined elastic modules. For our purpose here, it is easier and safer to use Voigt's notation only as an abbreviation for the indices in the elastic modules and stay with the standard tensor notation and summation otherwise. In the shorthand notation, the elastic tensor becomes a 6x6 symmetric tensor with 21 independent components, at most. The number of independent components is further reduced by the crystal symmetry. For crystals with cubic symmetry, the elastic tensor has only three independent components.

$$\begin{pmatrix} c_{11} & c_{12} & c_{12} & 0 & 0 & 0 \\ c_{12} & c_{11} & c_{12} & 0 & 0 & 0 \\ c_{12} & c_{12} & c_{11} & 0 & 0 & 0 \\ 0 & 0 & 0 & c_{44} & 0 & 0 \\ 0 & 0 & 0 & 0 & c_{44} & 0 \\ 0 & 0 & 0 & 0 & 0 & c_{44} \end{pmatrix} \quad (3.47)$$

It is easy to see that the elastic tensor must have this form, even without a formal proof. For example, the cubic axes are equivalent. Therefore, the diagonal components for normal and shear distortions must be equal ($c_{11} = c_{22} = c_{33}$ and $c_{44} = c_{55} = c_{66}$). A shear strain along one cubic axis cannot give rise to forces which would cause a shear along another cubic axis ($c_{45} = 0$, etc.). Furthermore, a shear cannot cause a normal stress ($c_{14} = 0$, etc.), and finally the forces perpendicular to a strain along one cubic axis must be isotropic ($c_{12} = c_{13}$, etc.).
For a hexagonal crystal, the elastic tensor has the components

$$\begin{pmatrix} c_{11} & c_{12} & c_{13} & 0 & 0 & 0 \\ c_{12} & c_{11} & c_{13} & 0 & 0 & 0 \\ c_{13} & c_{13} & c_{33} & 0 & 0 & 0 \\ 0 & 0 & 0 & c_{44} & 0 & 0 \\ 0 & 0 & 0 & 0 & c_{44} & 0 \\ 0 & 0 & 0 & 0 & 0 & c_{66} \end{pmatrix}. \quad (3.48)$$

A hexagonal crystal is elastically isotropic in its basal plane. The tensor component that describes the stress-strain relation for a shear distortion in the basal plane, $c_{66}$, is therefore related to the tensor components $c_{11}$ and $c_{12}$ by the *isotropy condition*

$$2c_{66} = c_{11} - c_{12} \,. \tag{3.49}$$

If the same condition would hold also for a cubic material, i.e. if

$$2c_{44} = c_{11} - c_{12} \tag{3.50}$$

the material would be elastically isotropic with only two independent elastic constants, the *Lamé-constants*

$$\mu = c_{44} \text{ and } \lambda = c_{12} \,. \tag{3.51}$$

In general cubic crystals are far from being isotropic (an exception is the element tungsten (see Table 3.3)!). Nevertheless, elasticity of interface systems is often studied assuming elastic isotropy, because this model provides analytical solutions for many essential problems. For an isotropic solid Hook's law (3.45) becomes

$$\tau_{ik} = \lambda \delta_{ik} \sum_i \varepsilon_{ii} + 2\mu \varepsilon_{ik} \tag{3.52}$$

Hook's law can also be written in its inverse form

$$\varepsilon_{k\ell} = \sum_{ij} s_{k\ell ij} \tau_{ij} \tag{3.53}$$

in which $s_{k\ell ij}$ is the tensor of elastic constants. The tensor $s$ has the same symmetry as the tensor $c$. The isotropy condition is

$$s_{44} = 2(s_{11} - s_{12}) \,. \tag{3.54}$$

For an isotropic solid the elastic tensor has the independent components

$$Y = 1/s_{11} \text{ and } \nu = -s_{12}/s_{11} \,. \tag{3.55}$$

$Y$ and $\nu$ are *Young's modulus* and *Poisson-number*, respectively, related to the Lamé-constants $\mu$ and $\lambda$ by

$$\nu = \frac{\lambda}{2(\lambda + \mu)} \qquad Y = \frac{\mu(2\mu + 3\lambda)}{\mu + \lambda} \tag{3.56}$$

For a proof of this relation and others see e.g. [3.9] 96ff.

While resorting to the isotropic solid is necessary for many problems, it is advantageous to stay with the anisotropic crystal in cases where the symmetry reduces the tensor relation to scalar ones. An example is the equation of motion for longitudinal waves along the axis of a cubic material. All components of the strain tensor except one, e.g. $\varepsilon_{11}$, vanish and Hook's law (3.45) reduces to

$$\tau_{11} = c_{11}\,\varepsilon_{11} = \frac{\partial u_1}{\partial x_1}. \tag{3.57}$$

The change in stress component $\tau_{11}$ on a length element $dx_1$ due to the strain is balanced by the inertial force so that

$$\rho\,\ddot{u}_1 = \frac{\partial \tau_{11}}{\partial x_1} = c_{11}\frac{\partial^2 u_1}{\partial x_1^2} \tag{3.58}$$

in which $\rho$ is the mass density. The velocity of sound for a longitudinal wave is therefore

$$v_{\mathrm{L}} = \sqrt{c_{11}/\rho}. \tag{3.59}$$

Just as easily one obtains the sound velocity of a shear wave as

$$v_{\mathrm{T}} = \sqrt{c_{44}/\rho}. \tag{3.60}$$

## 3.3.2 Elastic Energy in Strained Layers

Thin films systems offer the unique possibility to synthesize materials in a state in which the strain can have a magnitude that could not be realized for bulk materials. Such materials can have unusual and advantageous mechanical or electrical properties. An example is strained silicon in which electrons and holes possess a mobility twice as high as normally [3.12]. Another example is the negative-electron-affinity strained $GaAs_{0.95}P_{0.05}$ -photocathode that provides electrons of a particular high degree of spin polarization [3.13]. Technically the strain is realized by growing the material epitaxially on a substrate which has a larger lattice constant, e. g., Si on Ge or a SiGe-alloy. Under certain circumstances, Si grows then pseudomorphic with the larger lattice constant of the substrate. Growth and stability of such films are determined by the elastic energy stored in the film. We therefore consider the elastic energy in strained crystalline material in this section. As for the bulk elastic waves, the easiest access to the problem is achieved by considering the special case of a thin film with cubic structure with the film plane parallel to (001). We denote the axis perpendicular to the film as $x_3$ (=[001]) and take the [100] and the [010]-directions as the $x_1$ and the $x_2$-axes, respectively. In a zero pressure environment ($\tau_{33} = 0$) the state of strain is described by

$$\begin{aligned}
\varepsilon_{11} &= s_{11}\tau_{11} + s_{12}\tau_{22} \\
\varepsilon_{22} &= s_{12}\tau_{11} + s_{11}\tau_{22}
\end{aligned} \tag{3.61}$$

If the strain is isotropic $\varepsilon_{11}=\varepsilon_{22} = \varepsilon$, the stresses are also isotropic $\tau_{11} = \tau_{22} = \tau$, and the differential of the energy density $dU_{elast}$ is

$$dU_{elast} = 2\tau d\varepsilon = \frac{2\varepsilon\,d\varepsilon}{s_{11} + s_{12}} \tag{3.62}$$

The energy density per area $\gamma_{elast}$ is therefore

$$\gamma_{elast} = \frac{t\,\varepsilon^2}{s_{11} + s_{22}} = t\,\varepsilon^2\,\frac{Y}{1-v} \tag{3.63}$$

in which $t$ is the film thickness, and $Y$ and $v$ are Young's modulus and Poisson-number, respectively, as introduced in (3.55). Written in this form, the expression for the elastic energy density also applies to an isotropic solid.

If one has a one-dimensional strain ($\varepsilon_{22} = 0$, $\varepsilon = \varepsilon_{11}$) the elastic energy per area is

$$\gamma_{elast} = t\,\varepsilon^2\,\frac{Y}{2(1-v^2)}. \tag{3.64}$$

We note that in the framework presented here the energy associated with a two-dimensional strain (3.63) cannot be recovered by calculating the energy involved in two successive, one-dimensional strains orthogonal to each other. The reason is that the second stretch would be applied to the film already strained in the other direction. To calculate the energy associated with such an operation one would have to resort to strain tensors describing finite strains properly, instead of using the tensor of infinitesimal strain $\varepsilon_{ij}$.

Equations (3.63) and (3.64) can be used also for a (111) oriented film plane provided $Y$ and $v$ are the properly transformed quantities. For cubic crystals the relevant transformation relations are

$$\begin{aligned}
s'_{11} &= s_{11} - 2S\left(l_1^2l_2^2 + l_2^2l_3^2 + l_1^2l_3^2\right) & s'_{22} &= s_{22} - 2S(m_1^2m_2^2 + m_2^2m_3^2 + m_1^2m_3^2) \\
s'_{12} &= s_{12} + S\left(l_1^2m_1^2 + l_2^2m_2^2 + l_3^2m_3^2\right) & s'_{66} &= s_{44} + 4S(l_1^2m_1^2 + l_2^2m_2^2 + l_3^2m_3^2) \\
S &= s_{11} - s_{12} - \frac{1}{2}s_{44}
\end{aligned} \tag{3.65}$$

The coefficients $l_i$ and $m_i$ are the cosines of the angles of the new axes $x_1$ and $x_2$ with the cubic axes, respectively (Fig. 3.14).

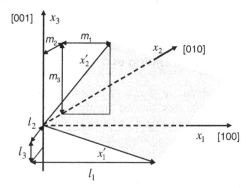

**Fig. 3.14.** Illustration of the cosines of the projection angles of the $x_i'$-axes onto the $x_i$-axes used in (3.64)

For the (111) surface, e. g., the Young's modulus and the Poisson ratio are

$$Y_{(111)} = \frac{4}{2s_{11} + 2s_{12} + s_{44}} \tag{3.66}$$

$$v_{(111)} = -Y_{(111)}(2s_{11} + 10s_{12} - s_{44})/12 \tag{3.67}$$

**Table 3.3.** Young's moduli (in $10^{10}$ N/m$^2$) and Poisson-numbers for the (100) and (111) planes of some cubic crystals.

| Material | $Y_{(100)}$ | $v_{(100)}$ | $Y_{(111)}$ | $v_{(111)}$ |
|---------|-------------|-------------|-------------|-------------|
| W | 39.5 | 0.287 | 39.4 | 0.287 |
| Fe | 13.0 | 0.364 | 21.4 | 0.383 |
| Cu | 6.66 | 0.42 | 9.51 | 0.361 |
| Ag | 4.37 | 0.428 | 8.35 | 0.514 |
| Au | 4.29 | 0.459 | 8.16 | 0.573 |
| Pt | 13.6 | 0.419 | 18.5 | 0.450 |
| Si | 13.0 | 0.279 | 16.9 | 0.262 |
| Ge | 10.3 | 0.273 | 13.8 | 0.252 |

Table 3.3 shows Young's moduli and the Poisson-numbers for the (100) and (111) planes for selected cubic crystals. With the exception of W, there is a considerable anisotropy in the elastic properties. The anisotropy is particular large for the noble metals Cu, Ag, Au. The large anisotropy is due to the particular shape of the Fermi-surface which has "neck- and belly states" at the boundary of the Brillouin-zone along the [111] direction that cause a high modulus of the (111) plane.

### 3.3.3 Thin Film Stress and Bending of a Substrate

The stress in thin films, which are grown on a wafer substrate, can cause a bending of the entire wafer. While this may be mostly an undesired effect, it can also be used to measure the stress in thin films during and after film growth or film processing. Such measurements serve to learn about growth mechanisms, the build-up of stress in epitaxial systems, and stress relaxation. Experiments of his kind have become rather popular lately and this section is devoted to the theoretical background of the experimental techniques [3.14]. An illustrative example is displayed in Fig. 3.15 [3.15]. Pseudomorphic silver films are grown on a Fe(100) substrate. Plotted is the integral over the stress in the silver film, which is the measured quantity, vs. the film thickness. The misfit between the lattice nearest neighbor distances on Ag(100) (fcc) and Fe(100) (bcc) is $\varepsilon_{mf} = 8 \times 10^{-3}$, the distance on Ag being smaller.

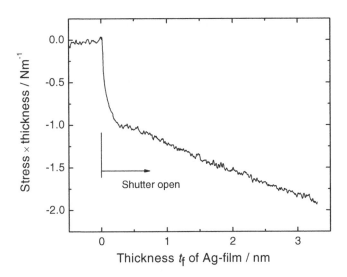

**Fig. 3.15.** Integral over the stress in a thin silver film deposited on a Fe(100) substrate at 150 K vs. the film thickness. After completion of the first monolayer the stress agrees with the stress calculated from the misfit of -0.8% (courtesy of Dirk Sander, [3.15]).

The stress calculated form this elastic deformation is $\tau = \varepsilon\, Y_{100}\,/(1-\nu_{100}) =$ $6.1\times10^{8}\ \mathrm{Nm^{-2}}$, in good agreement with the measured slope of the curve. The first monolayer, however, is determined by interfacial properties and does not fit into the scheme.

Frequently, the interface between the substrate and the growing film ranges beyond one monolayer. For the system Ag/Fe(100) this is the case when deposition is being performed at 300 K. The intermixing between Fe and Ag occurs in the first 5 monolayers, and is reflected in the stress curves. Thus, stress curves are a sensitive, yet not always easy to interpret tool to monitor structural changes during epitaxial growth.

We confine the discussion of the stress-induced bending of cantilevers to thin deposited films. In that case, one may perform the calculation of the bending caused by the film stress on one of the two sides of the cantilever as if the neutral plane of the cantilever (the plane that is neither stretched nor compressed) were in the center. The reason is that a stress on one side is equivalent to a combination of stresses of opposite sign and of equal sign on the two sides. The component of equal sign leads to an elongation of the cantilever, which, because of the linearity of the elastic equations, has no effect on the bending. The stress of opposite sign must lead therefore to the same bending, and for those stresses, the neutral plane is in the center. We choose the coordinate system such that the $x_3$-axis is perpendicular to the film and the cantilever and have its origin in the center plane (Fig. 3.16).

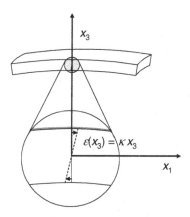

**Fig. 3.16.** Illustration of the coordinates and the strain in a cantilever, which bends due to a compressively stressed film on the upper side.

We first consider the case of free unsupported plates and deposited films which share a common set of in-plane principal axes (denoted as $x_1$ and $x_2$) of the stress/strain tensors, or for which a common set can be chosen because of in-plane isotropy. Then, the $x_1$ and $x_2$-axes are also the principle axes for the curvature. The corresponding curvatures are denoted as $\kappa_1$ and $\kappa_2$. The calculation proceeds following the simplifying assumption of Euler and Bernoulli [3.16] (valid if the thickness of the plate is much smaller then the lateral dimensions). According Euler and Bernoulli only the stresses, $\tau_{11}$ and $\tau_{22}$, along the direction of curvatures are important and the strains $\varepsilon_{11}$ and $\varepsilon_{22}$ are symmetric around a neutral plane in the center of the plate (Fig. 3.16). In other words, $\varepsilon_{11}$ and $\varepsilon_{22}$ are

$$\varepsilon_{11}(x_3) = -\kappa_1 x_3, \qquad \varepsilon_{22}(x_3) = -\kappa_2 x_3 \tag{3.68}$$

when $x_3$ is measured from the neutral plane. The sign convention used in (3.68) defines the curvatures as negative (downwards as shown in Fig. 3.16) for a compressive, i.e. negative stress in the film. The stress $\tau_{33}$ vanishes identically because of the boundary condition. The curvatures along the principal axes, $\kappa_1$ and $\kappa_2$, can be calculated by remembering that torques must vanish identically in the solid. An alternative pathway is to calculate the equilibrium shape of the plate as the shape of minimal elastic energy (see [3.17]). The torques for the $x_1$ and $x_2$-directions are

$$-\int_{-t/2}^{t/2}\tau_{11}(x_3)\,x_3\,\mathrm{d}\,x_3 = \frac{t}{2}\int_{t/2}^{t/2+t_f}\tau_{11}(x_3)\,\mathrm{d}\,x_3 = \tau_{11}^{(f)}t/2$$

$$-\int_{-t/2}^{t/2}\tau_{22}(x_3)\,x_3\,\mathrm{d}\,x_3 = \frac{t}{2}\int_{t/2}^{t/2+t_f}\tau_{22}(x_3)\,\mathrm{d}\,x_3 = \tau_{22}^{(f)}t/2 \tag{3.69}$$

Here, $t$ and $t_f$ denote the thickness of the cantilever and the film, respectively. Since only the integral over the stress components in the deposited film enter in the equations for the bending, it useful to define the integrals $\tau_{11}^{(f)}$ and $\tau_{22}^{(f)}$ as the "*film stresses*". If the films thickness is merely of the order of a monolayer or even less, it is still the integral over the stress that is measured. In Sect. 4.2.2 this integral will be discussed further as one of the surface excess quantities denoted as the *surface stress* $\tau_{ii}^{(s)}$. As a nonzero surface stress exists even on clean surfaces (comp. sect. 4.2.2) it is the change in the surface stress $\Delta\tau_{ii}^{(s)}$ that is being measured with the bending bar technique. The bulk stresses $\tau_{11}(x_3)$ and $\tau_{22}(x_3)$ can be expressed in terms of the strains $\varepsilon_{11}(x_3)$ and $\varepsilon_{22}(x_3)$ and the curvatures $\kappa_1$ and $\kappa_2$ using Hook's law (3.61) and (3.68).

$$\tau_{11}(x_3) = \frac{s'_{22}\,\varepsilon_{11}(x_3) - s'_{12}\,\varepsilon_{22}(x_3)}{s'_{11}\,s'_{22} - s'_{12}s'_{21}}$$

$$\tau_{22}(x_3) = \frac{s'_{11}\,\varepsilon_{22}(x_3) - s_{21}\,\varepsilon_{12}(x_3)}{s'_{11}\,s'_{22} - s'_{12}s'_{21}} \tag{3.70}$$

in which the $s'_{ij}$ are the elastic constants transformed into the plane spanned by $x_1$, $x_2$-axes. After inserting (3.70) and (3.68) into (3.69) and rearranging the terms one obtains a set of two equations relating the curvatures $\kappa_1$ and $\kappa_2$ to the film stresses $\tau_{11}^{(f)}$ and $\tau_{22}^{(f)}$.

$$\kappa_1 = \frac{6}{t^2}(s'_{11}\,\tau^{(f)}_{11} + s'_{12}\,\tau^{(f)}_{22})$$

$$\kappa_2 = \frac{6}{t^2}(s'_{21}\,\tau^{(f)}_{11} + s'_{22}\,\tau^{(f)}_{22})$$

(3.71)

For the special case of a system, which is elastically isotropic in the film plane and has an isotropic thin film stress $\tau^{(f)}$ (3.71) reduces to

$$\kappa = \frac{6(1-\nu')}{t^2 Y'}\tau^{(f)}.$$

(3.72)

This equation which was first derived for a one dimensional film stress (without the $(1-\nu')$-term) by Stoney [3.18] and is named after him. Eqs. (3.71) are therefore called generalized Stoney-equations.

Technically, the measurement of a cantilever bending requires that the plate serving as a cantilever be clamped along one edge, which keeps the curvature fixed to zero along that edge. Fixing of the curvature along one edge causes a significant perturbation effect on the bending of the entire sheet. The effect becomes marginal if the curvatures are measured at the end of a long sheet, which is clamped on one of the short sides. The magnitude of the disturbance induced by the clamping also depends on the method by which the bending is measured. Experiments which measure the bending using a capacitor at the end of the cantilever or the tip of a scanning tunneling microscope detect the deflection $\zeta$ along the $x_3$-axis. The curvature of the sample, as it appears in (3.71), denoted as $\kappa_\zeta$ is then calculated as

$$\kappa_\zeta \equiv 2\zeta(L)/L^2$$

(3.73)

in which $L$ is the distance between the point of measurement and clamping position. If the reflection angle of a light beam is used, the slope $\zeta'$ at the point of reflection is measured which defines

$$\kappa_{\zeta'} \equiv \zeta'(L)/L.$$

(3.74)

The change in the angle between two reflected beams determines the curvature $\kappa \cong \zeta''$ as the difference in the slope between two points. The perturbation induced by fixing the curvature at one end is the smallest on the curvature $\zeta''$ and the largest on the deflection $\zeta$. It is convenient to express the effect of clamping by the "dimensionality" D defined by rewriting the curvature $\kappa_1$ (3.72) as

$$\kappa_1^{(+)} = \frac{6 s_{11}'}{t^2} (1 + s_{11}' / s_{12}') \left[ 1 - (2 - D) s_{12}' / s_{11}' \right] \tau^{(f,+)}$$

$$\kappa_1^{(-)} = \frac{6 s_{11}'}{t^2} (1 - s_{12}' / s_{11}') \left[ 1 + (2 - D) s_{12}' / s_{11}' \right] \tau^{(f,-)}$$

(3.75)

in which $\tau^{(f,+)}$ and $\tau^{(f,+)}$ are the isotropic and antitropic part of the thin films stress defined by

$$\begin{pmatrix} \tau_{11}^{(f)} & 0 \\ 0 & \tau_{22}^{(f)} \end{pmatrix} \equiv \underbrace{\frac{1}{2} \left( \tau_{11}^{(f)} + \tau_{22}^{(f)} \right) \begin{pmatrix} 1 & 0 \\ 0 & 1 \end{pmatrix}}_{\tau^{(f,+)}} + \underbrace{\frac{1}{2} \left( \tau_{11}^{(f)} - \tau_{22}^{(f)} \right) \begin{pmatrix} 1 & 0 \\ 0 & -1 \end{pmatrix}}_{\tau^{(f,-)}} .$$

(3.76)

A corresponding equation holds for $\kappa_2$. Eq. (3.71) is recovered by setting $D = 2$. The effect of clamping on the bending is expressed by allowing for $D < 2$. Values for $D$ have been calculated for thin film stresses on (100) and (111) surfaces of cubic cantilever materials crystals using finite element methods [3.19, 20].

The results for various anisotropies $A' = 2(s_{11}' - s_{12}') / s_{44}'$ and Poisson numbers $v' = -s_{12}' / s_{11}'$ are shown in Fig. 3.17 for the curvatures $\kappa_\zeta$, $\kappa_{\zeta'}$ and $\kappa_{\zeta''}$ as a function of the aspect ratio $a$ defined as the length of the cantilever $L$ over the length of the clamped edge. The results were calculated under the assumption that the point of measurement is at the end of the cantilever. For (110) surfaces of cubic cantilevers, finite element calculations have to be performed for individual systems. The results of such calculations are conveniently expressed in terms of correction factors $v_{ij}$ in Eq. (3.71)

$$\kappa_1 = \frac{6}{t^2} (v_{11} s_{11}' \tau_{11}^{(f)} + v_{12} s_{12}' \tau_{22}^{(f)})$$

$$\kappa_2 = \frac{6}{t^2} (v_{21} s_{21}' \tau_{11}^{(f)} + v_{22} s_{22}' \tau_{22}^{(f)})$$

(3.77)

Table 3.4 shows sets of correction factors for the (110) faces of Cu, Mo, and Si for various aspect ratios. The correction factors depend on whether the deflection $\zeta$ or the slope $\zeta'$ is measured. The limiting value $v_{ij} = 1$ is approached very gradually if the deflection is measured. However, the factors also depend on the elastic properties. Deviations from $v_{ij} = 1$ are particularly large for elastically anisotropic materials, e.g. Cu.

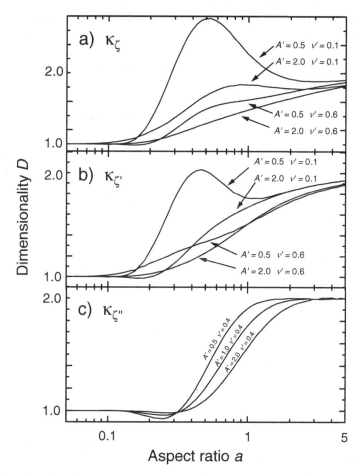

**Fig. 3.17.** Dimensionality as a function of the aspect ratio for cantilevers made from cubic crystalline materials with (100) or (111) surfaces [3.19]. Parameters are the elastic anisotropy $A' = 2(s'_{11} - s'_{12})/s'_{44}$ and the Poisson number $v' = -s'_{12}/s'_{11}$. The perturbation on the curvature is practically negligible if the distance between the point of measurement and the fixed edge exceeds the length of the fixed edge by a factor 3. The case of free bending is approached very slowly if the deflection is measured. Corrections have to be made in most cases when this detection scheme is used.

**Table 3.4.** Correction factors $v_{ij}$ for cantilevers with {110} surfaces (Klaus Dahmen, unpublished).

| | $\kappa_\zeta$ | | | | $\kappa_{\zeta'}$ | | | |
|---|---|---|---|---|---|---|---|---|
| $a$ | $v_{11}$ | $v_{12}$ | $v_{21}$ | $v_{22}$ | $v_{11}$ | $v_{12}$ | $v_{21}$ | $v_{22}$ |
| Cu | $[\bar{1}10]$ | | [001] | | $[\bar{1}10]$ | | [001] | |
| 1 | 0.791 | 0.401 | 0.492 | 0.823 | 0.790 | 0.397 | 0.538 | 0.839 |
| 2 | 0.853 | 0.580 | 0.660 | 0.881 | 0.884 | 0.668 | 0.763 | 0.917 |
| 4 | 0.907 | 0.735 | 0.798 | 0.929 | 0.942 | 0.834 | 0.882 | 0.959 |
| V | $[\bar{1}10]$ | | [001] | | $[\bar{1}10]$ | | [001] | |
| 1 | 0.971 | 0.701 | 0.698 | 0.970 | 0.965 | 0.649 | 0.625 | 0.963 |
| 2 | 0.976 | 0.754 | 0.744 | 0.974 | 0.982 | 0.815 | 0.799 | 0.980 |
| 4 | 0.985 | 0.846 | 0.835 | 0.984 | 0.991 | 0.907 | 0.899 | 0.990 |
| Cr | $[\bar{1}10]$ | | [001] | | $[\bar{1}10]$ | | [001] | |
| 1 | 1.00 | 0.979 | 1.021 | 1.001 | 0.995 | 0.682 | 0.666 | 0.994 |
| 2 | 0.997 | 0.837 | 0.834 | 0.997 | 0.997 | 0.817 | 0.797 | 0.996 |
| 4 | 0.998 | 0.867 | 0.859 | 0.997 | 0.998 | 0.908 | 0.897 | 0.998 |
| Mo | $[\bar{1}10]$ | | [001] | | $[\bar{1}10]$ | | [001] | |
| 1 | 0.989 | 0.796 | 0.796 | 0.990 | 0.983 | 0.666 | 0.639 | 0.982 |
| 2 | 0.989 | 0.781 | 0.773 | 0.988 | 0.991 | 0.819 | 0.799 | 0.990 |
| 4 | 0.992 | 0.855 | 0.843 | 0.992 | 0.995 | 0.909 | 0.899 | 0.995 |
| Si | $[\bar{1}10]$ | | [001] | | $[\bar{1}10]$ | | [001] | |
| 1 | 0.961 | 0.607 | 0.624 | 0.962 | 0.954 | 0.537 | 0.585 | 0.958 |
| 2 | 0.969 | 0.697 | 0.716 | 0.971 | 0.975 | 0.755 | 0.784 | 0.978 |
| 4 | 0.980 | 0.803 | 0.822 | 0.982 | 0.988 | 0.877 | 0.892 | 0.989 |

## 3.4 Elastic Interactions Between Defects

### 3.4.1 Outline of the Problem

The elastic energy in strained layers was calculated in Sect. 3.2.2 under the assumption that the layer is laterally extended and dislocation free, and therefore homogeneously strained. Finite size islands of lattice-mismatched layers, however, are strained inhomogeneously. Lately, ab-initio calculation of large non-periodic ensembles of atoms have become available and have been applied to the problem of strain in lattice mismatched heteroepitaxial islands [3.21, 22]. Figure 3.18 displays the qualitative picture that has emerged from these calculations. It is assumed that the unstrained bond lengths of the island atoms are 2% smaller than that of the substrate atoms, which corresponds to a Co-layer on Cu. The displacements in Fig. 3.18 are exaggerated by about a factor of 20. The relatively large vertical distortions reflect the fact that the surface is most easily deformed in this direction.

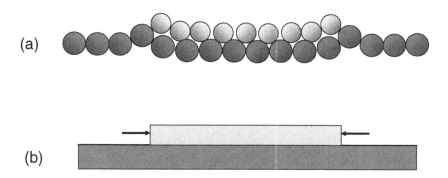

**Fig. 3.18**. (a) Distortion of the surface structure caused by the deposition of an island (light grey balls) with 2% smaller lattice constant, corresponding to Co deposited on Cu [3.21]. The atom displacements are exaggerated by about a factor 20. (b) The strain field originating from the island may be considered as arising from forces per length along the perimeter.

Surprisingly, the displacement pattern is very much the same, albeit smaller in magnitude, for homoepitaxial islands [3.22]. The reason is that the lower coordination of the surface atoms causes a reduction of the equilibrium bond distance between the atoms in the surface layer. Hence, even a homoepitaxial island has some lattice mismatch with the substrate and is under tensile stress.

Because of the inhomogeneity of the strain field, the elastic energy has contributions arising from the displacements of the atoms along the periphery that are not covered by (3.62). One might expect these contributions to scale as the length of the periphery; however, things are not so simple. The displacement pattern extends into the bulk and reaches out laterally. The longer the length of the periphery, the more extended is the displacement pattern. This gives rise to a characteristic additional logarithmic dependence of the elastic energy associated with boundary of strained islands. The atoms at the periphery of the islands have an even lower coordination than the surface atoms, which causes additional distortions with a characteristic displacement pattern that extends laterally and into the bulk of the substrate. The elastic distortions affect the island self-energy, but also induce interactions between islands. The same train of arguments applies to other defects on surfaces, be it straight steps on vicinal surfaces or point defects. The elastic self-energy with its curious scaling and the lateral interaction between islands or defects have far-reaching consequences on the equilibrium structure of patterned surfaces, on the stability of certain configurations, on phase transition and on nucleation phenomena. In order to treat the elastic energy arising from the local displacements of the atoms at defects one introduces a trick. Rather than beginning with the local displacement pattern as calculated from ab-initio methods and then look for the continuation of that pattern in an elastic continuum, one places forces at the site of the defect or at the island boundary and studies the strain field originating from these forces. The type of forces is chosen in accordance with the cause and type of the displacement pattern. For example, the long-range part of the displacement pattern of the strained island in Fig. 3.18a is the same as the one caused by a line of force monopoles along the perimeter. The forces point inwards for a tensile island strain (Fig. 3.18b). The far field of the displacement pattern arising from the atom displacements at the island periphery, which result from the reduced coordination, is the same as that of line force dipoles. The same holds for steps on vicinal surfaces. The strain field of point defects is again that of a force dipole (Fig. 3.19).

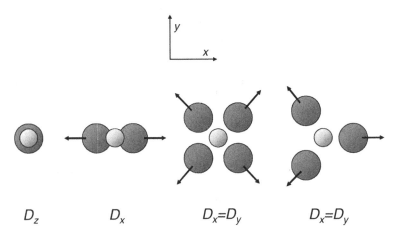

$$D_z \qquad D_x \qquad D_x{=}D_y \qquad D_x{=}D_y$$

**Fig. 3.19.** The long-range strain fields originating from defects in various sites is the same as the strain field of force dipoles placed on the surface at the site of the defect.

For a defect in atop position, the prevailing component of the force dipole is perpendicular to the surface. For a defect in a twofold bridge site, the force dipole has a significant component parallel to the surface along the direction of the bridge. Defects in fourfold or threefold sites bear isotropic parallel force dipoles. In all cases, the strength of the forces remains arbitrary within elasticity theory. If ab-initio calculations are available, the strength can be chosen such that the elastic distortions arising from the forces correspond to the calculated displacements. As mentioned, such calculation have become available only lately with the development of theoretical methods and the computing power to handle a very large number of atoms. Even without the knowledge of the absolute values of the forces, elasticity theory makes important prediction on the scaling of the self-energies and interaction energies with the size of the defects and the distance between effects. These relations are discussed in the following. The starting point is the strain field arising from a single point force placed at the surface. Within the framework of linear elasticity the displacement pattern for an arbitrary arrangement of forces, monopoles or dipoles, is simply a superposition of the displacements of individual forces. The energy $E$ associated with a strain field $u(r)$ and a force field $f(r)$ is

$$E = -\frac{1}{2V} \int_V dr\, f(r)\,u(r) \tag{3.78}$$

The factor 1/2 accounts for the fact that all products of forces and displacements are considered twice in the integration over the volume $V$. In the problem considered here the forces are situated at the surface $z = 0$ so that

$$f(r) = f(x, y)\delta(z). \tag{3.79}$$

The integral reduces correspondingly. An analytical solution for the displacement field $u(r)$ that arises from a point force $F$ at $x,y,z = 0$ exists for an elastically isotropic half space [3.23].

$$u_x(x, y) = \frac{1+v}{\pi Y} \frac{1}{r}\left[(1-v)F_x + \frac{vx}{r^2}(xF_x + yF_y) - \frac{(1-2v)x}{2r}F_z\right] \tag{3.80a}$$

$$u_y(x, y) = \frac{1+v}{\pi Y} \frac{1}{r}\left[(1-v)F_y + \frac{vy}{r^2}(xF_x + yF_y) - \frac{(1-2v)x}{2r}F_z\right] \tag{3.80b}$$

$$u_z(x, y) = \frac{1+v}{\pi Y} \frac{1}{r}\left[\frac{1-2v}{2r}(xF_x + yF_y) + (1-v)F_z\right] \tag{3.80c}$$

This set of equations is our starting point for the calculations of defect interactions.

### 3.4.2 Interaction Between Point and Line Defects

We consider the interaction energy between two point defects represented by a force dipole in $x$-direction at a distance $r$ (Fig. 3.20). The interaction energy $E_{\text{int}}$ is

$$
\begin{aligned}
E_{\text{int}} &= -F_x\left[2u_x(x,y) - u(x-d,y) - u(x+d,y)\right] \\
&= F_x d^2\frac{\partial^2 u_x}{\partial x^2} \\
&= D_x^2\frac{1+\nu}{\pi Y}\left\{-\frac{1-\nu}{(x^2+y^2)^{3/2}} + 3\frac{(1-6\nu)x^2}{(x^2+y^2)^{5/2}} + \frac{15\nu x^4}{(x^2+y^2)^{7/2}}\right\}
\end{aligned}
\tag{3.81}
$$

with $D_x = F_x d$.

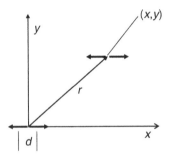

**Fig. 3.20.** Illustration of the geometry of two in-plane force dipoles placed at a distance $r$.

The interaction therefore scales as $r^{-3}$. Special solutions for $x = 0$ and $y = 0$ are

$$
\begin{aligned}
E_{\text{int}} &= -\frac{D_x^2(1-\nu^2)}{\pi Y r^3} \quad \text{for} \ \ x = 0 \\[2mm]
E_{\text{int}} &= \frac{4D_x^2(1+\nu)^2}{\pi Y r^3} \quad \text{for} \ \ y = 0
\end{aligned}
\tag{3.82}
$$

Depending on the orientation the interaction changes from attractive when dipoles are aligned along the $y$-axis to repulsive when they are aligned along the $x$-axis.

The method can likewise be applied to the problem of step-step interactions [3.24]. Due to the reduced coordination, step atoms have a reduced bond length to their neighbors, which causes an extended strain field around the step (Fig. 3.21a and b). The strain field is mimicked by force dipoles with $x$ and $z$-components on each step atom. In the framework of continuum theory this corresponds to a line density of force dipoles $D_x$ and $D_z$, which are assumed to sit in the surface plane at

$z = 0$. The actual geometric structure of the stepped surface is neglected in this approach. Suppose two steps of the same orientation run parallel along the $y$-axis at a distance $x = L$. The dipole in a length element $dy$ in one step interacts then with all dipoles of the other step. With the integrals

$$\int_{-\infty}^{\infty} \frac{dy}{(1+y^2)^{3/2}} = 2, \quad \int_{-\infty}^{\infty} \frac{dy}{(1+y^2)^{5/2}} = 4/3, \quad \int_{-\infty}^{\infty} \frac{dy}{(1+y^2)^{3/2}} = 16/15 \quad (3.83)$$

one obtains after some algebra the step-step interaction energy per unit length for a pair of steps at distance $L$.

$$E_{\text{step}} = \frac{2(1-v^2)}{\pi Y} \frac{1}{L^2} (D_x^2 \pm D_z^2) \quad (3.84)$$

**(a)**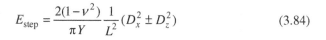

**(b)**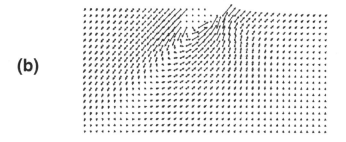

**Fig. 3.21. (a)** Schematic sketch of the atom displacements at a step edge. At large distance the displacement pattern can be mimicked by placing a density of force dipoles with $x$ and $z$ components on a semi-infinite elastic half-space. **(b)** Displacements calculated by Shilkrot and Srolovitz [3.25] for a stepped Au(100)-surface using the *embedded atom model* (EAM). The displacements are enlarged by a factor of 100.

This equation was first derived by Marchenko and Parshin [3.24]. The interaction is repulsive for the force dipole in the $x$-direction. For the $z$-component of the force dipole, the interaction is attractive for steps of opposite sign (Fig. 3.22a) and repulsive for steps of equal sign (Fig. 3.22b).

Attempts have been made to improve the classical relation of Marchenko and Parshin (eq. 3.79) by calculating the strain field in theoretical models [3.25, 26], by considering the realistic surface structure and by taking the elastic anisotropy into account. The net result is that the $L^{-2}$ -dependence is well preserved down to small distances between steps and is not changed by the inclusion of elastic anisotropy. The original proposal of Marchenko and Parshin that the perpendicular component of the force dipole should be given by the product of surface stress and step height cannot be held up, however[1]. This is not so surprising since the surface stress (Sect. 4.2.2) is a macroscopic quantity of the flat surface while the strain field of steps results from the change in the local bonding of step atom. Quantitative calculations of the forces dipoles associated with steps using model potentials yield numbers of the order of 0.2 nN for $D_x$ and $D_z$. However the use of model potentials is questionable since the same models produce too low numbers for the surface energy and the surface stress [3.28].

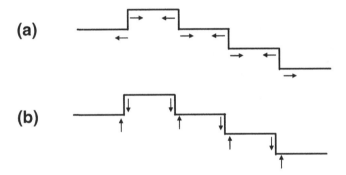

**Fig. 3.22.** The figure illustrates that the parallel force dipole leads to step-step repulsion regardless of the step orientation (**a**) while for the $z$-component of the force dipole is attractive interaction for steps of opposite sign since both displacement field operate in the same direction in that case (**b**).

### 3.4.3 Pattern Formation via Elastic Interactions

In 1988 Alerhand et al. predicted that one-dimensional stripes bearing different stresses should self-assembly into a periodic arrangement of stripes of a particular size [3.29]. With the reconstructed Si(100) surface in mind, Alerhand et al. concluded that the surface should even form monatomic up and down steps spontaneously, and that these steps should arrange with a periodicity given by the

[1] G. Prévot and B. Croset infered from their calculations that the Marchenko proposition is sound [3.27]. However, their reasoning is based on an artefact of the model potential. The surface stress produced by the model potential is far too low!

ratio of the step formation energy and the difference of the surface stresses on the stripes with the alternate orientation of the dimer reconstruction (Fig. 1.34). Eight years later, the spontaneous formation of the striped phase was indeed observed on boron doped Si(100) surfaces. Earlier, a striped phase with a characteristic periodicity was observed for adsorbed oxygen on the Cu(110) surface [3.30]. The effect was explained later by the same group in terms of the Alerhand-model [3.31]. Figure 3.23 displays an STM image of the striped phase of oxygen on Cu(110) together with a model of the local structure [3.32].

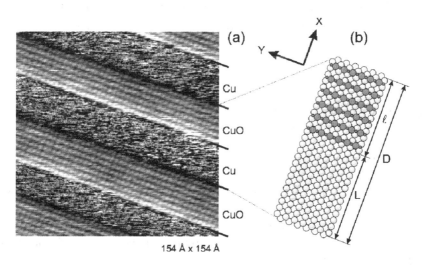

**Fig. 3.23.** (a) STM image of the striped phase of oxygen adsorbed on Cu(110) (courtesy of Peter Zeppenfeld, [3.32]). (b) Model of the structure showing that the oxygen atoms form chains with Cu-atoms in alternate sequence. The Cu-atoms are provided by nearby steps or they are taken Cu-surface atoms whereby vacancy islands are formed.

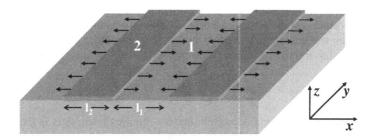

**Fig. 3.24.** Stripe domains of different stress on surface. The elastic strain field originating from the change in the stress can be mimicked by force monopoles per length of the boundary $f_x$.

The linear striped phases are the simplest examples of a large family of systems that show self-assembly into a one or two-dimensional periodic pattern. Well investigated examples are the checkerboard pattern of adsorbed nitrogen on Cu(100) [3.33, 34], the periodic arrangement of patches of a Pb/Cu surface alloy on Cu(111) [3.35], and the chevron superstructure on the reconstructed Au(111) surface (Fig. 1.13) [3.36]. In combination with nucleation kinetics in epitaxial growth an enormous wealth of patterns has been reported (see e.g. [3.37-39]). Here, we focus on the simple case of striped stress domains.

The stress can arise from the misfit strain in an epitaxial film. The strain field caused by the termination of the film stress $\tau_{xx}^{(f)}$ at the edges of the stripes (Fig. 3.24) is the same as the strain field caused by line force monopoles $f_x$ oriented perpendicular to the stripe and parallel to the surface

$$f_x = -\tau_{xx}^{(f)}(2) = -\int \tau_{xx}(2,z)dz . \qquad (3.85)$$

The stripes may also consist of differently oriented surface reconstructions, of stripes with and without an adsorbate as in Fig. 3.23, or of an alloy phase on the bare surface. In the latter cases, the line force monopole is given by the differences in the surface stresses (Sect. 3.3.3, see also Sect. 4.2.2, eq. 4.7)

$$f_x = \int [\tau_{xx}(1) - \tau_{xx}(2)]dz \equiv \tau_{xx}^{(s)}(1) - \tau_{xx}^{(s)}(2) . \qquad (3.86)$$

The strain energy per area produced by a periodic arrangement of line force monopoles of alternate sign was calculated by Alerhand et al. [3.29] for an elastically isotropic substrate.

$$\gamma_{\text{elastic}} = -\frac{f_x^2}{\pi} \frac{2(1-\nu^2)}{Y(l_1+l_2)} \ln\left\{ \frac{l_1+l_2}{2\pi a_c} \sin\left( \frac{\pi l_1}{l_1+l_2} \right) \right\} . \qquad (3.87)$$

Here, $Y$ and $\nu$ are Young's modulus and Poisson number, respectively, $l_1$ and $l_2$ are the width of the stripes, and $a_c$ is a cut-off distance of the order of an atom diameter. It is important that the system is periodic with alternate orientation of the force monopoles as the energy of just two interacting line monopoles diverges. The elastic energy has a (rather shallow) minimum when the periodicity length $L_{\min}$ is

$$L_{\min} = 2\pi a_c e / \sin(\pi \Theta_s) \quad \text{with} \quad \Theta = l_1 /(l_1 + l_2) \qquad (3.88)$$

If the formation of the stripe domain boundary requires energy, as is the case for the Si(100) surface where the phase boundaries are formed by steps, then the total energy contains an additional term

$$\gamma_{\text{boundary}} = \frac{2\beta}{l_1 + l_2} \tag{3.89}$$

and $L_{\text{min}}$ becomes

$$L_{\text{min}} = 2\pi a_c e^{(1+2\beta/C_2)} / \sin(\pi\Theta) \tag{3.90}$$

in which $C_2 = 2f_x^2(1-v^2)/[\pi Y(l_1 + l_2)]$. It is remarkable that there is a minimum in the total energy regardless of the magnitude of the domain boundary energy. With application to the Si(100) surface this has the interesting consequence that the Si(100) surface should form a periodic array of up and down steps spontaneously, regardless of the magnitude of the step energy and the difference in the stress tensor for the two differently oriented dimer reconstructions. However, this result is merely a theoretical one, insofar as the energy gained in the spontaneous formation of periodic steps becomes marginally small and $L_{\text{min}}$ becomes very large. Since the energy minimum exists only for ordered stripes, the natural morphological disorder on the surface as well as the kinetic barriers for atom transport hinder the formation of ordered stripe patterns.

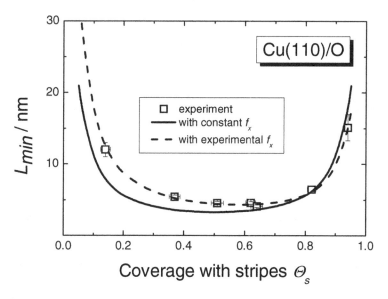

**Fig. 3.25.** Characteristic length $L_{\text{min}}$ in STM images of the striped phase of oxygen on Cu(10) vs. the coverage of the surface with stripes $\Theta_s$ [3.40]. The full line is a fit according to (3.90) with assumed constant monopole lines forces $f_x$. The dashed line is a fit with the actual line forces as measured by the bending bar technique.

Equation (3.90) is easily tested for the striped oxygen phase on Cu(110) surface since the fractional coverage with oxygen stripes is easily controlled by the oxygen exposure. Figure 3.25 shows experimental results Of Bombis et al. that was obtained from an analysis of STM-images [3.40]. While (3.85) predicts $L_{\min}(\Theta_s)$ to be symmetric around $\Theta_s = 0.5$ the experimental data show clearly an asymmetry. Bombis et al. performed stress measurements on the same system and found that the stress in the oxygen-covered stripes displays a significant dependence on $\Theta_s$. The stress in the oxygen-stripes and thus the strength of the line force monopole decreases with $\Theta_s$. The dashed line in Fig. 3.25 is a fit to (3.85) with the actually measured line force monopoles $f_x(\Theta_s)$. The asymmetry of the data is then well reproduced.

# 4. Equilibrium Thermodynamics

## 4.1 The Hierarchy of Equilibria

Equilibrium thermodynamics was developed in the second half of the 19th century. Its early extension to the description of interfaces by J. W. Gibbs [4.1] was developed at a time when scientists had little understanding of surfaces and interfaces and no knowledge of their atomic structure. The thermodynamics of interfaces was therefore formulated with a minimum set of assumptions, without taking into account any specifics of a particular surface or interface system. Instead, interface thermodynamics introduced some very general concepts, as e. g. the surface particle excesses, which - while being defined within the closed framework of thermodynamics - could not, or at least not easily, be identified with anything measurable outside the framework of thermodynamics. Furthermore, since, the early formulation of interface thermodynamics took place long before the concept of equilibrium in the electronic system of metals and semiconductors could be formulated with the help of Fermi-Statistics one could not conceptually connect the electronic properties of solid materials with electrolytes. Last, not least some of the basic assumptions or conventions in early interface thermodynamics such as the assumption of a global equilibrium and the definition of a dividing plane turned out to be imprudent in the light of our present understanding of interfaces on the atomic scale. It is probably therefore that current textbooks on surface science appear to treat interface thermodynamics more out of a sense of duty than with care, and, once done with it, seldom refer to that treatment in the remainder of the text. In recent years however, surface thermodynamics received considerably more attention, and rejuvenation at the same time. The revolution introduced by the discovery of the scanning tunneling microscope has, among other things, brought about the possibility to observe single atoms on surface and to track their motion. The same instrument permits the observation of epitaxial growth phenomena during or after deposition on the length scale of microns and beyond. We are therefore faced with the task to connect single atom properties and dynamics with macroscopic morphological features, thereby bridging 4-8 orders of magnitude on the length scale and even more orders of magnitude on the time scale. Just as it is impossible to observe the development of the morphology during an epitaxial growth process on the mm/cm-length scale by keeping track of all atom motions, it is also impossible to describe the spatiotemporal development of large-scale morphological features as a collection of single atom properties. Treating such multi-scale processes requires means to interconnect single atom properties with a coarse-grained description. It is here, where thermodynamics has found its present role.

Connecting single atom properties with large-scale surface features is greatly facilitated by the existence of hierarchy of quasi-equilibria. What is meant by this is best illustrated with the help of scanning tunneling microscope (STM) images. Fig 4.1 displays two STM images of a Cu(111) surface shortly after deposition of several monolayers of Cu-atoms, and the same surface after a period of 12 hours during which the surface was held at a constant temperature T = 314 K. The black cross marks the center of a stack of islands, which has moved from the right side of the image to the left due to some drift. During that long period of time a considerable coarsening of morphological features has taken place. Small islands have disappeared and steps have straightened. This coarsening is thermodynamically driven by the minimization of the step (free) energy.

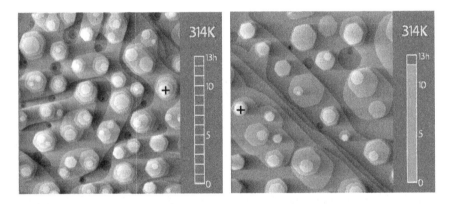

**Fig. 4.1.** STM image of a Cu(111) surface (left) shortly after deposition of several monolayers of Cu and (right) after 12 h (courtesy of Margret Giesen). Each contrast level corresponds to one monolayer. The black cross marks the same spot on the surface, which has drifted from left to right. During the time span of 12 h the surface features become larger and the surface flattens. The mean shape of all islands is the same and stays constant during the entire time: the islands are in equilibrium with themselves, while the surface is globally not in equilibrium.

Ultimately, all islands would disappear, and the surface would become flat, at least on the scale of the image. One feature, however, persists during the entire coarsening process: Save for some fluctuations, the shape of the islands stays the same, independent of their size. This is owed to the much faster diffusion of atoms along the perimeter of an island compared to the exchange of atoms with the terraces. In other words, each island is always in equilibrium with itself. Thus, equilibrium thermodynamics can be applied to the island shapes despite the fact that the surface is not in equilibrium on a larger scale. The two-dimensional equilibrium shape of the islands, e.g., can be obtained by averaging over a sufficiently large number of individual shapes. A more quantitative analysis of Fig. 4.1 would reveal that atom exchange between layers of different height is slow compared to the exchange between islands on the same terrace. Establishing equilibrium between the terraces of different height therefore takes a much longer time than the equilibration of islands on the same terrace. Consequently, the surface reaches a state in which the roughness consists merely of smoothly curved steps with no islands present any more. STM images such as displayed in Fig. 4.1 are typically observed in ultra-high vacuum with no Cu-vapor pressure present. The (at T = 314 K extremely) slow evaporation has no effect on the surface morphology. Hence, even in this simple case one has a hierarchy of equilibria in which each level of the hierarchy is established on a time scale, which differs by many orders of magnitude from the next. At each stage, equilibrium thermodynamics can be applied to some features, which are in (quasi-) equilibrium, and non-equilibrium thermodynamics can be applied to other features varying slowly on the time scale considered.

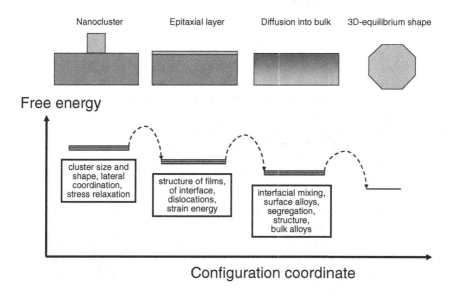

**Fig. 4.2.** The hierarchy of equilibria on surfaces (see text for discussion).

Coarsening on a Cu(111) surface is a very simple example for the hierarchy of equilibria. The hierarchy of equilibria can be substantially more complex if heterogeneous interfaces with their adjacent bulk phases are considered. Figure 4.2 displays an overview of various interface features that include 3D nanoclusters of a different material on a substrate. Each of these clusters may or may not display an equilibrium shape, the clusters may be in equilibrium or not with respect to their cluster size distribution, and the lateral distribution on the surface may or may not correspond to an equilibrium phase. Spreading the material laterally over the surface in the form of a thin film may lead to a state of higher or lower energy. Within the thin film system, one may have features, which display a local equilibrium shape. One may have dislocations in the films, and entire network of dislocations. Furthermore, there may be intermixing of material with bulk, a formation of surface and bulk alloys. All these features and aspects, of which we have named only a few, are amenable to a properly defined thermodynamic description, despite the fact that one is far away from the global equilibrium of the system. The treatise on interface thermodynamics in this section lays the foundation for that endeavor and commences with the thermodynamics of flat interfaces.

## 4.2 Thermodynamics of Flat Interfaces

### 4.2.1 The Interface Free Energy

According to the first law of thermodynamics, the sum of the heat $\delta Q$ and the work $\delta W$ applied to a system raises the internal energy $U$ of the system by an amount

$$dU = \delta Q + \delta W \tag{4.1}$$

with $U$ being a thermodynamic potential, which is a unique function of certain independent variables. The independent variables for the internal energy $U$ are the entropy $S$, the volume $V$, the number of particles of various types $n_i$, the electric and magnetic fields, and the electrical charge. The work

$$\delta W = \delta W_{\text{mech}} + \delta W_{\text{chem}} + \delta W_{\text{electr}} . \tag{4.2}$$

can be of mechanical, chemical, electro-magnetic or electrostatic nature. The electrostatic work is of particular interest for solids immersed in an electrolyte, and requires a particular careful consideration in conjunction with a specific experimental situation to which we turn a little later. The thermodynamic potential whose variation at constant temperature $T$ is equal to the applied work is the (Helmholtz) free energy

$$F = U - TS \tag{4.3}$$

With the supplied differential heat $\delta Q$ being $\delta Q = TdS$ the total differential of $F$ for a homogeneous system is

$$dF = -SdT + V\sum_{kl} \tau_{kl} \, d\varepsilon_{kl} + \sum_i \mu_i dn_i ... \tag{4.4}$$

Here, $\tau_{kl}$ and $\varepsilon_{kl}$ are the components of the stress and strain tensors and $\mu_i$ is the chemical potential of the particles of type $i$ in the system. The second term accounts for the fact that solids in a finite volume may have six independent components of the symmetric stress and strain tensors. It replaces the $-PdV$-term in the thermodynamics of gases and liquids. Note that the sign convention of the stress is opposite to that of the pressure! The third term describes the chemical work associated with bringing $dn_i$ particles with a chemical potential $\mu_i$ into the system.

The derivation of the thermodynamics of a surface or an interface requires careful thinking. One reason is that the introduction of an interface makes the system necessarily inhomogeneous. A second reason is that one would like to use thermodynamic arguments even when the system is not in complete equilibrium. A typical example for a partial equilibrium is a surface of a solid in equilibrium with a gas phase, however not in equilibrium with the bulk of the solid because of slow or negligible diffusion of adsorbed components into the bulk. Another example is encountered in segregation phenomena, where the surface concentration of some impurity may be in equilibrium with the bulk concentration, but not with the gas phase. In the first case, the chemical potential of particles at the surface would equal the chemical potential in the gas phase, but be different from chemical potential of the same particle dissolved in the bulk. In the second case the chemical potential of the segregating species at the surface equals that in the bulk, differs, however, from the chemical potential in the surrounding gas phase. These important and frequently encountered situations cannot be treated with the instrumentation and logical apparatus of conventional Gibbs thermodynamics, since Gibbs thermodynamics assumes global equilibrium.

For the definition of a free energy that is associated with an interface we consider two phases denoted by I and II. One may think of phase I as a solid and phase II as either another solid, a liquid, an electrolyte, a gas phase or vacuum. Inside the bulk, the phases shall be in equilibrium. In order to define a finite system without any undesired interfaces one may invoke the trick commonly used in Solid State Physics. We assume that phase I and II and the interfaces between them form an infinite periodic sequence along the $z$-axis (Fig. 4.3) and impose periodic boundary conditions. In the $x,y$-plane the system is assumed to be homogeneous. The thereby formed *supercell* has the length $L$ and the area $A$. The periodicity length $L$ of the supercell shall be large compared to the extension of the interfaces so that the bulk of phases II and I are homogeneous. As interface regions can extend up to μm length, the periodicity length $L$ may also be quite large. To make things easier we assume that the solid has not a polar axis so that the two interfaces between Phase I and II are identical. In keeping with typical situations encountered in surface physics, we assume that the external pressure is zero.

The supercell as depicted in Fig. 4.3a has a finite and fixed volume $V = AL$. The periodic boundary conditions ensure that the number of particles in the cell remains constant. The thermodynamic potential appropriate for that situation is the Helmholtz free energy. We can therefore can define a Helmholtz free energy for

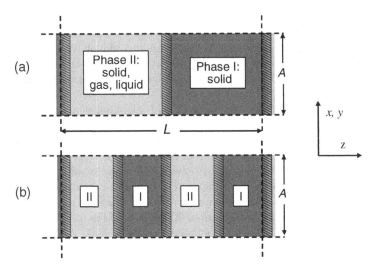

**Fig. 4.3.** Illustration on the thermodynamics of flat interfaces, see text for discussion.

one supercell with two interfaces, which we denote as $F^{(2)}(L)$. The free energy can be calculated as the partition function of the system, at least in principle. The supercell method described above is actually the standard procedure in ab-initio calculations of finite size systems. The system be in total equilibrium. For example, the chemical potential of a dissolved species can be different inside phase I and II. In particular the concentration of a species might be zero in one phase (corresponding to a chemical potential $\mu = -\infty$) even if it were finite, if small in equilibrium. In order to define the Helmholtz free energy associated with the formation of an interface we reduce the periodicity length by a factor of two such that we have four interfaces instead of two within the supercell of length $L$ (Fig. 4.3b). The free energy of that system is denoted as $F^{(4)}(L)$. The free energy associated with the introduction of the two additional interfaces is obviously $F^{(4)}(L)-F^{(2)}(L)$. We can therefore define an interface Helmholtz free energy $F^{(s)}$ and the area specific free energy $f^{(s)}$ as

$$f^{(s)} = F^{(s)} / A = (F^{(4)}(L) - F^{(2)}(L))/2A \qquad (4.5)$$

The definition with the help of a supercell and periodic boundary conditions has the advantage that there is no need to specify where one bulk phase ends and the next one begins as is done in Gibbs thermodynamics with the Gibbs dividing plane. The interface energy can be positive or negative. It is always positive for a free surface since otherwise the condensed phase would be an unstable form of matter. The interface energy between two solids may be negative if the bonds between the atoms of the two phases are stronger than between the atoms in each phase. Still, energy would be gained in that case by creating more interfaces between the two phases, but the system may be frozen into a particular state. Currently interface science deals more often than not with such metastable systems, which is one more reason to go beyond Gibbs thermodynamics.

We note that no satisfactory experimental method is available to determine the surface free energy of crystal faces. The existing methods concern either the liquid phase or the amorphous state at high temperature. Theory has however advanced to a stage where surface energies can be calculated with fair reliability.

The Helmholtz free energy $F$ is a thermodynamic potential with respect to the variables temperature, strain, and particle numbers. We now need to define the corresponding variables for the interface free energy $F^{(s)}$. We begin with the discussion of strain and stress.

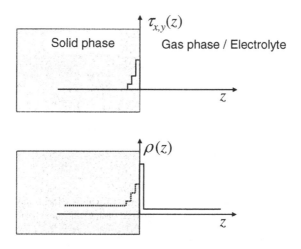

**Fig. 4.4.** Illustration of the stress $\tau$ and the particle density at an interface. Due to a redistribution of the electronic charge, a nonzero stress parallel to the surface may exist in the first layers of the solid. Gas phase molecules or atoms dissolved in the bulk may accumulate in a dense layer on the surface because of either segregation or adsorption.

The bulk components of the stress tensor vanish, both in the bulk of the gas/liquid and the solid phase, because of the homogeneity and the condition of zero external pressure. Near the surface the stress may have non-vanishing components $\tau_{kl}(z)$ within the surface plane of a solid (Fig. 4.4). The interface stress or surface stress $\tau_{kl}^{(s)}$ is defined as

$$\tau_{kl}^{(s)} = \int_{\text{interface}} \tau_{kl}(z)\,\mathrm{d}z \tag{4.6}$$

where the indices $k,l = 1,2$ denote the in-plane components of the stress tensor. The surface stress in the plane of drawing in Fig. 4.4 is a force per length with the length along the axis perpendicular to the plane of drawing. It can also be defined as the $z$-integral of all forces per length needed to keep the material unstrained if in a *Gedanken* experiment the interface is cut vertically to the surface with the material on the right hand side being removed. We note that the surface stress $\tau^{(s)}$ need not be positive! A positive surface stress means that work is required to stretch the surface elastically. This stress is also called a *tensile surface stress*. If the surface stress is negative, it is called *compressive*. The intrinsic surface stress of clean metals seems to be always tensile, but may become compressive upon chemisorption of electronegative adsorbates [4.2]. For semiconductors surfaces the surface stress may be compressive even when clean [4.2, 3]. The tensile surface stress on clean metal surfaces can be understood to be a consequence of the redistribution of the bonding charge.

**(a)** $\tau^{(s)} > 0$          **(b)** $\tau^{(s)} < 0$

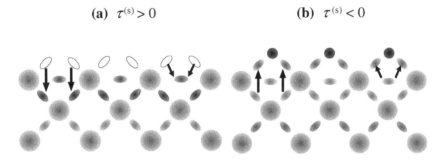

**Fig. 4.5.** Illustration of the charge redistribution at metal surfaces and its effect on the surface stress for **(a)** a free surface and **(b)** a surface with electronegative adsorbates.

The charge that is not required for bonding because of the existence of a surface redistributes to enhance the charge between the first and second layer. This causes the typical contraction of the interlayer distance. The charge also increases the charge density between the surface atoms to cause an attractive force to the positively charged cores of the atoms in the surface layer (Fig. 4.5a). The atoms cannot yield to that force since they are hold in registry by the substrate; however, a tensile surface stress arises from the charge redistribution. Upon adsorption of electronegative atoms, some of the electron charge between the surface atoms is removed to cause a compressive surface stress (Fig. 4.5b). Adsorption of electropositive adsorbates increases the tensile stress.

Typical surface stresses of solids are of the order of 1 N/m. As the surface stress originates from a redistribution of the electronic charge at the surface which is confined to a distance of about 1nm a surface stress of 1 N/m corresponds to a bulk stress of 1 Gpa. Neglecting external pressures on the surface thermodynamics of solids is therefore justified as long as the pressures are small compared to 1 GPa.

When additional interfaces are inserted into the supercell in the procedure described above components of bulk constituents may flow towards or away from the interface to change the particle density there. For the interface free energy, the variable "particle number" in of total free energy is to be replaced by the *surface excess numbers* $n_i^{(s)}$, which are defined as the integral over the $z$-dependent densities $\rho_i(z)$ minus the bulk densities

$$n_i^{(s)} = A \int_{interface} [\rho_i(z) - \rho_{I,II}] \, dz . \tag{4.7}$$

In order to subtract the bulk contributions correctly one has to define where exactly the bulk of the solid ends and the gas/liquid phase begins. Considering the case of adsorption from the gas/liquid phase, it is useful to place the dividing plane such that all atoms which belong to the solid state and remain within the (unaltered) solid state under the given circumstances fall on one side, while atoms belonging to the gas/liquid phase and adsorbed atoms fall on the other side. If segregation is considered, the dividing plane would be placed best between the segregated surface phase and the gas/liquid. The exact position of the interface is not important in either case, since the concentration is zero in one of the adjacent bulk phases[2]. As a function of the independent variables the interface free energy $F^{(s)}(T, \varepsilon_{kl}, n_i^{(s)})$ can be written as

$$F^{(s)}(T, \varepsilon_{kl}, n_i^{(s)}) = U^{(s)} - S^{(s)}T + A \sum_{kl} \tau_{kl}^{(s)} \varepsilon_{kl} + \sum_i \mu_i n_i^{(s)} . \tag{4.8}$$

We note that the indices $kl$ denote merely the $x$ and $y$ components of the surface stress and strain tensors. There is no contribution to the mechanical work from the $z$-component since we have assumed that the pressure be zero.
The total differential of the interface free energy is

$$dF^{(s)} = -S^{(s)} \, dT + A \sum_{kl} \tau_{kl}^{(s)} \, d\varepsilon_{kl} + \sum_i \mu_i \, dn_i^{(s)} . \tag{4.9}$$

In surface science, one often deals with adsorption phenomena where the adsorption process saturates after a monolayer coverage. The concentration $\rho_i(z)$ can be written as

---

[2] Gibbs has placed the dividing plane such that $\sum_i \mu_i n_i^{(s)} = 0$. However, this is inexpedient for solid surfaces since the position of the dividing plane would change with the concentration of adsorbed atoms. Nevertheless this convention is frequently cited in many textbooks which is very confusing as one attempts to apply a thermodynamics based on the Gibb's convention to concrete situations (see [4.4])

$$\rho_i(z) = n_i^{(s)} \delta(z) / A .\tag{4.10}$$

Divided by the area which corresponds to one surface atom $\Omega_s$, $n_i^{(s)}$ defines a fractional coverage $\Theta_i$

$$\Theta_i = n_i^{(s)} / \Omega_s .\tag{4.11}$$

## 4.2.2 Charged Interfaces

To develop the thermodynamics of electrified interfaces between a solid and an electrolyte we need to admit the contribution of electrical energy. The cell doubling method requires that the bulk phases be in equilibrium with themselves within their bulk region. This ensues that there are no electric fields inside the bulk phases as otherwise the chemical potential of electrons and ions would depend on the position, which constitutes a non-equilibrium situation. Furthermore, the integral of the charge density over the supercell must be zero to avoid infinite self-energy terms. Since the Helmholtz free energy has the charge as the independent variable, we need to consider the charge distribution inside the interface for the definition of an interface free energy. Because of the required neutrality, the charge distribution involves a charge on the solid and a countercharge of opposite sign and the same magnitude inside the electrolyte. If the solid is a metal, the charge on the solid is localized within the screening length that amounts to about one tenth of the Fermi-wavelength (Sect. 3.2.1, Fig. 3.5). The countercharge in the electrolyte extends over the Debye-length (Table 3.2). The resulting charge distribution in the supercell is depicted in Fig. 4.6a. The charge distribution causes a potential difference between the solid and the electrolyte (Fig. 4.6b). Homogeneity in the $x,y$-plane requires that the charge density and the potential depend on $z$ only. The electric work term in the free energy is therefore calculated as for a parallel plate capacitor. The energy of each interface capacitor is directly the electric contribution to the interface free energy. The total free energy $F^{(s,tot)}$ is therefore

$$F^{(s,tot)}(T, \varepsilon_{kl}, n_i^{(s)}, \sigma, ...) = F^{(s)}(T, \varepsilon_{kl}, n_i^{(s)}) + \phi^{(M)} q\tag{4.12}$$

where $q$ is the total charge and $\phi^{(M)}$ is potential on the metal. Without loss of generality, we set the potential of the electrolyte bulk phase as zero[3]. Hence, for the geometry and the side conditions considered one can write $dF^{(s)}$ as

---

[3] If one thinks of electrons as the carriers of the charge $q$ and treats the electrons as particles one may replace the term $\phi^{(M)} q$ by $-\tilde{\mu}_e n_e$ in which $\tilde{\mu}_e$ is the electrochemical potential of electrons. This notation is typically used in electrochemistry.

$$dF^{(s)} = -S^{(s)} \, dT + A \sum_{kl} \tau_{kl}^{(s)} \, d\varepsilon_{kl} + \sum_i \mu_i \, dn_i^{(s)} + \phi^{(M)} \, dq \, . \tag{4.13}$$

The differential form of the area specific free energy is obtained by observing that

$$dF^{(s)} = A \, df^{(s)} + f^{(s)} \, dA \tag{4.14}$$

so that

$$df^{(s)} = -s^{(s)} \, dT + \sum_{kl} (\tau_{kl}^{(s)} - f^{(s)} \delta_{kl}) \, d\varepsilon_{kl} + \sum_i \mu_i \, d\Gamma_i^{(s)} + \phi^{(M)} \, d\sigma \, . \tag{4.15}$$

Here, we have introduced the surface specific entropy $s^{(s)} = S^{(s)} / A$, the surface particle excess (which frequently is the surface coverage per area!) $\Gamma_i = n_i^{(s)} / A$, the specific charge $\sigma = q / A$, and we have taken into account that

$$dA = A \sum_k d\varepsilon_{kk} = A \sum_{kl} \delta_{kl} \, d\varepsilon_{kl} \tag{4.16}$$

with $\delta_{kl}$ the Kronecker symbol.

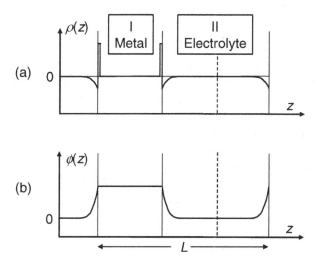

**Fig. 4.6.** A supercell with charged interfaces; **(a)** shows the charge density and **(b)** the potential. The mean charge is zero. On the metal surface, the charge is localized within the Thomas-Fermi screening length. In the electrolyte, the characteristic decay length is the Debye-length.

For a surface at constant temperature $T$, constant particle excesses $\Gamma_i$ and constant surface charge $\sigma$ one has a relation between the surface stress and the specific surface free energy

$$\tau_{kl}^{(s)} = f^{(s)} \delta_{kl} + \left. \frac{\partial f^{(s)}}{\partial \varepsilon_{kl}} \right|_{T,\Gamma_i,\sigma} . \tag{4.17}$$

This relation has been derived by Shuttleworth [4.5] and is therefore known as the *Shuttleworth relation*. As this particular Shuttleworth relation refers to an experimental situation in which surface coverage and charge density are held constant, this form is useful for a surface in a gas phase at temperatures were the exchange of adsorbate atoms with the gas phase is slow so that the coverage would not change upon application of a strain to the solid.

We conclude this section by considering the derivative of the free energy with respect to the charge at the point of zero charge for a surface in vacuum

$$\left. \frac{\partial f^{(s)}}{\partial \sigma} \right|_{T,\Gamma_i,\varepsilon_{kl},\sigma=0} = \phi_0^{(M)} , \tag{4.18}$$

which is formally a potential according to (4.15). The question is what is the meaning of that potential? By definition, the variation of the surface free energy is the work per area on the surface required to bring a charge per area from the solid into the vacuum. Suppose the charge would be that of one electron and the charge would be removed from the solid in the form of an electron. The electron would be brought from the Fermi-level up to the vacuum potential whereby a work amounting to the *work function* would be performed on the solid. The quantity $e\phi_0^{(M)}$ should therefore be the work function. It is a well-known result of density functional theory that only the charge density at the surface within some screening length is changed by removing an electron from a material [4.6]. Hence, the work required to remove an electron from the bulk of a material is entirely acting upon the surface and thereby changes only the free energy of the surface and not the free energy of the bulk! The work $\Phi = e\phi_0^{(M)} = e\partial f^{(s)} / \partial\sigma$ is therefore the work function of the material in its given state, that is, with possibly some adsorbate coverage. For surfaces in an electrolyte $\Phi = e\partial f^{(s)} / \partial\sigma \big|_{T,\Gamma_i,\varepsilon_{kl},\sigma=0}$ is the work function of the metal in contact with an electrolyte in the state of zero charge, however with a Stern layer of bonded water and other molecules on the surface.

### 4.2.3 Charged Surfaces at Constant Potential

The area specific Helmholtz free energy $f^{(s)}$ is the isothermal work per area to create a surface while keeping the particle excesses $\Gamma^{(s)}$, the strain and the charge density constant. It is therefore the appropriate thermodynamic potential for a surface in vacuum, which is typically uncharged and on which the number of adsorbed particles is fixed. To solid-state physicists $f^{(s)}$ is frequently known as the *surface tension*. Metallurgists who often work with materials under high pressure tend to take the Gibbs surface free energy per area as the *surface tension* [4.4, 7], as this so defined surface tension is the appropriate surface property to consider when the partial pressures i.e. the chemical potentials in the surrounding gas phase are being held constant. Electrochemists, finally, are interested in (highly) charged surfaces in equilibrium with electrolytes of defined concentration and at a constant potential with respect to a reference electrode. They need again a different specific surface thermodynamic potential referring to the chemical potentials $\mu_i$ of species in the electrolyte (as determined by the concentrations of neutral and ionic species) and the electric potential $\phi$ as independent variables; a quantity, which is likewise called *surface tension*. This can create quite a bit of confusion unless the independent variables are clearly stated.

The transformation to the surface tension $\gamma$, which has the temperature $T$, the strain $\varepsilon_{kl}$, the chemical potentials $\mu_i$, and the electric potential $\phi$ as independent variables is performed via the Legendre-transformation

$$A\gamma = F^{(s)} - \sum_i \mu_i n_i - \phi q \, . \tag{4.19}$$

We eliminate the superscript (s) in $\gamma$ as this symbol is used universally for the area specific surface tension. By invoking 4.13 and 4.14 one obtains the differential form of the surface tension $\gamma$ as

$$d\gamma(T, \varepsilon_{kl}, \mu_i, \phi) = -s^{(s)} \, dT + \sum_{kl} (\tau_{kl}^{(s)} - \gamma \delta_{kl}) d\varepsilon_{kl} - \sum_i \Gamma_i \, d\mu_i - \sigma \, d\phi \tag{4.20}$$

The *Shuttleworth relation* for $\gamma$ then reads

$$\tau_{kl}^{(s)} = \gamma \delta_{kl} + \left. \frac{\partial \gamma}{\partial \varepsilon_{kl}} \right|_{T, \mu_i, \phi} \tag{4.21}$$

We note that the charge density $\sigma$ in (4.20) is the charge density measured in an experiment with electrodes held at constant potential. This experimental charge density may have two components, one from loading the electrochemical double layer, the other from the adsorption of ions, which are unloaded on the surface by forming a chemical bond with the substrate. These *specifically adsorbed* ions contribute to the surface excesses $\Gamma_i$. Experimentally, one cannot distinguish between a current stemming from charging the double layer or from an unloading of ions. For electrodes kept at constant potential, the interface system remains neutral even with specific adsorption. The charge provided by the specifically adsorbed ions is transported away as a current by the external power supply that keeps the electrode potential constant.

A typical experiment in electrochemistry is isothermal ($dT = 0$) and involves a surface on a bulk material of considerable thickness which keeps the surface strain $\varepsilon_{kl}$ constant ($d\varepsilon_{kl} = 0$). In that case, one has

$$d\gamma = -\sum_i \Gamma_i \, d\mu_i - \sigma \, d\phi . \tag{4.22}$$

This equation is known as the *electrocapillary equation* or *Lippmann equation*[4]. Because of (4.22) the surface tension $\gamma$ has a maximum at the potential where the surface is uncharged. This potential is called the potential of zero charge (*pzc*) and is denoted as $\phi_{pzc}$ in the following. The value of $\phi_{pzc}$ can be related to the work function if the surface excesses $\Gamma_i$ are zero or constant. Suppose the work function of the surface would change by some amount $\Delta\Phi$, e.g. by a change in the surface structure. Then by virtue of (4.18) and (4.19)

$$\Delta\Phi = e\,\Delta\left\{ \frac{\partial f^{(s)}}{\partial \sigma}\bigg|_{\sigma=0} \right\} = e\,\Delta\left\{ \frac{\partial \gamma}{\partial \phi}\frac{\partial \phi}{\partial \sigma}\bigg|_{\sigma=0} + \sigma\frac{\partial \phi}{\partial \sigma}\bigg|_{\sigma=0} + \phi\big|_{\sigma=0} \right\}. \tag{4.23}$$

The first two terms vanish and $\phi\big|_{\sigma=0} \equiv \phi_{pzc}$ by definition, so that

$$\Delta\Phi\big|_{\Gamma_i} = e\,\Delta\phi_{pzc}\big|_{\Gamma_i} , \tag{4.24}$$

which is a rather important result! Note, however, that $\Delta\Phi$ is the change of the work function of the substrate when in contact with the particular electrolyte! By integration of (4.22) one obtains

$$\gamma = \gamma(\phi_{pzc}) - \int_{\phi_{pzc}}^{\phi}\sigma(\phi')d\phi' = \gamma(\phi_{pzc}) - \int_{\phi_{pzc}}^{\phi} d\phi' \int_{\phi_{pzc}}^{\phi'} C(\phi'')d\phi'' \tag{4.25}$$

with C the differential interface capacity

$$C = \frac{\partial \sigma}{\partial \phi}\bigg|_{\mu_i} . \tag{4.26}$$

---

[4] Sometimes $\partial\gamma/\partial\phi\big|_{\mu_i,T} = -\sigma$ is called the Lippmann equation.

The capacity $C$ can be expanded into a Taylor series

$$C(\phi) = C(\phi_{pzc}) + C_1(\phi - \phi_{pzc}) + \frac{1}{2}C_2(\phi - \phi_{pzc})^2 ... , \qquad (4.27)$$

so that

$$\gamma(\phi) = \gamma(\phi_{pzc}) - \frac{1}{2}C(\phi_{pzc})(\phi - \phi_{pzc})^2 ... \qquad (4.28)$$

As shown in section 3.2.3 the interface capacity has a minimum at *pzc* for dilute electrolytes. The next term in the expansion (4.28) is therefore the forth order term. The dependence of the surface tension $\gamma$ on the potential is illustrated with the example of the Au(100) surface. Under vacuum conditions the Au(100) surface is reconstructed. The surface layer contains about 25% more atoms, which are arranged in a quasi-hexagonal, incommensurate structure (Sect. 1.2.1). For

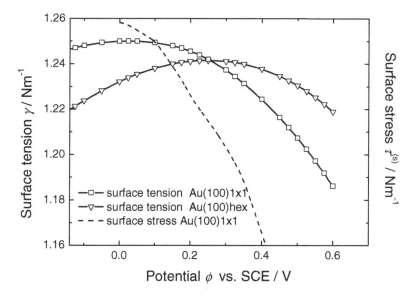

**Fig. 4.7.** Surface tension of Au(100) surfaces in 0.01M $HClO_4$ [4.8]. The absolute value of the surface tension is taken from theory [4.9]. The reconstructed surface is stable for potentials below about 0.25 V vs. SCE (*Saturated Calomel Electrode*), while above 0.25 V the unreconstructed surface is stable. The lifting of the reconstruction for positive potentials is indeed observed in scanning tunneling microscopy [4.10, 11]. The variation of the surface stress as determined experimentally [4.11] is plotted as a dashed line on the same scale for comparison.

Au(100) surfaces in contact with an electrolyte both the reconstructed and the unreconstructed phase may be stable, depending on the potential.

As the capacity $C(\phi)$ and the potential of zero charge $\phi_{pzc}$ can be determined from electrochemical experiments, the potential dependence of the surface tension can be calculated from experimental data. The results shown in Fig. 4.7 refer to the unreconstructed and reconstructed (hex) Au(100) surfaces in 0.01M $HClO_4$ [4.8] (1 M is one mole of the species in one mole of $H_2O$). The lower *pzc* of the unreconstructed Au(100) surface is due to the lower work function of that surface (4.24). The work function of the unreconstructed surface is lower because the work function scales with the density of surface atoms. The relative values of the surface tensions at their respective *pzc* can approximately be obtained from the observation of the stability range of the two phases. Merely for the absolute value of the surface tension one has to resort to theory [4.9, 12]). The variation of the surface stress with applied potential has been measured for Au(100) in the same electrolyte using the bending bar technique [4.11] and is plotted as a dashed line in Fig. 4.7 for comparison (the absolute value is arbitrary). The surface stress changes more rapidly with the potential and does not have a maximum at *pzc*.

## 4.2.4 Maxwell Relations and Their Applications

Since $f^{(s)}$ and $\gamma$ are state functions, second derivatives with respect to their independent variables do not depend on the sequence in which the derivatives are made. This gives rise to the so-called Maxwell relations. Consider e.g. the surface tension with the total differential as given by (4.20). The trivial equality $\partial^2 \gamma / \partial\phi\partial\mu = \partial^2 \gamma / \partial\mu\partial\phi$ leads to the not so trivial Maxwell relation

$$\left.\frac{\partial \Gamma_i}{\partial \phi}\right|_{\mu_i;T,\dots} = \left.\frac{\partial \sigma}{\partial \mu_i}\right|_{\mu_{j\neq i},\phi,T\dots} . \tag{4.29}$$

Suppose one had means to determine the dependence of the surface charge density $\sigma$ on the chemical potential $\mu_i$ and the electric potential $\phi$, then one could obtain the surface excess $\Gamma_i$ of the species $i$ or surface coverage $\Theta_i$ as a function of $\phi$ for all values of $\mu_i$ by integration

$$\Gamma_i(\phi,\mu_i) = \int_{-\infty}^{\phi} \left(\frac{\partial \sigma(\phi')}{\partial \mu_i}\right) d\phi' . \tag{4.30}$$

The lower boundary $-\infty$ stands for a potential where the species is not adsorbed so that $\Gamma_i(-\infty) = 0$. Of course, the integrand is also zero there. Alternatively, one may take the full coverage as a reference where the surface charge is again independent of the chemical potential so that the integrand is zero. This method to determine surface coverages on surfaces in an electrolyte is called *chronocou-lometry*[5]. While the theoretical foundation of the method is simple, the experiments are cumbersome and require considerable skill, care and diligence. An example from a group, which has cultivated such measurements is shown in Fig. 4.8 [4.13]. The example concerns the adsorption of $SO_4^-$-ions on Au(111) electrodes in a supporting electrolyte of 0.05 M $KClO_4$+0.02 M $HClO_4$ with varying small concentrations of $K_2SO_4$ ranging between $5\times10^{-6}$ M and $5\times10^{-3}$ M.

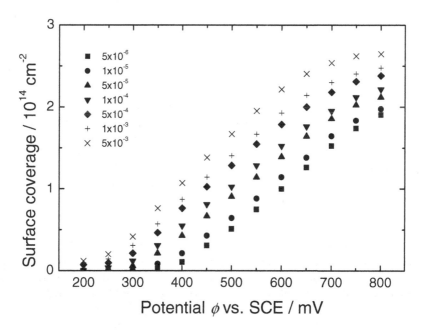

**Fig. 4.8.** Surface coverage with $SO_4^-$-ions on a Au(111) electrode surface as a function of the electrode potential measured with respect to SCE. Parameter is the concentration of $SO_4^-$-ions in the electrolyte, in other words the chemical potential of the electrolyte with respect to $SO_4^-$-ions (After [4.13]).

The supporting electrolyte serves to keep the conductance of the electrolyte constant. It is assumed that ions of the supporting electrolyte do not adsorb on the surface. The concentration of $SO_4^-$-ions in the liquid is proportional to the molality of $K_2SO_4$, and for the dilute concentrations (see Sect. 6.2.4) one has the

---

[5] The term *chronos* (Greek "time") presumably refers to the method of measuring charges by integrating currents obtained in potential sweeps.

relation between the chemical potential and the concentration

$$\mu_{SO_4^-} = \mu_0 + k_B T \ln(\rho_{SO_4^-}) . \qquad (4.31)$$

The higher the concentration and thereby the chemical potential the lower the positive potential at which a particular coverage of the surface with $SO_4^-$-ions is obtained. A more detailed analysis of the adsorption isotherms is performed in Sect 6.2.5.

A further useful Maxwell relation is derived from (4.15). Assuming an isotropic surface stress for simplicity one obtains

$$\left.\frac{\partial \mu_i}{\partial \varepsilon}\right|_{\Theta,T...} = \left.\frac{\partial \tau^{(s)}}{\partial \Gamma_i}\right|_{\varepsilon,T...} = \frac{1}{\Omega_s}\left.\frac{\partial \tau^{(s)}}{\partial \Theta_i}\right|_{\varepsilon,T...} \qquad (4.32)$$

in which $\Omega_s$ is the area covered by one molecule of the adsorbed species. The coverage dependence of the surface stress $\tau^{(s)}$ can be measured, i.e. by the bending bar technique [4.2], so that (4.32) provides information on the strain dependence of the chemical potential of the adsorbed phase. The strain dependence of the chemical potential is approximately equal to the strain dependence of the heat of adsorption $Q$ [4.14]. Hence, the not easily measured strain dependence of the heat of adsorption can be obtained from the more accessible coverage dependence of the surface stress. A few examples are shown in Fig. 4.9.

As a final example, we consider a Maxwell-relation between the dependence of the surface stress on the charge density and the strain dependence of the work function $\Phi$. For simplicity, we assume that the surface stress is isotropic. From (4.15) one obtains a Maxwell-relation, which relates the charge dependence of the surface stress with the strain dependence of the potential $\phi$. The only way the potential can change upon a strain is when the work function changes. We can therefore replace the potential $\phi$ by the work function $\Phi$ and obtain

$$\left.\frac{\partial \tau^{(s)}}{\partial \sigma}\right|_{T,\Gamma_i,\sigma=0} = -\frac{1}{e}\left.\frac{\partial \Phi}{\partial \varepsilon}\right|_{T,\Gamma_i,\sigma=0} . \qquad (4.33)$$

The variation of the work function with strain is material specific and in general not known. For a qualitative picture one may resort to the jellium model (see Sect. 3.2.1) in which the work function is a monotonous function of the electron density. If the density is described in the typical way by the radius $r_s$ in units of Bohr (0.529 Å) of a sphere which has the volume taken by one electron, the dependence of the work function on the electron density can be described by [4.15]

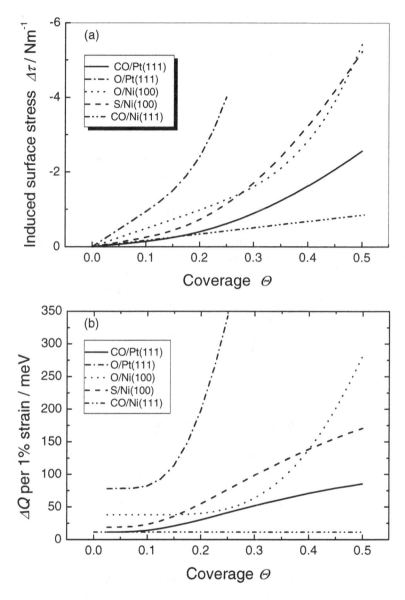

**Fig. 4.9**: **(a)** Surface stress as a function of coverage for Pt(111), Ni(100) and Ni(111) surfaces. The results were obtained using the bending bar technique [4.2]. **(b)** The variation of the heat of adsorption with strain calculated from the coverage dependence of the surface stress [4.14].

$$\Phi / eV = 4.63 - 0.38 \, r_s \, . \tag{4.34}$$

After some simple algebra one obtains for the strain derivative of the work function

$$\left. \frac{\partial \Phi}{\partial \varepsilon} \right|_{T, \Gamma_i, \sigma = 0} / eV = -0.38 \frac{1}{3} r_s \cong -0.13 \, r_s \tag{4.35}$$

The variation of the surface stress with the charge is therefore necessarily negative, which is in agreement with the existing experimental data (cf. Fig. 4.7). Gold has $r_s = 3.0$, and therefore $\partial \tau / \partial \sigma \cong -0.38 \, V$ . The experimental values are -0.9 V and $-1.6V$ on Au(111) and Au(100) in 0.1 M HClO$_4$, respectively [4.16]. While the simple model predicts the sign of the effect correctly, it fails to account for the effect quantitatively. One should not be surprised by that, however: The jellium model underestimates, the surface energy, the surface stress and the work function of 5d-metals by far [4.15, 17]. This is partly because localized d-electrons are not considered in the jellium model, partly because these quantities are intimately related to inhomogeneity of the electronic charge distribution due to the atomic structure (see Fig. 4.6) which is likewise not included in the jellium model. Furthermore, (4.33) is valid around *pzc* and for constant $\Gamma_i$. In the experiments on the surface stress of gold surfaces in an electrolyte, the chemical potentials $\mu_i$ were kept constant, which does make a difference, in particular as a specific adsorption of ions around *pzc* from the HClO$_4$-electrolyte cannot be excluded.

### 4.2.5 Solid/Solid and Solid/Liquid Interfaces

As an application of the thermodynamics of flat surfaces and interfaces, we consider phases, which have a line of contact in common (Fig. 4.10a) which - without loss of generality- may be assumed normal to the plane of drawing. The condition of equilibrium requires a relation between surface tensions of the interfaces and the angles. We assume for the moment that the interface tensions are independent of the orientations of the interfaces. The interfaces are in equilibrium if the total interface energy is stationary against a variation of the contact angles. In order to make that comparison one introduces a "virtual" displacement $\vec{s}$ of the line of contact in any arbitrarily chosen direction (arrow in Fig. 4.10a). The condition that the total interface energy of the three phases considered in Fig. 4.10a be stationary with respect to the virtual displacement $\vec{s}$ requires that

$$\gamma_1 \cos \alpha_1 + \gamma_2 \cos \alpha_2 + \gamma_3 \cos \alpha_3 = \sum_i \gamma_i \cos \alpha_i = 0 \tag{4.36}$$

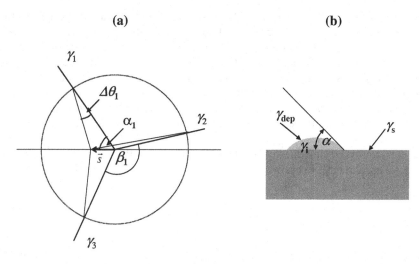

**Fig. 4.10. (a)** Suppose three (or more) phases share a common line, which is assumed perpendicular to the plane of drawing. Equilibrium requires that the interface energy is stationary with respect to any arbitrarily oriented virtual in-plane displacement $\vec{s}$ of the common line. The interfaces $i$ change length by the amount $s\cos\alpha_i$, with $\alpha_i$ the angle between the displacement vector $\vec{s}$ and a vector pointing from the center along the interface $i$. For solids, the interface energies change also because of the rotation by the amount $\Delta\theta_i$ since the interface tensions then depend on the orientation. **(b)** A special application of the equilibrium condition is the contact angle of a solid cluster or a liquid droplet on the flat surface of a substrate.

with $\alpha_i$ the angles between the interface $i$ and the direction of the virtual displacement[6].

Equation (4.36) is occasionally interpreted as a "balance of forces" acting upon the common line. This is a gross misconception: The virtual displacement $\vec{s}$ compares different sets of contact angles; it is not an elastic deformation. If (4.36) were to describe a balance of forces upon an elastic distortion, it would involve the interface stresses $\tau_i^{(s)}$ and not the interface tensions $\gamma_i$! According to the *Shut-*

---

[6] Equation (4.36) can also be written in terms of the tensions $\gamma_i$ and the angles $\beta_i$ subtended by the respective two other interfaces $j\neq i$ (Fig. 4.8a). This operation is achieved by first writing (4.36) three times each with one of the three interfaces as the plane of projection for the angles $\alpha_i$. After converting the angles $\alpha_i$ into the angles $\beta_i$ and some algebra one obtains the relation

$$\frac{\gamma_1}{\sin\beta_1} = \frac{\gamma_2}{\sin\beta_2} = \frac{\gamma_3}{\sin\beta_3} \tag{4.36a}$$

that is well known in the material science of grain boundaries.

*tleworth relation* (4.21) the stress and tension differ by the strain derivative of the tension which is an intrinsic property of solids involved. The difference between the interface stress and tension may have an arbitrary positive or negative value. If (4.36) holds for the tensions it cannot hold for the stresses, except incidentally. In general, the forces exerted by the interface stresses on the common line are therefore not balanced! The unbalanced forces along the common line give rise to long-range elastic deformations of the phases (cf. Sect. 3.4), which are difficult to handle theoretically and therefore constitute one of the unsolved problems in the physics of mixed phases.

For crystalline solids, the surface tension depends on the orientation. This contributes a second term to each variation in the interface tensions of the form

$$\Delta \gamma_i = \frac{\partial \gamma_i}{\partial \theta_i} \Delta \theta_i \tag{4.37}$$

where $\Delta \theta_i$ is the rotation of the orientation of the interface $i$ due to the displacement vector $\vec{s}$ and the derivative is to be taken at the orientation of the interface $i$. The condition that the interface energy be stationary is then

$$\sum_i \gamma_i \cos \alpha_i + \frac{\partial \gamma_i}{\partial \theta_i} \sin \alpha_i = 0 . \tag{4.38}$$

It will be shown in section 4.3 that the derivative of the interface tension with respect to the orientation is of the same order of magnitude as the interface tension when the interface is a low index crystallographic direction and forms a "facet". It is small only if the surface is thermodynamically "rough" (see section 4.3).

Another application of the principle of stationary interface energy is the calculation of the contact angle of a solid or liquid deposit on a flat surface (Fig. 4.10b). In that case, (4.36) turns into the Young-Dupré equation

$$\gamma_i + \gamma_{dep} \cos \alpha = \gamma_s \tag{4.39}$$

in which $\gamma_s$, $\gamma_{dep}$ and $\gamma_i$ are the surface tensions of the substrate and deposit and the interface tension between substrate and deposit, respectively (Fig. 4.10b). For simplicity, we have assumed that the surface of the deposit is "rough" near the line of contact with the substrate so that $\partial \gamma_{dep} / \partial \theta_{dep} \ll \gamma_{dep}$.

If the sum of the interface tension and the surface tension of the deposit is smaller than the surface tension of the substrate,

$$\gamma_i + \gamma_{dep} < \gamma_s , \tag{4.40}$$

then the condition (4.39) cannot be fulfilled for any contact angle. Rather one has a complete wetting of the substrate by the deposit.

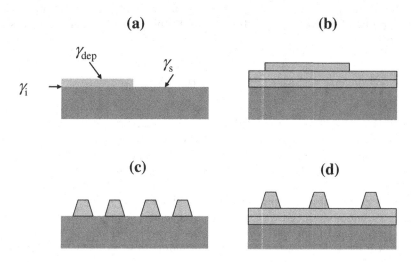

**Fig. 4.11.** (a) Definition of surface and interface tensions for heteroepitaxial growth; (b) layer-by-layer growth, or *Frank-van-der-Merwe growth*; (c) 3D-cluster or *Vollmer-Weber growth*; (d) growth with one or more wetting layers continued by 3D-cluster growth, known as *Stranski-Krastanov growth*.

The condition (4.40) and its opposite

$$\gamma_i + \gamma_{dep} > \gamma_s \tag{4.41}$$

constitute an important criterion for the preferred growth mode in heteroepitaxial growth. The criterion is named after E. Bauer [4.18]. If (4.40) holds, heteroepitaxial growth proceeds in a layer-by-layer manner. One monolayer is completed before the next one begins to grow (ideally). This growth mode is also called *Frank-van-der-Merwe growth* (Fig.4.11b). If (4.41) holds, the growth mode is in the form of 3D-clusters. This growth mode is *called Vollmer-Weber growth* (Fig.4.11c).

Heteroepitaxial growth is frequently pseudomorphic which means that the deposit grows with a lattice constant matched to the substrate. Since the natural lattice constants of the deposited film $a_f$ and the substrate $a_s$ may differ, the deposited film is in a state of strain if the growth is pseudomorphic. The misfit strain $\varepsilon_{mf}$ is

$$\varepsilon_{mf} = \frac{a_f - a_s}{a_s} . \tag{4.42}$$

For simplicity, we assume that the pseudomorphic strain and the elastic properties of the deposited film are isotropic. Because of the misfit, the film carries an elastic energy of

$$\gamma_{elast} = \frac{t}{2}\frac{Y}{1-\nu}\varepsilon_{mf}^2 + \Delta\tau^{(s)}\varepsilon_{mf} \qquad (4.43)$$

per area. Here, $Y$ is the *Young modulus* of the deposited film, $\nu$ the *Poisson ratio* and $t$ the thickness of the deposited film. The change in the surface stress $\Delta\tau^{(s)}$ is the total change in the surface stress due to film deposition. Depending on the sign of $\Delta\tau^{(s)}$ and the sign of the misfit, the second term can be negative or positive, stabilizing or destabilizing *Frank-van-der-Merwe growth*. For larger film thickness, the elastic energy always works in favor of 3D-clusters, hence *Vollmer-Weber growth*. Pseudomorphic growth therefore frequently begins as *Frank-van-der-Merwe growth* and turns into *Vollmer-Weber growth* after one or more layers. This growth mode is known as *Stranski-Krastanov growth* (Fig. 4.11d). Alternatively, the elastic energy in pseudomorphic films may relax by the formation of dislocations (Sect. 1.3.1) after a *critical thickness* $t_c$.

## 4.3 Curved Surfaces and Surface Defects

### 4.3.1 The Crystal Equilibrium Shape

For crystalline materials, the surface tension is a function of the orientation of surface. The equilibrium shape minimizes the total surface energy for a given volume. Which thermodynamic potential is the relevant surface energy depends on the environment and the parameters that are kept constant during the equilibration process, and whether with the environment equilibrium is established at all. For a crystal in vacuum, e.g., the equilibrium shape is established via surface or bulk diffusion. The number of atoms in crystals remains constant. The relevant energy is then the Helmholtz surface free energy $f^{(s)}$. For a crystal held in equilibrium with its own vapor phase the relevant surface energy is the surface tension $\gamma$. Since typically a crystal in that case will be uncharged, the Legendre-transformation with respect to the surface charge and potential (4.19) is irrelevant. One may also ask for the equilibrium shape of a crystal in a vapor phase of a gas, which can absorb on the surface but contains no atoms of the crystal substrate. Let us assume further that the temperature is low enough so that there is no evaporation of the substrate material. In that case, the number of substrate atoms is kept constant while vapor phase atoms or molecules may adsorb on the crystal surface to build up surface excesses. In that (not untypical) experimental situation the appropriate surface energy is a particular surface tension for which the Legendre transformation over the species $i$ is performed for all but the species which make up the bulk of the substrate. It is not sufficient to treat this case by making the surface excess of the substrate species zero by definition: with the equilibrium vapor pressure being zero for substrate atoms their chemical potential would diverge to minus infinity, leaving the product of surface excess and chemical potential undefined, unless special models are invoked describing the mathematics of this limit. One may also consider a crystal in equilibrium with an aqueous electrolyte with some finite concentrations of ions, however with zero concentration of ions of the substrate material. With regard to the substrate atoms, the case is equivalent to a crystal in vacuum at moderate temperatures. The gold surface in a $H_2SO_4$- electrolyte (Fig. 4.8) is an example. Here, the appropriate surface tension $\gamma$ is one where the Legendre transformation (4.19) is performed with respect to $SO_4^-$- ions and the surface charge, however not for the gold atoms. In summary, the appropriate surface energy function depends entirely on the details of the experimental conditions. In what follows we denote the surface tension that suits the specific experiment by the symbol $\gamma$ and keep in mind that $\gamma$ is not the same in each case. With that important caveat, the condition for the equilibrium shape is that

$$\Pi^{(s)} = \int \gamma(\vec{n}) \, dS \qquad (4.44)$$

should be minimal for a given total volume of the crystalline material. Here, $\vec{n}$ denotes a unit vector perpendicular to the surface and $dS$ is an element of the surface. For the special case of crystal in vacuum, $\Pi^{(s)}$ is the Helmholtz surface free energy $F^{(s)}$. If the surface tension is isotropic as for a liquid, $\Pi^{(s)}$ is minimized by a spherical shape. For a general anisotropic surface tension, the equilibrium shape is obtained from the Wulff-construction, which is illustrated in Fig. 4.12 (For the nontrivial proof of the Wulff-construction see e.g. [4.4, 19]): On each ray connecting the origin with the point of $\gamma(\theta)$ in a polar plot (dotted lines in Fig. 4.12) perpendicular lines are constructed (dashed lines). The area inside the ensemble of all the dashed lines marks the equilibrium shape of the crystal in the plane of drawing. Figure 4.12 shows two examples. In the first one (upper Wulff-plot), $\gamma(\theta)$ is a continuous differentiable function everywhere and the equilibrium shape has a finite nonzero curvature.

Surfaces, which have a finite curvature on the equilibrium shape, are called *rough* surfaces. The equilibrium shape has a corner at $\theta = 45°$, which means that certain orientations of high $\gamma(\theta)$ are missing. In the lower half of Fig. 4.12 $\gamma(\theta)$ has a cusp at $\theta = 0°$, in other words, $\gamma(\theta) \propto |\theta|$. The equilibrium shape is then a flat plane with a particular lateral extension, which depends on the azimuthal orientation. These areas are called *facets* or *singular surfaces*. Physical realizations of facets are the low index crystal planes.

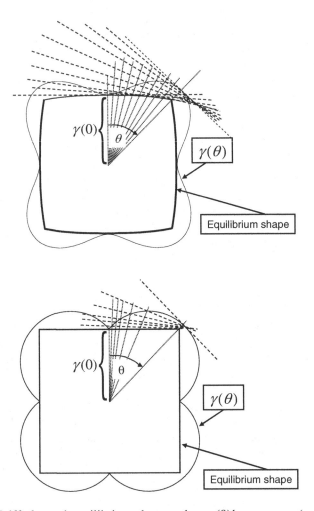

**Fig. 4.12.** Wulff plot and equilibrium shapes when $\gamma(\theta)$ has no cusp (upper panel) and when $\gamma(\theta)$ has a cusp at $\theta = 0$ (lower panel).

In Fig. 4.13 a low index $(h\,k\,l)$ surface is depicted together with crystal planes, which are slightly tilted to the $(h\,k\,l)$ plane by an angle $\pm\theta$. Tilted planes are composed of a regular sequence of $(h\,k\,l)$-terraces separated by steps of height $h$. One may think of these steps as being one atom layer high as this would be a typical realization of a tilted surface. However, thermodynamics can be formulated without having a specific atomic model in mind. The ensemble of $(hkl)$-terraces has the same energy as on the non-tilted surface. The creation of steps requires additional

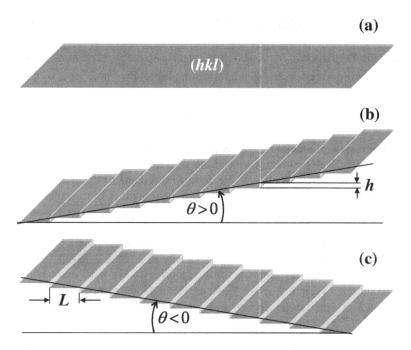

**Fig. 4.13.** Surfaces, which are tilted with respect to a low index plane ($h k l$) **(a)** by an angle $\theta$ **(b, c)** have a higher surface tension due to the step line tension $\beta$.

work, which is inversely proportional to the distance $L$ between the steps (Fig. 4.13b,c). By expressing the step density in terms of the step height $h$ one has

$$\gamma_p(\theta) = \frac{\gamma(\theta)}{\cos\theta} = \gamma_0 + \frac{\beta^{(\pm)}}{h}|\tan\theta| \quad . \tag{4.45}$$

With $\gamma_p(\theta)$ we have introduced a surface tension defined with respect to a length scale projected on to the surface with the angle $\theta = 0$. The quantity $\beta^{(\pm)}$ is the *step line tension* of steps of height $h$. The plus or minus sign refers to the sign of the tilt angle. In case of centrosymmetric surfaces, e.g. a {100} surface of an fcc-structure, $\beta^{(+)} = \beta^{(-)}$. On {111}-surfaces of fcc crystals, the A- and B-steps (Sect. 1.3.1) have a different structure and thus a different energy so that $\beta^{(+)} \neq \beta^{(-)}$. For uncharged surfaces in vacuum, the line tension is equal to the step free energy, which in turn at $T = 0$ K is equal to the step energy. At moderate temperatures, the step free energy is a little lower, but still roughly equal to the step energy since the step entropy is small (cf. Section 5.2.1).

Equation (4.45) can be considered as the first step of an expansion in powers of $p = \tan\theta$

$$\gamma_p(p) = \gamma_0 + \gamma_1 p + \gamma_2 p^2 + \gamma_3 p^3 + \gamma_4 p^4 ... \tag{4.46}$$

in which the higher order terms correspond to step-step interactions. The term $W_2^{(int)} = \gamma_2 p^2$ represents a step-step interaction proportional to $L^{-1}$ and so forth.

$$W_n^{(int)} \propto L^{-(n-1)} \quad n \geq 2 \tag{4.47}$$

Models for step-step interactions are considered in Sect. 3.4.2 and 5.2.2. Here we are concerned with the consequences of the various terms on the crystal equilibrium shape in the vicinity of a facet. We describe the shape function around a facet as $z(x)$ in which the $z$-axis is parallel to the orientation which corresponds to $\theta = 0$ and the $x$-axis is orthogonal to the $z$-axis in the direction of an advancing $\theta$ (Fig. 4.14). Both the $x$- and the $z$-axis have their origin at the center of the equilibrium shape, which is also the center of the polar $\gamma$-plot (Wulff-point).

The shape function $z(x)$ is obtained most easily in a parameterized form [4.20] by making use of the Wulff-construction. According to Fig. 4.14 one has the following relation between the surface tension $\gamma(\theta)$ and the distance of a point on the shape curve $R$ in which the slope is $\theta$ and the origin.

$$\begin{aligned}\frac{\gamma(\theta)}{\gamma_0} &= \frac{R}{z_0}\cos(90° - \theta - \alpha) = \frac{R}{z_0}\sin(\theta + \alpha) \\ &= \frac{R}{z_0}(\sin\alpha\cos\theta + \cos\alpha\sin\theta) = \frac{1}{z_0}(z\cos\theta + x\sin\theta)\end{aligned} \tag{4.48}$$

In the reduced coordinates

$$\hat{x} = x\gamma_0/z_0 \quad \hat{z} = z\gamma_0/z_0 \tag{4.49}$$

equation (4.48) becomes

$$\gamma_p(p) = \hat{z} + \hat{x} p . \tag{4.50}$$

The parameterized form of the shape function is therefore

$$\begin{aligned}\hat{x} &= \gamma_p'(p) = \gamma_1 + 2\gamma_2 p + 3\gamma_3 p^2 + 4\gamma_4 p^3 ... \\ \hat{z} &= \gamma_p(p) - \gamma_p'(p) \, p = \gamma_0 - \gamma_2 p^2 - 2\gamma_3 p^3 - 3\gamma_4 p^4 ...\end{aligned} \tag{4.51}$$

in which $\gamma_p'(p)$ is the derivative of $\gamma_p(p)$ with respect to $p$. Assuming that all coefficients $\gamma_n$ are positive (repulsive step-step interactions) the slope of the shape curve is negative definite and approaches zero continuously as $x$ decreases.

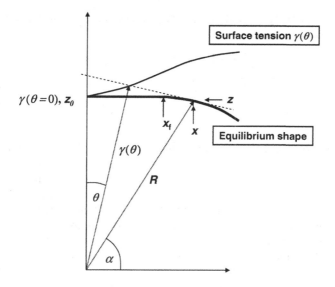

**Fig. 4.14.** Illustration for the derivation of the shape function $z(x)$ (see text for details). Note that $\gamma(\theta)$ and $R(\theta)$ represent a different metric.

The slope becomes zero at a finite value of $\hat{x}$ which marks the extension of the facet $x_f$

$$\hat{x}_f = \gamma_1 = \beta / h \qquad x_f = \frac{\beta z_0}{h \gamma_0} .$$

$$\hat{z}_f = \gamma_0 \qquad z_f = z_0 \qquad\qquad\qquad (4.52)$$

The size of the facet is therefore proportional to the step line tension. As a general trend, the step line tension decreases with rising temperature because of the increasing vibration and configuration entropy. Eventually, the line tension of steps and thus the facet may even vanish completely prior to melting of the crystal. This phase transition is called the *roughening transition,* and the temperature at which this happens is called the *roughening temperature* $T_R$. Models for the temperature dependence of the step line tension based on statistical physics are considered in Sect. 5.2.4. The dependence of the step line tension on the electrode potential for surfaces in an electrolyte is discussed in section 4.3.5.

Standard models for the step-step interactions consider elastic interactions (Sect. 3.4.2) and entropic interactions (Sect. 5.2.2). Both are repulsive and depend on the step-step distance $L$ as $L^{-2}$, and therefore contribute to a positive $\gamma_3$-term. The corresponding projected surface tension is plotted in Fig. 4.15a as a dashed line. Many qualitative and quantitative experiments support the understanding that $L^{-2}$-interactions are dominating in general [4.21, 22]. The solution for the shape function for $\gamma_1$, $\gamma_3 > 0$; $\gamma_2 = \gamma_4 = 0$ is therefore of particular practical importance. The resulting shape function

$$z = z_0 - \frac{2}{3}\sqrt{\frac{\gamma_0}{3z_0\gamma_3}}(x-x_f)^{3/2} \qquad x \geq x_f$$

$$z = z_0 \qquad\qquad\qquad\qquad\qquad x \leq x_f$$

(4.53)

is known as a Pokrovsky-Talapov or Gruber-Mullins shape [4.23, 24] (dashed line in Fig. 4.15b). The shape function changes quantitatively but not qualitatively for nonzero, positive $\gamma_2$ and $\gamma_4$ [4.20]. Qualitative changes of the equilibrium shape are introduced by attractive step-step interactions. The transition between the facet and the rough surface becomes abrupt with a finite contact angle. Effective attractive step-step interactions and finite contact angles have been observed, e.g., for Au(100) and Au(111) surfaces [4.25]. In that case, they can be attributed to the reconstruction of gold surfaces. The reconstruction tends to make particular step-step distances energetically favorable over others. It is, however, inappropriate to describe the surface tension of these surfaces by an expansion in powers of the inverse distance. Rather one has pronounced minima in the energy relative to the expansion in powers of $p$ for certain "magic" orientations for which the terraces sizes fit to integral numbers of reconstruction cells [4.26].

Stepped surfaces in an electrolyte experience attractive interactions of the $\gamma_2 < 0$ type (Sect. 4.3.5). We therefore take that case as an example and calculate the contact angle at the facet boundary from (4.51) by looking for nontrivial solutions of $z(x) = z_0$ for $\gamma_2 < 0$.

$$z = z_0: \quad p^2(\gamma_2 + 2\gamma_3 p) = 0$$

(4.54)

In addition to the facet solution $p = 0$ one has a solution for $p > 0$ corresponding to a finite contact angle $\theta_f$ between the facet and the rough section

$$\theta_f = \arctan(-\gamma_2/2\gamma_3).$$

(4.55)

In case of attractive interactions, the size of the facet is no longer described by (4.53). The size does not only depend on the step line tension either. By inserting (4.55) into (4.51) one obtains

$$x_f = \frac{\beta z_0}{h\gamma_0} - \frac{1}{4}\frac{z_0\gamma_2^2}{\gamma_3\gamma_0}.$$

(4.56)

In case of partly attractive interactions, the facet size shrinks compared to purely repulsive interactions! The effect of a negative $\gamma_2$-term on the projected surface

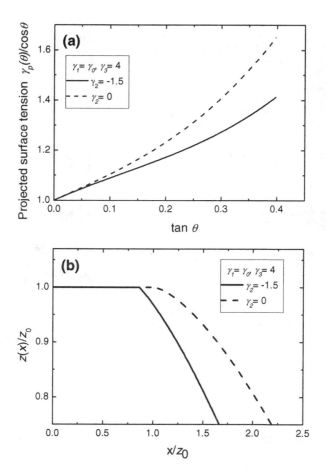

**Fig. 4.15.** (a) Projected surface tension $\gamma(\theta)/\cos\theta$ as function of $\tan\theta$ for attractive interactions $\gamma_2$ and repulsive interactions $\gamma_3$ and for repulsive interactions $\gamma_3$ only, solid and dashed line respectively. (b) The equilibrium crystal shape.

tension $\gamma_p(\theta)$ and on the equilibrium shape is illustrated by the solid lines in Fig. 4.15a and 4.15b, respectively.

Experimental studies on crystal equilibrium shapes are tedious and therefore rare. Figure 4.16a displays an STM image of a lead particle which was deposited on a Ru(0001) surface at 420 K after equilibration for about 20 h. Various facets are clearly seen. In images with enhanced contrast, the first step marking the boundary is clearly discernible Fig. 4.16b. With the boundary of the facet exactly fixed accurate experimental results are achieved on the transition region between the facet and the rough surface as well as on shape function (for further details the reader is referred to the review of H. P. Bonzel [4.20]).

**(a)**                                                    **(b)**

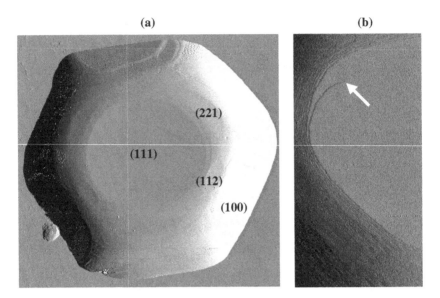

**Fig. 4.16.** STM images of lead particles equilibrated on a Ru(0001) substrate (courtesy of Christian Bombis). The particles grow well with their (111) face on the basal plane of Ru since the lattice mismatch (4.22) is only about +4.7%. **(a)** Total view of the particle imaged in reduced contrast. The step marking the facet boundary is clearly seen in images with enhanced contrast **(b)**. A step originates where a dislocation line of a dislocation with a screw component penetrates the surface (white arrow in **(b)**).

## 4.3.2 Rough Surfaces

We consider a rough surface of a solid in an experimental situation where the number atom of the solid remains fixed, so that, concerning the atoms of the solid, the Helmholtz free energy is the appropriate thermodynamic potential. The solid may nevertheless be in contact with an electrolyte or a vapor phase so that ions or gas atom/molecules of different nature than those making up the solid substrate can absorb on the surface. The appropriate thermodynamic potential is then one in which the $n_i \, \mu_i$ -Legendre transformation is performed for all but the substrate species. As with regard to the latter, the surface tension keeps the property of the specific Helmholtz free energy, namely that the chemical potential of the substrate atoms $j$ is obtained by the derivative of the Helmholtz free energy of the solid with respect to the number of atoms $j$: $\mu_j = \partial F / \partial n_j$. We are now interested in the work required to create a small extension of the area of curved surface of the solid. Let A and B two points on the surface on the surface, with A defining the origin on the x-axis and B a second point on the $x$-axis at the distance $L_x$ from the origin (Fig. 4.17a). Between the points A and B the surface profile has a contour

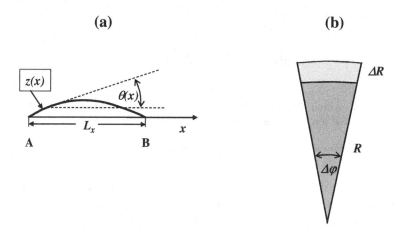

**Fig. 4.17.** Illustrations for the derivation of **(a)** the surface stiffness and **(b)** the chemical potential of a curved surface.

described as $z(x)$ with a total length $L$. Orthogonal to the plane of drawing the profile is assumed flat over a length unit $L_y$. We calculate the thermodynamic potential $\Pi^{(s)}$ as a function of the length of the contour $L$.

$$\Pi^{(s)} / L_y = \int_A^B \gamma(\theta(x))\mathrm{d}s \; , \; L = \int_A^B \mathrm{d}s \; . \tag{4.57}$$

The integrals are to be taken along the surface contour $z(x)$. The surface tension depends on $z(x)$ only via the angle $\theta(x)$. The integral is converted into an integral along the $x$-axis by replacing the length element $\mathrm{d}s$ along the contour by $\mathrm{d}x = \cos\theta\,\mathrm{d}s$. By expanding the surface tension and $\cos^{-1}\theta$

$$\gamma(\theta) = \gamma(0) + \gamma'(0)\theta + \frac{1}{2}\gamma''(0)\theta^2 \ldots$$

$$\frac{1}{\cos\theta} = 1 + \frac{1}{2}\theta^2 \ldots \tag{4.58}$$

and by observing that the integral over $\theta(x)$ vanishes one obtains

$$\Pi^{(s)} / L_y = \int_A^B \frac{\gamma(\theta(x))}{\cos\theta(x)}\mathrm{d}x \cong \gamma(0)L_x + \left(\gamma(0) + \gamma''(0)\right)\int_A^B \frac{1}{2}\theta^2(x)\,\mathrm{d}x$$

$$\cong \gamma(0)L_x + \left(\gamma(0) + \gamma''(0)\right)\left(L - L_x\right) \tag{4.59}$$

since

$$L = \int_0^{L_x} dx / \cos\theta(x) \cong \int_A^B \left\{ 1 + \frac{1}{2}\theta^2(x) \right\} dx .  \tag{4.60}$$

The quantity, which describes the variation of the thermodynamic potential $\Pi^{(s)}$ per area with the length of the contour $L$, is known as the *stiffness of the surface*

$$\tilde{\gamma}(\theta) = \gamma(\theta) + \gamma''(\theta) .  \tag{4.61}$$

With the help of the surface stiffness $\tilde{\gamma}$, one can calculate the chemical potential of a body that is enclosed in a curved contour and is in equilibrium with the surface. Suppose one has an element of the surface with a particular radius of curvature $R$ (Fig. 4.17b). By expanding the radius by $\Delta R$ while keeping the curvature constant the surface area increases by $L_y \Delta L = L_y \Delta R \Delta\varphi$. The number of particles in the body increases by $\Delta N = L_y \Delta R \Delta\varphi R / \Omega$ with $\Omega$ the volume of an atom. Because of the homogeneity with respect to the surface area, equilibrium between bulk and surface requires that $\mu \Delta N = \tilde{\gamma} L_y \Delta L$ and thus

$$\mu = \Omega \tilde{\gamma} / R .  \tag{4.62}$$

The sign of the chemical potential of the surface is positive if $R$ has its center inside the body (concave surface). When the center is outside (convex surface) the number of particles in the body is shrinking for an increasing contour length and the chemical potential is negative. We therefore describe the local surface contour $\kappa_x$ at any point of the surface as positive for a concave surface. With curvatures orthogonal to the plane of drawing denoted as $\kappa_y$ the chemical potential is

$$\mu = \Omega\left( \tilde{\gamma}(\theta_x)\kappa_x + \tilde{\gamma}(\theta_y)\kappa_y \right).  \tag{4.63}$$

This equation is referred to as the *Herring-Mullins equation*. The derivation of (4.63) requires merely that bulk and surface be in local equilibrium in the particular area considered. With the understanding that the coordinates $x$, $y$ are defined on a *coarse scale* so that each "infinitesimal" element $dx$, $dy$ contains many atoms one may therefore define a *position dependent chemical potential of a surface* $\mu(x, y)$. The Herring-Mullins equation then serves as a basis for a local thermodynamic description of a surface, thereby providing an important link between the macroscopic world and quantum physics. Equation (4.63) is a useful starting point for the description of surface self-diffusion (Sect. 10.2.2), e.g., or for the process of two- and three- dimensional Ostwald ripening (Sect. 10.4).

If surface and bulk are in global equilibrium with each other then the sum of the products of surface stiffness and the curvature is a constant all over the surface contour. Applied to a liquid droplet one has $\tilde{\gamma} = \gamma$ and thus

$$\mu_{\mathrm{liq}} = \frac{2\Omega\,\gamma}{R} \tag{4.64}$$

and since $\mu \propto \ln p$

$$p = p_\infty \mathrm{e}^{\frac{2\Omega\,\gamma}{R}} \tag{4.65}$$

with $p_\infty$ the vapor pressure of the flat surface. Equation (4.63) may therefore be considered as a generalization of the *Gibbs-Thomson equation* to anisotropic solids.

The principle of the derivation of the Herring-Mullins equation may also be applied to derive a useful relation between the distance of a facet from the center of a crystal equilibrium shape, the surface tension of the facet and the chemical potential of the crystal. Let $z_0$ be the distance of a particular facet from the center and $\gamma_0$ the surface tension of the same facet. Consider a uniform expansion of the crystal with $z_0$ being expanded by $\Delta z_0$. The area of the facet is $A_\mathrm{f} = c\,z_0^2$ with $c$ a constant depending on the facet shape. The area of the facet then varies upon an expansion by $\Delta A_\mathrm{f} = c\,2z_0\,\Delta z_0$ and the number of particles changes by $\Delta N = A_\mathrm{f}\,\Delta z_0 / \Omega$. Equating the change in the thermodynamic potentials of surface and bulk as above leads to the equation

$$\mu = 2\Omega\frac{\gamma_0}{z_0}, \tag{4.66}$$

which holds for any arbitrary facet (Gibbs-Wulff Theorem).

## 4.3.3 Line Tension and Stiffness

With (4.45) we have introduced the step line tension as the additional contribution to the surface tension due to the presence of steps. Accordingly, the step line tension may be calculated formally as the difference between the surface tension of stepped and flat surfaces per step and step length, and this is how one actually proceeds in total energy calculations of the step energy. Since the definition, meaning and value of the surface tension depends on the thermodynamic boundary conditions, the step line tension likewise refers to specific boundary conditions. This is of particular importance for surfaces held at constant potential in an electrolyte. The same remark applies to other surface defects. These aspects are covered in Sect. 4.3.5. Here, we focus on the step line tension of surfaces in vacuum where it is identical to the step free energy per length. While the step line tension in a coarse-grained description can be defined for steps of arbitrary height and orientation, we now use the term in a more restricted sense as the line tension of a monolayer high step. Steps of this kind are found experimentally on the low index surface orientations (Sect. 1.3). On any given surface orientation, the step line tension depends on the orientation of the step on that surface. On the (100) surface of an fcc-crystal, e.g., step atoms have the highest coordination (coordination number = 7) for steps oriented along the [011] direction. Steps running in this direction have therefore the lowest energy of all orientations on the (100) surface. Since entropic contributions to the free energy of steps along the [011] direction are small at moderate temperatures, the [011] orientation is also the direction of lowest free energy. Deviation from the [011] orientation require the formation of kinks in the steps which costs energy because kink atoms have a lower coordination number (= 6). Hence the energy increases with increasing deviation from the [011] orientation, regardless which way the deviation goes. One might therefore think that the angular dependence of the step line tension $\beta(\theta)$ near the direction of dense packing of atoms (defined as $\theta = 0°$) should be described by an equation analogous to (4.45) with a term proportional to $|\tan\theta|$, the proportionality constant being the free energy of kinks. The Wulff-construction of the equilibrium shape of a two-dimensional island would then reveal facets at $\langle 011 \rangle$ orientations. This is not so! Two-dimensional islands have no facets. In other words, steps are thermodynamically rough for temperatures above $T = 0$ K and the free energy of kinks vanishes. This is a special consequence of a general theorem that there are no phase transitions in one-dimensional systems at finite temperature for interactions decaying faster than $1/x^2$ when $x$ measures the distance along the chain. The physical reason beyond the formalities is that the fluctuations in 1D-objects simply are too large. Steps may be considered as 1D-objects, mostly at least. Exceptions may be steps on surfaces with reconstructions, such as on the clean Au(111), Au(100), Ir(100), Pt(100) surfaces. Experiments indicate that the unit cells of the reconstruction are structurally correlated with the steps. In that case, steps are no longer 1D-objects and monolayer high islands may have facets. The same may apply to surfaces, which are reconstructed because of adsorbate layers. So far, however, no case of a faceted monolayer island has been reported. This may be partly owed to the difficulty to distinguish between a faceted edge and an edge, which has an extremely small, but finite curvature.

Figure 4.18 displays experimental island equilibrium shapes for the Cu(100), Ag(111), and the Pt(111) surface for two temperatures each [4.27, 28]. The higher the temperature, the more roundly the islands are shaped. This implies that the entropy of steps is larger for steps with orientations off the direction of dense packing. It will be shown in sections 5.2.1 and 5.3.2 that this larger entropy is a configurational entropy due to the presence of many kinks in the steps. For temperatures approaching $T = 0$K the equilibrium shapes eventually become polygons. For (100) surfaces, e.g., the equilibrium shape would be a square or a square with truncated edges, an octagon. For (111) surfaces, the $T = 0$ K

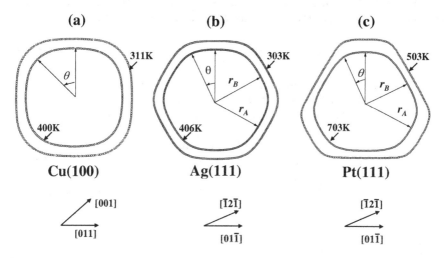

**Fig. 4.18.** Experimentally determined equilibrium island shapes on **(a)** Cu(100), **(b)** Ag(111), and **(c)** Pt(111) [4.27, 28]. The higher the temperature the rounder is the shape, eventually approaching a circle. For Ag(111), and also for Cu(111) A- and B-steps have nearly identical step line tensions, for Pt(111) the line tensions for A- and B-steps differ.

equilibrium shape is a truncated triangle or a hexagon if A- and B-steps have the same formation energy. As for rough surfaces, one may calculate the change in the free energy of a curved step upon elongation of the step contour. This leads to the definition of a step stiffness $\tilde{\beta}$, which depends on the orientation of the step. Analogous to (4.61) one finds

$$\tilde{\beta}(\theta_s) = \beta(\theta_s) + \partial^2 \beta(\theta_s)/\partial\theta_s^2 \tag{4.67}$$

in which $\theta_s$ denotes the orientation of the step (as opposed to the polar angle) pointing towards the step denoted as $\theta$ (Fig. 4.18)). The relation between the stiffness, curvature and chemical potential is (4.63)

$$\mu = \Omega \tilde{\beta}(\theta_s) \kappa(\theta_s). \tag{4.68}$$

Since the chemical potential is constant along the periphery of an island that has its equilibrium shape, the stiffness varies along the periphery to keep the product of curvature and stiffness constant. For the, at low temperatures, nearly straight sections of the steps in the ⟨011⟩ directions of close packing this means that the stiffness becomes very large and diverges as $T$ approaches zero. The divergence is inversely proportional to the concentration of thermally excited kinks (Sect. 4.3.7). With (4.66) a relation between the ratio of the surface tension of a facet and its distance from the center to the chemical potential was derived. Although steps have no facets, one may nevertheless apply the equivalent of (4.66) to 2D-islands

$$\mu = \Omega_s\, \beta_0 / y_0 . \tag{4.69}$$

Here, $\Omega_s$ is the area of an atom and $\beta_0$ and $y_0$ are the line tensions of the steps at the points of minimal curvature and the distance from the center, respectively. For islands on the {111} surface the index 0 stands for either A- or B-steps. Equation (4.69) can be proved also directly by assuming the $T = 0$ K polygonal shapes. For very high temperatures for which all equilibrium shapes converge to a circle (4.69) is just the Gibbs-Thomson equation in two dimensions. The size dependence of the chemical potential of islands on surfaces gives rise to a 2D-Ostwald ripening: larger islands grow at the expense of smaller ones. Eventually all islands disappear and the atoms migrate to straight steps which typically exist in sufficient abundance on surfaces.

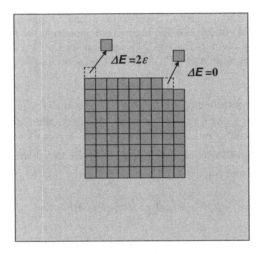

**Fig. 4.19.** The chemical potential of a square island in the Kossel-model. The free energy of the island does not change upon removal of all but the last atom in an edge. The Gibbs-Thomson chemical potential of the island is entirely due to that last atom.

It is interesting to contemplate the microscopic origin of the Gibbs-Thomson equation. For that purpose, we consider a simple model for the bonding of atoms: The atoms are represented by cubes that bond to each other via the faces of the cubes. This model is known as a Kossel-model. For islands on a (100) surface of an fcc-crystal, it is equivalent to a nearest neighbor bond model (Fig. 4.19). The total energy associated with the perimeter of an island of $N$ atoms is $E = 4\varepsilon N^{1/2}$. The chemical potential in the continuum approximation is

$$\mu = \frac{\partial E}{\partial N} = \frac{2\varepsilon}{\sqrt{N}} \tag{4.70}$$

As seen from Fig. 4.19 the microscopic length of the perimeter does not change with the removal of a corner atom or atoms from kink sites. The total energy of the island does not change in this case and the "chemical potential" for these atoms is zero. Only the removal of the last atom in a row reduces the length of the perimeter by two units. The chemical potential of that last atom is therefore $2\varepsilon$. The probability for an atom to be the final one in the row is $N^{-1/2}$, so that on the average $\mu = 2\varepsilon N^{-1/2}$. The continuum equation is thereby recovered. The fact that only a fraction of atoms carries the message of a higher chemical potential of an island does not hurt as all thermodynamic quantities are defined as averages over large ensembles. This has to be seen apart from the question whether deviations from the Gibbs-Thomson equation exist for small islands. Such deviations would arise, e.g., from next-nearest neighbor or long-range interactions.

We conclude this section with the remark that the concepts of line tension and line stiffness also apply to other line defects. Domain walls between ordered adsorbate regions are of particular interest.

### 4.3.4 Point Defects

The hierarchy of equilibria on surfaces is established by the difference in the speed of transport processes along step edges, on terraces and across step edges. The carriers of the transport processes are single atoms (or molecules for molecular crystals), but also single atom vacancies. Occasionally the random motion of units of several atoms may contribute to the mass transport, however in most cases the contribution of adatoms or single atom vacancies prevails by far. The transport is driven by local gradients in the chemical potential of transporting species. For the transport along a step edges the local chemical potential is given by the curvature of the steps. Diffusion along steps can thus be described by considering the local curvature and certain transport coefficients (Sect. 10.2). For the purpose of a quantitative description of atom transport across terraces in the continuum approach one needs to know the chemical potential of the transporting species on terraces. In most practical cases, the concentration of the species is very small, so that interactions between them can be neglected. The diffusing species then form a dilute 2D, and therefore non-interacting lattice gas (cf. Sect. 5.4). Each site on the surface may or may not be occupied by the defect; hence, one has occupation numbers 0 and 1. The appropriate statistics is the Fermi-statistics. By inverting (3.7), one obtains the chemical potential as a function of the fractional coverage $\Theta$ of a site as

$$\mu = \mu_0 + k_B T \ln[\Theta/(1-\Theta)] \cong \mu_0 + k_B T \ln \Theta . \tag{4.71}$$

Here, $k_B$ is the Boltzmann constant and $\mu_0$ is the ground state energy of the defect. In order to calculate the equilibrium coverage $\Theta_{eq}$ we need to recall the special property of a kink site, namely that it repeats itself after removing or adding an atom to a step the kink (Fig. 4.19). The equilibrium coverage is therefore given by

$$\Theta_{eq} = e^{-\frac{W_d}{k_B T}} \qquad (4.72)$$

with $W_d$ the work required to generate the defect from the kink site. The work is equal to the change in the appropriate thermodynamic potential of the surface. For uncharged surfaces in vacuum, this is the change in the Helmholtz free energy, which can be approximated by the change in energy, such as can be calculated in total energy calculations. For surfaces in an electrolyte environment $W_d$ is a different quantity (Sect. 4.3.6). It is useful to consider the order of magnitude of the equilibrium coverage $\Theta_{eq}$. The adatom formation energy on Cu and Ag surfaces is about 0.7 eV, hence $\Theta_{eq} \cong 10^{-12}$ at 300K. The neglect of interactions in (4.71) is thus well justified. Despite the smallness of the equilibrium concentration, transport process can be quite rapid on these surfaces even around room temperature.

## 4.3.5 Steps on Charged Surfaces

The surface tension of stepped surfaces differ from the flat surface for two reasons: one is the additional Helmholtz free energy required to generate steps on surfaces. A second reason, which comes to bear only for charged surfaces, is that surface steps carry a dipole moment with the positive end pointing outwards. The dipole moment originates from the non-perfect screening of the sharp contour of the positively charged ion cores. This effect, the Smoluchowski effect [4.29] was discussed in Sect. 3.2.1. The step dipole moments reduce the work function of vicinal surfaces, and, since a shift in the work function causes a corresponding shift in the potential of zero charge (4.24), steps shift the potential of zero charge towards negative potentials. The relation between the shift in the work function and the step dipole moment is

$$\Delta\Phi / e = \phi_{pzc,\theta} - \phi_{pzc,0} = -\frac{p_z}{\varepsilon_0\, a_\parallel\, L}, \qquad (4.73)$$

in which $p_z$ denotes the dipole moment per step atom, $a_\parallel$ is the length of an atomic unit at the steps (the atom diameter for a densely packed step), $\varepsilon_0$ is the absolute dielectric permeability, and $L$ is the (mean) distance between steps.

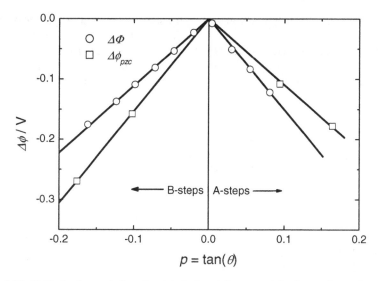

**Fig. 4.20.** Shifts in the work function (circles) and the potential of zero charge (squares) for stepped Au(111) surfaces [4.30, 31].

**Table 4.1.** Dipole moments per step atom $p_z$ for steps on silver, gold and platinum surfaces. Data refer to a non-aqueous electrolyte, an aqueous electrolyte and to surfaces in vacuum.

| Surface | Step type | $p_z/e\text{Å}$ | Environment | Reference |
|---------|-----------|-----------------|-------------|-----------|
| Ag(111) | [1$\bar{1}$0] (111), B-Step | 0.0138 | non-aqueous | [4.32] |
| Ag(111) | [1$\bar{1}$0] (100), A-Step | 0.0238 | non-aqueous | [4.32] |
| Ag(100) | [1$\bar{1}$0] (111) | 0.0054 | non-aqueous | [4.32] |
| Au(111) | [1$\bar{1}$0] (111), B-Step | 0.06 | aqueous | [4.31] |
| Au(111) | [1$\bar{1}$0] (111), B-Step | 0.041 | vacuum | [4.30] |
| Au(111) | [1$\bar{1}$0] (100), A-Step | 0.042 | aqueous | [4.31] |
| Au(111) | [1$\bar{1}$0] (100), A-Step | 0.056 | vacuum | [4.30] |
| Pt(111) | [1$\bar{1}$0] (111), B-Step | 0.184 | vacuum | [4.30] |
| Pt(111) | [1$\bar{1}$0] (100), A-Step | 0.094 | vacuum | [4.30] |

The dipole moments of step atoms need not be the same in vacuum and in an electrolyte. For aqueous electrolytes, e.g., the arrangement of the dipolar water molecules around the step atoms may differ from the flat surface, which may cause either a reduction or an enhancement of the dipole moment. Figure 4.20 displays a comparison of the shift in the work function and the potential of zero charge as a function of the step density [4.31]. Interestingly the dipole moment of steps in an aqueous electrolyte is higher than for steps in vacuum in case of B-type steps, however lower in case of A-type steps. Table 4.1 summarizes a few data on the dipole moments of steps as determined from the shift in the *pzc* for silver surfaces in non-aqueous electrolyte, for gold surfaces in an aqueous electrolyte and from the change of the work function of gold and platinum in vacuum.

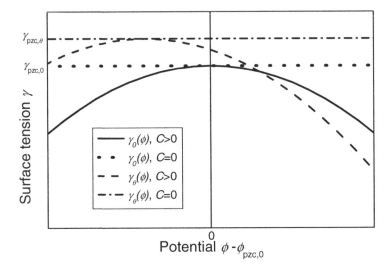

**Fig. 4.21.** Surface tensions (schematic) for flat and stepped surfaces, with and without charge ($C > 0$ and $C = 0$) as function of the potential (solid line: flat surface, $C > 0$; dotted line: flat surface, $C = 0$; dashed line: stepped surface, $C > 0$; dash-dotted line, $C = 0$).

Figure 4.21 illustrates the effect of the shift of the *pzc* on the surface tension. The surface tensions for a stepped and a flat surface in an electrolyte are plotted schematically as dashed and dotted lines, respectively. It is assumed that the interface capacity is the same in both cases, so that the parabolae (4.28) have the same curvature. Merely the potential of zero charge is shifted towards negative potentials for the stepped surface. The apex of the parabola representing the stepped surface is shifted upwards compared to the flat surface. The amount of this upwards shift must correspond to the difference in the specific Helmholtz free energy of the two surfaces which is the appropriate surface tension for uncharged surfaces. This can be seen by the following argument: Suppose the capacities $C$ are made zero by going to an infinitely dilute electrolyte, then the parabolas degenerate to horizontal lines through the apices. With the capacity equal zero, the surfaces become uncharged, regardless of the potential and the difference in the surface tension is the difference in the specific Helmholtz free energy.

For an arbitrary potential $\phi$ and capacitance $C$, the step line tension is the difference between surface tension of the stepped and the flat surface multiplied with the length $L$ between the steps. Since we are interested in the line tension of isolated steps, we take the difference in the limit of large $L$.

$$
\begin{aligned}
\beta(\phi) &= \lim_{L\to\infty} L\left\{ \gamma_{\text{pzc},\theta} - \int_{\phi_{\text{pzc},\theta}}^{\phi}\sigma_\theta(\phi')\mathrm{d}\phi' - \gamma_{\text{pzc},0} + \int_{\phi_{\text{pzc},0}}^{\phi}\sigma_0(\phi')\mathrm{d}\phi' \right\} \\
&= \lim_{L\to\infty} L\left\{ \beta_{\text{H}}/L - \int_{\phi_{\text{pzc},0}}^{\phi}(\sigma_\theta(\phi')-\sigma_0(\phi'))\mathrm{d}\phi' - \int_{\phi_{\text{pzc},\theta}}^{\phi_{\text{pzc},\theta}+\Delta\phi_{\text{pzc}}}\sigma_\theta(\phi')\mathrm{d}\phi' \right\} \qquad (4.74) \\
&= \beta_{\text{H}} - \lim_{L\to\infty} L\left\{ \int_{\phi_{\text{pzc},0}}^{\phi}(\sigma_\theta(\phi')-\sigma_0(\phi'))\mathrm{d}\phi' \right\}
\end{aligned}
$$

The second integral vanishes in the large $L$ limit since its value is proportional to $(\Delta\phi_{\text{pzc}})^2 \propto L^{-2}$. With (4.74) the potential dependence of the step line tension is expressed in terms of the potential dependence of the difference in the charge densities on the stepped and flat surfaces. A significant contribution to the difference in the charge densities, which exists under all circumstances, arises from the shift in the *pzc* by $\Delta\phi_{\text{pzc}}$. In addition to the effect of the *pzc*-shift, the charge density on stepped surfaces may differ because of a change of the interface capacity due to the presence of steps, and because of a specific adsorption of ions, which may adsorb in different quantities and at different potentials at step sites compared to the flat surface. The effect on the capacity has again two contributions; one is due to the enhanced polarizability of the substrate electron system near steps and tends to increase the interfacial capacitance [4.33]. The capacity is reduced by the screening of the part of the terrace surface adjacent to the steps due to the geometric structure of the step [4.34]. Both effects are very small and partly compensate each other so that a variation of the capacity due to steps can be neglected to a good approximation [4.33]. In the absence of specific adsorption or in case the specific adsorption at steps is as on the flat surface, the shift in *pzc* is the therefore prevailing effect on the charge density. Figure 4.22 illustrates the variation in the charge density in such a case. The charge density for the stepped surface with the *pzc* shifted by $\Delta\phi$ can be written as an expansion in powers of $\Delta\phi$ in terms of the charge density of the flat surface.

$$
\begin{aligned}
\sigma_\theta(\phi) &= \sigma_0(\phi+\Delta\phi) + \Delta\sigma_\pm(\phi) \\
&= \sigma_0(\phi) + \frac{\partial\sigma_0}{\partial\phi}\Delta\phi... + \Delta\sigma_\pm(\phi)
\end{aligned} \qquad (4.75)
$$

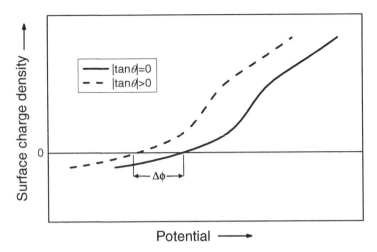

**Fig. 4.22.** Surface charge density vs. potential for flat and stepped surfaces (schematic). A rise in the slope at a certain potential and a leveling off is typical if ions are specifically adsorbed in that potential range. The dashed line representing the stepped surface is merely shifted horizontally, which is equivalent to the assumption that the specific adsorption of ions is not altered near step sites.

The term $\Delta\sigma_\pm(\phi)$ stands as a reminder that there could be further contributions due to specific adsorption at step sites or a change in the capacity due to a different polarizability of the stepped interface. Neglecting those one obtains for the step line tension the remarkably simple equation

$$\beta(\phi) = \beta_{\mathrm{H}} - \lim_{L\to\infty} L\{\Delta\phi\,\sigma_0(\phi)\} = \beta_{\mathrm{H}} - \frac{p_z}{\varepsilon_0 a_\parallel}\sigma_0(\phi)\,. \tag{4.76}$$

The potential dependence of the line tension can therefore be expressed in terms of the step dipole moment $p_z$ (which may be potential dependent also) and the charge density of the flat surface. For small deviations of the potential from $\phi_{pzc}$, the contribution arising from $\Delta\sigma_\pm(\phi)$ can be expressed in terms of the difference in the capacitance per step length $\Delta\tilde{C}$ of the stepped and non-stepped surface as $\Delta\tilde{C}(\phi-\phi_{\mathrm{pzc}})^2/2$. The latter contribution is, however, small, at least in the absence of specific adsorption and for low to moderate electrolyte concentrations. This was confirmed by a microscopic theory in which the charge distribution was calculated by a simultaneous solution of the jellium model for the solid and the Poisson-Boltzmann equation for the electrolyte [4.33]. The charm of equation (4.76) lies in the fact that the surface charge of the flat surface at any given potential and the dipole moment $p_z$ (via the change in *pzc*) can be measured independently, so that the potential dependence of the line tension can be determined from independent experimental data. If the surface charge is measured for both the stepped and the non-stepped surface, the potential dependence of the step line tension can be determined from the experiment even without any further assumption (first line in (4.74)).

An interesting aspect of (4.76) is that it predicts the existence of an *electrochemical roughening transition* for positive potentials when $\beta(\phi) = 0$. By taking approximate numbers for $\beta_{pzc}a_\parallel$ from clean surfaces in vacuum as 0.22 eV, 0.25 eV, and 0.34 eV one calculates critical charges of +0.37, +0.17, and +0.07 $e$/atom for the (111) surfaces of Ag, Au, and Pt, respectively. These charges are within the experimentally accessible range.

In section 4.3.1 we have considered the expansion of the projected surface tension in powers of $p = \tan\theta$ and the effect of the various terms on the crystal equilibrium shape. Here, we show that charging of surfaces necessarily leads to an effective attractive step-step interaction term of the $\gamma_2$-type. The projected potential dependent surface tension is

$$\gamma_p(\theta,\phi) \equiv \frac{\gamma(\theta,\phi)}{\cos\theta} = \frac{1}{\cos\theta}\left\{\gamma_{pzc,\theta} - \int_{\phi_{pzc,\theta}}^{\phi}\sigma_\theta(\phi')\,d\phi'\right\}. \tag{4.77}$$

By inserting (4.75) one obtains

$$\gamma_p(\theta,\phi) = \gamma_{pzc,0} + \frac{\beta_H}{h}p - \frac{1}{\cos\theta}\left\{\int_{\phi_{pzc,\theta}}^{\phi}\left(\sigma_0(\phi') + \frac{\partial\sigma_0}{\partial\phi'}\Delta\phi_{pzc}\right)d\phi'\right\}$$

$$\cong \gamma_{pzc,0} + \left(\frac{\beta_H}{h} - \frac{p_z}{\varepsilon_0 a_\parallel h}\sigma_0(\phi)\right)p$$

$$-\left(1+\frac{1}{2}p^2\cdots\right)\left\{\frac{1}{2}C_{pzc}\left(\frac{p_z}{\varepsilon_0 a_\parallel h}\right)^2 p^2 + \int_{\phi_{pzc,0}}^{\phi}\sigma_0(\phi')d\phi'\right\} \tag{4.78}$$

$$\gamma_p(\theta,\phi) = \gamma_0(\phi) + \frac{\beta(\phi)}{h}p - \frac{1}{2}\left\{C_{pzc}\left(\frac{p_z}{\varepsilon_0 a_\parallel h}\right)^2 + \int_{\phi_{pzc,0}}^{\phi}\sigma_0(\phi')d\phi'\right\}p^2\cdots$$

Here, $\gamma(\phi)$ and $\beta(\phi)$ are the surface tension of the flat surface and the step line tension (4.76), respectively, $h$ is the step height and $C_{pzc}$ is the (mean) capacitance in the range $\phi_{pzc,\theta} < \phi < \phi_{pzc,0}$. The last term in (4.78) has a negative sign and is proportional to $p^2$. It therefore corresponds to an attractive step-step interaction proportional to $L^{-1}$. According to (4.55) the equilibrium crystal shapes in an electrolyte has therefore always a *finite contact angle* between the facet and the rough surface! Consider Ag(111) surfaces as an example: the repulsive interaction term

$\gamma_3$ is about 1.75 meV/$\text{Å}^2$ [4.21]. With the data from Table 4.1 one calculates $\gamma_2 = -1.1$ meV/$\text{Å}^2$ and $-3.4$ meV/$\text{Å}^2$ and thus contact angles of 33° and 62°, for A- and B-steps respectively. The contact angle increases with the potential further away from *pzc* when the second contribution to the $p^2$-term gains importance.

Interactions between steps are observable directly via the step-step distance distribution on vicinal surfaces. Vicinal surfaces in vacuum display a Gaussian distribution of step-step distances due to repulsive interactions. The width of the Gaussian reflects the strength of the interaction (Sect. 5.3.2). Attractive interactions, on the other hand, are evidenced by the formation of step bunches. The formation of step bunches with time on a surface, which initially, due to the surface preparation, displayed a regular step distance distribution, has been observed on Ag(19 19 17) surfaces in a 1 mM $CuSO_4$+0.05 $H_2SO_4$ electrolyte [4.35].

## 4.3.6 Point Defects on Charged Surfaces

The hierarchical structure of equilibration processes on surfaces brings about local varying chemical potentials, which act as driving forces for ripening process on surfaces. Atom transport in these processes is either via single adatoms or via single atom vacancies, rarely also via small clusters as the diffusion species. The thermodynamics of these defects is therefore important for the potential dependence of ripening effects as well as for surface diffusion processes in general. It is a common observation that the speed of surface self-diffusion processes increases at positive potentials, occasionally also at negative potentials. The term *electrochemical annealing* has been cast to stress an apparent analogy to the speed-up of equilibration processes at higher temperatures. As is shown in this section the expression is misleading insofar as the increase in the speed is due to a lowering of the activation barriers rather than being akin to a raise in temperature.

The activation energies in surface transport process involve the work required for the formation of the atom transporting species and the activation energies for diffusion. For uncharged surfaces in vacuum the formation energy is the Helmholtz free energy $\Delta F_{\text{def}}$ required for the creation of an adatom or a vacancy from a kink site (Sect. 4.3.4, see also Fig. 4.19). For charged surfaces kept at constant potential, the making of, e.g., an adatom from a kink does not change the charge distribution at the step, as the kink reproduces itself in the creation of an adatom (Fig. 4.19). However, the newly created adatom on the surface gives rise to a flow of charge in response to the locally varied potential around the adatom. The effect of this variation in the charge distribution on the thermodynamics of defect creation is calculated by the same method as for the step line tension on charged surfaces, namely as the difference of the surface tension of surfaces with and without the defect. By writing (4.74) for point defects rather than for a line defect one obtains

$$
\begin{aligned}
E_{\text{def}}(\phi) &= \lim_{\rho \to 0} \frac{1}{\rho} \left\{ \gamma_{pzc,\rho} - \int_{\phi_{pzc,\rho}}^{\phi} \sigma_\rho(\phi')\mathrm{d}\phi' - \gamma_{pzc,0} + \int_{\phi_{pzc,0}}^{\phi} \sigma_0(\phi')\mathrm{d}\phi' \right\} \\
&= \Delta F_{\text{def}} - \lim_{\rho \to 0} \frac{1}{\rho} \left\{ \int_{\phi_{pzc,0}}^{\phi} (\sigma_\rho(\phi') - \sigma_0(\phi'))\mathrm{d}\phi' \right\} .
\end{aligned}
\tag{4.79}
$$

With the expansion (4.75) the defect energy becomes

$$
E_{\text{def}}(\phi) = \Delta F_{\text{def}} - \frac{p_z}{\varepsilon_0} \sigma_0(\phi) .
\tag{4.80}
$$

Here, $p_z$ is the dipole moment of the adatom or the vacancy, and $\sigma_0(\phi)$ is the surface charge density of the surface without defects as before. We can also immediately write down the potential dependence of an activation energy, e.g. in a diffusion process as

$$
E_{\text{act}}(\phi) = E_{\text{act}}(\phi_{pzc}) - \frac{\Delta p_z}{\varepsilon_0} \sigma_0(\phi)
\tag{4.81}
$$

with $\Delta p_z$ the difference between the dipole moment in the activated transition state and the initial state. As the surface charge depends linearly on the potential $\phi$ (at least in a small range) the two equations (4.80) and (4.81) state that to first order formation energies of defects and activation energies of defect migration depend linearly on the potential. All processes of surface self-diffusion should therefore increase exponentially for positive potentials. This is the physical reason for a multitude of qualitative observations on electrochemical surfaces, which have been summarized under the name *electrochemical annealing*. A few quantitative studies which confirm the exponential increase of coarsening processes have also been reported lately [4.21].

## 4.3.7 Equilibrium Fluctuations of Line Defects and Surfaces

This section considers the spatial equilibrium fluctuations of steps or other linear systems and of rough surfaces for which the surface tension is a continuously differentiable function. Temporal aspects of these fluctuations will be considered in Sect. 10.5. We begin with the fluctuations of a line, and generalize the result to the two-dimensional case later. The line profile is described as a position $x(y)$ so that $y$ represents the mean direction of the line. The work associated with the fluctuations of the line profile is

$$\Delta E = \tilde{\beta}\left(\int_0^{L_0} ds - L_0\right) \cong \tilde{\beta}\int_0^{L_0}\left(\frac{dx}{dy}\right)^2 dy , \tag{4.82}$$

in which $\tilde{\beta}$ is the line stiffness (4.67) for the $y$-direction and $L_0$ is the length along this direction. To avoid finite size effects $L_0$ can be chosen as the length unit of periodic boundary conditions. The position $x(y)$ is then expanded into the Fourier-series of partial waves of wave vector $q = 2\pi n/L_0$ with an integer number running from $-\infty$ to $+\infty$

$$x(y) = \frac{1}{\sqrt{L_0}}\sum_q \eta_q e^{iqy} . \tag{4.83}$$

In thermal equilibrium, each capillary wave of amplitude $\eta_q$ carries the energy $k_B T/2$. The thermal fluctuations of a line can therefore be calculated by summing up the contribution of all capillary waves. Because of

$$\sum_q e^{i(q-q')y} = \delta_{qq'} \tag{4.84}$$

the expansion has the inversion

$$\eta_q = \frac{1}{\sqrt{L_0}}\int_0^{L_0} dy\, x(y)\, e^{-iqy} . \tag{4.85}$$

After differentiation of (4.83), insertion into (4.82), by using the identity

$$\int_0^{L_0} dy\, e^{i(q+q')y} = L_0 \delta_{q,-q'} \tag{4.86}$$

and

$$\eta_q^* = \eta_{-q} , \tag{4.87}$$

which is a consequence of the reality of $x(y)$, one obtains the energy in terms of the amplitudes of the partial waves as

$$\Delta E = \frac{1}{2}\tilde{\beta}\sum_q |\eta_q|^2 q^2 . \tag{4.87}$$

According to the equipartition principle each capillary wave contributes $k_B T/2$ to the total energy so that

$$\left|\eta_q\right|^2 = \frac{k_B T}{\tilde{\beta} q^2}. \tag{4.88}$$

We are now interested in the mean square deviation of the amplitude $x(y)$

$$G(y, y') = \left\langle \left(x(y) - x(y')\right)^2 \right\rangle = 2\left\langle \left(x(y)\right)^2 \right\rangle - 2\left\langle x(y)x(y') \right\rangle. \tag{4.89}$$

$G(y, y')$ is frequently called the correlation function of the line system, although, strictly speaking, the correlation function is only the second term on the right hand side of (4.89). The first term on the right hand side is obtained by inserting the Fourier-expansion (4.83) and by using (4.84) as

$$2\left\langle \left(x(y)\right)^2 \right\rangle = \frac{2}{L_0} \sum_q \left|\eta_q\right|^2 = \frac{2k_B T}{\tilde{\beta} L_0} \sum_q \frac{1}{q^2}. \tag{4.90}$$

The second term is

$$2\left\langle x(y)x(y') \right\rangle = \frac{2}{L_0} \left\langle \sum_{qq'} \eta_q e^{iqy} \eta_{-q'} e^{-iq'y'} \right\rangle$$
$$\equiv \frac{2}{L_0} \left\langle \sum_{qq'} \eta_q \eta_{-q'} e^{i(q-q')y} e^{iq'(y-y')} \right\rangle, \tag{4.91}$$

which by using (4.84) becomes

$$2\left\langle x(y)x(y') \right\rangle = \frac{2}{L_0} \left\langle \sum_q \left|\eta_q\right|^2 e^{iq(y-y')} \right\rangle = \frac{2k_B T}{\tilde{\beta} L_0} \sum_q \frac{1}{q^2} e^{iq(y-y')} \tag{4.92}$$

With the standard substitution of the discrete variable $q$ by a continuous variable $k$

$$\frac{1}{L_0} \sum_q \rightarrow \frac{1}{2\pi} \int dk \tag{4.93}$$

one obtains

$$\left\langle \left(x(y) - x(y')\right)^2 \right\rangle = \frac{2k_B T}{\tilde{\beta}} \frac{1}{2\pi} \int_{-\infty}^{+\infty} dq \frac{1}{q^2} \left(1 - \cos q(y - y')\right) \tag{4.94}$$

By solving the integral one arrives at the desired expression for the mean square fluctuations

$$\left\langle \left( x(y) - x(y') \right)^2 \right\rangle = \frac{k_B T}{\tilde{\beta}} \left| y - y' \right|. \qquad (4.95)$$

The term $k_B T / \tilde{\beta}$ is called the *diffusivity* of the line. The spatial fluctuations of a linear system, e.g. a step, therefore increase linearly with the distance from the origin, and they are inversely proportional to the stiffness $\tilde{\beta}$, which is an intuitively appealing result.

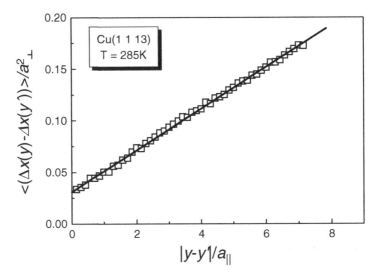

**Fig. 4.23.** Spatial correlation function for steps on the Cu(1 1 13) surface at 285 K. In order to avoid the problem of the absolute reference direction the difference in the position of two adjacent steps $\Delta x(y)$ is plotted. A kink energy of $\varepsilon_k = 0.128$ eV is obtained from the slope [4.36]. The offset at $\left| y - y' \right| = 0$ is due to random noise in the STM-image.

Figure 4.23 shows experimental data of the step correlation function on a Cu(1 1 13) surface which confirms the linear dependence [4.36]. The steps on this surface vicinal to the (100) plane run along the densely packed [011]-direction. In this case, the diffusivity has a straightforward microscopic interpretation (see also Sect. 5.2.1). At any atom position there is a finite probability $P_k$ for having a kink in the positive or negative direction. If the temperature is not too high $P_k \ll 1$ and kink length larger than one atom are thereby excluded. The probability to find a kink of one atom length unit $a_\perp$ is

$$P_k = 2e^{-\frac{\varepsilon_k}{k_B T}} \tag{4.96}$$

with $\varepsilon_k$ the energy required to make a kink in the step. The factor of 2 arises because kinks can be in the positive or negative direction. Due to these kinks the step engages in a random walk concerning the $x$-direction and the mean square displacement in that random walk is

$$\left\langle (x(y) - x(y'))^2 \right\rangle / a_\perp^2 = P_k |y - y'| / a_\parallel . \tag{4.97}$$

Here, $a_\parallel$ is the atomic length unit along the $y$-direction (for the (100) surface $a_\parallel = a_\perp$). The line stiffness of a step on a (100) surface along the [011]-direction is therefore

$$\tilde{\beta} = \frac{k_B T a_\parallel}{2 a_\perp^2} e^{\frac{\varepsilon_k}{k_B T}} . \tag{4.98}$$

For low temperatures, the stiffness is much higher than the line tension $\beta$ (4.67) and diverges as $T \rightarrow 0$ K. For the equilibrium shape of an island this means that the curvature of steps along the [110]-direction approaches zero (4.68, 4.69).

Steps and domain walls are frequently pinned at point defects, e.g. impurity atoms on the surface. In that case, one is interested in the fluctuations of a line over a distance $L$ between two fixed points. Obviously, the fluctuations are zero at the pinning defects and have their maximum at midpoint between the pinning centers. Rather than the magnitude of the fluctuation as a function of the distance, one takes the average of the fluctuations over the entire length $L$ as a measure of the intensity of the fluctuations. The natural capillary wave expansion for this problem is the expansion into functions $\sin qx$ with $q = n\pi/L$ with $n = 1,2,3...$

$$x = \frac{2}{\sqrt{L}} \sum_{n>0} \eta_n \sin(\pi n y / L) . \tag{4.99}$$

Writing the fluctuation energy in terms of the capillary waves and applying the equipartition principle now leads to amplitudes $\eta_n$

$$\eta_n^2 = \frac{k_B T L^2}{2\pi^2 \tilde{\beta}} \frac{1}{n^2} . \tag{4.100}$$

The mean square of the fluctuations is then

$$\left\langle x^2 \right\rangle = \frac{1}{L} \int_0^L x^2(y)\,dy = \frac{k_B T}{6\tilde{\beta}} L \tag{4.101}$$

With this result, the line tension of step can be determined from the fluctuations of steps between two pinning centers. The method has been applied to determine the step stiffness on Si(111) surfaces at 900 °C [4.37].

Equation (4.94) can be generalized to the two-dimensional spatial fluctuations of a thermodynamically rough surface. The profile of the surface be described by the height function $h(r)$ with $r$ a two-dimensional vector. The generalization of (4.94) to the 2D-case is

$$\left\langle (h(r) - h(r'))^2 \right\rangle = \frac{2k_B T}{\tilde{\gamma}} \frac{1}{(2\pi)^2} \int_{-\infty}^{+\infty} dq \frac{1}{q^2} (1 - \cos q \cdot (r - r')) \tag{4.102}$$

with $\tilde{\gamma}$ the surface stiffness. By introducing polar coordinates $dq = q\,dq\,d\varphi$ this becomes

$$\left\langle (h(r) - h(r'))^2 \right\rangle = \frac{k_B T}{\tilde{\gamma}} \frac{1}{\pi} \int_0^{q_{max}} dq \frac{1}{q} (1 - J_0(q|r - r'|))$$

$$= \frac{k_B T}{\tilde{\gamma}} \frac{1}{\pi} \int_0^{|r-r'|q_{max}} \frac{dx}{x} (1 - J_0(x)) \tag{4.103}$$

in which $J_0$ is the Bessel function and $q_{max}$ is a cut off wave vector which is of the order of the inverse of an atomic distance $a$. The integrand is free of poles in the entire range. For small arguments, $J_0$ approaches one and the integrand vanishes. We are interested in the correlation function for large distances, in particular in the mathematical form of the divergence. In the limit of large arguments $x = q|r-r'|$, the Bessel function $J_0$ oscillates symmetrically around zero and its contribution to the integral vanishes therefore. The height correlation function therefore diverges as

$$\left\langle (h(r) - h(r'))^2 \right\rangle = \frac{k_B T}{\tilde{\gamma} \pi} \ln \left( \frac{|r - r'|}{a} \right) \tag{4.104}$$

for large distances. We see that for a 2D-system the correlation function diverges logarithmically, significantly slower than for a linear system. The atomic distance $a$, and thus the absolute value of the roughness, remains undefined within the continuum theory. From the physics point of view, a cut-off wave-vector should have been introduced into the continuum theory of linear fluctuations as well. It is, however, mathematically not necessary there since the integral converges nicely. Furthermore, the microscopic result from random walk theory yields exactly the same result for all length scales. Equation (4.97) therefore holds for all distances on the y-axis.

## 4.3.8. Island Shape Fluctuations

This section considers the thermal fluctuations of steps forming the perimeters of homoepitaxial monolayer islands or any other closed loops of a linear boundary that possess a line tension. An example for the shape fluctuations of a monolayer island is shown in Fig. 4.24 which displays the equilibrium shape on a Cu(100) surface at $T = 408$ K [4.38] (solid line) together with the shape observed in a single STM-image (dashed line). The systematic measurement of shape fluctuations provides an elegant method to determine absolute values for the step line tension. The theory behind the method is akin to the theory of the fluctuations of linear systems; it deviates, however, in detail from the theory for infinitely long lines or lines extending from point A to a point B. The differences arise from different boundary conditions, from the different energy functional, and from a different normal mode expansion. The boundary condition here is that the number of atoms in the islands stays constant. The energy functional is

$$E = \oint \beta(n)\mathrm{d}s \,, \tag{4.105}$$

in which $n$ is a vector normal to step orientation. For simplicity we consider the mathematical simple case of an orientation independent line tension $\beta(n) = \beta$. The theory nevertheless applies also to equilibrium shapes such as displayed in Fig. 4.24 as long as the orientation dependence of $\beta$ is not too strong (for highly anisotropic systems see [4.39]). For an angle independent line tension $\beta$, the equilibrium shape is a circle of radius $R$. Because of the fluctuations, the actual radius in any instantaneous image $j$ of the island is different at each point of the perimeter, which is denoted by a radius $r(\theta, j)$ with $\theta$ the polar angle. The relative variation $g(\theta, j)$ defined as

$$g(\theta, j) = \frac{r(\theta, j) - R}{R} \tag{4.106}$$

can be expanded in a Fourier series

$$g(\theta, j) = \sum_{n=-\infty}^{\infty} g_n(j)\mathrm{e}^{in\theta} \tag{4.107}$$

with $g_n(j)$ the Fourier coefficients.

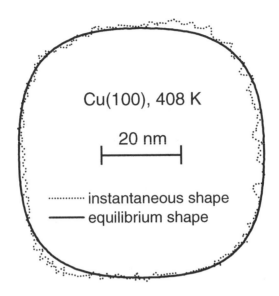

**Fig. 4.24.** Equilibrium shape of an island of one monolayer of atoms on a Cu(100) surface at a temperature of $T = 408$K (solid line). The equilibrium shape is obtained experimentally by averaging over several hundred STM-images. In each of these images, the shape of the island deviates from the equilibrium shape due to thermal fluctuations. One individual shape is shown as a dashed line [4.38].

The side condition of constant area reads then

$$R^2 = \frac{1}{2\pi} \int_0^{2\pi} r^2(\theta, j) \, d\theta = R^2 + \frac{R^2}{2\pi} \int_0^{2\pi} d\theta \left[ 2g(\theta, j) + g^2(\theta, j) \right]. \quad (4.108)$$

The identity of the left and right side of the equation requires that

$$\int_0^{2\pi} g(\theta, j) \, d\theta = -\frac{1}{2} \int_0^{2\pi} g^2(\theta, j) \, d\theta. \quad (4.109)$$

The length element d$s$ in (4.105) expressed in terms of $g(\theta, j)$ is

$$ds = R \, d\theta \left( \left[ 1 + g(\theta, j) \right]^2 + g'^2(\theta, j) \right)^{\frac{1}{2}} \quad (4.110)$$

in which $g' = \partial g / \partial \theta$. For small fluctuations g<<1 and d$s$ is approximately

$$ds \cong R\,d\theta + \left( g(\theta, j) + \frac{1}{2}g'^2(\theta, j) \right) d\theta \tag{4.111}$$

With (4.109) and (4.111) the energy (4.105) is

$$E = E_0 + \frac{\beta R}{2} \int_0^{2\pi} \left[ g'^2(\theta, j) - g^2(\theta, j) \right] d\theta . \tag{4.112}$$

The second term is the additional energy due to excursions from the equilibrium shape. This term is now expanded into the normal modes (4.107).

$$\int_0^{2\pi} g^2(\theta, j)\,d\theta = \sum_n \sum_m g_n g_{-m} \int_0^{2\pi} e^{i(n-m)\theta}\,d\theta$$

$$= 2\pi \sum_n \sum_m g_n g_{-m} \delta_{nm} = 2\pi \sum_n |g_n|^2 \tag{4.113}$$

$$\int_0^{2\pi} g'^2(\theta, j)\,d\theta = 2\pi \sum_n n^2 |g_n|^2$$

We have made use of the fact that $g(\theta, j)$ is a real function so that $g_{-n} = g_n^*$. In terms of the amplitudes of the normal modes the energy E is

$$E = E_0 + \pi \beta R \sum_n |g_n(j)|^2 (n^2 - 1) \tag{4.114}$$

The contribution of the term with $n = 0$ to the energy is zero. The reason is that this term does not represent a shape fluctuation but rather a motion of the center of mass, which does not change the energy. The fluctuation associated with $g_0 \neq 0$ corresponds to the Brownian motion of the entire island. As before, the equipartition principle requires that in the mean over a large ensemble each of the normal modes carries the energy $k_B T/2$ so that

$$\left\langle |g_n|^2 \right\rangle = \frac{k_B T}{2\pi \beta R(n^2 - 1)} \qquad n \neq 0 \tag{4.115}$$

The experimental data concern the ensemble average of the fluctuation function $G(j)$ defined as

$$G(j) = \frac{R^2}{2\pi} \int_0^{2\pi} g^2(\theta, j)\,d\theta = R^2 \sum_n |g_n(j)|^2, \quad n \neq 0 . \tag{4.116}$$

The term with $n = 0$ is excluded if the origin of the radius $r(\theta, j)$ lies in the center of gravity so that the integral $\int r(\theta, j) \, d\theta$ vanishes. With (4.115) the ensemble average of the fluctuation function $<G>$ becomes

$$\left\langle G(j) \right\rangle_j = \frac{3k_B T}{4\pi\beta} R \qquad (4.117)$$

Thus, $\left\langle G(j) \right\rangle_j$ is proportional to the mean radius $R$ of the island. A similar equation can be derived for non-circular equilibrium shapes [4.40]. The numerical factor changes slightly and the line tension $\beta$ in (4.117) is to be replaced by a complicated mean over the angle $\theta$. The mean can be disentangled to obtain the full dependence of the line tension on the angle with the help of the equilibrium shape (see [4.40, 41] for details). As experimental data on the perimeter position have a certain noise, which is independent of the radius R, the line tension $\beta$ is best obtained from the slope of a plot of $\left\langle G(j) \right\rangle_j$ versus the radius $R$, or versus the product of the radius $R$ and the temperature $T$.

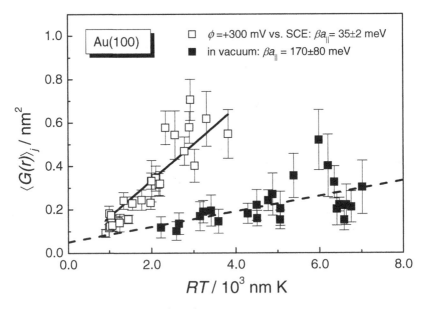

**Fig. 4.25.** Mean square fluctuations $\left\langle G(r) \right\rangle$ of Au(100) islands in a 50 mM $H_2SO_4$ electrolyte at a potential of +300 mV vs. SCE and of the same surface in vacuum (open and solid squares, respectively).

Figure 4.25 shows two examples of such plots, one referring to Au(100) in vacuum (solid squares), the other to the same surface in an $H_2SO_4$ electrolyte (open squares). The fluctuations are larger in the electrolyte and the line tension is therefore lower. This may be partly due to the fact that the Au(100) is unreconstructed in the electrolyte at +300 meV and reconstructed in vacuum. The largest contribution to the difference, however, is presumably owed to the positive surface charge in the electrolyte (Sect. 4.3.5). A summary of step line tensions obtained from island shape fluctuations is shown in Table 4.2.

**Table 4.2.** Step line tensions obtained from island shape fluctuations.

| Surface | $\beta$ | comment | reference |
|---------|---------|---------|-----------|
| Cu(100) | 220 ±11 meV/atom | UHV | [4.38] |
| Cu(111) | 256 ±22 meV/atom | UHV | [4.38] |
| Ag(111) | 233 ±13 meV/atom | UHV | [4.38] |
| Au(100) | 170 +80/-17 meV/atom | UHV | [4.41] |
| Au(100) | 40-60 meV/atom | in 0.05M $H_2SO_4$ | [4.42] |
| TiN(111) | 210±40/290±60 meV/Å | strong anisotropy | [4.43] |

# 5. Statistical Thermodynamics of Surfaces

The advantage of the continuum thermodynamics of surfaces in the coarse-grained view is that the results are exact and model independent. On the other hand, the quantities defined in the coarse-grained description do not translate into an atomistic picture unless models for the surface structure and the interactions between the atoms are invoked. This section discusses the most important atomic scale models that are used in statistical thermodynamics. We begin with the definitions of the standard thermodynamic potentials within the framework of quantum statistics.

## 5.1 General Concepts

### 5.1.1 Internal Energy and Free Energy

The internal energy $U$ of system is by definition

$$U = \frac{\sum_i E_i e^{-\frac{E_i}{k_B T}}}{\sum_i e^{-\frac{E_i}{k_B T}}} \equiv k_B T^2 \frac{\partial}{\partial T} \ln \sum_i e^{-\frac{E_i}{k_B T}} . \tag{5.1}$$

The summation index $i$ counts all discernable quantum states of the system. Degenerate states are counted according to their degeneracy. In general, $i$ is a multiple infinite index number, and the sum (5.1) is an exceedingly complex one that can be solved only in a few simple cases. These cases and the results derived for the simple systems form the core of our understanding of the thermodynamic properties of large ensembles of atoms.

Phenomenological thermodynamics proves that the isothermal mechanical work $\delta W$ executed on a system (leaving the particle number, volume, charge, etc. constant) enhances the Helmholtz free energy $F$ of the system by the amount $\delta F = \delta W$. Phenomenological thermodynamics shows further that the free energy can be expressed in terms of the internal energy as

$$F = -T \int_0^T \frac{U(T')}{T'^2} \, \mathrm{d}T' . \tag{5.2}$$

Inserting (5.1) leads to the well-known result

$$F = -k_{\mathrm{B}} T \ln \sum_i \mathrm{e}^{-\frac{E_i}{k_{\mathrm{B}} T}} \equiv -k_{\mathrm{B}} T \ln Z \tag{5.3}$$

in which $Z$ is the *partition function* of the system with a fixed number of particles, the *microcanonical ensemble*.

### 5.1.2 Application to the Ideal Gas

The partition function is easily calculated for an ensemble of $N$ independent particles having merely translation degree of freedoms along the $x$, $y$, and $z$-axes. The energy levels of a free particle are

$$E = \frac{\hbar^2}{2m} (k_x^2 + k_y^2 + k_z^2) \tag{5.4}$$

where $k_x$, $k_y$, and $k_z$ are the components of the $k$-vector and $m$ is the mass of the particle. The system is assumed to be infinitely large. The eigenstates are nevertheless countable by assuming that the system is periodic with the periodicity length $L$ in all three directions.

$$k_{x,y,z} = 2\pi n_{x,y,z} / L \qquad -\infty < n_{x,y,z} < \infty \tag{5.5}$$

The partition function for one particle $Z_1$ is the product of the partition functions of the independent translations $Z_x$, $Z_y$, $Z_z$

$$Z_x = \sum_{k_x} \mathrm{e}^{-\frac{\hbar^2 k_x^2}{2mk_{\mathrm{B}}T}} = \sum_{-\infty}^{\infty} \mathrm{e}^{-\frac{h^2 n_x^2}{2mk_{\mathrm{B}}TL^2}} = \int_{-\infty}^{\infty} \mathrm{e}^{-\frac{h^2 x^2}{2mk_{\mathrm{B}}TL^2}} \, \mathrm{d}x = \frac{L}{h}(2\pi mk_{\mathrm{B}}T)^{1/2} \tag{5.6}$$

The partition function for $N$ independent particles is

$$Z = (Z_x Z_y Z_z)^N / N! = \left( V(2\pi mk_{\mathrm{B}}T / h^2)^{3/2} \right)^N / N! \tag{5.7}$$

in which $V = L^3$ is the volume. The division by $N!$ eliminates the identical configurations obtained by changing the enumeration of particles. With the Stirling-approximation

$$\ln N! \cong N \ln N - N \tag{5.8}$$

the free energy $F$ becomes

$$F = k_B T N \left( -1 + \ln \frac{N}{V} (h^2 / 2\pi m k_B T)^{3/2} \right) \tag{5.9}$$

Writing (5.9) in terms of the pressure $p$ of an ideal gas $p = N k_B T / V$ one obtains

$$F = k_B T N \left( -1 + \ln \frac{p}{k_B T} (h^2 / 2\pi m k_B T)^{3/2} \right). \tag{5.10}$$

If the particles are molecules then the partition functions for the rotational and vibrational degrees of freedom $Z_{rot}$ and $Z_{vib}$ contribute to the free energy. With the energy levels of the harmonic oscillator

$$E_i = (n + 1/2)\hbar\omega_i \tag{5.11}$$

the vibrational partition function becomes

$$Z_{vib} = \prod_i e^{-\hbar\omega_i / 2k_B T} (1 - e^{-\hbar\omega_i / k_B T})^{-1}. \tag{5.12}$$

The rotational energy levels are

$$E_{rot} = \frac{\hbar^2}{2I} J(J+1) \tag{5.13}$$

with $I$ the moment of inertia. Considering the $(2J+1)$ degeneracy of the rotational energy levels, the partition function is calculated to

$$Z_{rot} = 2 I k_B T / \hbar^2. \tag{5.14}$$

We finally calculate the chemical potential of the ensemble of independent particles, i.e. an ideal gas of molecules

$$\mu = \frac{\partial F}{\partial N}\bigg|_{V,T...} = k_B T \ln \frac{p}{k_B T} \left( \frac{h^2}{2\pi m k_B T} \right)^{3/2} \frac{1}{Z_{rot} Z_{vib}}. \tag{5.15}$$

Note that the derivative is to be taken at constant volume $V$, which means $\mu$ is the derivative of (5.9) with respect to $N$.

### 5.1.3 The Vapor Pressure of Solids

As a first application of (5.15), we calculate the equilibrium vapor pressure of a solid. The condition for equilibrium of two phases is that the chemical potentials be equal. We therefore equate the chemical potential of the vapor phase with the chemical potential of the solid. We consider the case of an elemental solid and assume that the vapor phase consist of single atoms, which is typical for most metals. In Sect. 1.3.2, it was argued that the cohesive energy of a solid is the binding energy of an atom in a *Halbkristallage*, a kink site in a densely packed row of atoms. The energy of an atom in that position with respect to the ground state in vacuum is therefore $-E_{coh}$. The chemical potential of the atom in the kink site has a contribution from the vibrational partition function. For simplicity, we replace the vibration spectrum by three Einstein-oscillators per atom with a frequency $\omega = k_B \Theta_D / \hbar$. In the high temperature limit the chemical potential of the solid $\mu_s$ with reference to the ground state level of the vapor phase is

$$\mu_s = -E_{coh} - 3k_B T \ln(T / \Theta_D). \tag{5.16}$$

Equating the chemical potentials of the solid (5.16) and the vapor phase (5.15) yields the equilibrium vapor pressure $p_{vap}$

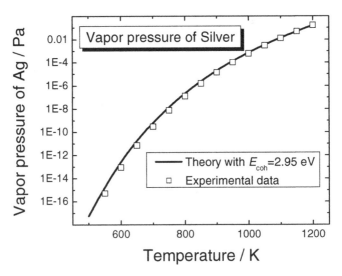

**Fig. 5.1.** Vapor pressure of silver according to (5.17) ($E_{coh} = 2.95$ eV, $\Theta_D = 225$ K) and experimental data (solid line and open circles, respectively) [5.1].

$$p_{\text{vap}} = \left( \frac{2\pi m k_B T}{h^2} \right)^{3/2} k_B T \left( \frac{\Theta_D}{T} \right)^3 e^{-\frac{E_{\text{coh}}}{k_B T}} \tag{5.17}$$

Fig. 5.1 shows the vapor pressure of silver (solid line) together with the experimental numbers. Equation (5.17) may be compared with the Clausius-Clapeyron equation of phenomenological thermodynamics

$$\frac{1}{p} \frac{\partial p}{\partial T} = \frac{Q}{k_B T} \tag{5.18}$$

in which $Q$ is the heat of vaporization per atom. The comparison shows that the heat of vaporization

$$Q = E_{\text{coh}} - \frac{1}{2} k_B T \tag{5.19}$$

is but for a trifle equal to the cohesive energy.

## 5.2 The Terrace Step Kink (TSK) Model

### 5.2.1 Basic Assumptions and properties

The *Terrace Step Kink* (TSK) model of surfaces assumes that the surface consists of flat terraces separated by steps in such a way that an unequivocal numbering of the steps by an index $n$ is possible. The steps are allowed to have kinks, but no overhangs so that the position of a step is uniquely described by one coordinate $x_n(y)$. Figure 5.2 shows a top view on step positions on a surface with square symmetry (e.g. $\langle 110 \rangle$ steps on $\{100\}$ surfaces of an fcc-material). The terrace heights to the left and right of steps differ by the step height $h$. We remark that this latter fact is of no consequence for the statistical thermodynamics of steps unless the height of the surface at a particular position is addressed, as e.g. in the height-height correlation function. The results of this section therefore apply also to other ensembles of linear systems such as domain walls in adsorbate phases (Sect. 1.3.1) or in thin film magnetism (see e.g. [5.2]).

In Sect. 4.3.7, we have considered the low temperature limit for the probability $P_k$ that the step has a kink of one atom length (4.96)

$$P_k = 2\exp(-\varepsilon_k / k_B T) \tag{5.20}$$

where $\varepsilon_k$ is the energy required to generate the kink (cf. Sect. 1.3.2). We now expand the considerations to arbitrary temperature. The probability to find a kink of the length of $n$ atomic units $a_\perp$ is

$$P_k(n) = \frac{2e^{-\frac{\varepsilon_k(n)}{k_B T}}}{1 + 2\sum_{n=1}^{\infty} e^{-\frac{\varepsilon_k(n)}{k_B T}}}. \tag{5.21}$$

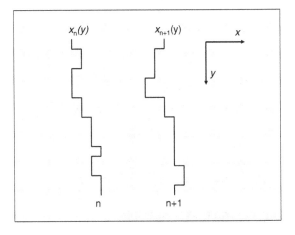

**Fig. 5.2.** Top view on step positions in the TSK-model. The position of a step $n$ is uniquely described by $x_n(y)$. Kinks have the length of one or more atom units.

The mean square of the length of a kink in units of $a_\perp$ is therefore

$$\langle n^2 \rangle = 2 \frac{\sum_{n=1}^{\infty} n^2 e^{-\frac{\varepsilon_k(n)}{k_B T}}}{1 + 2\sum_{n=1}^{\infty} e^{-\frac{\varepsilon_k(n)}{k_B T}}}. \tag{5.22}$$

Due to the equal probability for the existence of kinks of positive and negative sign, the steps undertake a random walk with respect to the $x$-coordinate as the step progresses along the $y$-direction. The randomness of the walk entails that the mean square displacement of the step in the $x$-direction is a linear function of the $y$-coordinate, the proportionality factor being $\langle n^2 \rangle$.

$$\left\langle \left(x_n(y)-x_n(y_0)^2\right\rangle/a_\perp^2 = \left\langle n^2\right\rangle \left|y-y_0\right|/a_\parallel \tag{5.23}$$

Here, $a_\parallel$ is the atomic length unit along $y$-direction. Equation (5.21) is the TSK-equivalent of (4.95), which was derived in the continuum approximation. Hence, the step stiffness $\tilde{\beta}$ in the TSK-model is

$$\tilde{\beta} = k_B T a_\parallel / \left\langle n^2\right\rangle a_\perp^2 . \tag{5.24}$$

Further evaluation requires the knowledge of the dependence of the kink energy as a function of the step length. Experimental data on probability to find kinks of a particular length on Si(100) showed that $\varepsilon_k(n)$ is well described by a corner energy and a linear dependence on the length [5.3]

$$\varepsilon_k(n) = \varepsilon_c + \varepsilon_1 n . \tag{5.25}$$

For steps on Si(100) $\varepsilon_c$ is about three times $\varepsilon_1$. This entails that once the temperature is high enough to have thermally generated single atom kinks one has also a large probability for kinks of multiple length. For metal surfaces on the other hand, the reverse is true. The step energy per atomic length unit is typically about twice as high as the energy of a one atom long kink. A kink of a length $n > 2$ consists of an inner corner, an outer corner and step of length $(n-2)$. While there appears to be no information available on the energy of a kink with length of two atom units, the energy of kinks with $n > 2$ fits to (5.25) with a negative corner energy $\varepsilon_c = \varepsilon_k - a_\parallel\beta$ and $\varepsilon_1 = a_\parallel\beta$. Using (5.25) with arbitrary values of $\varepsilon_c$ and $\varepsilon_1$ we calculate $\left\langle n^2\right\rangle$ as

$$\left\langle n^2\right\rangle = \frac{2cq(1+q)}{(1-q)^2(1-q+2cq^2)} \quad c = e^{-\frac{\varepsilon_c}{k_B T}} \quad q = e^{-\frac{\varepsilon_1}{k_B T}}, \tag{5.26}$$

which reduces to

$$\left\langle n^2\right\rangle = \frac{2}{q+q^{-1}-2} \tag{5.27}$$

if the corner energy $\varepsilon_c$ vanishes.

The denominator in (5.22) is the partition function for a step in the TSK-model with no step-step interactions. The free energy per step atom of a step is therefore

$$F \equiv \beta(T)a_{\parallel} = \beta(0)a_{\parallel} - k_{\mathrm{B}}T \ln(1 + 2c \sum_{n=1}^{\infty} q^n)$$

$$= \beta(0)a_{\parallel} - k_{\mathrm{B}}T \ln(1 + \frac{2cq}{1-q})$$

(5.28)

with $\beta(T=0)a_{\parallel}$ the step free energy per atom. Equation (5.28) has the low temperature expansion

$$\beta(T)a_{\parallel} = \beta(0)a_{\parallel} - 2k_{\mathrm{B}}T\, \mathrm{e}^{-\frac{\varepsilon_k}{k_{\mathrm{B}}T}}$$

(5.29)

in which $\varepsilon_k = \varepsilon_c + \varepsilon_1$ is the energy to create a single kink. The second term in (5.28)
and (5.29) represents the configuration entropy that results from the thermal me-
andering in space. With increasing temperature, the entropic term can become as
large as the step energy so that the free energy of the step vanishes. At this temperature, steps are spontaneously created and the surface becomes
thermodynamically rough (Sect. 4.3.1).

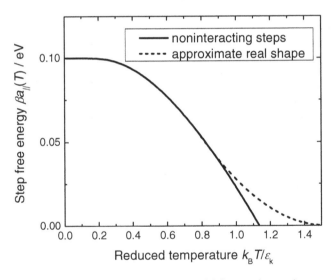

**Fig. 5.3.** Step free energy according to the TSK-model for non-interacting steps (solid line)
for $\beta(0)a_{\parallel} = \varepsilon_1 = 0.1\mathrm{eV}$. The corner energy is assumed to be zero so that the energy of a
single atom kink is $\varepsilon_k = \varepsilon_1$. According to the model, the step free energy becomes zero at
the roughening temperature $T_R = 1.13\varepsilon_k/k_{\mathrm{B}}$. In reality, the roughening transition is approached very gradually because of step-step interactions (dashed line).

According to the TSK-model of non-interacting steps the roughening temperature should be $T_R = 1.13\ \varepsilon_k/k_B$ when $\beta(0)a_\| = \varepsilon_k$. However, the TSK-model predicts an unrealistic value for the slope of the step free energy at the roughening temperature where steps are created spontaneously. Actually, steps repel each other by elastic as well as entropic interactions (to be studied in the next section). Because of the repulsive interactions between the steps, the energy cost for the creation of new steps increases with the density of already existing steps. Consequently, the free energy does not dive into the zero line as the TSK-model of non-interacting steps suggests, but rather approaches the zero level gradually with even a zero slope at the roughening temperature $T_R$.

### 5.2.2 Step-Step Interactions on Vicinal Surfaces

We now study the meandering of steps on vicinal surfaces with step-step interactions taken into account. We consider merely repulsive interactions so that vicinal surfaces are stable. Repulsive interactions arise from two sources. One is the elastic repulsive interaction as investigated in Sect. 3.4.2. The second interaction is of entropic nature and arises from the fact that steps are in each other's way so that their meandering is hindered. The following analysis requires that the vicinal surfaces are thermodynamic rough (Sect. 4.3.7). Later, Sect. 5.2.4 shows when this requirement is fulfilled.

The elastic interaction considered in Sect. 3.4.2 is non-local as each step at position $x_n$, $y$ interacts with all other steps $x_m$ at any $y'$. The simple $L^{-2}$ dependence of (3.80) was derived for straight steps separated by a constant distance $L$, whereas we now wish to consider explicitly meandering steps for which the distance to next step varies along the $y$-axis. The problem is not solvable in full generality. One usually reduces the interaction to a local pair interaction potential between steps $v[x_{n+1}(y)-x_n(y)]$. With this interaction potential and the further assumption that the kink energy be proportional to the kink length (no corner energy) the Hamiltonian is

$$H = \sum_{n,y} \varepsilon \left| x_n(y) - x_n(y+1) \right| + v\left[ x_{n+1}(y) - x_n(y) \right] \qquad (5.30)$$

We have dropped the index in the notation of the kink energy $\varepsilon$. The variables $x$ and $y$ are now discrete and in units of the atomic length $a_\perp$ and $a_\|$, respectively. The free energy of the system is given by the grand partition function

$$F = -k_B T \ln\left[ \text{Tr} \exp(-H\{x_n(y)\}/k_B T) \right] \qquad (5.31)$$

in which Tr denotes the trace. This harmlessly looking expression is in general a rather unapproachable beast. For the ansatz (5.30) the Hamiltonians in each line $y$ are independent of each other and the partition function separates line wise so that

$$Z = Z_y^N = \left(\mathrm{Tr}\exp\left[-H_y(x_n)/k_B T\right]\right)^N \qquad (5.33)$$

with $N$ number of $y$-values, i.e. the step length in atomic units. We see that the Hamiltonian (5.30) corresponds to $N$ independent systems, each possessing $p$, in general interacting particles (steps). Analytical solutions require still further simplifications concerning the interaction potential. The simplest case is no interaction at all. Another solvable case is a mean field approach in which the potential $V = \sum_{n,y} v[x_{n+1}(y) - x_n(y)]$ is replaced by a potential $V(x_n(y))$ that is the same for all steps, regardless their position with respect to the other steps.

We consider first steps with no explicit interactions. An elegant way to solve (5.33) is to map the problem onto an equivalent quantum mechanical problem. The method is called *transfer matrix method*. To this end, we describe the system with $p$ steps at position $y$ by a state vector

$$|\psi\rangle = |x_1, x_2, ... x_p\rangle \qquad (5.34)$$

where the $x_i$ denote the positions of the steps. Steps in the state vector $|\psi\rangle$ are created by the usual creation operators $c^+$ so that

$$|\psi\rangle = |x_1, x_2, ... x_p\rangle = c_{x_1}^+ c_{x_2}^+ ... c_{x_p}^+ |0\rangle \qquad (5.35)$$

Since no two steps exist at the same place, the occupation number for a step at $x_k$ is either 0 or 1[7]. The operators $c_{x_k}^+$ and $c_{x_k}$ are therefore fermion creators and annihilators, respectively. The operator $c_{x_k}^+ c_{x_k}$ has the eigenvalue 0 or 1, depending on the existence of a step at $x_k$. Kinks are created by removing a step at position $x$ and creating another one at $x$-1 or at $x$+1; i.e. kink are created by the pair of operators $c_{x-1}^+ c_x$ and $c_{x+1}^+ c_x$. We consider the matrix element $\langle \psi_y | Z_y | \psi_{y+1} \rangle$ which transfers the system from line $y$ to $y$+1. If there is a kink in the step it must have the value $\exp(-\varepsilon/k_B T)$ in the low temperature limit. Hence, $Z_y$ can be written as

$$Z_y = 1 + e^{-\varepsilon/k_B T} \sum_{x_k} c_{x_k}^+ c_{x_k+1} + c_{x_k}^+ c_{x_k-1} \cong \exp\left[e^{-\varepsilon/k_B T} \sum_{x_k} c_{x_k}^+ c_{x_k+1} + c_{x_k}^+ c_{x_k-1}\right]. (5.36)$$

The Hamiltonian $H_y$ in (5.33) that yields the correct partition function is therefore

---

[7] This condition is also referred to as the *no-crossing rule* for steps.

$$H_y = -k_B T e^{-\varepsilon/k_B T} \sum_{x_k} c_{x_k}^+ c_{x_k+1} + c_{x_k}^+ c_{x_k-1} . \tag{5.37}$$

For a large ensemble $N$, the value of the partition function is determined by its maximum eigenvalue, hence by the minimum eigenvalue of $H_y$, which is the ground state. The solution we are looking for is therefore the ground state of the Hamiltonian (5.37) to which one might add a mean field potential

$$V_{\text{meanfield}} = \sum_{x_k} V(x_k) c_{x_k}^+ c_{x_k} . \tag{5.38}$$

We now move from the particle representation to the space representation in which the states are described by a many particle wave function $\psi(x_1, x_2, \ldots x_p)$ that has the usual meaning of a probability-amplitude to find the steps at the positions $x_k$. We furthermore represent the many body state $\psi(x_1, x_2, \ldots x_p)$ by the product of single particle wave-functions $\psi(x_k)$, which is equivalent to the single electron approximation for a general fermion gas. This last approximation is not necessary if there is no interaction as the Hamiltonian of a non-interacting Fermi-gas can be solved exactly, but it simplifies the analysis considerably. Mapped onto the step problem this means we now consider the behavior of a single step in a mean field potential provided by all the other steps. The Schrödinger equation for the single step wave function is then

$$-k_B T e^{-\varepsilon/k_B T} [\psi(x+1) + \psi(x-1)] + V(x)\psi(x) = f\psi(x) \tag{5.39}$$

We have dropped the index $k$. The eigenvalue $f$ is the free energy of the step per atom length, which follows from the general definition of the free energy (5.3) and (5.33). Moving over to a coarse-grained description one may consider $x$ as a continuous variable and return to the normal metric by replacing $\psi(x+1) + \psi(x-1)$ first by $\psi(x+a_\perp) + \psi(x-a_\perp)$ and then by $a_\perp^2 \psi''(x) + 2\psi(x)$. Our tour de force through the transfer matrix method applied to steps is then finally rewarded with a nice equation for the probability amplitude $\psi(x)$ to find a step at position $x$.

$$-k_B T e^{-\varepsilon/k_B T} a_\perp^2 \psi''(x) + V(x)\psi(x) = (f + 2k_B T e^{-\varepsilon/k_B T})\psi(x) . \tag{5.40}$$

## 5.2.3 Simple Solutions for the Problem of Interacting Steps

### The Gruber-Mullins model

We consider two solutions of (5.40). The first is the particle-in-a-box solution, which here means that a step is assumed to meander freely between the boundary set by two neighboring steps at their mean distance $-L$ and $+L$. Evidently, the solution is

$$\psi(x) = \frac{1}{\sqrt{L}} \cos(x\pi/2L) \qquad (5.41)$$

The probability to find a step at $x$ is therefore

$$P(x) = \frac{1}{L} \cos^2(x\pi/2L) \qquad (5.42)$$

Equation (5.42) is known as the *Gruber-Mullins solution* of the step-step interaction problem [5.4]. By observing that the step position $x = 0$ corresponds to a terrace width $L = <L>$ the normalized terrace width distribution is

$$P(s) = \sin^2(\pi s/2), \qquad (5.43)$$

in which $s = L/\langle L \rangle$. The terrace width distribution is plotted in Fig. 5.4. Inserting the wave function into (5.40) yields the free energy per atom length as

$$f = -2k_B T e^{-\varepsilon/k_B T} + k_B T \frac{\pi^2 a_\perp^2}{4L^2} e^{-\varepsilon/k_B T}. \qquad (5.44)$$

Evidently, the energy of the ground state at $T = 0$ comes out as zero. To make contact with the real world we have to add the internal energy. The final result in our standard notation is therefore

$$\beta(T)a_\parallel = \beta(0)a_\parallel - 2k_B T e^{-\varepsilon/k_B T} + k_B T \frac{\pi^2 a_\perp^2}{4L^2} e^{-\varepsilon/k_B T}. \qquad (5.45)$$

Compared to the solution for a single step (5.29) equation (5.45) contains an additional $L$-dependent term, which represents the entropic repulsion of steps. However, the hard wall model overestimates the entropic repulsion. In reality, the neighboring steps also wander about which reduces the entropic repulsion somewhat. In the exact result obtained from the free fermion solution the 4 in the denominator of the third term in (5.45) is replaced by a 6 [5.5].

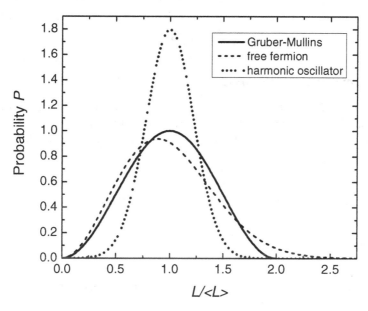

**Fig. 5.4.** Normalized terrace width distribution according to the Gruber-Mullins model, the free fermion model, and the harmonic oscillator model (solid, dashed and dotted line, respectively).

With (5.45) we have a microscopic solution for the third term of the expansion of the projected surface tension $\gamma_p(p)$ (4.46) in which $p$ was defined as $|\tan \theta|$ with $\theta$ the angle of inclination of the surface with respect to a low index surface. With the correct factor 6 inserted in the denominator of the last term of (5.45) one obtains for $\gamma_3$ in (4.46)

$$\gamma_3 = \frac{k_B T \pi^2 a_\perp^2}{6h^2} e^{-\varepsilon/k_B T} \tag{5.46}$$

in which $h$ is the step height. By introducing the line stiffness $\tilde{\beta}$ (5.24) the free energy is expressed in terms of macroscopically defined quantities.

$$\beta(T) a_\parallel = \beta(0) a_\parallel - \frac{(k_B T)^2 a_\parallel}{\tilde{\beta} a_\perp^2} + \frac{(\pi k_B T)^2 a_\parallel}{6 \tilde{\beta} L^2} \tag{5.47}$$

and $\gamma_3$ becomes

$$\gamma_3 = \frac{(\pi k_B T)^2 a_\parallel}{6 \tilde{\beta} h^3} \qquad (5.48)$$

### The harmonic oscillator model

The second solution of (5.40) that can be expressed in analytical form is obtained for elastic interactions (Sect. 3.4.2). In keeping with the definitions in Sect. 3.4.2 we write the elastic step-step interaction for one atomic unit step length as $a_\parallel A/L^2$. The single step considered meanders therefore in a potential

$$V(x) = a_\parallel A \sum_{n=1}^{\infty} \left( \frac{1}{(nL-x)^2} + \frac{1}{(nL+x)^2} \right)$$

$$\cong \frac{a_\parallel A \pi^2}{6L^2} + \frac{a_\parallel A \pi^4}{15L^4} x^2 \dots \qquad (5.49)$$

Here, $x$ denotes the deviation from the central position between two neighboring steps. It is also assumed that all other steps on the vicinal surface are at their mean position. The sum of the interactions to the next-nearest steps and all steps farther away adds a factor of $\pi^4/90 = 1.082$ to the next-nearest neighbor interaction. The resulting potential is harmonic as long as $x \ll L$. With (5.49) equation (5.40) becomes the Schrödinger equation of a harmonic oscillator. The probability to find a step at position $x$ is a Gaussian

$$\left| \psi^2(x) \right|^2 = \frac{1}{S\sqrt{2\pi}} e^{-\frac{x^2}{2S^2}} \qquad (5.50)$$

with

$$S = Lw = L \left( \frac{15 a_\perp^2 k_B T}{4 a_\parallel A \pi^4 \exp(\varepsilon/k_B T)} \right)^{1/4} \qquad (5.51)$$

Transformed into the normalized terrace width distribution $P(s)$ (5.50) becomes

$$P(s) = \frac{1}{w\sqrt{2\pi}} e^{-\frac{(s-1)^2}{2w^2}} \qquad (5.52)$$

Figure 5.4 shows this distribution as a dotted line with data for $A$ and $\varepsilon$ that correspond to Cu(100) vicinal surfaces ($A = 7$ meVÅ, $\varepsilon = 126$ meV [5.6]).

As for the Gruber-Mullins model, one can calculate the step free energy and obtains

$$\beta(T)a_\parallel = \beta(0)a_\parallel - 2k_BT\,e^{-\varepsilon/k_BT} + \frac{\pi^2 a_\parallel A}{6L^2}\left(1 + \sqrt{\frac{12a_\perp^2 k_BT \exp(-\varepsilon/k_BT)}{5a_\parallel A}}\right). \quad (5.53)$$

As expected, the free energy rises with the strength of the step-step interaction. The largest contribution is the temperature independent part of the third term, which simply arises from the constant term in the interaction potential (5.49).

Both the Gruber-Mullins and the harmonic oscillator solution yield a symmetric terrace width distribution. Experimental terrace width distributions are typically skewed with a higher weight for larger step-step distances. Einstein and Pierre-Louis [5.7 962] have proposed a general representation of the terrace width distribution based on a generalized Wigner surmise on random matrix theory, which reads

$$P(s) = a_\rho s^\rho e^{-b_\rho s^2}. \quad (5.54)$$

The parameter $\rho$ is determined by the interaction constant $A$

$$A = \frac{2a_\perp^2 e^{-\varepsilon/k_BT} k_BT}{a_\parallel} \frac{\rho(\rho-2)}{4}. \quad (5.55)$$

The other constants are determined by normalization and the requirement that $P(s)$ must have a unit mean. Equation (5.54) is exact for $\rho = 1$, 2, and 4. The case $\rho = 2$ is equivalent to $A = 0$, the case of non-interacting fermions [5.8]. This distribution, originally proposed heuristically [5.9] to describe the result of a complex approximate form derived by Joós et al. [5.8], is shown in Fig. 5.4 as a dashed line. Experimental data on the terrace width distribution seem to have the general form (5.54) and step-step interactions have been determined using (5.54) and (5.55) [5.10, 11].

### 5.2.4 Models for Thermal Roughening

In Sect. 4.3.1 the concept of thermodynamically rough surfaces and the roughening transition was introduced. Within the framework of the continuum theory, it was shown that the height-height correlation function diverges logarithmically above the roughening temperature $T_R$ (4.104).

$$\lim_{|r-r'|\to\infty}\left\langle(h(r)-h(r'))\right\rangle = \left|\begin{array}{ll} = const & T < T_R \\ \propto \ln|r-r'| & T > T_R \end{array}\right. \quad (5.56)$$

Microscopic models with analytical solutions for the roughening transition also exist. However, the models require drastically simplifying assumptions about the solid so that they have little predictive value for the roughening transition of real solids. Villain et al. [5.12 717] introduced the most realistic model. It concerns the special case of the roughening transition of vicinal surfaces. The model considers the Hamiltonian

$$H = \varepsilon \sum_{n,y} \left[ x_n(y+1) - x_n(y) \right]^2 + W_{el} \sum_{n,y} \left[ x_{n+1}(y) - x_n(y) \right]^2 . \qquad (5.57)$$

The second terms describes the step-step interaction as quadratic in the distance. With reference to (5.49) $W_{el}$ can be expressed by the interaction constant $A$ as

$$W_{el} = \frac{a_{\parallel} a_{\perp}^2 \pi^4 A}{15 L^4} . \qquad (5.58)$$

The quadratic dependence of the kink energy on the kink length assumed with the first term in (5.57) is unrealistic but hurts little as the roughening transition occurs at temperatures well below $\varepsilon / k_B$. The model predicts the roughening temperature at

$$\frac{W_{el}}{k_B T_R} e^{-\varepsilon / k_B T_R} = \frac{\pi^2}{2} . \qquad (5.59)$$

The step-step interaction constant $W_{el}$ has been determined experimentally from the terrace width distribution for a number of surfaces [5.6, 13]. A rather complete set of data exists for the Cu(11n) surfaces [5.9]. For this series, the roughening temperatures are calculated in Table 5.1. With the exception of the (113) surface, all the Cu vicinals should be rough above room temperature. This is satisfying since the random walk model for steps as explicated above is applicable only to thermodynamic rough surfaces.

**Table 5.1.** Roughening temperatures $T_R$ for Cu(1 1 n) surfaces, calculated using (5.59) with $A = 7.1$ meVÅ and the kink energy $\varepsilon_k = 128$ meV.

| Surface | (1 1 3) | (1 1 5) | (1 1 7) | (1 1 9) | (1 1 11) | (1 1 13) | (1 1 15) | (1 1 17) |
|---------|---------|---------|---------|---------|----------|----------|----------|----------|
| $T_R$/K | 388 | 270 | 223 | 197 | 180 | 168 | 159 | 152 |

For the roughening transition of facets, two analytically solvable models exist. Both models involve severely simplifying assumptions. In one model, the solid is assumed to consist of columns of height $h_i$ with the atoms in the form of cubes (Kossel crystal, cf. Sect. 4.3.3) and the Hamiltonian is expressed in terms of the nearest neighbor difference in the columnar height

$$H = \frac{\varepsilon}{2}\sum_{i,\delta}\left|h_i - h_{i,\delta}\right| \tag{5.60}$$

in which $\varepsilon$ corresponds to the energy required to "break a bond". The sum over $\delta$ extends over the four nearest neighbor columns. The factor 1/2 accounts for the fact that each "bond" appears twice in the sum. The model is referred to as *Absolute (height difference) Solid On Solid* (ASOS) model. The model possesses a roughening transition at

$$T_R(ASOS) = 1.24\varepsilon/k_B \tag{5.61}$$

In the other solvable model the absolute value of the height difference is replaced by the square

$$H = \frac{\varepsilon}{2}\sum_{i,\delta}\left(h_i - h_{i,\delta}\right)^2 . \tag{5.62}$$

The model is called *Discrete Gaussian Solid On Solid* (DGSOS) model. Its roughening temperature

$$T_R(DGSOS) = 1.46\varepsilon/k_B \tag{5.63}$$

is somewhat higher as for the ASOS-model since the energy increases more rapidly for larger differences in the columnar height. The models are useful insofar as their study has revealed the nature of the roughening transitions to be of the Kosterlitz-Thouless type. However, the predicted roughening temperatures are excessively high. Furthermore, the models provide no clue to the dependence of the roughening temperature on the crystallographic orientation of the facet, which is the most interesting aspect of roughening, at least from the experimental side.

### 5.2.5 Phonon Entropy of Steps

As important as the configuration entropy of steps is for the terrace width distribution and the roughening transition, the prevailing entropic term at room temperature stems from the phonon spectrum. Due to the different bonding of the atom at step sites, the local vibration spectrum around step atoms from the spectrum of flat surfaces. The phonon contribution to the free energy of surfaces and steps has been addressed by the Rahman group in a series of papers [5.14-18]. The authors find that the change in local bonding not only affects the vibration spectrum of the step atoms but also that of their neighbors. The vibrational contribution to the step free energy per step atom is

$$f_{\text{vib}} = k_{\text{B}} T \int_0^{\infty} \ln(2 \sinh \frac{\hbar \omega}{2 k_{\text{B}} T}) \big[ n_{\text{stepped}} (\omega) - n_{\text{flat}} (\omega) \big] d\omega . \qquad (5.64)$$

Here, $n_{\text{stepped}}(\omega)$ and $n_{\text{flat}}(\omega)$ denote the spectral densities per atom for a sample with and without a step. As an example, the free energy of steps on copper surfaces are shown in Fig. 5.5 [5.18]. The differences between the various steps are quite remarkable. For the steps on the Cu(100) surface, the vibrational free energy is marginally small, however steps on Cu(111) have vibrational free energy that amounts to 5-8% of the step energy (Table 4.2) at 300 K. It has been proposed by this author to estimate the step free energy by invoking an Einstein oscillator model [5.19]. Based on the scaling properties of a Morse-potential the oscillator frequencies $\omega_{\text{E}}$ were proposed to be proportional to the square root of the coordination number. According to this crude model, the vibration spectrum of a surface atom would be characterized by an Einstein temperature ($\Theta = \hbar \omega / k_{\text{B}}$) $\Theta_{\text{surf}} = \sqrt{8/12} \, \Theta_{\text{bulk}}$ and $\Theta_{\text{surf}} = \sqrt{9/12} \, \Theta_{\text{bulk}}$ for the (100) and (111) surfaces, respectively. The Einstein temperatures for the step atoms would be $\Theta_{\text{step}} = \sqrt{7/12} \, \Theta_{\text{bulk}}$ for all steps considered here.

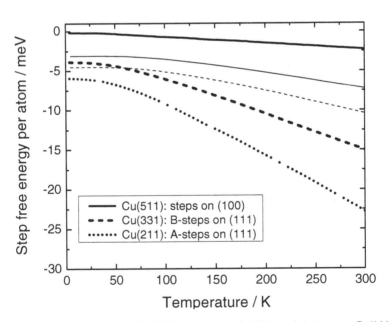

**Fig. 5.5.** Free energy of steps on Cu(001), B-steps on Cu(111) and A-steps on Cu(111) are shown as fat solid, dashed and dotted lines, respectively [5.18]. The thin solid and dashed lines represent calculations for the (100) and (111) surfaces, respectively, using an Einstein oscillator model with a simple scaling of the characteristic frequencies.

In this model, the vibration free energy would amount to

$$f_{\text{vib}} = 3k_B T \ln\left(\sinh(\Theta_{\text{step}}/2T)/\sinh(\Theta_{\text{surf}}/2T)\right) \tag{5.65}$$

The high temperature limit of (5.65) has a particular easy form, which is

$$f_{\text{vib}}(100) = 3k_B T \ln(7/8)^{1/2} \tag{5.66}$$

$$f_{\text{vib}}(111) = 3k_B T \ln(7/9)^{1/2} \tag{5.67}$$

for steps on the (100) and (111) surfaces, respectively. With $\Theta_{\text{bulk}} = 343K$, $\Theta_{\text{step}}$ and $\Theta_{\text{surf}}$ inserted, the thin solid and dashed lines in Fig. 5.5 are obtained for steps on Cu(100) and Cu(111). By construction, the model cannot account for a difference between A- and B-steps. The model also fails in any other quantitative sense as it underestimates the step free energy on Cu(111) and overestimates on Cu(100). Despite this obvious failure, the model has been (ab)used several times in the analysis of crystal and island equilibrium shapes where quantitative values for the step free energy enter crucially [5.20, 21].

## 5.3 The Ising-Model and the Crystal Equilibrium Shape

### 5.3.1 The Model and the Shape Function

The Ising-model is popular in many fields of physics because it is comparatively simple and can be solved analytically in two dimensions. The Hamiltonian is

$$H_{\text{Ising}} = \sum_{ij} J_{ij} s_i s_j \tag{5.68}$$

in which the indices $i, j$ run over the nearest neighbors $J_{ij}$ is the interaction strength and $s_i = \pm 1$ are spin variables. In the context of surface problems, one writes the Hamiltonian in terms of nearest neighbor occupation numbers

$$n_i = (s_i + 1)/2 = 0,1 \quad . \tag{5.69}$$

With a nearest neighbor interaction energy $V_0$ and after introducing a nonzero chemical potential the Ising-Hamiltonian becomes

$$H_{\text{Ising}} = \frac{1}{2} V_0 \sum_{i,j} n_i n_j - \mu \sum_i n_i \quad . \tag{5.70}$$

In this form, the Hamiltonian is used as a starting point for the theoretical analysis of two-dimensional phases and phase transitions. A special mean-field solution will be discussed in Sect. 5.4.4. Here, we address the problem of the crystal equilibrium shape in two dimensions. Analytical solutions have been derived for the square and the honeycomb lattice [5.22, 23]. These solutions have found application in the description of the equilibrium shape of two-dimensional islands to which problem we now attend.

**(a)**                                              **(b)**

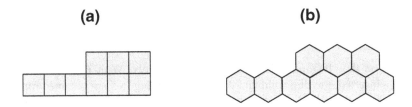

**Fig. 5.6.** Kinked steps in the square **(a)** and hexagonal lattice **(b)**. In the square lattice, step atoms have one broken bond, kink atoms two. The kink energy is therefore equal to the step energy per atom. In the hexagonal lattice, each step atom has two broken bond and the kink energy is one half of the step energy per atom.

In the Ising-model, the energy of a step is proportional to its microscopic length. For the square lattice, this means that the kink energy is equal to the step energy per atom. On the honeycomb lattice (henceforth named "hexagonal"), the energy per atom of the densely packed step is twice as large as the kink energy (Fig. 5.6).

We denote the energy parameter in the Ising-model that corresponds to the kink energy as $\varepsilon$. The equilibrium shapes are given by implicit expressions. For the square lattice the shape is described by [5.22]

$$\cosh\left[\varepsilon(x-y)/2k_{B}T\right]\cosh\left[\varepsilon(x+y)/2k_{B}T\right]=A_{sq},\qquad(5.71)$$

with

$$A_{sq}=\frac{1}{2}\cosh(\varepsilon/k_{B}T)\coth(\varepsilon/k_{B}T).\qquad(5.72)$$

For the hexagonal lattice the shape is [5.23]

$$\cosh(2y\varepsilon/k_{B}T)+\cosh\left((\sqrt{3}x+y)\varepsilon/k_{B}T\right)$$
$$+\cosh\left((\sqrt{3}x-y)\varepsilon/k_{B}T\right)=A_{hex}\qquad(5.73)$$

with

$$A_{\text{hex}} = \frac{\cosh^3(2K^*) + \sinh^3(2K^*)}{\sinh(2K^*)}$$

$$(5.74)$$

$$\tanh(K^*) = \exp(-2\varepsilon/k_B T) \ .$$

The coordinates $x$, $y$ are chosen such that for both lattices the nearly straight sections at low temperature are oriented parallel to the $x$-axis. The scaling of the cartesian coordinates is so that $y(x=0) = \pm 1$ in the limit $\exp(-\varepsilon/k_B T) \ll 1$, and that the size of the islands described by (5.71) and (5.73) remains approximately constant with temperature.

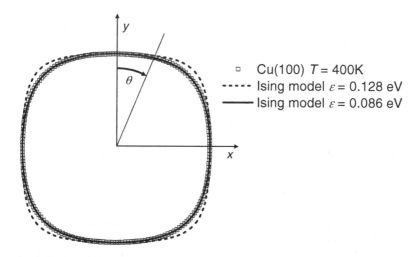

**Fig. 5.7.** Comparison of the experimental equilibrium shape of islands on Cu(100) at 400K with the Ising-shape (5.67). The Ising-model describes the experiment well if the parameter $\varepsilon$ is fitted. Using the experimental kink energy $\varepsilon = 0.128$ eV makes the island too squared.

The Ising-shapes describe the experimentally observed equilibrium shapes of 2D-islands quite well if the parameter $\varepsilon$ is fitted to the experiment. Figure 5.7 shows a comparison of experiment [5.24] and theory for the equilibrium shape of one atom layer high islands on Cu(100). The experimental shape is missed when the Ising-parameter is equated with the experimentally measured kink energy of $\varepsilon = 0.128$ eV [5.24, 25]. The reason is that the experimental step energy per atom $a_\parallel \beta$ is about twice as high as the kink energy ($a_\parallel \beta = 0.22$ eV [5.24]) and not equal to the kink energy as the Ising-model for a square lattice has it.

## 5.3.2 Further Properties of the Model

For moderately low temperatures, i.e., in the limit $\exp(-\varepsilon/k_B T) \ll 1$, $A_{sq}$ (5.72) and $A_{hex}$ (5.74) are approximated by

$$A_{sq} = \frac{1}{4} e^{\varepsilon/k_B T} \tag{5.75}$$

$$A_{hex} = \frac{1}{2} e^{2\varepsilon/k_B T} \; . \tag{5.76}$$

With these approximations, the distances from the center of the island to the point of minimum curvature $y(x = 0)$ are

$$y_{sq}(x = 0) = 1 - \frac{2k_B T}{\varepsilon} e^{-\varepsilon/k_B T} \tag{5.77}$$

$$y_{hex}(x = 0) = 1 - \frac{k_B T}{\varepsilon} e^{-\varepsilon/k_B T} \; . \tag{5.78}$$

Because of the mechanics of the Wulff-construction, the minimum and maximum distances from the center of the island possessing a tangent orthogonal to the radius vector are proportional to the step free energy. After multiplication with the step energies in terms of the parameter $\varepsilon$ one obtains for both lattices

$$a_\parallel \beta(T) = a_\parallel \beta(T = 0) - 2k_B T \, e^{-\varepsilon/k_B T} \; . \tag{5.79}$$

We have thereby recovered the low temperature approximation to the free energy of a step running along the direction of dense atom packing (5.29).

By using Wulff's theorem in two dimensions (4.66) and the chemical potential $\mu$ in terms of the step energy at $\theta = 0$ (4.69)

$$\mu = \Omega_s \tilde{\beta}(\theta) \kappa(\theta) = \Omega_s \beta(\theta = 0) / y(x = 0) \; , \tag{5.80}$$

one can calculate the step stiffness $\tilde{\beta}(\theta = 0)$. With the product of curvature $y''(x = 0)$ and $y(x=0)$

$$y y''\big|_{sq,x=0} k_B T \cong 2\varepsilon e^{-\varepsilon/k_B T} \left( 1 - \frac{2k_B T}{\varepsilon} e^{-\varepsilon/k_B T} \right) \cong 2\varepsilon e^{-\varepsilon/k_B T} \tag{5.81}$$

$$y y''\big|_{hex,x=0} k_B T \cong 3\varepsilon e^{-\varepsilon/k_B T} \left( 1 - \frac{k_B T}{\varepsilon} e^{-\varepsilon/k_B T} \right) \cong 3\varepsilon e^{-\varepsilon/k_B T} \tag{5.82}$$

one obtains

$$\tilde{\beta}_{sq}(\theta = 0) = \frac{k_B T a_{\parallel}}{2a_{\perp}^2} e^{\varepsilon / k_B T} \tag{5.83}$$

$$\tilde{\beta}_{hex}(\theta = 0) = \frac{k_B T a_{\parallel}}{a_{\perp}^2} e^{\varepsilon / k_B T} \tag{5.84}$$

To arrive at (5.83) we have inserted the scaling units $a_{\parallel}$ and $a_{\perp}$. In order to obtain (5.84) one has to consider that on the hexagonal lattice kinks involve an advancement along the step direction by $3/2a_{\parallel}$ and that the step energy that corresponds to this advancement is $3\varepsilon$. Equation (5.83) is the same as (4.98), and (5.80) can be obtained accordingly from (4.97) when the metric of the hexagonal lattice is taken into account. We have therefore recovered the TSK-expressions for the stiffness.

The TSK-stiffness in combination with 5.80 can be used to determine the kink energy from an Arrhenius-plot of the curvature multiplied by $k_B T$ [5.24]. Because of the considerable difficulties to determine the curvature exactly at $\theta = 0$ the method is inferior to the use of the step-step correlation function (cf. Fig. 4.23).

Of interest are also the aspect ratios, defined as the ratio of the radii to the "corners" and the "straight" sections $r_{45°}/r_{0°}$ and $r_{30°}/r_{0°}$ for the square and hexagonal lattices, respectively. These ratios represent the ratios of the free energies of the steps where, forced by the orientation, every atom is a kink atom (100% kinks) to the free energy of the steps with thermal kinks only. Here, the Ising-results are

$$\frac{r_{45°}}{r_{0°}} \cong \sqrt{2}\left(1 - \frac{\ln 2}{\varepsilon} k_B T\right) \Big/ \left(1 - \frac{2k_B T}{\varepsilon} e^{-\varepsilon / k_B T}\right) \tag{5.85}$$

$$\frac{r_{30°}}{r_{0°}} \cong \frac{2}{\sqrt{3}}\left(1 - \frac{\ln 2}{2\varepsilon} k_B T\right) \Big/ \left(1 - \frac{k_B T}{\varepsilon} e^{-\varepsilon / k_B T}\right) . \tag{5.86}$$

The leading terms for the temperature dependencies of the ratios are the linear terms in the numerators. The denominators in (5.85) and (5.86) are proportional to the free energies of the straight steps. While (5.85) and (5.86) are derived in the low temperature limit, the equations represent a good approximation also for moderately high temperatures. Figure 5.8 compares the approximation with the exact (numerical) solution of the Ising-shape. The approximation agrees well with the exact result up to $k_B T = 0.4\,\varepsilon$ and $0.5\,\varepsilon$, for the square and hexagonal lattice, respectively. This condition corresponds, for example, to temperatures of 700 K, 680 K, and 580 K for Cu(100), Cu(111) and Ag(111), respectively. At these temperatures, islands disappear very quickly due to diffusion. Experimental observations on the equilibrium shapes of these islands therefore concern lower temperatures.

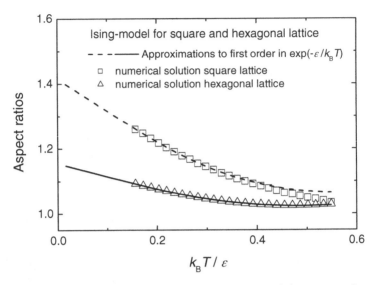

**Fig. 5.8.** Comparison of the low temperature approximation of the aspect ratios and the exact solution of the Ising-model.

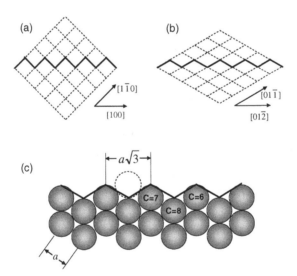

**Fig. 5.9.** The various possible paths of a step with a mean direction corresponding to the 100% kinked step on the square and hexagonal lattice, **(a)** and **(b)**, respectively. Only the paths closest to the center path have sufficient statistical weight to survive in the macroscopic limit. **(c)** A second path of significant statistical weight corresponds to adding the dashed atom. Boundary atoms with coordination numbers C = 6 and C = 8 are thereby replaced by two atoms with C = 7.

The entropic $\ln 2$ -terms in the numerator of (5.85) and (5.86) can also be derived directly by considering the free energy of the 100% kinked steps on a square and hexagonal lattice. Figure 5.9 displays the possible paths of the steps in the two cases. The entropy arises from the various configurations of the steps which all have a mean orientation along the direction of the 100% kinked step and are energetically equivalent in the Ising-model. These various paths are illustrated by dashed lines in Fig. 5.9a and b, for the square and hexagonal lattice, respectively. The configuration entropy is calculated easily by making contact with gambling theory [5.26]: The number of possibilities in a coin tossing game with $N$ trials to arrive at $N/2$ "head" and $N/2$ "tale" results is $N!/[(N/2)!]^2$. The entropy is therefore

$$S = k_B \ln \frac{N!}{[(N/2)!]^2} , \tag{5.87}$$

which by virtue of Stirling's formula becomes

$$S = N k_B \ln 2 \tag{5.88}$$

in the macroscopic limit $(N \to \infty)$. Hence, the entropy per atom on the kinked step is $k_B \ln 2$ and the partition function per atom is $Z = 2$. This means that in the macroscopic limit only two alternative paths per atom survive. Those ones stick closest to the center path and correspond to adding or removing one atom to the step as illustrated in Fig. 5.9c. All other paths have a statistical weight lower then $e^N$ and vanish therefore in the macroscopic limit. With the entropy (5.88), the free energy per atom for the 100% kinked step becomes

$$a_k \beta(\theta_k, T) = a_k \beta(\theta_k, T=0) - k_B T \ln 2 . \tag{5.89}$$

Here, $a_k$ denotes the length per atom of the kinked step which is $a_{\parallel}/\sqrt{2}$ and $a_{\parallel}\sqrt{3}/2$ for the square and hexagonal lattice, respectively, and $\theta_k$ is the angle of the 100% kinked step with respect to the "straight" steps. With (5.89) one calculates the ratio of the free energies of the 100% kinked step and the "straight" steps and thereby the aspect ratios of the islands as defined before

$$\frac{r_{\theta_k}}{r_0} = \frac{a_{\parallel}\beta(\theta_k, T)}{a_{\parallel}\beta(0, T)} = \frac{a_{\parallel}\beta(\theta_k, 0) - \dfrac{a_{\parallel} \ln 2}{a_k} k_B T}{a_{\parallel}\beta(0,0)\left(1 - \dfrac{2 k_B T}{a_{\parallel}\beta(0,0)} e^{-\varepsilon_k / k_B T}\right)} . \tag{5.90}$$

This equation is equivalent to (5.85 and 5.86, see also Fig. 5.9) since

$$\frac{a_{\parallel}}{a_k} = \frac{\beta(\theta_k,0)}{\beta(0,0)} \ . \tag{5.91}$$

Equation (5.90) offers an interesting way to determine the step energy $a_{\parallel}\beta(0,0)$ from experimental island equilibrium shapes by fitting the experimentally observed aspect ratios as a function of temperature to (5.90). Since the kink energy can be determined in independent experiments, this fit involves matching two parameters to the experiment.

One may relax the requirement that the energies of all paths depicted in Fig. 5.9a,b be equal, and make allowance for a (small) energy difference $\Delta E_b$ between the two paths next to the center path. The two paths correspond to adding or removing the dashed atom in Fig. 5.9. Boundary atoms with coordination numbers $C = 6$ and $C = 8$ are thereby replaced by two atoms with $C = 7$. In the Ising model, the energies of the two states are the same. However, also in more realistic models with a curved form of the energy vs. coordination number, the energy difference is small as the energy vs. coordination number is rather linear around $C = 7$ (Fig. 1.45).

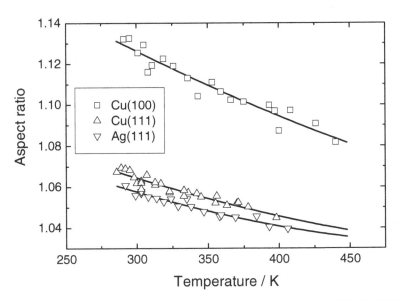

**Fig. 5.10.** Experimental data on the aspect ratio vs. temperature for Cu(100), Cu(111), and Ag(111) [5.24]. The solid lines are fits with $a_{\parallel}\beta = 0.22\pm0.02$ eV, $0.27\pm0.03$ eV, and $0.25\pm0.03$ eV, respectively.

With the energy difference $\Delta E_b$ the partition function per atom is

$$Z = 2\cosh(\Delta E_b / 2k_B T) \cong 2\left[1 + \frac{1}{2!}\left(\frac{\Delta E_b}{2k_B T}\right)^2 + \frac{1}{4!}\left(\frac{\Delta E_b}{2k_B T}\right)^4 \cdots\right]. \quad (5.92)$$

The additional terms can be ignored as long as $(\Delta E_b / 2k_B T)^2 \ll 1$. If that condition is not fulfilled, the absolute value of the partition function is affected by $\Delta E_b$, and the partition function becomes temperature dependent. The experimental aspect ratio then obeys

$$\frac{r_{\theta_k}}{r_0}\left(1 - \frac{2k_B T}{a_\parallel \beta(0,0)}e^{-\varepsilon_k / k_B T}\right) = \frac{\beta(\theta_k,0)}{\beta(0,0)} - \frac{a_\parallel \ln Z(T)}{a_k a_\parallel \beta(0,0)}k_B T. \quad (5.93)$$

The three unknown parameters in (5.93), $\beta(0,0)$, $\Delta E_b$, and the ratio $\beta(\theta_k,0)/\beta(0,0)$, can be determined from a self consistent fit of the temperature dependence of the aspect ratio. This three-parameter fit requires that the data have sufficiently low noise over a wide temperature range.

Figure 5.10 shows experimental data on the aspect ratios of 2D-island equilibrium shapes for Cu(100), Cu(111), and Ag(111) [5.24]. The solid lines are fit with (5.86) from which the data in Table 5.2 were obtained. The data agree well with those obtained from island shape fluctuations (Sect. 4.3.8).

**Table 5.2.** Step free energies and the ratio of the step energy in the corners and the straight sections at $T = 0$ K obtained from the aspect ratios of 2D-islands vs. temperature [5.24].

| Surface | $a_\parallel \beta$ /eV | $\beta(\theta_k)/\beta(0°)$ /eV |
| --- | --- | --- |
| Cu(100) | 0.22±0.02 | 1.24±0.01 |
| Cu(111) | 0.27±0.03 | 1.138±0.008 |
| Ag(111) | 0.25±0.03 | 1.136±0.009 |

# 5.4 Lattice Gas Models

## 5.4.1 Lattice Gas with No Interactions

The interaction potential of atoms (or molecules) that are chemically bonded to the surface varies strongly with the lateral position of the adsorbed species. Such *chemisorbed* atoms primarily reside in defined surface sites. On crystalline surfaces, these sites have a periodic structure described by a translation lattice. Any statistical model of chemisorbed adsorbate layers must account for this basic property. If adsorbates assume a random position in the available sites (which they obviously can do only if the fractional coverage is less than unity), they are said to form a *lattice gas*. The alternative is the formation of highly coordinated ensembles or even ordered structures. In a more general sense, all models dealing with the occupation of a periodic arrangement of sites, stochastically or ordered, are called *lattice gas models*.

The simplest approximation one can make is to assume only one type of sites, and further that the adsorption energy for a site is independent on the occupation of neighboring sites. This model is called the *non-interacting lattice gas*. The model is easily amended by introducing a mean field interaction. In taking the most important property of an adsorbate layer into account, the model is a good approximation to the real world with respect to segregation and adsorption isotherms. It fails (as all mean field models do) in the description of the evolution of different phases as a function of the adsorbate coverage and the transitions between various phases. In the following, we study the basic properties of the noninteracting lattice gas. To be able to consider the equilibrium with species dissolved in the bulk or with a surrounding gas phase we calculate the chemical potential as a function of fractional surface coverage.

The allowed occupation numbers for each site $i$ are $n_i = 0$ and $n_i = 1$. For the non-interacting lattice gas, the occupation statistics is therefore the same as for electrons in the free electron approximation and we can borrow the result for the mean occupation number $\langle n \rangle$ per site from there,

$$\langle n \rangle \equiv \Theta_{\mathrm{ad}} = \left(1 + \exp\left[(E_{\mathrm{ad}} - \mu_{\mathrm{ad}})/k_{\mathrm{B}}T\right]\right)^{-1}. \tag{5.94}$$

In the context of adsorption, the mean occupation number per site is called *fractional coverage*, or simply *coverage*, which we denote as $\Theta_{\mathrm{ad}}$. $E_{\mathrm{ad}}$ and $\mu_{\mathrm{ad}}$ are the energy of the adsorbate and its chemical potential, respectively. Solving (5.94) for the chemical potential yields

$$\mu_{\mathrm{ad}} = E_{\mathrm{ad}} + k_{\mathrm{B}}T \ln \frac{\Theta_{\mathrm{ad}}}{1 - \Theta_{\mathrm{ad}}}. \tag{5.95}$$

Adsorbed atoms and molecules may possess low-lying vibrational energy levels so that the vibrational partition function (5.12) differs from unity. Adding the corresponding term yields

$$\mu_{\mathrm{ad}} = E_{\mathrm{ad}} + k_{\mathrm{B}}T \ln \frac{\Theta_{\mathrm{ad}}}{1 - \Theta_{\mathrm{ad}}} - k_{\mathrm{B}}T \ln Z_{\mathrm{vib,ad}}. \tag{5.96}$$

The vibration partition function contains a factor with the ground state energies $(1/2)\hbar\omega_i$ of the harmonic oscillators in the exponent (5.12). The sum of these ground state energies adds a temperature independent contribution to $E_{\mathrm{ad}}$. The remaining contribution of the vibrational partition function is small for most chemisorbed species since their vibration quanta are typically of the order of 300 K $k_B$ or higher. There is a noteworthy exception, however. The vibrational degrees of freedom that correspond to translations parallel to the surface can have very low frequencies of the order of a few meV. For example, the vibration frequencies of the CO translation modes on Ni(100) are $\hbar\omega_{top} = 3.2$ meV and $\hbar\omega_{bridge} = 3.7$ meV, for the atop site and the bridge site respectively [5.27]. The atop-site translational mode is twofold degenerate. This site has therefore the larger entropy. The higher vibration entropy reduces the free energy level so that at room temperature occupation of the atop site is preferred over the bridge site. At low temperatures, the bridge site with its larger binding energy wins. An entropy-induced conversion of adsorption sites for CO on Ni(100) [5.28, 29] is the consequence.

Molecules frequently dissociate upon adsorption into atoms or other fragments. In case of adsorption of a diatomic molecule like $H_2$, $N_2$ or $O_2$ the atoms reside in the same type of sites with the same energy. The chemical potential of that adsorbate phase is

$$\mu_{ad} = 2E_{ad} + 2k_BT \, \ln\frac{\Theta_{ad}}{1-\Theta_{ad}} - 2k_BT \ln Z_{vib,ad} \qquad (5.97)$$

An attractive or repulsive interaction between the adsorbed species can be mimicked in a mean field approach by adding a coverage dependent energy term $W(\Theta_{ad})$ to (5.96). The mean field approximation works particularly well for isotherms. In the context of adsorption, this mean field approximation goes under the name *Bragg-Williams approximation*. The approximation is equivalent to the molecular field approximation in magnetism.

## 5.4.2 Lattice Gas or Real 2D-Gas?

As an alternative model to the lattice gas model, one might neglect the lateral corrugation of the potential completely and assume that the atoms are bonded to the surface in a one-dimensional trough. Atoms can then be treated as two-dimensional van-der-Waals gas in which the atoms interact merely via a hard-core repulsive pair potential and can move about freely otherwise. If $N$ denotes the number of atoms in the surface area $A$ and $A_{ad}$ the area occupied by a single atom the total available area is $A-NA_{ad}$. Writing (5.9) for two dimensions and replacing the total area by the available area leads to an expression for the free energy of the 2D-van-der-Waals gas. After differentiation with respect to the number of particles at constant area, one obtains the chemical potential as

$$\mu_{2DvdW} = E_{ad} + k_BT \, \frac{\Theta_{ad}}{1-\Theta_{ad}} + k_BT \ln \frac{h^2}{2\pi m k_BT} \frac{1}{A_{ad}} \frac{\Theta_{ad}}{1-\Theta_{ad}} \qquad (5.98)$$

in which we have introduced the coverage $\Theta_{ad} = NA_{ad}/A$.

It is an interesting question whether an adsorbate is better described by the lattice gas or by the van-der-Waals gas model. The answer must depend on the magnitude of the lateral corrugation of the potential as well as on the temperature. Figure 5.11 shows an illustration of the potential (solid line) and the localized states of adsorbates in the potential as dotted lines. The energy levels near the minimum are approximately described by a harmonic oscillator. With rising energy, the levels moves closer together and eventually merge into a continuum of states. The lattice gas approximation assumes that the atoms sit in the ground state with some small occupation of the higher vibrational levels if one adds the corresponding term to the chemical potential (5.97). The 2D-gas model describes the chemical potential of atoms in the continuum of states above the potential maxima. For high enough temperatures, the atoms occupy primarily the continuum and the 2D-gas model should apply, whereas at low temperatures the atoms should reside in the ground state and form a lattice gas. The question is whether one can find a criterion for the transition temperature. The issue was already addressed in 1946 by Hill [5.30] and later again by Doll and Steele [5.31] who included the vertical motion of the adsorbate atoms in their statistical model. For a 2D-cosine potential, they found that the transition between the lattice gas and the free gas takes place at $T_c = 0.2 \, V_0 / k_B$, in which $V_0$ is the difference between the top and the bottom of the potential, i.e. approximately the activation energy for diffusion $E_{diff}$[8]. A qualitative, model independent estimate of the transition temperature may be obtained from the comparison of the chemical potentials of the independent lattice gas $\mu_{lat}$ (5.95) and the chemical potential of the 2D-gas $\mu_{2D}$ (5.98) considering the ground state of the latter as being shifted upwards by the amount $V_0$. The atoms prefer the state of lower chemical potential. Hence, the lattice gas model should be appropriate if $\mu_{lat} < \mu_{2D}$. This condition is equivalent to

$$V_0 > k_B T \left\{ \ln \frac{2\pi m k_B T \, A_{ad}}{h^2} - \frac{\Theta_{ad}}{1 - \Theta_{ad}} \right\} \tag{5.99}$$

According to (5.99) the atoms should always prefer the localized sites for higher coverages: in the 2D van-der-Waals gas, the atoms would be so much in each other's way that the chemical potential of the 2D-gas increases steeply.

In the low coverage limit, there is a temperature dependent critical depth of the potential $V_c$ beyond which the system behaves like a lattice gas. Figure 5.12 shows the critical potential $V_c$ for an atom of mass 16 and a hard core area $A_{ad}$ of 4 Å$^2$ for coverages between $\Theta_{ad} = 0$ and 0.8 in steps of 0.1. According to Fig. 5.12, strongly chemisorbed atoms like oxygen, carbon, and nitrogen should always form a lattice

---

[8] Doll and Steele discuss this result as being identical to the earlier result of Hill obtained without considering the vertical movement of the vibrating atoms. However, Fig. 1 of Hill's paper [5.30] clearly shows that Hill has the transition temperature a factor of four lower than the value quoted by Doll and Steele.

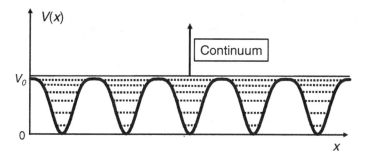

**Fig. 5.11.** Illustration of the lateral corrugation of the surface potential (solid line), the discrete energy levels of localized adsorbates (dotted lines), and the continuum of states at higher energies.

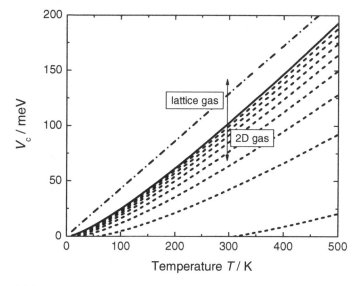

**Fig. 5.12.** Critical value of the potential corrugation in the limit of zero coverage (solid line) and coverages $\Theta_{ad} = 0.1, 0.2, \ldots, 0.8$ for an atom mass m=16 with a hard-core area $A_{ad} = 4$ Å$^2$. The dash-dotted line is the result of the solution for a 2D-cosine potential [5.31].

gas. Weakly chemisorbing species such as CO, in particular on metals with a filled d-shell, are closer to the limit. CO molecules do reside in defined sites; however, their chemical potential is affected by lowering the vibrational levels of the hindered translations. For higher coverages, CO is displaced from the low coverage sites by the lateral interactions and incommensurate or high order commensurate lattices are formed (see also Frenkel-Kontorova model, Sect. 1.3.1). Physisorbed layers are even more in the realm of 2D-gases, however without really conforming to the simple 2D-van-der-Waals models since the corrugation of the potential is still too strong for that. In summary, many systems are well described by the lattice gas model; hardly anyone conforms to the 2D-gas-model.

### 5.4.3 Segregation

A simple and illustrative example for the application of the lattice gas model is the problem of segregation. Bulk single crystal materials always contain impurities from the refining or the growth process. Some of these impurities do not form a substitutional alloy with the material, but rather they are dissolved in interstitial sites. The binding energy in these interstitial sites is frequently lower than on the surface, simply because there is more room at surfaces and the surface atoms of the substrate have free bonds ready to become engaged in bonding. If the temperature is high enough for effective diffusion, the bulk impurity atoms will segregate to the surface and form an adsorbate layer there. Typical examples are carbon and sulfur in metal crystals. In Section 2.2.3, we have remarked that segregation can be a significant problem for the preparation of clean surfaces. With the help of the lattice gas model, we show what precisely the nature of the problem is. Since the concentration of the impurities in the substrate is very small, we can write for the chemical potential $\mu_s$ of the dissolved impurities

$$\mu_s = E_s + k_B T \ln c_s \qquad (5.100)$$

in which $E_s$ is the energy of the impurity in the solid solution and $c_s$ is the concentration per available site in the bulk. By equating $\mu_{ad}$ and $\mu_s$ one obtains for the equilibrium surface coverage $\Theta_{ad}$

$$\Theta_{ad} = \frac{1}{1 + c_s^{-1} e^{\frac{E_{ad} - E_s}{k_B T}}} \cdot \qquad (5.101)$$

Figure 5.13 shows a typical example. It is assumed that the initial bulk impurity concentration is 10 ppm ($10^{-5}$). The difference in the binding energies at the surface and in the interstitial site is assumed to be $E_{ad} - E_s = -0.7$ eV. The solid line in Fig. 5.13 is the equilibrium surface concentration for that case. However, below a certain temperature bulk diffusion stops so that the equilibrium coverage is not reached. In Fig. 5.13, the temperature below which there is no further segregation is assumed to be $T_{min} = 500$ K. For the initial bulk concentration, the surface

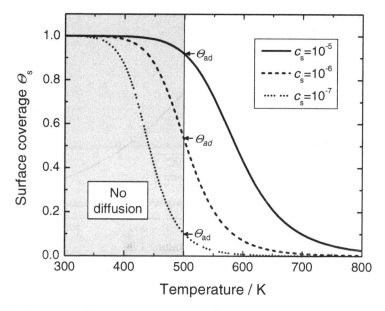

**Fig. 5.13.** Surface equilibrium concentration of a segregating impurity as a function of temperature for three different bulk concentrations $c_s$. It is assumed that the difference in the binding energy at the surface and in the interstitial site is $E_{ad} - E_s = -0.7$ eV. The shaded area marks the region where there is no diffusion, so that the actual surface coverages found experimentally after annealing are the ones marked at the upper diffusion limit, which is assumed to be at 500 K.

concentration is nevertheless quite high. This surface concentration can be removed effectively by sputtering the surface with noble gas ions (Sect. 2.3.2). In order to heal the surface damage the substrate must be annealed, whereupon further impurity atoms segregate to the surface. By repetitive sputter-annealing cycles, the bulk impurity concentration is reduced. This in turn reduces also the surface concentration but the removal of impurities by sputtering becomes less and less effective as the leaching process advances. Since many experiments require a surface cleanness of $10^{-4}$ to $10^{-6}$, surface preparation can become quite tedious.

Figure 5.14 illustrates the basic principle of segregation and the experimental problem it might cause. Actual experimental data on segregation rarely fit the simple isotherm so perfectly as the typical segregating impurities carbon and sulfur form various ordered phases as a function of coverage. Carbon may even form a graphitic overlayer. Figure 5.14 shows the segregation of carbon on Ni(100) [5.32]. The sample was intentionally doped with carbon. The segregation curve is fitted to (5.101) with $E_s\text{-}E_{ad} = 0.35$ eV and a carbon bulk concentration of 0.5%. The fit is not unique, however, since the experimental data do not conform to a simple isotherm.

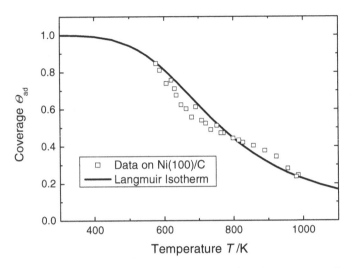

**Fig. 5.14.** Equilibrium carbon coverage on Ni(100) surface due to segregation from the bulk [5.32]. Coverage $\Theta_{ad} = 1$ is defined with respect to the maximum coverage with a c(2×2) layer. The solid line is a fit to (5.101) with a carbon bulk concentration of 0.5% and a segregation energy of 0.35eV. The fit is not unique, however.

## 5.4.4 Phase Transitions in the Lattice Gas Model

The field of 2D-phase transitions is enormously rich, in experiment as well as in theory. Even comparatively simple adsorption systems display complex patterns of various phases, commensurate, incommensurate or disordered ones as a function of coverage and temperature. Transitions between phases, especially the order-disorder transition as a function of temperature have been studied extensively using diffraction techniques. Particular attention was paid to the critical exponents, e.g. the temperature dependence of the intensity of a diffracted beam as the system moves from order to disorder. These critical exponents where discussed in terms of *universality classes* of particular theoretical models. Other interesting aspects concern the spatial distribution of coexisting phases and the domain walls separating the phases (Sect. 1.3.2). The material has been reviewed by Persson [5.2] and Patrykiejew et al. [5.33].

In the field of phase transitions, it proved to be more difficult than in most other areas of surface science to make theory match the experimental data. The theory of phase transitions, in particular if concerned with the critical exponents assumes a perfect homogeneity on the surface. By definition, critical exponents reveal themselves close to the phase transition when the spatial fluctuations of the system involve large surface areas. The closer the system is to a phase transition the more extended are the fluctuations. Small amounts of impurities or defects have therefore drastic effects on the behavior of the system. The intensity of a diffracted beam in an order-disorder transition, e.g. which according to theory should dive into the zero line as $(1-T/T_c)^{2\beta}$ with $\beta$ the critical exponent characteristic for a particular universality class, in reality displays a rounded shape and approaches zero rather gradually. This may render an unequivocal determination of the critical exponent impossible, likewise the unambiguous determination of the type of lateral interactions between the adsorbates. We exemplify these statements with the thoroughly and carefully studied order-disorder transition of hydrogen on Pd(100) [5.34, 35].

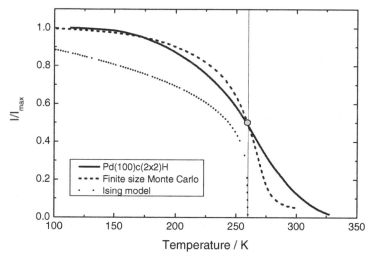

**Fig. 5.15** Normalized intensity of the (1/2, 1/2) diffraction peaks. Solid line: experimental result for Pd(100)c(2×2)H with $\Theta_{ad} = 0.5$ [5.34]. Dashed line: Monte Carlo simulation with repulsive nearest neighbor and attractive next-nearest neighbor interactions [5.35]. The dotted line is the intensity according to the Ising model with nearest neighbor interactions.

Figure 5.15 shows the experimental intensity of the (1/2, 1/2) spot for the c(2×2) hydrogen overlayer [5.34] as a function of temperature (solid line). As an example for the result of analytically solvable models, the intensity according to the Ising model with nearest neighbor interactions is plotted as a dotted line. The insufficient modeling of the lateral interactions causes the deviation between experiment and theory at lower temperatures. The qualitatively different behavior near the phase transition however is caused by inhomogeneity of the surface in the experiment. The inhomogeneity is crudely simulated by performing the calculation on a small lattice. The dashed line is the result of a Monte-Carlo simulation on a 40×40 lattice with periodic boundary conditions by Binder et al. [5.35] with repulsive nearest neighbor and attractive next-nearest neighbor interactions. The simulation matches the low temperature regime quite well and displays a tail beyond the transition temperature, in qualitative agreement with the experiment.

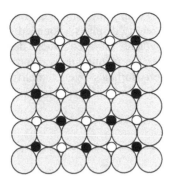

**Fig. 5.16.** Adsorbates in a square lattice with nearest-neighbor repulsive interactions. Black and white circles represent the A- and B-sites.

In order to elucidate the relation between the qualitative features of the phase diagram and the lateral interactions between the atoms we consider the Ising Hamiltonian (5.70) on a square lattice. We assume repulsive interactions $V_0$ between the nearest neighbors. Suppose one has a fractional coverage of $\Theta_{ad} = 0.5$. The system is perfectly ordered in a c(2×2) pattern if only every other site is occupied (black circles in Fig. 5.16). At the same time, the energy is minimal as no nearest neighbor sites are occupied. We denote the sublattice with the black circles by the letter A and the other one by B. Perfect order is then characterized by the fractional coverages

$$\Theta = \Theta_A / 2, \quad \Theta_B = 0 . \tag{5.102}$$

Here, $\Theta$ denotes the fractional coverage with respect to all sites. At higher temperatures, some of the nearest neighbor sites may become occupied, whereby the system becomes disordered. Half order diffraction spots persist, albeit with a lower intensity. Above a particular transition temperature $T_c$, the half order diffraction peaks vanish completely and the system is disordered. The state of disorder is characterized by an equal occupation of the sublattices A and B.

$$\Theta = \Theta_A = \Theta_B \tag{5.103}$$

We want to establish the relation between the transition temperature $T_c$ and $V_0$. A state close to the order-disorder transition is described by

$$\Theta_A = \Theta + \Delta, \quad \Theta_B = \Theta - \Delta \tag{5.104}$$

We assume now a mean field model in which the actual interaction $V_i$ of an atom in a particular site $i$ of the sublattice A with its neighboring atoms, which would depend on the actual occupation of the neighboring sites, is replaced by the interaction with the mean occupation of the neighboring sites.

$$V_{A,i} = \sum_{\delta=1}^{4} V_0 \Theta_{i,\delta} \Rightarrow V_A = 4V_0 \Theta_B \,. \tag{5.105}$$

The sum is over the four nearest neighbors. The mean interaction potential $V_A$ renormalizes the energy of all sites by the same amount. The occupation statistics remains Fermi-statistics. With (5.104) the fractional coverage $\Theta_A$ becomes

$$\Theta_A = \Theta + \Delta = \frac{1}{e^{\frac{4V_0(\Theta - \Delta) - \mu}{k_B T}} + 1} \tag{5.106}$$

Expanding (5.106) for small $\Delta$ and considering the limit $\Delta \to 0$ leads to the self-consistency equation for the transition temperature $T_c$

$$T_c = 4\Theta(1 - \Theta)V_0 / k_B \,. \tag{5.107}$$

This mean field results for the phase diagram is plotted in Fig. 5.17. The maximum value of $T_c$ occurs at $\Theta = 1/2$ and is

$$T_{c,\text{meanfield}} = V_0 / k_B \,. \tag{5.108}$$

This is considerably higher than the exact solution of the Ising-model

$$T_{c,\text{Ising}} \cong 0.57 V_0 / k_B \,. \tag{5.109}$$

The reason that $T_c$ is so much higher in the mean field model is the total neglect of fluctuations. The fluctuations are also responsible for the fact that in the Ising model the ordered c(2×2) is confined to a narrow coverage range between 0.37 and 0.63 (Fig. 5.17, dotted line). The inclusion of additional next-nearest neighbor attractive interaction stabilizes the c(2×2) structure in a wider coverage range. The dashed lines in Fig. 5.17 mark the range of stability for a ratio between next-nearest neighbor interactions $V_{nnn}$ to nearest neighbor interactions $V_{nn}$ of -1/2. For $V_{nnn} < 0$ the phase diagram shows a coexistence of the c(2×2) structure with a dilute and dense lattice gas phase, at low and high coverages respectively. Tricritical points occur at $\Theta \approx 0.31$ and $\Theta \approx 0.69$. All theoretical phase diagrams are symmetric around $\Theta = 0.5$ (*particle-hole symmetry*), whereas the experimental phase diagram displays some asymmetry. Such asymmetry may come about by three-body forces [5.35] or by energy associated with the displacement in the atom positions due to the lateral interactions. Such displacements occur if the nearest or the next-nearest neighbor coordination are not symmetric [5.2]. As shown by Persson

[5.2], the consideration of the lateral relaxation in the atom positions skews the phase boundary towards the left, in agreement with the experimental data (Fig. 5.17).

At the time when the phase diagram Fig. 5.17 was first investigated experimentally, the distribution of atoms on the available sites could merely be inferred from the diffraction data or obtained from Monte-Carlo simulations under the assumption of a particular interaction potential. After the advent of the scanning tunneling microscope, direct observations of the position of atoms became possible. The coexistence of ordered phases with a lattice gas was observed in STM-images e.g. by Wintterlin et al. [5.36].

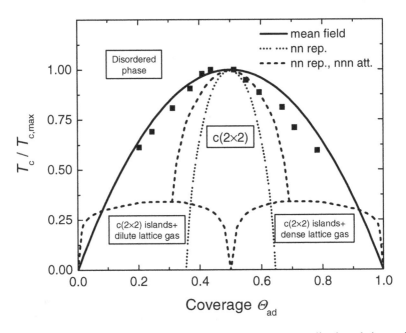

**Fig. 5.17.** Phase diagram for a square lattice. All curves are normalized to their specific maximum transition temperature $T_c$ at $\Theta = 1/2$. The solid line is (5.108) obtained for the mean field model with repulsive nearest-neighbor interactions. The dotted line is obtained from a Monte Carlo simulation with the same type of interaction [5.35]. The dashed lines mark the phase boundaries for a system with additional attractive next-nearest neighbor interactions. The data points (black squares) refer to the Pd(100)c(2×2)H system of Behm et. al [5.34].

# 6. Adsorption

Surface Science is an interdisciplinary field. This is particularly true for the science associated with adsorption and desorption. Physics and chemistry of localized and delocalized bond formation, thermodynamics and statistical physics, molecular dynamics as well as electrochemistry and catalysis meet here. Starting from a general discussion of bonding mechanisms, this sections deals with the thermodynamics of equilibrium phases and the kinetics of adsorption and desorption. The section concludes with a discussion of chemistry and the physical properties of the most common adsorbates.

## 6.1 Physisorption and Chemisorption – General Issues

The nature of the bonding that is involved in adsorption is addressed by the somewhat antiquated terms *physisorption* (= *physical adsorption*) and *chemisorption* (= *chemical adsorption*). The terms are not well defined since in the older literature the distinction between physisorption and chemisorption was made according to the adsorption energy, with physisorption denoting the realm of lower adsorption energies. There are, however, weak chemical interactions as well. In the following, we confine the term physisorption strictly to adsorption mediated by van-der-Waals forces. Van-der-Waals forces originate in the ground state fluctuations of the electronic charge of an atom, which generates a dynamic dipole moment $p_{fluct}$. The electric field emerging from this fluctuating dipole at the position of another atom at distance $r$ is proportional to $-p_{fluct}/r^3$. The field induces a dipole moment $p_{ind}$ of strength $p_{ind} \propto p_{fluct}/r^3$ in the second atom. The energy of the induced dipole $p_{ind}$ in the electric field of the original atom is negative (attractive interaction) and proportional to $r^{-6}$. Consequently, atoms attract each other even in the absence of chemical bonding. At smaller distances, Pauli-repulsion between closed shells eventually balances the attractive van-der-Waals interaction. Pauli-repulsion is proportional to the overlap of wave functions and increases therefore exponentially with decreasing distance. For convenience, the exponential dependence is traditionally replaced by an $r^{-12}$-dependence in analytical calculations. The result is the Lennard-Jones potential

$$V(r) = V_0 \left\{ \left( \frac{r_0}{r} \right)^{12} - 2 \left( \frac{r_0}{r} \right)^6 \right\}, \tag{6.1}$$

in which $-V_0$ is the potential at the equilibrium distance $r_0{}^9$. For a physisorption system, this potential is the only interaction between the adsorbate and the atoms of the solid phase. In contrast to chemical interactions, the van-der-Waals interaction is with <u>all</u> atoms of the solid. The interaction potential between an atom at $r_{atom}$ and the surface is therefore the sum over all two-body potentials (6.1),

$$V_{surf}(r_{atom}) = V_0 \sum_{n,\alpha} \left\{ \left( \frac{r_0}{|r_{atom} - r_{n,\alpha}|} \right)^{12} - 2 \left( \frac{r_0}{|r_{atom} - r_{n,\alpha}|} \right)^6 \right\}. \qquad (6.2)$$

The sum goes over all unit cells denoted by the triplet $n = (n_1, n_2, n_3)$ and the atoms in the unit cell denoted by $\alpha$. For the repulsive part only the nearest neighbors matter, for the attractive part, however, the summation has important consequences for the functional dependence of the physisorption potential on the distance $z$ from the surface. For not too small distances, the sum can be replaced by the integral

$$V(z) = \rho \, r_0^6 V_0 \int_0^\infty \frac{2\pi r \, dr \, dz'}{\left( (z+z')^2 + r^2 \right)^3} = \frac{\pi r_0^6 \rho V_0}{3 z^3} \qquad (6.3)$$

in which $\rho$ is the density of atoms. The integration over three dimensions reduces the $r^{-6}$ power law to a $z^{-3}$ dependence of the potential. Figure 6.1 displays the resulting potential for an fcc-structure when the van-der-Waals pair equilibrium distance $r_0$ is 2.6 in units of the substrate bond distance $a_s$, which is chosen to approximately represent the case of Xe on Pt(100). The lateral variation of the minimum in the potential (6.2) is illustrated in Fig. 6.2. Shown is the variation from the atop-position to the fourfold site where the potential has its minimum. The potential is rather anharmonic around the minimum. If represented by a two-dimensional Fourier-expansion, higher order Fourier-components must be employed. The $z$-position of the minimum potential varies with the lateral coordinate. The eigenstates of the physisorbed atoms are therefore not simply eigenstates of a two-dimensional oscillator (see also Sect. 5.4.2).

As straightforward as it is to write down an atom/surface potential for physisorption, the result is meaningful only in the limit of large distances from the surface. As the atom approaches the surface, other, chemical interactions come into play even in case of rare-gas adsorption. A well-studied example is the adsorption of rare-gases on transition metals. Experiments show a considerably higher heat of adsorption than expected for physisorption [6.1, 2]. The large reduction of the work function ($> 0.5$ eV) is indicative of a considerable charge transfer from the rare-gas atom into the solid. More importantly, the description of

---

[9] Occasionally, $V(r)$ is expressed in terms of a hard-wall distance. Then the factor 2 in the second term vanishes.

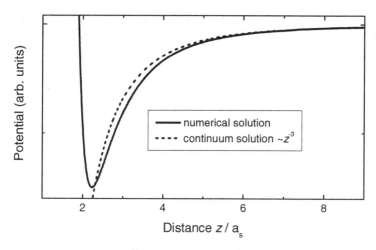

**Fig. 6.1.** Van-der-Waals bonding to a surface of an fcc-crystal. The parameters are chosen to represent Xe on Pt(100). The distance $z$ is in units of the surface lattice constant $a_s$. Solid line is the numerical solution for a pair-wise Lennard-Jones Potential. The dashed line is the continuum solution for the attractive part (6.3).

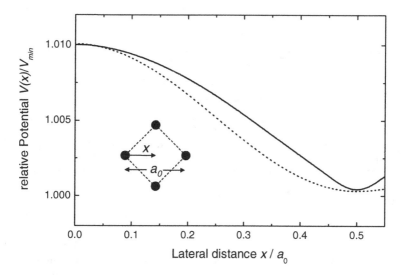

**Fig. 6.2.** Relative corrugation of the minimum in the physisorption potential (solid line). The potential deviates significantly from a simple cosine-function (dotted line).

the interaction of rare-gas atoms with surfaces in terms of van-der-Waals interactions yields a qualitatively wrong picture of the preferred adsorption site, the lateral interactions between the adsorbates and the vibrational states.

An early attempt to invoke density functional theory for the description of rare-gas adsorption was made by N. D. Lang [6.3] who studied rare-gases on a jellium substrate. J. E. Müller investigated the bonding of rare-gas atoms to transition metals [6.4]. Figure 6.3 shows the schematic picture developed by Müller for the example Xe on Pt. The overlap of the occupied Xe5p-states with the occupied Pt5d-states leads to bonding and anti-bonding states (dashed lines in Fig. 6.3). As both, the bonding and antibonding states are occupied the interaction is repulsive (Pauli-repulsion). The mixing with the unoccupied Pt5d states (named *polarization states* by Müller) leads to an overall downshift of the bonding and antibonding states, and thereby to a weak chemical bonding. Transferred into spatial coordinates Müller found that the polarization of the metal causes a charge increase where the Coulomb-potentials of the metal and the Xe-atoms overlap. This effect is stronger when the Xe-atom sits in an atop position! This is contrary to what the van-der-Waals bonding predicts (Fig. 6.2). Needless to say, that also the magnitude of the lateral corrugation of the potential as well as the curvature of the potential in the z-direction is not at all represented by the Lennard-Jones potentials! It is therefore no wonder that earlier attempts to describe experimental

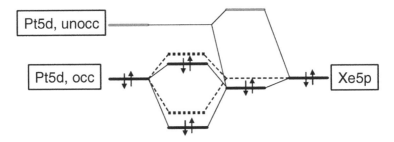

**Fig. 6.3.** Schematic picture of the bonding of rare-gas atoms on a transition metal for the example of Xe on Pt: The coupling of the occupied Xe5p-states with the occupied Pt5d-states leads to occupied bonding and anti-bonding states (dashed lines) and thereby to Pauli-repulsion. Mixing with the unoccupied Pt5d states (*polarization states*) leads to a considerable charge transfer and to an overall downshift of the electron states, and thereby to a weak chemical bonding.

observations concerning the adsorption energies, lateral interaction energies and the vibration modes of the Xe-atoms in terms of Lennard-Jones potentials failed. The theory provides furthermore an understanding of what causes the large work function shift. The unoccupied d-states are localized on the substrate. The participation of these states in the bonding requires a charge transfer from the Xe-atom to the surface. The associated dipole moment with the positive end pointing away from the surface causes the reduction of the work function. The charge transfer is nicely exemplified with charge density contours. Figure 6.4 displays the contour lines of the charge density $\rho(r)$ in a $Pt_{22}$-cluster with one Xe-atom adsorbed. In order to emphasize the changes brought about by the bonding the charge density of the bare cluster and the bare Xe-atom is subtracted, so that $\rho(r) = \rho(XePt_{22}) - \rho(Xe) - \rho(Pt_{22})$ is plotted in Fig. 6.4. The dashed and solid lines correspond to charge deficit and surplus, respectively. The charge transfer from the Xe-atom to the Pt-cluster is clearly visible. A significant part of the charge is located not on the Pt-atom to which the Xe-atom is bonded but rather in a ring around the Pt-atom. This has consequences for the bonding of the adjacent Xe-atom residing on the next-nearest neighbor Pt-atom (dashed circle in Fig. 6.4) since this second Xe-atom finds the states into which it would donate charge as a single atom already occupied by the first Xe-atom. The charge transfer to the substrate is therefore hindered. This causes a repulsive interaction term that partly compensates the attractive van-der-Waals interaction. The interaction remains attractive (18 meV), but is significant lower than a pure van-der-Waals interaction (30 meV) [6.4]. The hindered charge transfer for a neighboring Xe-atom causes also a considerable downshift in the frequency of the vibration perpendicular to the surface from $\hbar\omega = 8.5\,meV$ for a single Xe-atom to $\hbar\omega = 3.7\,meV$ for a full monolayer or a Xe-island [6.5].

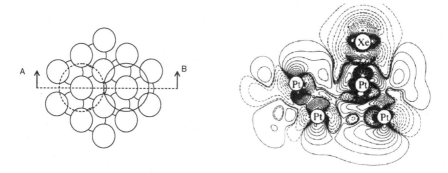

**Fig. 6.4.** Left panel: top view on the $Pt_{22}$-cluster with an adsorbed Xe-atom. The neighboring Xe-atom is drawn as a dashed circle. Right panel: charge density difference $\rho(r) = \rho(XePt_{22}) - \rho(Xe) - \rho(Pt_{22})$ along the intersection AB shown on the left. Dashed and solid contour lines indicate charge deficit and surplus, respectively. Note the net charge transfer into the substrate!

The specific bonding described above predicts that the strength of the Xe-metal interaction scales with the deepness of the metal potential, which in general correlates with the density of states at the Fermi-level. The bonding is particular strong on transition metals, but it should exist also for sp-metals and even for semiconductor surfaces and graphite, because of the unoccupied conduction band states.

It should be noted that the picture developed here is based on a localized, quantum-chemical approach to the problem. In that approach, the natural basis-set of wave functions are the eigenstates of individual atoms and the discussion is in terms of local charges and bonds. A somewhat different picture emerges from theories that start from periodic lattices and describes the results in terms of the global density of states [6.6]. Neither one picture can be proven right or wrong. They are just different, more or less appealing ways to rationalize a result of quantum mechanics.

While the existence of a genuine van-der-Waals bonding on any of the surfaces that are in the mainstream of interest remains elusive, van-der-Waals forces dominate at large distance from the surface. The $z^{-3}$-dependence of the attractive van-der-Waals interaction (dashed line in Fig. 6.1) exists for all molecules approaching the surface. The van-der-Waals attraction together with a polarization interaction and a Pauli-repulsion at close distances is therefore a basis for the discussion of the adsorption of chemically saturated molecules to which we turn now. Cases of particular interest are the dissociative adsorption of the diatomic molecules $H_2$, $O_2$, and $N_2$. These molecules, when they approach the surface, experience a potential similar to the one drawn in Fig. 6.1: a van-der-Waals attraction at large distances and a Pauli-repulsion at short distances, if the molecule does not dissociate. In addition, a state of molecular bonding may exist. The $O_2$ molecule, e.g., binds strongly as a molecule.

Figure 6.5 illustrates the energetics of the adsorption process. The energy[10] of the undissociated molecule as a function of distance from the surface $z$ is plotted as a solid line. This may be the physisorption energy or the energy of the chemical bonding of the undissociated molecule with the surface. The figure also shows two examples of the energy variation of the atoms when the molecule is dissociated in the gas phase (dashed and dash-dotted lines). At large distances from the surface, the curves begin at half the dissociation energy of the molecule $E_{diss}$. Evidently, dissociation upon adsorption occurs only if the energy gained in bonding is larger than the dissociation energy. Depending on the dissociation energy, of the adsorption energy, on the equilibrium distances in the bound states, and on the overall shape of the potential curves, the crossover point between molecular and dissociative adsorption may or may not lie above zero (with reference to the ground state of the molecule in the gas phase). If the crossover point is above zero, the molecule has to overcome an activation barrier for dissociative adsorption.

The one-dimensional model for dissociative adsorption as displayed in Fig. 6.5 grossly oversimplifies the problem. The molecule has internal degrees of freedom (vibration, rotation, equilibrium distance between the atoms), the adsorbed atoms and the solid have vibrational degrees of freedom, and the molecule approaches the surface from different angles and strikes the surface at a different position with respect to the surface structure. For each vibration or rotational eigenstate of the molecule, for each angle of approach, each point of contact, one has a different energy/distance relation. Furthermore, while the molecule is on its course of approach, the bond length, the vibrational levels, the rotational energies respond to the interaction with the solid, the occupation of the energy levels in the molecule

---

[10] For simplicity and generality we use the term *energy*. The term stands for the correct thermodynamic potential according to the thermodynamic boundary conditions (Chapt. 4).

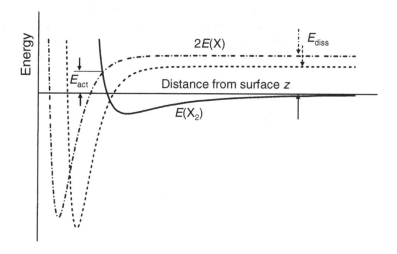

**Fig. 6.5.** Grossly simplified illustration of the energetics of the adsorption process for diatomic molecules: The solid line represents the physisorption potential or the potential of whatever bond the molecule may establish as a whole with the surface. The dashed and dash-dotted lines show two possible cases for the energy of the dissociated molecule. Depending on the energy of dissociation in the gas phase, the equilibrium bond distance and the adsorption energy for the atoms, there may be a barrier for dissociation or not.

changes, rotational energy is transferred into vibrational energy and kinetic energy is converted into phonons. Even electronic states may be excited, so that one leaves the realm of the Born-Oppenheimer approximation. Adsorption induced electronic excitations can give rise to chemo-luminescence [6.7] and hot electron emission [6.8]. In recent years, laser-technology has developed to a point where one is able to control and select the exact eigenstate of molecules in the gas phase. The energy transfer to the solid can be investigated in femtosecond pump-probe experiments, which has opened the new field of state selective chemistry [6.9].

Because of its extreme complexity and its multidimensionality, the problem of dissociative chemisorption is not amenable to a full quantum theoretical treatment. Currently, calculations involving 7 degrees of freedom are state of the art. A classical approach using molecular dynamics in connection with a potential derived from ab-initio calculations is however possible (see e.g. A. Gross [6.10])

Without straining mental and computer capacities too much, one may move up one step from the simple scheme in Fig. 6.5 and consider the approach of a diatomic molecule with a fixed orientation with respect to the surface. Vibrations and rotations as well as the structure of the surface are neglected. As an example, we consider the approach of a molecule, which has its axis parallel to the surface. Far away from the surface, the distance between the atoms is at its gas-phase equilibrium value $r_m$. Figure 6.6 displays the equipotential contour lines of the molecule as it approaches the surface. It is assumed that no molecular adsorption state (as typical for CO or $O_2$) exists. The molecule travels along the path of minimum energy (dotted line). For smaller distances from the surface, the wave functions of the molecule and the solid begin to overlap and the equilibrium bond distance between the molecule increases. The molecule may or may not encounter an activation barrier; eventually the potential becomes completely flat with respect to the distance between the atoms: the molecule is dissociated. For a structured surface, one would have a different 2D-potential for each $x$, $y$-starting position in the gas-phase with respect to the structured surface and each initial orientation of the molecule. The potential would also depend on the initial angle of the molecular trajectory with respect to the surface. The orientation of the molecule would change as it approaches the surface. In total, the model would involve six independent coordinates.

While calculations that include vibrational and rotational excited states can be performed nowadays, one would like to explore what can be said about the ther-

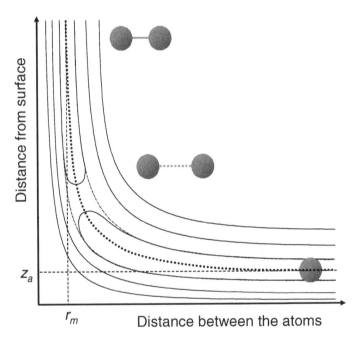

**Fig. 6.6.** Schematic drawing of the contour lines of the atom potential for a molecule approaching a flat surface ("*elbow plot*"). For large distances, the potential has a minimum at the gas-phase equilibrium bond distance. As the molecule draws closer to the surface, the bond distance increases. Eventually the potential is independent of the atom-atom distance. The molecule may or may not encounter an activation barrier along the path (solid and thin dash-dotted contour lines).

modynamics of dissociative adsorption without resorting to sophisticated theory. With respect to the probability of dissociative adsorption, a simple approach is to postulate that the molecule travels on the hypersurface of coordinates along the path of minimal energy so that the effective activation barrier is the minimum barrier along this *reaction coordinate*. The probability of dissociative adsorption, the *sticking coefficient,* can then be expressed in terms of certain properties of the transition state, which leads to the transition state theory of rates (Sect. 10.2).

The complexity of the molecule-surface interaction is also reflected in the desorption process. Dissociatively adsorbed molecules recombine during desorption. If the molecules travel along a path that involves an activation barrier, they must gain kinetic energy. Concerning the velocity component perpendicular to the surface, the desorbing molecule must have a kinetic energy that reflects the height of the activation barrier. Part of the energy may also go into vibrational and rotational excitations. Even in the absence of an activation barrier, the occupation of rotational and vibrational states is in general not given by the thermal equilibrium distribution that would correspond to the substrate temperature $T_s$. One might think of building a clever *perpetuum mobile* (perpetual motion engine) by adsorbing molecules at surfaces and receiving hyperthermal molecules desorbing in return. Alas, the *principal of detailed balance* ensures that the second law of thermodynamics is not violated. For each direction of the impinging molecule, each velocity, each point of impact, each internal excitation, the sticking coefficient is such that the differential fluxes in and out are identical in equilibrium. Alternatively, one may argue that the principle of detailed balance is the reason why desorbing molecules have a nonthermal velocity distribution and a nonthermal distribution on the vibrational and rotational levels since evidently these molecules in general should have a different sticking probability.

Detailed balance is an extremely powerful principle. It can be used to perform calculations on the kinetics starting at either end of a process and obtain information on the reverse process. It is also an indispensable check on quantum statistic calculations that detailed balance is obeyed. Not all models or assumptions one might be inclined to make or to employ for the sake of easing the task fulfill this requirement. On a more elementary level, the principle of detailed balance can be applied to express the rate of desorption into a particular angle in terms of the sticking coefficient of molecules traveling the reverse direction and the chemical potential of the adsorbed phase. This is helpful insofar as the sticking probability is a measurable quantity, and it is often close to unity in case of non-activated adsorption.

## 6.2 Isotherms, Isosters and Isobars

### 6.2.1 The Langmuir Isotherm

We consider the equilibrium between an ideal gas and an adsorbed phase described by the non-interacting lattice gas model. By equating the chemical potentials (5.15) and (5.96, 5.97) one obtains

$$\left(\frac{\Theta_{ad}}{1-\Theta_{ad}}\right)^{\alpha} = e^{\frac{E_g - \alpha E_{ad}}{k_B T}} \frac{h^3}{k_B T (2\pi m k_B T)^{3/2}} \frac{1}{Z_r} \frac{Z_{vib,ad}^{\alpha}}{Z_v} p .$$  (6.4)

Here, $E_g$ is the ground state energy level of the gas phase and $\alpha$ denotes the number of (equal) atoms into which in molecule dissociates upon adsorption, hence $\alpha$ is either 1 or 2. For $\alpha = 1$, equation (6.4) represents the *Langmuir Isotherm*. The isotherm describes the equilibrium between two phases, the gas phase and the adsorbate phase. It is therefore, that only the properties of the two phases enter, and not the pathway by which the equilibrium is established.

This statement may warrant a further remark. In the chemical literature, the Langmuir Isotherm (and other equilibrium properties) is almost exclusively derived by considering adsorption and desorption <u>kinetics</u> and by equating the adsorption rate with the desorption rate to allow for equilibrium. This treatment introduces (within that framework) undefined kinetic parameters, and thereby obscures the true nature of the Langmuir Isotherm. Furthermore, the dependence on $\Theta_{ad}$ (the left hand side of (6.4)) is introduced as a property of the adsorption kinetics, namely that the dependence of the sticking probability on the coverage should be proportional to $1 - \Theta_{ad}$, which is rarely if ever the case. Furthermore, according to this "derivation", the isotherm would depend on the coverage dependence of the sticking probability, which is incorrect.

Solving (6.4) for the coverage $\Theta_{ad}$ yields

$$\Theta_{ad} = \frac{(K_{eq}(T) p)^{1/\alpha}}{1 + (K_{eq}(T) p)^{1/\alpha}} .$$  (6.5)

The equilibrium constant $K_{eq}(T)$ is

$$K_{eq}(T) = e^{\frac{E_g - \alpha E_{ad}}{k_B T}} \frac{h^3}{k_B T (2\pi m k_B T)^{3/2}} \frac{1}{Z_r} \frac{Z_{vib,ad}^{\alpha}}{Z_v} .$$  (6.6)

Figure 6.7 displays the isotherms for $\alpha = 1$ and 2. For non-dissociative adsorption, the coverage rises linearly with the pressures in the small coverage/pressure regime to level off as the coverage increases. For dissociative adsorption of a di-

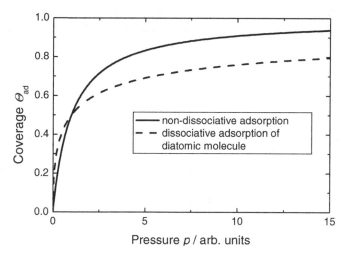

**Fig. 6.7.** Adsorption isotherms in the non-interacting lattice gas model for non-dissociative adsorption (Langmuir Isotherm) and the dissociative adsorption of a diatomic molecule.

atomic molecule, the coverage initially increases proportional to the square root of the pressure, and levels off more quickly. Full coverage may not be reached at all for kinetic reasons. As a rule, dissociative adsorption requires two empty nearest-neighbor sites rather than two empty sites somewhere on the surface. As the coverage approaches unity, the number of nearest neighbor pairs reduces more rapidly than the number of pairs at some arbitrary distance. Eventually, only single sites with no nearest-neighbor empty site around will remain on the surface. The full coverage is therefore not reached at all. Rather the adsorption stops at about 95% coverage [6.11]. Note, however, that this is a kinetic argument stating that equilibrium may not be reached; it does not concern the equilibrium itself. Given enough diffusion and long enough time, the dispersed empty sites will disappear.

## 6.2.2 Lattice Gas with Mean Field Interactions – the Fowler-Frumkin Isotherm

Experimental isotherms rarely conform to the Langmuir Isotherms because of the lateral interactions between adsorbed species. Figure 6.8 shows three isobars for the molecular adsorption of CO on a Ni(111)-surface as an example [6.12]. The pressure is varied by an order of magnitude. The decay of the coverage with increasing temperature is much slower than for the Langmuir Isotherm (dotted line). The slower decay can be interpreted in various ways. The binding energy of CO to the Ni-substrate might decrease when neighboring sites become occupied, different sites may become occupied, or there may be a direct repulsive interaction between the CO-molecules. Isotherms do not distinguish between these possibilities.

To account for a variation of the mean adsorption energy with coverage one may write an additional coverage dependent term into the chemical potential of a lattice gas (5.96).

$$\mu_{ad} = \alpha\left(E_{ad} + k_B T \ln\frac{\Theta_{ad}}{1-\Theta_{ad}} - k_B T \ln Z_{vib,ad}\right) + W(\Theta_{ad}). \qquad (6.7)$$

The factor $\alpha$ is $\alpha = 1$ for atom adsorption and $\alpha = 2$ for the dissociative adsorption of diatomic homonuclear molecules. The expression is easily generalized to diatomic molecules consisting two different atoms. The term $W(\Theta_{ad})$ describes the change in the adsorption energy with coverage. It can also be understood as a mean field approach to describe lateral interaction between the adsorbed species. In the context of adsorption, this mean field approximation is called *Bragg-Williams approximation* (cf. Sect. 5.4.1). The approximation is equivalent to the molecular field approximation in magnetism (Sect. 9.5)

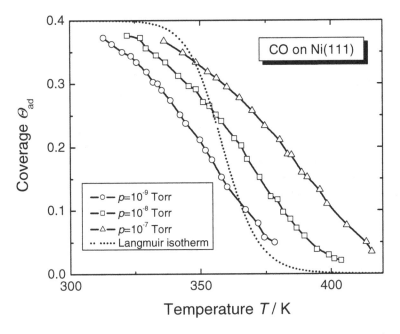

**Fig. 6.8.** Adsorption isobars for CO on Ni(111) [6.12]. The decay of the coverage is much slower than in the Langmuir Isotherm since the effective binding energy increases as the coverage becomes smaller. The coverage is here defined as the number of CO-molecules per surface atom.

With the mean field interaction added, the isotherm assumes the form

$$\left(\frac{\Theta_{ad}}{1-\Theta_{ad}}\right)^{\alpha} e^{\frac{W(\Theta)}{k_BT}} = K_{eq}(T)\ p\ . \tag{6.8}$$

The equilibrium constant $K_{eq}(T)$ is as in (6.6). A positive $W(\Theta_{ad})$ stands for an increasing energy level of the adsorbed state for larger coverages, hence for repulsive interactions between the adsorbed species. Negative $W(\Theta_{ad})$ stand for attractive interactions. Equation (6.8) holds for arbitrary analytical forms of $W(\Theta_{ad})$. Consider for example repulsive dipole/dipole interactions: The dipoles can be electric dipoles or elastic dipoles (3.81). The interaction energy scales with the distance $r$ as $r^{-3}$ and $W(\Theta_{ad})$ becomes

$$W(\Theta_{ad}) = w\Theta_{ad}^{3/2}\ . \tag{6.9}$$

Dipole interactions are of the order of a few meV and are therefore small compared to other interactions or to the variations in the adsorption energy. To account for the latter, frequently a linear variation of $W(\Theta_{ad})$ with $\Theta_{ad}$ is assumed,

$$W(\Theta_{ad}) = w\Theta_{ad}\ . \tag{6.10}$$

Depending on the scientific community, the resulting isotherms are named *Fowler-isotherms* (physics) [6.13] or *Frumkin-isotherms* (electrochemistry) [6.14].

Figure 6.9 shows a set of isotherms for non-dissociative adsorption and various values of $w$ in units of $k_BT$. The coverage is now plotted vs. the logarithm of the pressure, which is proportional to the chemical potential of the gas-phase. The case $w = 0$ represents the Langmuir Isotherm. For positive values of $w$, the equilibrium pressure is higher for a given coverage, the coverage increases more slowly with rising pressure. For $w < 0$, the isotherm is steeper. All isotherms are symmetric around $\Theta_{ad} = 0.5$. The slope at this point is

$$\frac{\partial \Theta_{ad}}{\partial \ln p} = \frac{1}{4+w/k_BT}\ . \tag{6.11}$$

A rough estimate of the interaction energy is therefore obtained from the slope at $\Theta_{ad} = 0.5$,

$$w = k_BT\left(\frac{\partial \Theta_{ad}}{\partial \ln p}\right)^{-1} - 4k_BT\ . \tag{6.12}$$

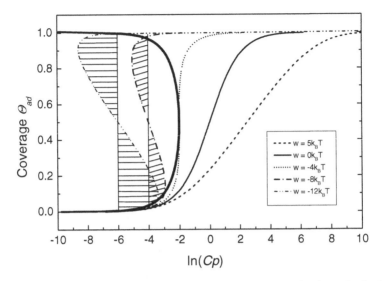

**Fig. 6.9**. Adsorption isotherms with lateral interactions between adsorbates in the Bragg-Williams approximation. The coverage is plotted vs. the logarithm of the pressure, i.e. vs. the chemical potential of the gas phase. The thin solid line ($w = 0$) is the Langmuir Isotherm. For $w > 0$, the coverage rise more gradually with $\ln p$ ; the chemical potential of the gas phase has to be larger to overcome the repulsive interactions in the adsorbed phase. For attractive interactions, the uptake is faster. Eventually, the adsorbate phase condenses into a *"lattice liquid"*. The thick solid lines separates stable from instable regions.

Similarly, from the slope of isobars at $\Theta_{ad} = 0.5$ the lateral interaction is obtained as

$$w = -Q\left( T\frac{\partial \Theta_{ad}}{\partial T} \right)^{-1} - 4k_B T \ . \tag{6.13}$$

At a critical value $w_c = -4k_B T$ the slope of the isotherm becomes infinite at $\Theta_{ad} = 0.5$, and below $-4k_B T$ one obtains formally three different coverages for a given pressure. This means that the system is instable and not adequately described by the continuous curves. As in the case of the van-der-Waals gas, the true curve is given by the Maxwell construction of a vertical line leaving the same area to the calculated curves on both sides of the line (thin solid lines in Fig. 6.9). The resulting isotherms then call for an increasing coverage in the low-pressure regime. At a particular critical pressure, which corresponds to the critical coverage given by the fat solid line in Fig. 6.9, more atoms are adsorbed without a pressure increase until an upper coverage limit given by the same fat line is reached. From thereon the coverage increases again according to the rising pressure. The system behaves very much like a condensation into a liquid state, although the adsorbate remains in a lattice gas state. We remark that negative values of $w$ stand for attractive interactions between the adsorbed species. The microscopic structure of the condensed adsorbed phase consists therefore of (large) close-packed adsorbate islands and uncovered areas on the surface.

The interesting isotherms produced by attractive interactions are typical for rare-gas adsorption, where the adsorbed atoms attract each other through van-der-Waals forces. As discussed in Sect. 6.1 this attraction may be partly balanced by repulsive chemical interactions. However, even then the interaction remains attractive. Figure 6.10 shows the adsorption isotherms of xenon on graphite for three different temperatures as measured by Suzanne et al. [6.15]. Beyond a particular pressure, which depends on the temperature, the coverage rises abruptly from a lower to an upper critical value (compare Fig. 6.9). The process repeats for further monolayers at higher pressures until the effect of the substrate vanishes and the condensed phase grows indefinitely. From the isotherms, one obtains a heat of adsorption of about 0.24 eV/atom. This value is comparable to the heat of adsorption of xenon on platinum [6.2]. This and the fact that further layers condense at higher pressures is indicative of a chemical interaction between graphite and xenon as discussed in Sect. 6.1 for xenon on platinum.

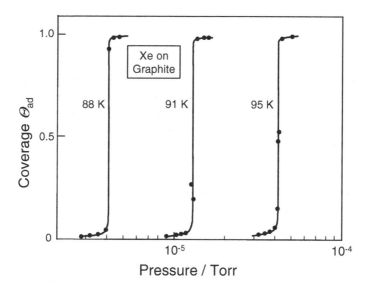

**Fig. 6.10** Isotherms at three different temperatures for the first layer adsorption of xenon on graphite (after ref. [6.15]). Above a particular pressure, the coverage rises abruptly from the lower critical coverage to the higher critical coverage. The curve repeats itself for the second and further monolayers until the influence of the substrate vanishes and the condensed phase grows infinitely. A heat of adsorption of about 0.24 eV/atom is derived from the isotherms.

Lateral interactions between chemisorbed species are frequently attractive at long distances and always repulsive at short distances. However, the attractive part does not necessarily show up in the isotherms, as in that case the chemisorbed species tend to grow in islands of a particular structure. For denser layers, the adsorption energy typically decreases as more substrate surface bonds become engaged in the bonding to the adsorbate. While this effect is not a lateral interaction between adsorbates, it has the effect that the energy level of the adsorbate rises with the coverage. Isotherms, which are less steep than the Langmuir Isotherm, are therefore typical for chemisorption. In terms of the Bragg-Williams model, this means positive values of $w$. With a proper choice of $w$, or at least with a particular function $W(\Theta_{ad})$ a quantitative agreement with experimental isotherms or isobars is nearly always obtained. Since isotherms represent an integral property of the gas-surface interactions, fine details of the lateral interactions such as phase transitions in adsorbed layers are hardly visible in the experimental data due to the limited precision of the measurements.

## 6.2.3 Experimental Determination of the Heat of Adsorption

Experimentally, the ambient pressure is not as easily varied as the temperature. Adsorption equilibria are therefore mostly investigated with isobars: Here, the coverage is measured as a function of the surface temperature while the ambient pressure is held constant. Figure 6.11 displays a set of isobars that were calculated from the Langmuir Isotherms. The parameters are chosen such that the curves could represent very crudely CO-adsorption on transition metals in ambient pressures ranging from $10^{-9}$ to $10^{-5}$ mbar. The temperature dependence of the vibrational and rotational partition functions is neglected. From curve to curve, the pressure is varied by one order of magnitude. If it were not for the temperature dependent prefactor factors in (6.4), each order of magnitude in the pressure would displace the isotherm by the same amount. Experimental data on a set of isobars as displayed in Fig. 6.11 can be used to calculate the isosteric heat of adsorption $Q(\Theta_{ad})$ by employing the Clausius-Clapeyron equation,

$$Q(\Theta_{ad}) = -\frac{\partial \ln p_{eq}}{\partial(1/k_B T)} .$$ 

(6.14)

This heat of adsorption is nearly but not completely equal to the difference in ground state energy of the gas-phase $E_g$ and $E_{ad}$. Neglecting again the contributions from the rotational and vibrational partition functions in (6.4) one obtains

$$Q(\Theta_{ad}) = E_g - E_{ad} + \frac{5}{2} k_B T .$$ 

(6.15)

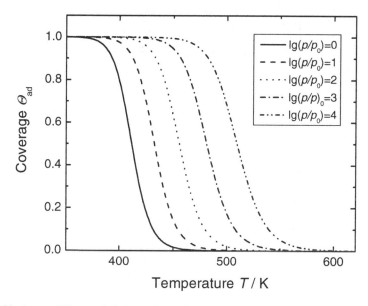

**Fig. 6.11.** A set of Langmuir Isobars for pressures varying by an order of magnitude. Because of the temperature dependent prefactor in (6.4) the Langmuir Isobars are shifted by an amount which increases with the temperature.

For chemisorbed systems, the energy of adsorption $\Delta E_{ad} = E_g - E_{ad}$ is of the order of one eV or more, so that the difference between the heat of adsorption $Q$ and $\Delta E_{ad}$ becomes marginal. For weakly chemisorbed species and isobars obtained at higher pressures, and comparatively high temperatures, the difference could matter. In principle, the temperature dependence of the heat of adsorption could also be determined from the isobars using the Clausius-Clapeyron equation. In reality, the data rarely cover a sufficiently large $p/T$-range to do so. Figure 6.12 displays the typical Arrhenius-plot for $\ln p_{eq}(\Theta_{ad})$ vs. $T^{-1}$ that would be used to determine the heat of adsorption. Even for the pseudo-experimental data calculated from the Langmuir Isotherm, which is free of experimental errors, the deviations of the data points from a straight line (solid line in Fig. 6.12) are hardly visible. Therefore, experiments cannot resolve the temperature dependence of the prefactor and of the heat of adsorption.

It is important to realize in this context that nominal equilibrium experiments on single crystal surfaces in vacuum are strictly speaking not completely in equilibrium insofar as only the temperature of the sample is varied. The temperature of the gas phase is determined by the walls of the vacuum chamber and therefore remains at about 300 K. This means that only the temperature in the temperature dependent terms of the chemical potential of the adsorbed state is varied [6.16]. With respect to the isotherms, these are the temperatures in the exponential term and in the vibrational partition function of the adsorbed state.

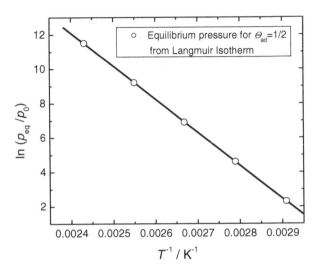

**Fig. 6.12.** Arrhenius plot of the equilibrium pressure for which the coverage is $\Theta_{ad} = 0.5$ as calculated from the Langmuir Isotherm (6.4) (Fig. 6.11). The variation in the slope caused by the temperature dependent prefactor is hardly visible, despite the fact that the pressure range covers four orders of magnitude. Given the experimental errors, the temperature dependence of the heat of adsorption and therefore the temperature dependence of the prefactor cannot be obtained from experimental data.

Neglecting the vibrational excitations, the heat of adsorption obtained from iso-bars measured in an UHV-chamber is

$$Q_{UHV}(\Theta_{ad}) = E_g - E_{ad} .\tag{6.16}$$

As an example we show the heat of adsorption of Xe on Pt(111) as a function of coverage in Fig. 6.13. The coverage was measured using the elastic scattered intensity of a thermal beam of He-atoms [6.2]. The method is extremely sensitive and does not perturb the adsorbed layer. Up to a coverage of about $\Theta = 1/3$ the heat of adsorption rises because of the attractive interactions between the Xe-atoms. After that, the hard-core repulsion reduces the heat of adsorption.

For a long time, Arrhenius-plots as shown in Fig. 6.12 and thermal desorption spectra (Sect. 6.3) were the only way to obtain information on the heat of adsorption on well-defined single crystal surfaces. Direct calorimetric measurements were not feasible because of the very small heat released in an adsorption process in relation to the heat capacity of a bulk crystal. The development of an extremely sensitive single crystal calorimeter in combination with a molecular beam adsorption by D. A. King and collaborators was a major step forward [6.17, 18]. The

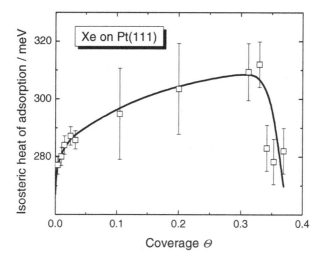

**Fig. 6.13.** Isosteric heat of adsorption of Xe on Pt(111) (after Kern et al. [6.2]). The coverage $\Theta$ is the fractional coverage in relation to the number of surface atoms. For small coverages, the heat of adsorption increases with coverage due to attractive interactions. At about $\Theta = 1/3$, the hard-core repulsion sets in and the heat of adsorption drops sharply. The solid line is a heuristic fit with $Q/\text{meV} = 267 + 65\Theta^{1/3} - 2 \times 10^{10}\Theta^{20}$. The fit is used later in the context of thermal desorption.

equipment is shown schematically in Fig. 6.14. The metal single crystals have a thickness of merely 0.2 μm. The crystal is prepared by epitaxial growth on a water dissolvable single crystal salt (e.g. NaCl). The supporting crystal is dissolved afterwards to obtain a freestanding film, which is then mounted on a support ring. The gas is dosed from a molecular beam source. The source is calibrated employing the spinning rotor gauge (Sect. 2.2.1). The gauge sits in a housing featuring a small tubular orifice, which can be moved into the beam. The equilibrium pressure inside the housing when the beam is directed into the tubular orifice can be converted into the number of particles in the beam per second, since the pressure and the conductance of the tube yield the flux out of the housing. The sticking coefficient on the sample is measured by observing the signal of reflected molecules from the beam using a mass spectrometer. If the signal is as high as when an inert gold surface is moved into the beam instead of the sample then the sticking coefficient is zero. It is one, when no beam molecules are reflected from the sample. The heat of adsorption increases temporarily the temperature of the sample. The increase is measured by an infrared detector. Light pulses of defined energy from a He-Ne Laser calibrate the IR-bolometer. The results for the heat of adsorption obtained by the direct measurements agree in general rather well with data obtained from isobars [6.18].

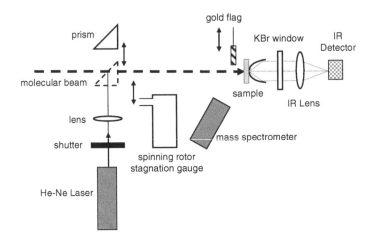

**Fig. 6.14.** Set-up for a direct measurement of the heat of adsorption after Stuck et al. [6.18].

## 6.2.4 Underpotential Deposition

The term *underpotential deposition* (upd) refers to the deposition of metal ions with a charge of $+ze$ ( $Me_A^{z+}$ ) on a substrate of a different material, mostly another metal $Me_S$. If the equilibrium potential at which this deposition occurs is positive of the equilibrium potential for the formation of the solid phase of the deposited metal, then this phenomenon is called underpotential deposition and the corresponding difference is the upd-shift. The reason for the existence of upd is the same as for the filling of a first monolayer (or the sequential filling of more layers) of a rare-gas on a substrate at a pressure below the equilibrium pressure of its solid phase. The binding energy between the rare-gas atoms and the substrate is larger than the binding energy in a kink site of the crystalline phase of the rare-gas. For rare-gas/solid interactions, the difference in the binding energy could amount to a factor of two. For metal-on-metal deposition, the differences are smaller. Because of this smallness, other factors such as the structure of the upd-layer as a function of the density, or a possible stabilization of the layer by ions from the electrolyte play an important role for the magnitude of the upd-shift and as to whether upd occurs at all.

The processes, which are compared, are illustrated in Fig. 6.15. The equilibrium between the solid phase and the electrolyte concerns the equilibrium between a metal atom at a Halbkristallage (kink site in half crystal position, see Sect. 1.3.2) and the metal-ion in the electrolyte (Fig. 6.15a). The reaction between the two states of the metal atom can be written as

$$Me \Leftrightarrow Me^{z+} + ze^-$$ (6.17)

The energies involved in the reaction are the cohesive energy, the energy required to remove an electron from the bare atom, the energy to bring the electron from the vacuum level into the solid, and the solvation energy of the metal ion. The electrode potential at which this reaction is in equilibrium is the Nernst-potential.

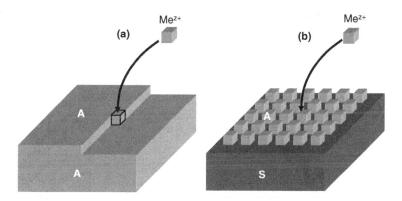

**Fig. 6.15.** **(a)** Electrochemical deposition of bulk material A and **(b)** of a layer of the material A on a substrate S of a different material. If the latter case **(b)** occurs at a potential that is positive of the Nernst-potential of material A it is called _underpotential deposition_ (upd). The upd-layer is stable in the potential range between the upd-potential and the Nernst-potential. Underpotential deposition requires that the mean binding energy for an atom in the upd-layer of material A is larger than the cohesive energy (binding energy to a kink site) of material A.

The upd-reaction can be written as

$$Me_S + Me_A \Leftrightarrow Me_S + Me_A^{z+} + ze^- \qquad (6.18)$$

The equilibrium electrode potential depends on the structure of the substrate and the adsorbed layer. Because of the lateral interactions in the upd-layer, the equilibrium potential also depends on the coverage. Phase transitions in the upd-layer affect the upd-potential as well as the possible co-adsorption of anions from the electrolyte and the formation of a compound structure with these anions. In other words, upd-layers are even more complex than the typical adsorbate layers in UHV-physics. However, just as for adsorbates in vacuum one does not need to have the full understanding of the complexity of upd-layers to obtain a qualitative picture of the behavior of isotherms.

In order to describe the equilibrium thermodynamics of upd we need expressions for the chemical potentials of the various phases involved. We begin with the electrolyte. We assume that the electrolyte is sufficiently dilute so that ions do not interact with each other. In that case, their partition function is that of independent particles like for the ideal gas. From the free energy $F$ of the ideal gas (5.9) one obtains for the chemical potential in terms of the density $\rho$,

$$\mu = \mu_0 + k_B T \ln\left\{\rho (h^2 / 2\pi m k_B T)^{3/2} /(Z_{\text{vib}} Z_{\text{rot}})\right\} = \mu_0(T) + k_B T \ln \rho . \quad (6.19)$$

Here, we are interested only in the concentration dependence. We have therefore stored all other terms in a temperature dependent chemical potential $\mu_0(T)$. For sufficiently dilute electrolytes, the chemical potential of the ions is therefore proportional to the logarithm of their concentration. The concentration is now defined by the molar ratio with respect to the solvent. In terms of this concentration $\rho_{\text{Me}^{z+}}$ the chemical potential of the ions $\text{Me}^{z+}$ in a dilute electrolyte is

$$\mu_{\text{Me}^{z+}} = \mu_{\text{Me}^{z+},0}(T) + k_B T \ln \rho_{\text{Me}^{z+}} . \quad (6.20)$$

For larger concentrations, (6.20) is no longer valid and the chemical potential becomes a complex function of the concentration of the ions, the concentration of all other ions in the electrolyte and the properties of the solvent. Debye and Hückel [6.19] derived a mean field solution to the problem. However, in general the chemical potential can only be determined experimentally via reaction equilibria. Chemists like to keep the simple functional form of (6.20) and hide the complexity of the functional dependence on the concentration by introducing an *activity* $a(\rho_{\text{Me}^{z+}},....)$ that is defined by the chemical potential of ions of a particular concentration.

$$\mu_{\text{Me}^{z+}}(\rho_{\text{Me}^{z+}}) = \mu_{\text{Me}^{z+},0}(T) + k_B T \ln a(\rho_{\text{Me}^{z+}},...) \quad (6.21)$$

Underpotential deposition is defined with respect to the equilibrium electrode potential for bulk deposition. We therefore consider this case first. In Sect. 5.1.3 the chemical potential of the solid phase with reference to the vacuum level was derived as (5.16)

$$\mu_s = -E_{\text{coh}} - 3k_B T \ln(T/\theta) . \quad (6.22)$$

As in (6.20) we put terms arising from vibrational partition function into a temperature dependent ground state chemical potential $\mu_{s,0}(T)$. In the course of the reaction (6.17) z-electrons are left behind at the Fermi-level of the metal. To consider equilibrium of the reaction (6.17) their energy can either be added to the chemical potential of the electrolyte or be subtracted from the chemical potential of the solid. We choose the latter option to stress the point that the electrons sit on the metal. Subtracting the energy of z-electrons $-ze(\phi-\phi_{ref})$ with reference to an arbitrary reference potential $\phi_{ref}$ one obtains

$$\mu_s = \mu_{s,0}(T) + ze(\phi - \phi_{ref}) . \tag{6.23}$$

Equilibrium between the ions in the electrolyte and the solid requires that

$$\mu_{Me^{z+}} = \mu_s = \mu_{Me^{z+},0}(T) + k_B T \ln \rho_{Me^{z+}} = \mu_{s,0}(T) + ze(\phi_{eq} - \phi_{ref}) . \tag{6.24}$$

The equilibrium potential is therefore

$$\phi_{eq} = \phi_{ref} + \frac{\mu_{Me^{z+},0}(T) - \mu_{s,0}(T)}{ze} + \frac{k_B T}{ze} \ln \rho_{Me^{z+}} . \tag{6.25}$$

Apart from a constant that is characteristic for the system, the equilibrium electrode potential is therefore proportional to the logarithm of the ion concentration. For each order of magnitude in the concentration the equilibrium potential shifts by

$$\Delta\phi_{eq} = \frac{k_B T}{ze} \ln 10 = 58.2 / z \quad \text{meV at } 20°C . \tag{6.26}$$

The corresponding equilibrium condition for an upd-layer can be obtained if one hides the complexity of a real system in a mean field approach (Sect. 6.2.2) and writes for the upd-phase

$$\mu_{upd} = \mu_{0,upd}(T) + w\Theta_{ad} + ze(\phi - \phi_{ref}) + k_B T \ln \frac{\Theta_{ad}}{1 - \Theta_{ad}} . \tag{6.27}$$

The coverage $\Theta$ is defined as the fraction of the saturation coverage in the upd-layer. The equilibrium potential is now

$$\phi_{eq,upd} = \phi_{ref} + \frac{\mu_{Me^{z+},0}(T) - \mu_{upd,0}(T)}{ze}$$
$$- \frac{1}{ze}\left( w\Theta_{ad} + k_B T \ln \frac{\Theta_{ad}}{1 - \Theta_{ad}} \right) + \frac{k_B T}{ze} \ln \rho_{Me^{z+}} . \tag{6.28}$$

In experiments, the temperature and the electrolyte concentration is kept fixed. Equation (6.28) then relates potential and the coverage in the upd-layer. Figure 6.16 shows isotherms for the upd of a $Me^{2+}$ metal as a function of the potential for various values of $w/k_B T$. In all cases the potential is referred to the potential at which $\Theta_{ad} = 0.5$. Depending on the sign and magnitude of the interaction constant $w$ one has a steep or a gradual transition region. For $w/k_B T < -4$ one obtains a sudden change in the coverage as indicated by the vertical short-dotted line in Fig. 6.16. From general reasoning, one would expect the interaction between the atoms in a metal upd-layer to be strongly attractive. STM studies show that metal upd-layers grow as a dense layer in islands or from steps. An abrupt change in the coverage at a particular electrode potential is therefore expected for metal upd. However, because of the experimental difficulty to measure static charges, upd is almost never studied in equilibrium. Rather one uses voltammograms, where the potential is swept at a particular rate (Sect. 2.3.2). A few equilibrium isotherms on single crystal surfaces of silver and copper were reported in the late seventies [6.20]. They show abrupt as well as smooth transitions within a potential range of 50 mV. A smooth transition, however, can also result from surface inhomogeneities. As these early studies had not STM-control over the surface quality and since it does not take much of structural inhomogeneity to produce a 50 mV shift in binding energy, it remains open whether upd-isotherms of metals are abrupt or smooth.

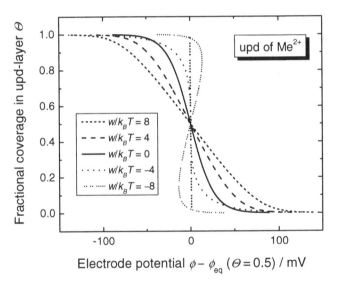

**Fig. 6.16.** Model isotherms for underpotential deposition. Parameter is the mean-field interaction constant $w$. For metal upd, one would expect strong attractive lateral interactions since the atoms in the upd-layer are densely packed even at partial coverages. The upd-transition should therefore be sharp as indicated by the dotted vertical line.

According to the model, the upd-deposition saturates for negative potentials, since the chemical potential becomes infinitely large at $\Theta = 1$. Real isotherms tend to display a slow further uptake of atoms as the upd-layer becomes compressed. At more negative electrode potentials equilibrium with bulk crystal deposition is eventually reached. Then, the growth continues indefinitely if the concentration of ions is kept constant.

From (6.28) and (6.25) one obtains the difference in the equilibrium potentials for upd and bulk deposition, which is noted as the upd-shift $\Delta\phi_{upd}$ It is convenient to define the upd-potential as the potential at the point of inflection ($\Theta = 0.5$) and take as the ground state chemical potential $\mu_{upd,0}(T)$ the chemical potential of the upd-layer at that coverage.

$$\Delta\phi_{upd} \equiv \phi_{eq,upd} - \phi_{eq} = \frac{1}{ze}\left(\mu_{s,0}(T) - \mu_{upd,0}(T, \Theta_{ad} = 0.5)\right) \qquad (6.29)$$

Equation (6.29) is the recipe to calculate, at least approximately, the upd-shift $\Delta\phi_{upd}$: Lacking better knowledge, one disregards the vibrational partition functions and calculates the difference in the ground state energies per atom for the bulk of the deposit (the negative of the cohesive energy) and the ground state energy of a surface covered with the deposit. This way the problem is amenable to treatment by standard total energy calculations. If the structure of the upd-layer is known at a particular coverage, e.g. to be a compressed layer at saturation, one may furthermore neglect the entropic factors from site occupation and calculate the upd-shift directly for the particular structure [6.21, 22].

### 6.2.5 Specific Adsorption of Ions

The term *specific adsorption of ions* refers to a situation where ions from an electrolyte form a chemical bond with the surface as an adsorbed layer, in the same way as atoms or molecules chemisorbed from the gas phase might do. As for chemisorption, the term *specific adsorption* implies that a surface compound and not a bulk compound is formed in the process. Hence, specific adsorption and chemisorption concern the same final product, a layer of molecules or atoms chemically bonded to the substrate surface. The difference is in the initial state, which is more complex in the case of an electrolyte as it consists of an ensemble of solvent molecules, the solvation shells of the ions and counter ions in the electrolyte. The process of specific adsorption therefore involves at least a partial stripping of the solvation shell. Some kind of a solvation shell may remain as the chemisorbed layer may include H-bonded water molecules, $OH^-$, or $H^+$. The ions in the electrolyte carry a positive or negative charge that is quantized in units of the elementary charge. In forming the chemical bond with the surface, the charge is transferred to the solid. The bond may retain a partial ionic character (Sect. 3.1.3); However, this merely means that the bond with the substrate atoms is polarized, not ionic as a whole. Insofar the term "*specific adsorption of ions*" is somewhat misleading, especially when the adsorbed species is addressed in the notation for ions, e.g. as $SO_4^{2-}$ or $HSO_4^-$. Specific adsorption of ions is a ubiquitous phenomenon, which has been studied extensively using the traditional methods of electrochemistry such as voltametry and chronocoulometry (Sect. 4.2.4) as well as infrared spectroscopy, x-ray diffraction and tunneling microscopy. Well studied is the relatively simple adsorption of halogens on the coinage metals, in particular gold.

The adsorption isotherms for specific adsorption are derived the same way as the isotherms for underpotential deposition. The adsorbed phase is treated in the mean field approximation. The relation between the coverage $\Theta_{ad}$ and the electrode potential $\phi_{el}$, the Frumkin-isotherm, is obtained by equating the chemical potential of the adsorbed phase with the chemical potential of the ions in solution.

$$\phi_{el} = \phi_0 \mp \frac{1}{ze}\left( w\,\Theta_{ad} + k_B T \ln \frac{\Theta_{ad}}{1-\Theta_{ad}} \right) + \frac{k_B T}{ze}\ln \rho_{I^{z\pm}} . \tag{6.30}$$

The reference potential $\phi_0$ is a function of the binding energy of the ions at the surface and the solvation energy. The sign of the second term depends on the charge of the ions in solution, negative for a positively charge ions, and vice versa. As an example, Fig. 6.17 shows the adsorption isotherm for iodine on a gold film [6.23]. The coverage was determined by chronocoulometry and, ex situ, by x-ray photoemission spectroscopy (XPS). The coverage vs. potential fit to the Frumkin-isotherm (6.30) with $w/k_B T = 12$, which indicates a relatively strong repulsive interaction between the iodine atoms. Figure 6.17 seems to indicate a complete saturation at $\phi_{sat} = -0.2$ V. By applying more positive potentials on the electrode, one may nevertheless compress the iodine layer even against the Pauli-repulsion between the ions because of the large energy $e(\phi - \phi_{sat})$ per atom provided by the

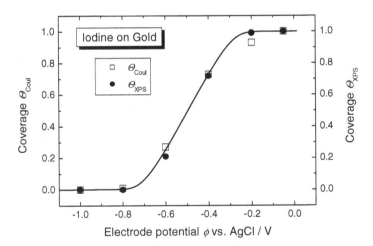

**Fig. 6.17.** Fractional coverage of iodine on a Gold film. The coverages where determined using chronocoulometry and x-ray photoemission (XPS) (after Bravo et al. [6.23]). The solid line is a fit to the Frumkin-isotherm with $w/k_B T = 12$.

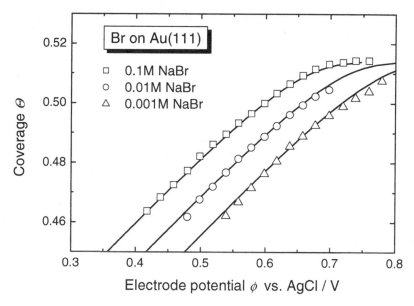

**Fig. 6.18.** Coverage of electrodeposited bromine on Au(111) (after Magnussen et al. [6.25]. The coverage is in Br-atoms per surface Au-atoms. The solid lines are Frumkin-isotherms shifted with respect to each other by $k_B T/e$ ln10.

electrode potential. This *electrocompression* can lead to a sequence of phases as a function of coverage [6.24]. Electrocompression of compact halogen layers was studied by Magnussen et al. [6.25]. Figure 6.18 shows the Br-coverage of a Au(111)-surface in the regime of a uniformly compressed hexagonal bromine layer. The layer is incommensurate with the substrate in the entire range. The solid lines in Fig. 6.18 are Frumkin-isotherms that are displaced along the x-axis with respect to each other according to the change in the concentration of Br⁻-ions in the solution by

$$\Delta\phi = \frac{k_B T}{ze}\ln 10 \tag{6.31}$$

with z = 1. The fact that the charge of one electron appears in the shift is consistent with the reasoning above that the ions of the solution have to give up one electron in forming the surface bond. Experimentally the observed shift with concentration do not always conform to (6.31) with integer multiples of the electron charge. This happens if the chemisorbed layer of ions involves the incorporation of a fractional monolayer of other ions e.g., OH⁻, or H⁺. To describe this effect heuristically by conventional thermodynamics the term *electrosorption valency* $\gamma'$ was introduced. With reference to (6.30) $\gamma'$ can be defined as

$$\gamma' = \frac{k_{\mathrm{B}}T}{e}\left(\frac{\partial \phi}{\partial \ln \rho}\bigg|_{\Theta}\right)^{-1} \qquad (6.32)$$

The incorporation of non-stoichiometric amounts of OH$^-$ and H$^+$ is presumably the reason why the isotherms for sulfate adsorption display a non-integer electrosorption valency. Figure 6.19 shows isotherms for sulfate adsorption on Au(111) as reported by Shi et al. [6.26]. The data was already discussed in Sect. 4.2.4 (Fig. 4.8) in the context of Maxwell relations and chronocoulometry. The solid lines in Fig. 6.19 are again fits to the Frumkin-isotherm, now with $w = 13.5k_{\mathrm{B}}T$. The curves are rigidly shifted with respect to each other assuming an electrosorption valency of one. For low coverages, this shift is approximately in agreement with experiment, indicating that sulfate ions in aqueous solution are monovalent, solvated HSO$_4^-$-ions. The experimental data deviate substantially for larger potentials and non-saturated coverages. A convincing interpretation in terms of a structural model for the disordered sulfate adlayer is still lacking.

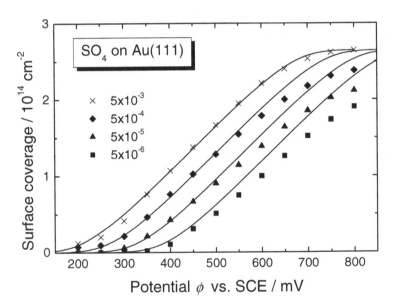

**Fig. 6.19.** Surface coverage of Au(111) with SO$_4$ (after Shi et al. [6.26]) for various concentrations of sulfate in a supporting electrolyte (0.05M KClO$_4$+0,02M HClO$_4$+xM K$_2$SO$_4$). The solid lines are Frumkin-isotherms with $w = 13.5k_{\mathrm{B}}T$. The isotherms are shifted with respect to each other assuming an electrosorption valency of one.

## 6.3 Desorption

### 6.3.1 Desorption Spectroscopy

*Thermal Desorption Spectroscopy* (TDS) is one of the oldest techniques in surface science [6.27, 28]. In early work, it was used to study desorption from ribbon-shaped tungsten filaments which were rapidly heated by passing an electric current through the filament. Differently strong adsorbed species would thereby desorb in the sequence of their binding energies, giving rise to pressure bursts that were detected by a pressure gauge. TDS is therefore a spectroscopy of the activation energies for desorption. If the adsorption process does not involve an activation energy, as is frequently the case, then the activation energies for desorption roughly correspond to the heats of adsorption (Fig. 6.5 and eqs. (6.15, 16)). TDS is then a spectroscopy of the heats of adsorption. Because of the use of filaments, the method was also named *flash filament technique.*

Modern versions of the technique use quadrupole mass spectrometers (Sect. 2.2.1) for detection and simultaneous chemical analysis of the desorbing species. Single crystal surfaces other than tungsten are not easily prepared as ribbons. Rather they come in the form of disks or as bead crystals (Sect. 2.1). Heating such a crystal in vacuum, inevitably causes desorption also from those surface areas that are neither properly prepared nor of the desired crystallographic orientation. It is therefore necessary to ensure that only the species desorbing from the surface of interest are probed. This is best achieved by placing the mass spectrometer into a separately pumped housing that is connected to the main chamber via a tube with a diameter smaller than the single crystal surface area. The orifice of the tube is brought close to the crystal, so that the species desorbing from the sides of the crystal are not in line-of-sight of the mass spectrometer. The desorption-signals from these species are sufficiently suppressed if the pumping speed of

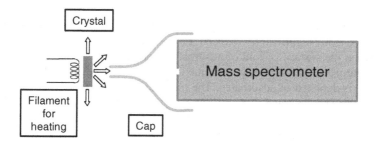

**Fig. 6.20.** To avoid interference from species desorbing from the sides of the crystal the entrance of the mass spectrometer is covered by a glass cap. The inside of the cap should be gold plated and electrically grounded to prevent distortion of the electric fields inside the mass spectrometer.

the main chamber is high enough so that the pressure increase during desorption is small. Simple, but rather effective is also to cover the opening of the ionization chamber of the mass spectrometer with a glass cap that ends into a tube pointing towards the sample (Fig. 6.20). The cap should be gold-coated on the inside with the gold film electrically grounded to avoid charging of the inner walls. Alternatively, the cap can be made from stainless steel [6.29].

In 1962 Redhead proposed to turn flash desorption into a quantitative method for the determination of activation energies for desorption [6.30] by raising the temperature linearly in time.

$$T(t) = T_0 + \alpha t \tag{6.33}$$

This has become the standard procedure in TDS. Since the temperature is increased according to a certain program, thermal desorption spectroscopy is occasionally also called $\underline{T}$emperature $\underline{P}$rogrammed $\underline{D}$esorption (TPD). To calculate the pressure increase as function temperature Redhead made a simple ansatz for the desorption rate

$$r_{des} = n_{ad} \Theta_{ad}^n \nu_0 e^{-\frac{E_{act}}{k_B T}} . \tag{6.34}$$

Here, $r_{des}$ is number of desorbing species per time and surface area, $E_{act}$ is the activation energy for desorption, $n_{ad}$ is the number of adsorbate sites per area, $\nu_0$ is a rate constant, $n$ is the "order of the reaction" and $\Theta_{ad}$ is again the fractional coverage of adsorbate sites. The order is $n = 1$ for the direct desorption of the adsorbed species. If the desorption requires a recombination of two adsorbed atoms then it seems reasonable to assume that the rate is proportional to $\Theta^2$, hence desorption should be of second order. Zero order desorption should occur if the desorption product results from an autocatalytic surface reaction. However, we shall see shortly that this interpretation of experimentally determined exponents is too simplistic.

With rising temperature, the desorption-rate increases exponentially as long as the coverage is not significantly reduced. For zero-order desorption, the rate drops to zero when the surface is void of adsorbates. For first and second order desorption, the rate passes through a smooth maximum to become eventually zero when the surface is depleted of adsorbates. The maximum of the desorption-rate is easily calculated. The coverage changes with time as

$$\frac{d\Theta_{ad}}{dt} = -\frac{r_{des}}{n_{ad}} . \tag{6.35}$$

The maximum in the desorption rate corresponds to the point of zero slope of $r_{des}$ and therefore to the zero of the second derivative of the coverage. After inserting (6.33) and (6.35) in (6.34) and solving for the zero of the second derivative of $\Theta_{ad}$

with respect to time one obtains an implicit solution for the temperature $T_m$ at which the maximum occurs

$$\frac{\alpha E_{act}}{k_B T_m^2} = \begin{cases} \nu_0\, e^{-\dfrac{E_{act}}{k_B T_m}} & \text{for} \quad n = 1 \\[2ex] \nu_0\, 2\Theta_{ad,m}\, e^{-\dfrac{E_{act}}{k_B T_m}} & \text{for} \quad n = 2 \end{cases}. \tag{6.36}$$

Here, $\Theta_{ad,m}$ is the coverage at the maximum of the desorption rate which is approximately equal to half the initial coverage $\Theta_{ad,in}$. For second order desorption, the maximum shifts with the initial coverage (to lower temperatures), while the maximum is independent of coverage for first order desorption. The numerical solution to (6.36) is well described by

$$T_{max}\, /\,K = 10 + \left\{ 347 - 20\lg(10^{-13}\,K\nu_0\Theta_{ad,in}^{n-1}\,/\,\alpha) \right\} E_{act}\,/\,eV . \tag{6.37}$$

The equation can be used to roughly estimate the activation energy for desorption by assuming a value for the rate constant $\nu_0$. Both, the rate constant and activation energy can be determined from experiment if desorption spectra are measured with different heating rates. A variation of the rate by two orders of magnitude is however required to keep the error reasonably low. Figure 6.21 displays the complete desorption spectra calculated from the Redhead-ansatz (6.34) for $n = 0, 1, 2$. The heating rate is set to $\alpha = 1$ K/s. The activation energy and rate constant were chosen as 1.5 eV and $10^{13}$ s$^{-1}$, respectively, and the initial coverage is varied from $\Theta = 0.1$ to 1.0 in steps of 0.1. Zero order spectra are characterized by an exponential increase in the rate, followed by a sudden drop to zero. The drop-off occurs the earlier the smaller the coverage is. First order desorption spectra are also somewhat skewed to the low temperature side. The peak position is independent of coverage. In second order desorption, the peak position shifts to lower temperatures with increasing initial coverage. Common to all spectra is that the peak position depends essentially linear on the activation energy and somewhat on the rate of the temperature increase.

With his ansatz for the desorption rate, Redhead made three assumptions, neither one is fulfilled in reality. Most importantly, the activation energy changes with coverage. The rate constant $\nu_0$ is to be replaced by a temperature dependent prefactor, and the coverage dependence of the rate can be significantly more complicated than assumed in (6.34). The most important consequence of these complications is that the simple classification of the spectra according to the order of desorption cannot be upheld. To understand these effects, we need to develop a more detailed understanding of the desorption process.

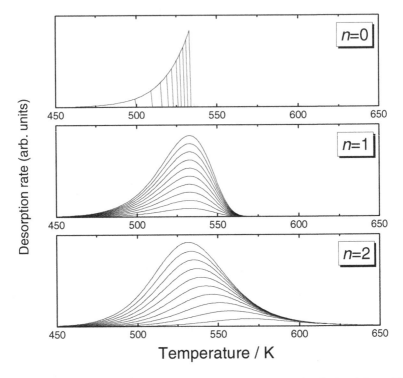

**Fig. 6.21.** Desorption spectra for the desorption order $n = 0$, 1, 2 calculated from (6.34) under the assumption of a constant heating rate $\alpha = 1$ K/s, a rate constant $\nu_0 = 10^{13}$ s$^{-1}$, and an activation energy $E_{\text{act}} = 1.5$ eV. The initial coverage is varied from 0.1 to 1 in steps of 0.1.

## 6.3.2 Theory of Desorption Rates

The theory of kinetic processes is significantly more sophisticated than the theory of equilibria, because of the many channels, which couple the degrees of freedom of the adsorbed species to the degrees of freedom of the desorbing species. It is however, possible to derive an expression for the rate of desorption that contains mostly equilibrium properties of the adsorbed phase and the gas phase and a single parameter, which account for the kinetics. This parameter is the sticking probability for a gas phase species (see also Sect. 6.1). This quantity is amenable to experimental determination, and is often near unity. The calculation of the desorption rate is based on the fact that in equilibrium the adsorption and desorption rates per area are equal. The flux $F$ of molecules impinging on the surface from the gas phase is known from kinetic gas theory (2.3) as

$$F = \frac{p}{(2\pi m k_B T)^{1/2}} . \tag{6.38}$$

The adsorption rate per area $r_{ad}$ is then

$$r_{ad} = s(\Theta_{ad}, T) \frac{p}{(2\pi m k_B T)^{1/2}} . \tag{6.39}$$

The sticking probability $s(\Theta_{ad}, T)$ depends on the surface coverage $\Theta_{ad}$ and the temperature $T$. In equilibrium, the rate of desorption and the rate of adsorption are equal and the pressure is the equilibrium pressure $p_{eq}$.

$$r_{des} = r_{ad} = s(\Theta_{ad}, T) \frac{p_{eq}}{(2\pi m k_B T)^{1/2}} . \tag{6.40}$$

The equilibrium pressure can be expressed in terms of the chemical potential of the gas phase (5.15), and in equilibrium this is equal to the chemical potential of the adsorbed phase $\mu_{ad}(\Theta_{ad}, T_s)$. The desorption rate is therefore

$$\frac{r_{des}}{n_{ad}} = s(\Theta_{ad}, T) \frac{k_B T}{h} \frac{2\pi m k_B T}{n_{ad} h^2} Z_{rot} Z_{vib} \, e^{-\frac{E_g - \mu_{ad}(\Theta_{ad}, T)}{k_B T}} . \tag{6.41}$$

While this is the desorption rate in equilibrium one can argue that the rate is not affected when the gas phase is taken away: adsorption and desorption rates per atom site are of the order of $1 \text{ s}^{-1}$. The internal relaxation times of a solid-state system are at most of the order of picoseconds. Therefore removing the gas phase cannot lead to a redistribution of energy over the internal degrees of freedom of a solid-state system and hence does not lead to a change in its thermodynamic properties. Equation (6.41) therefore describes the desorption rate even when no gas phase is present.

The rate contains an exponential term and various temperature dependent prefactors. The first one is the sticking coefficient, which contains all kinetic aspects of the problem. In particular, the sticking coefficient may involve an activation energy. We note that, the temperature in (6.41) is the crystal temperature, even if in an actual desorption experiment the crystal temperature differs from the temperature of the ambient gas phase. This follows from the fact that (6.41) was derived from an equilibrium situation. The temperature dependence of the sticking coefficient as an experimental quantity must be measured also in an equilibrium situation. In reality, the sticking coefficient is mostly measured with the gas phase at room temperature and the crystal at higher or lower temperature. In the case of activated adsorption, this so-measured sticking coefficient could deviate substantially from the sticking coefficient in equilibrium, since the probability to overcome the activation barrier should depend mostly on the kinetic energy of the molecules in the gas phase. The second term $k_B T/h$ is familiar from transition state theory (Sect. 10.1.3). It has the dimension of a frequency and amounts to $6.25 \; 10^{12} \; s^{-1}$ at 300 K. This frequency is often, erroneously, confused with the vibration frequency of an atom in the potential well and misinterpreted as an "*attempt frequency*". The third term is the ratio of density of the atoms in the phase space in vacuum and on the surface. Depending on the mass of the desorbing species and the density of sites on the surface, it can amount to a factor of hundred or thousand. Even for atom desorption the prefactor can therefore be as large as of $10^{15}$. Of the gas phase partition functions $Z_{vib}$ and $Z_{rot}$ only the latter contributes a larger factor. For CO, e.g., the rotational partition function is 120 at 300K.

In order to discuss (6.41) further we insert the mean field solution (6.7) for the chemical potential of the adsorbed phase. We consider explicitly three cases, (I) desorption of rare-gases, (II) desorption of diatomic molecules with particular attention to CO, and (III) desorption of a diatomic molecule that is dissociated in the adsorbed state.

*Case I: desorption of rare-gases*

The partition function in the gas phase contains only translations. The vibrational frequencies in the adsorbed state are low, so that there is a contribution from there. The rate of desorption is

$$\frac{r_{des}}{n_{ad}} = -\frac{d\Theta_{ad}}{dt} = s(\Theta_{ad}, T) \frac{\Theta_{ad}}{1-\Theta_{ad}} \frac{k_B T}{h} \frac{2\pi m k_B T}{n_{ad} h^2} \frac{1}{Z_{v,ad}} e^{-\frac{E_g - E_{ad} - W(\Theta_{ad})}{k_B T}} . \quad (6.42)$$

We discuss this equation with the system Xe on Pt(111) in mind, for which the thermodynamic data as well as vibration frequencies have been measured [6.2]. The heat of adsorption was shown in Fig. 6.13 and was fitted by a heuristic function. To conform to (6.42) the coverages in Fig. 6.13 have to be scaled to a saturation coverage for which we take $\Theta_{sat} = 0.37$. To calculate the desorption spectrum one needs the sticking coefficient. As an approximation, we assume that $s(\Theta_{ad}, T) = s_0(1-\Theta_{ad})$ with $s_0 = 1$. This cancels the $(1-\Theta_{ad})$-term in the denominator of (6.42). The vibration quantum $\hbar\omega$ for the vertical motion is about 3.5 meV [6.2]. We assume that the parallel vibration frequencies have the same value, so that the vibration partition function for the adsorbed state becomes (zero point energy neglected).

$$Z_{vib,ad} = \left(\frac{k_B T}{\hbar\omega}\right)^3 . \quad (6.43)$$

Finally, we convert the difference $E_g - E_{ad}$ into the heat of adsorption by using the definition (6.14) and the equilibrium condition between the chemical potential of the gas-phase (5.15) and the adsorbate phase (6.7).

$$E_g - E_{ad} = Q + \frac{1}{2}k_B T \qquad (6.44)$$

Equation (6.42) then becomes

$$r_{des} = n_s \Theta_{ad} v_0(T) e^{-\frac{Q}{k_B T}}, \quad \text{with}$$

$$v_0(T) = \frac{m\omega^3}{4\pi^2 n_{ad} k_B T} e^{-1/2} \cong \frac{5.6 \times 10^{15}}{T} \text{Ks}^{-1} . \qquad (6.45)$$

The resulting desorption spectra for a heating rate of $\alpha = 1$ Ks$^{-1}$ are displayed in Fig. 6.22 for the coverages 0.1 to 1.0 (solid lines). The compressive interaction between the adsorbed Xe-atoms at full coverage cause the early desorption at low temperatures. The low temperature tail vanishes for $\Theta_{ad} = 0.9$ and lower. Without that tail, the spectra resemble those of zero-order desorption (Fig. 6.21) although

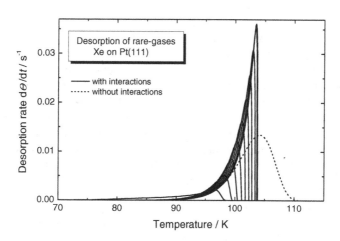

**Fig. 6.22.** Desorption spectra calculated for Xe on Pt(111) for coverages $\Theta_{ad} = 0.1$ to 1 in steps of 0.1. The spectra are typical for rare-gas desorption from transition metals. The spectra look as if the desorption were of zero order (Fig. 6.21) because of the attractive interaction between the adsorbate atoms. The long tail at low temperatures marks the early desorption of species from the compressed layer when repulsive interactions prevail. Without interaction the full coverage desorption spectrum would look as indicated by the dotted line.

the desorption is first order. This seeming zero-order shape results from the attractive interactions between the adsorbed Xe-atoms. The dashed line in Fig. 6.22 is the spectrum calculated for zero interactions, with the heat of adsorption set to its maximum value of 310 meV (Fig. 6.13).

### Case II: desorption of CO

Thermodynamics and kinetics of adsorption and desorption of carbon monoxide has been studied on many surfaces, in particular on the transition metals. There is consensus that the heat of adsorption decreases with coverage, although different studies come to different conclusions concerning the magnitude and the functional dependence on the coverage. For adsorption of CO on Pd(100), e.g., Behm et al. find that the heat of adsorption stays nearly constant almost up to saturation coverage and then drops sharply to 2/3 of its initial value [6.31]. Yeo et al. find a continuous, almost linear decrease [6.32]. The authors furthermore disagree on the coverage dependence of the sticking coefficient. This may be an indication that the method matters by which a result is obtained. Behm et al. measured the sticking coefficient by exposing the surface from the gas phase ambient pressure at 350 K and the heat of adsorption via isobars. Yeo et al. measured the sticking coefficient using a normal incidence CO beam, and the heat of adsorption was determined by calorimetry at 300 K. The sticking coefficient may depend on the orientation of the incoming CO molecules and the coverage dependence of the heat of adsorption could depend on whether the CO-layer is ordered (Yeo et al.) or disordered because of the higher temperature (Behm et al.).

For a survey on the qualitative features of CO-desorption spectra, these subtleties need not be taken into account. We model the spectra by assuming a heat of adsorption of 1.65 eV and a linear decrease down to 2/3 of the initial value at saturation coverage. This corresponds roughly to the coverage dependence of the heat of adsorption measured by Yeo et al. [6.32]. The saturation coverage on Pd(100) is 0.56 CO atoms per surface atom [6.31]. For our purpose, the sticking coefficient is sufficiently well described by $s(\Theta_{ad}, T) = 1 - \Theta_{ad}$ [6.31, 32]. The prefactor contains the rotational partition function of the gas phase molecule. CO on Pd adsorbs in a bridge site. The species has therefore one low lying vibrational mode from the hindered translation. We assume the frequency to be as for Ni(100) ($\hbar\omega = 3.7$ meV, Sect. 5.4.1). The desorption rate is then

$$\frac{r_{des}}{n_{ad}} = -\frac{d\Theta_{ad}}{dt} = \Theta_{ad} \nu_0(T) \, e^{-\frac{Q(1 - 2/3\Theta_{ad})}{k_B T}} . \qquad (6.46)$$

Here the conversion of the difference $E_g - E_{ad}$ into the heat of adsorption by using the definition (6.14) and the equilibrium condition (6.4) adds a factor $\exp(-5/2)$ to the prefactor $\nu_0(T)$, which thereby becomes

$$\nu_0(T) = \frac{k_B T}{h} \frac{2\pi m k_B T}{n_{ad} h^2} \frac{2I k_B T}{\hbar^2} \frac{\hbar \omega_{tr}}{k_B T} e^{-5/2} = 3.5 \times 10^{15} s^{-1} \frac{T^2}{(300K)^2} \ . \quad (6.47)$$

The calculated value is in reasonable agreement with the measured value of Behm et al. ($3\times10^{16}$ s$^{-1}$ at 500 K [6.31]).

The calculated desorption spectra for a heating rate of 14 K/s are displayed in Fig. 6.23. Because of the reduction of the heat of adsorption, the spectra shift towards lower energies for higher coverages. The overall width of the individual spectra reflects the amount by which the heat of adsorption changes with coverage for a given initial coverage. According to Fig. 6.23, desorption from a fully covered surface begins already below 300 K. A surface dosed at 350 K with CO with the CO-gas pumped off afterwards would therefore not display the low temperature part of the set of spectra shown in Fig. 6.23 [6.31]. CO-desorption spectra obtained after dosing a Pt(111) surface with CO at 100 K are very similar to the spectra shown in Fig. 6.23 [6.29].

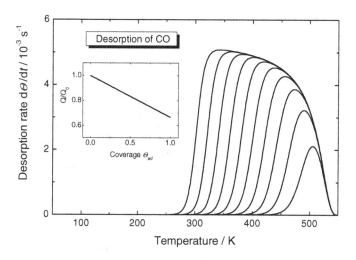

**Fig. 6.23.** Calculated desorption spectra for CO-desorption from a Pd(100) surface for coverages $\Theta_{ad}$ = 0.1-1 in steps of 0.1. The insert shows the assumed dependence of the heat of adsorption on the coverage. The broad appearance of the peak for high coverages is caused by the dependence of the heat of adsorption on the coverage.

It is also instructive to look at a set of desorption spectra calculated with the assumption that the heat of adsorption drops sharply beyond a particular coverage, e.g. because sites with a lower binding energy becomes occupied. Figure 6.24 displays a set of desorption spectra for that case. The heat of adsorption is assumed to drop down to 90% of its initial value at $\Theta_{ad}$ = 2/3 (see insert in Fig. 6.24). Otherwise, the data are as for Fig. 6.23.

## Case III: desorption of a dissociated diatomic molecule

Diatomic molecules like $N_2$, $O_2$, $H_2$, CO and NO, frequently dissociate upon adsorption. On transition metals, oxygen and hydrogen dissociate without an activation barrier. Desorption is therefore again determined by the heat of adsorption. As the molecules recombine in the process, desorption should be of second order. A second order process also follows from the simplest possible ansatz for the chemical potential of the adsorbed phase (5.97). Using (5.97) one obtains for the desorption rate

$$\frac{r_{des}}{n_{ad}} = -\frac{d\Theta_{ad}}{dt} = s(\Theta_{ad}, T)\left(\frac{\Theta_{ad}}{1-\Theta_{ad}}\right)^2 v_0(T)\, e^{-\frac{Q-W(\Theta)}{k_B T}} . \tag{6.48}$$

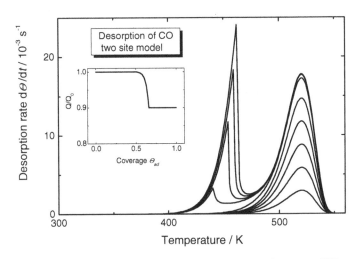

**Fig. 6.24.** Desorption spectra when the adsorbate fills sequentially two different sites of different binding energy. The coverages is varied between $\Theta_{ad} = 0.1$ and $0.9$ in steps of $0.1$. The insert shows the assumed shape of the functional dependence of the heat of adsorption on the coverage. Otherwise, the data are as for Fig. 6.23.

As an example we consider desorption of $H_2$ from a Pd(100)-surface. Adsorption and desorption of hydrogen, the sticking coefficient and the heat of adsorption for this surface has been studied by Behm et al. [6.33]. The vibration levels of H in the adsorbed state are too high to contribute to the partition function. The same holds for the molecular vibration in the gas phase. The prefactor $v_0(T)$ is now

$$v_0(T) = \frac{k_B T}{h}\frac{2\pi m k_B T}{n_{ad}h^2}\frac{2Ik_B T}{\hbar^2}e^{-7/2} = \frac{8\times10^{12}\,\mathrm{s}^{-1}T^3}{(300\,\mathrm{K})^3} . \tag{6.49}$$

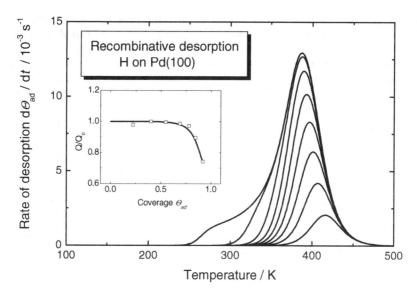

**Fig. 6.25.** Calculated desorption spectra for hydrogen on Pd(100) for coverages $\Theta_{ad} = 0.1\text{-}1$ in steps of 0.1. The coverage dependence of the heat of adsorption (solid line in insert) is taken from the experimental data (open squares [6.33])

The heat of adsorption stays constant up to about 70% of the saturation coverage which is at $\Theta = 1.3$ hydrogen atoms per Pd-surface atoms (see insert in Fig. 6.25). The zero coverage sticking coefficient is $s_0 = 0.5$ which indicates non-activated adsorption. The sticking coefficient decreases for higher coverages. For our purpose, the decrease is well enough described by $s(\Theta_{ad}, T) = 1 - \Theta_{ad}$. Behm et al. find for the prefactor in desorption at about 400 K the value $1.4 \times 10^{13}$ s$^{-1}$ [6.33]. Our calculated value $1.9 \times 10^{13}$ s$^{-1}$ at 400 K compares well with that experimental value. The calculated desorption spectra are displayed in Fig. 6.25. The desorption temperature and the overall shape of the curves agree favorable with the experiment [6.33], but the hump around 250 K carries more weight in the experiment. This weight depends on the shape of curve describing the heat of adsorption vs. coverage (see insert in Fig. 6.25). If the heat of adsorption falls off at a lower coverage, then the fraction of molecules desorbing in the low temperature regime increases. If the spectra display a hump, then the heat of adsorption stays approximately constant in a certain coverage range (compare Fig. 6.24).

*Summary*

In the early days of surface physics the various humps and peaks in desorption spectra were addressed as "states". These states were denoted by Greek letters and numbered in the sequence of their position on the temperature scale; $\alpha_1$, $\alpha_2$, ... would denote weakly bound, e.g. physisorbed species, $\beta_1$, $\beta_2$, ...would denote chemisorbed species. The understanding was that the peaks should correspond to individual species of different nature, for example species adsorbed in different sites. We have seen in the preceding section that humps and peaks can also arise from lateral interactions between the adsorbed species, which may or may not be concomitant with a change in the adsorption sites or a restructuring of the surface. Hydrogen on Pd(100), e.g. occupies only bridge sites [6.33]. Carbon monoxide on the other hand frequently changes the preferred site on the surface as a function of coverage. An example is CO on Pt(111) where CO first adsorbs in the atop sites and later in bridge sites. Despite the change in site, the desorption spectra look very similar to the set of curves displayed in Fig. 6.22 [6.34]. On the other hand, a quantitative description of the desorption spectra for the same system, self-consistent with isotherms, the sticking coefficients, and the partial coverages in the two sites is a formidable task that involves sophisticated and detailed considerations (for a very lucid discussion see [6.35]). In summary, one may state that desorption spectra while providing quick and qualitative information on the adsorption system are difficult to interpret in detail since equilibrium properties and kinetic effects are intertwined.

## 6.4 The Chemical Bond of Adsorbates

### 6.4.1 Carbon Monoxide (CO)

Carbon monoxide has been the drosophila (fruit fly) of Surface Science. Not only is adsorbed CO the by far most studied molecule; it has also been the molecule of choice to test new methods of surface analysis. There are reasons why CO became so popular. It adsorbs dissociatively and in a molecular state. As a molecule, CO can occupy different surface sites and may form complex surface lattices, depending on the coverage and the type of substrate. Research on CO adsorption and surface reactions involving CO was stimulated in the 70ties by the development of exhaust catalysts for the automotive industry. More recently, interest in CO catalysis was renewed in the context of methane reformation in fuel cells.

   Among the scientific questions, that evolved in connection with CO adsorption was whether and on which materials CO would adsorb dissociatively. Desorption spectroscopy is not conclusive in that regard as CO would always desorb as a molecule even if adsorbed as separate oxygen and carbon atoms. It was one of the great successes of surface vibration spectroscopy using inelastic electron scattering (Sect. 7.4). that, with respect to the tungsten surface, the question could be decided in favor of dissociative adsorption [6.36]. From systematic studies, the picture emerged that all transition metals in the Periodic Table left of a boundary between iron and cobalt dissociate CO upon chemisorption at room temperature. After completion of a monolayer, additional CO molecules bind to the surface in molecular form if the surface is at low temperatures. Elements to the right of the boundary bond CO as a molecule. For the transition metals near the boundary, dissociation depends on the surface orientation and on the presence of defects on the surface. Surface atoms with a lower coordination such as kink atoms or step atoms dissociate CO more easily.

Molecular bonding of CO involves the **Highest Occupied Molecular Orbital (HOMO)** and the **Lowest Unoccupied Molecular Orbital (LUMO)**. For CO, the HOMO is the $5\sigma$-orbital, which is a bonding orbital for the CO-molecule. The LUMO is the $2\pi^*$-orbital which is antibonding with respect to the CO molecular bond. The energies of both orbitals with reference to the vacuum level lie near the Fermi-level of transition metals. Coupling with the metal electrons broadens the molecular levels. The $5\sigma$-orbital forms bonding and antibonding combinations with the unoccupied metal states and thereby establishes a chemical bond to which the CO-molecule contributes two electrons. Thereby charge is donated to the surface. The $5\sigma$-orbital carries the largest weight on the backside of the carbon atom (Fig. 6.26). The CO-molecule therefore bonds in an upright position with the carbon pointing towards the surface.

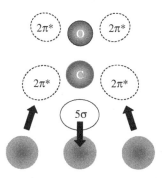

**Fig. 6.26.** Orbital scheme of the CO-molecule. As the largest weight of the $5\sigma$-orbital is on the backside of the carbon atom, CO bonds with the carbon atom pointing towards the surface.

The $2\pi^*$-orbitals form bonding and antibonding combinations with the occupied metal states. Electrons from the metal are back donated into the molecule. Since the $2\pi^*$-orbital is an antibonding orbital for the CO-molecule this back donation weakens the internal chemical bond of the CO molecule. The bonding mechanism for CO was first proposed by Blyholder [6.37]. Because of the described bonding mechanism, CO binds effectively to transition metals with their high density of states at the Fermi-level. The binding energy is around 1.5 eV. The binding energy with the coinage metals Cu, Ag, Au is significantly lower. The overlap of the $5\sigma$ orbital with the metal orbitals is best for the a-top site. For the $2\pi^*$-orbital, the bridge site between two metal atoms provides the best overlap. Both effects approximately compensate each other so that the binding energy of CO is not very site specific. The preferred binding sites may therefore depend on the coverage.

On Pt(111) e.g., adsorption is first in the atop site up to a coverage of 1/3 of a monolayer. Further CO molecules are then adsorbed in the two-fold bridge site. Adsorption of CO in various sites is easily distinguished by vibration spectroscopy. Bridge bonding to two or more surface atoms leads to a larger overlap with the $2\pi^*$-orbital and therefore to a larger back donation of electrons into the antibonding orbital of CO. This weakens the molecular bond. The simplest indicator of bond weakening is the CO-stretching vibration. Adsorption in bridge site leads to a lower CO-stretching vibration. The reasoning based on vibration spectroscopy is particular convincing if sites are occupied in sequence.

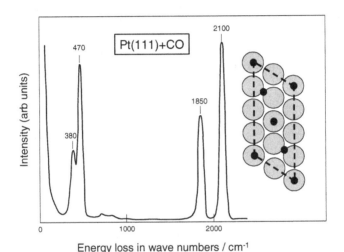

**Fig. 6.27.** Electron energy loss spectrum of 2 eV electrons backscattered in specular reflection from a CO-covered Pt(111) surface [6.29]. The energy loss is given in spectroscopic units (8.065 cm$^{-1}$ = 1 meV). The mode pair 2100 cm$^{-1}$/470 cm$^{-1}$ belongs to CO in the a-top site, the pair at 1850 cm$^{-1}$/380 cm$^{-1}$ to CO in the two-fold bridge site. The spectrum corresponds to half a monolayer coverage for which the system realizes a c(4×2) structure (see insert).

Figure 6.27 shows an electron energy loss spectrum (Sect. 7.2.2) for 0.5 monolayers of CO on Pt(111) [6.29]. The frequencies are given in wave numbers (cm$^{-1}$). The CO stretching vibration at 2100 cm$^{-1}$ and the associated metal-carbon vibration at 470 cm$^{-1}$ (corresponding to the vibration of the entire molecule against the surface) belong to CO in the a-top site. This pair of modes is the only one seen at low coverages. The CO stretching mode at 1850 cm$^{-1}$ and the metal-carbon mode at 380 cm$^{-1}$ belong to CO in the bridging site. At half monolayer coverage, the CO-molecules form a c(4×2) overlayer (insert in Fig. 6.27).

Substantially larger back donation than on the bridge sites occurs when electron-donating hydrocarbons [6.38] or alkali metals are coadsorbed. For CO coadsorbed with potassium on Pt(111), CO frequencies as low as 1380 cm$^{-1}$ have been observed [6.39, 40]. Note that the c(4×2) overlayer can also be realized with CO only in bridge sites by shifting the CO-unit cell with respect to the substrate. This was the structure originally proposed for CO on the Ni(111)-surface as there only one species was found for the ordered c(4×2) structure [6.41]. Later, however, LEED and photoelectron diffraction showed that CO occupies the hcp and fcc threefold hollow sites on Ni(111) [6.42, 43]. The Ni(111) surface thereby becomes rumpled and the CO-molecules are slightly tilted with respect to the (111)-orientation.

Tilted configurations are also realized for the compressed (2×1) p1g1 structure of CO on Pd(110) [6.44] and the (2×1) p2mg structures of CO on Ni(110) [6.45] and Rh(110) [6.46]. In all these cases, CO occupies the two-fold bridge site. On the Cu(110) surface, however, CO sits in the a-top site and is perpendicular oriented [6.47].

c(4x2) CO/NO                    c(4x2) NO

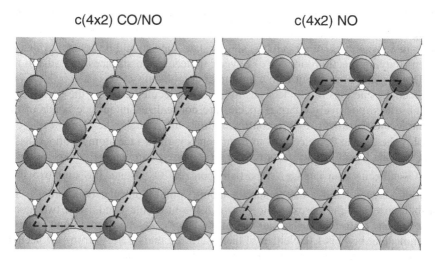

**Fig. 6.28.** (*left*) c(4×2) structure of CO [6.42, 43]. The same structure was proposed by Materer et al. for the c(4×2) structure of NO on Ni(111) [6.48]. An alternative proposition for the c(4×2) cell of NO on Ni(111) with more strongly and differently tilted NO molecules is shown on the *right* [6.49].

### 6.4.2 Nitric Oxide (NO)

The electronic structure of NO is similar to CO, except that now the $2\pi^*$ orbital is occupied by one electron. NO bonds always with the nitrogen atom and mostly in upright position. Because of the lower lying, partially occupied $2\pi^*$ orbital NO has a larger tendency to bond in bridging configurations. In Fig. 6.28 two alternatives for the c(4×2) structures for NO on Ni(111) are compared.

## 6.4.3 The Oxygen Molecule

Oxygen is a rather peculiar molecule. The formal double bond results from the occupation of three bonding and one antibonding orbital formed by the four p-electrons per atom. The splitting between the bonding and antibonding $5\sigma$ orbitals is large because of the large overlap of the $p_z$ atomic orbitals oriented along the molecular axis. The antibonding $5\sigma^*$ orbital therefore remains unoccupied. The $p_{x,y}$ atomic orbitals which are oriented perpendicular to the molecular axis form the degenerate $2\pi$ bonding orbitals and the $2\pi^*$ antibonding orbitals. The $2\pi$ orbitals are occupied with two electron pairs, the $2\pi^*$ orbital with one electron pair. If the molecule is ionized by removing one electron from the $2\pi^*$ orbital the bond strength increases. Even the doubly ionized $O_2^{2+}$ molecule is stable and so is the $O_2^-$-ion. The ground state of the oxygen molecule is a spin polarized triplet $^3\sum_g^-$-state (of the $2\pi^*$ electrons). By photoexcitation, the molecule can be brought into the more reactive singlet $^1\Delta_g$-state, which exists at a 0.98 eV higher energy. Because of the presence of hot filaments, a fraction of the oxygen molecules in a vacuum chamber is in the singlet state. One should be aware of this fact when studying reaction kinetics of oxygen molecules on surfaces that adsorb the triplet oxygen molecule but with a low sticking probability.

Since the energies of all electrons in the $O_2$-molecule lie below the Fermi level of substrates, positively charged species are not realized in surface bonds of the $O_2$-molecule. However, by a reconfiguration of the orbitals, the $O_2$-molecule can employ the $2\pi$ and $2\pi^*$ electrons to form bonds with surface atoms without dissociating. This flexibility accounts for the large variety of chemical bonds in which the $O_2$-molecule can engage: A partially ionized $O_2^{\delta-}$-state, the superoxo-state with the oxygen bonded to one surface atom (formal molecular bond order is 1.5 in that case) and the peroxo-state in which the molecule bonds with two atoms either to one or to two surface atoms. In the latter case, the formal bond order of the molecular bond is 1.0. Neither of these surface bonds is very strong. Molecular forms of adsorbed oxygen therefore exist only at low temperatures. Figure 6.29 shows an electron energy loss spectrum of the Pt(110) surface after adsorption of $O_2$ at 30 K [6.50]. In addition to physisorbed $O_2$ (1553 cm$^{-1}$) the spectrum shows three more losses which are associated with the $O_2$ stretching vibration: the $O_2^-$-state (1262 cm$^{-1}$) and two peroxo-states with the $O_2$-molecule bonding to one or two surface atoms (934 cm$^{-1}$ and 863 cm$^{-1}$, respectively). Oxygen dissociates on platinum upon annealing so that at 300 K only the spectral features of oxygen atoms survive.

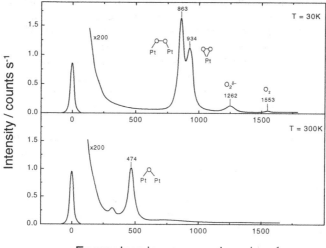

**Fig. 6.29.** Electron energy loss spectrum of the Pt(110) surface after deposition of molecular oxygen at 30 K (*upper panel*). Four different forms of $O_2$ are discernible. After heating to 300 K, the oxygen molecule dissociates and the spectrum displays the spectral features of adsorbed oxygen atoms (*lower panel*) [6.50].

### 6.4.4 Water

The importance of water/surface interactions for many areas of physics, chemistry, including electrochemistry, and biology can hardly be overestimated. One might therefore think that surface scientist should have responded to the challenge by making adsorption of water a prime objective of their research activities. It is, however, only lately that water adsorption has found the deserved attention, primarily for experimental reasons. Most experimental tools for structure analysis and spectroscopy employ electrons. Water has an extremely high cross section for electron-stimulated dissociation and desorption, rendering most electron based techniques useless, except if one can work with very low doses. Moreover, water interacts weakly with surfaces and therefore adsorbs only at low temperatures. The stable form of a layer of water molecules on a surface involves hydrogen bonds with other water molecules in the hexagonal configuration of a double layer resembling the structure of ice (Fig. 1.51). These double layers have been considered in Sect. 1.5 in the context of the solid/electrolyte interface. Here we are concerned with the adsorption of water monomers on surfaces. Adsorbed monomers of water exist at very low temperatures ($T \approx 20K$) as metastable species [6.51].

To prepare for the understanding of the water/surface bond we consider first the bonding in the molecule itself. The two hydrogen atoms in the water molecule are bonded by the oxygen $p_x$ and $p_y$ orbitals. The bond is strongly polarized with the charge shifted towards the oxygen atom, which gives rise to the large dipole moment of the water molecules (1.85 Debye, 1 Debye = 1/4.80 eÅ). The bond angle (104.5°) is larger than the 90° subtended by the unperturbed p-orbitals because of hybridization with the s-orbital and the Coulomb-repulsion between the hydrogen atoms that are partly divested of their electrons. The remaining four of the six valence electrons of oxygen form two *lone pair orbitals*, each occupied by two electrons. The chemical bond with the substrate is established by the formation of bonding and antibonding orbitals between an occupied lone pair orbital and the unoccupied substrate states. The bonding of the water molecules therefore involves a charge donation into substrate states. The polarity of the bond causes a reduction of the work function. The work function is further reduced by the dipole moment of the water molecule itself. The total work function shift is therefore described by attributing a dipole moment to each water molecule, which consists of two components one from the dipole moment of water $p(H_2O)$ and one from the dipole moment associated with the bonding to the surface $p_0$. The total dipole moment $p$ is $p = p_0 + p(H_2O)\cos \alpha$ with $\alpha$ the angle between the HOH-plane and the surface normal [6.53].

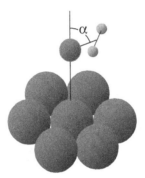

**Fig. 6.30.** The binding of a single water molecule on a Pt(111) surface [6.52, 53].

Theory has shown that the molecule binds best in the on-top sites. As the highest occupied (lone pair) orbital, the HOMO, is perpendicular to the plane of the molecule, the molecule is approximately planar to the surface [6.52, 53] (Fig. 6.30). The flat configuration of the molecule in an on top position was rediscovered in 2003 and it was shown that the angles $\alpha$ range between 75° and 84° [6.54] (Table 6.1). On some surfaces, water dissociates partially into H and OH. The oxygen pre-covered platinum is one case [6.55, 56]. OH-groups also play an important role as an intermediate in the famous catalytic reaction of oxygen and hydrogen to water [6.58] which was the first catalytic reaction ever reported (*Döbereiner lighter* [6.57]).

**Table 6.1.** Binding energies and orientation of single water molecules on {0001} and {111} metal surfaces after Michaelides [6.54]. The surface bond does practically not affect the HOH-bond angle.

| Substrate | Ru | Rh | Pd | Pt | Cu | Ag | Au |
|---|---|---|---|---|---|---|---|
| Energy /eV | 0.38 | 0.42 | 0.33 | 0.35 | 0.24 | 0.18 | 0.13 |
| Angle α | 84° | 81° | 83° | 83° | 75° | 81° | 67° |

Another case of partial dissociation which is of considerable technological importance is the Si(100) surface. This surface dissociates water with a sticking coefficient near unity even at 90 K and bonds the OH-group to one of the dangling bonds of a dimer and the hydrogen on the other to produce an ordered (2×1) structure [6.59, 60]. The asymmetric dimer of the clean surface (Fig. 1.24) becomes nearly symmetric. Because of the OH-groups, the surface is hydrophilic (Sect. 2.1.2). The hydrogen desorbs upon annealing, leaving the surface partly oxidized. Thermal oxidation of the Si(100) surface is therefore facilitated by adding water to the oxygen atmosphere (*wet oxidation*).

### 6.4.5 Hydrocarbons

We begin the discussion with the saturated hydrocarbons such as methane ($CH_4$), ethane ($C_2H_6$), propane ($C_3H_8$) etc., or cyclohexane ($C_6H_{12}$). Saturated hydrocarbons can establish a bond with the surface atom only via one or more hydrogen atoms in the form of a hydrogen bond, similar to one form of bonding of a network of water molecules with the surface as considered in Sect. 1.5. This somewhat unusual and a priori unexpected hydrogen bond was discovered in 1978 by vibration spectroscopy [6.61]. Figure 6.31 shows one of the original electron energy loss spectra representing the dipole active (Sect. 7.4.2) vibration modes of $C_6H_{12}$ adsorbed on Pt(111). Most of the vibration modes are at least approximately where the gas phase molecule has them. The corresponding gas phase modes are indicated in the figure together with their enumeration $v_i$ according to the chemical nomenclature. The broad intense loss in the range of the CH-stretching modes, but significantly downshifted compared to typical CH-stretching frequencies has no counterpart in the gas phase spectrum. The large intensity of the feature indicates that the mode has a large dynamical dipole moment. Together with the large width, this calls for the hydrogen atoms engaged in hydrogen bonding with the surface.

An unperturbed cyclohexane molecule may engage up to three hydrogen atoms in hydrogen bonding (Fig. 6.32). There are two principally different possibilities for hydrogen bonding with three H-atoms. One form would involve the H-atoms of three immediate carbon neighbors, the H-atoms labeled 1, 2 and 3 in Fig. 6.32. The alterative is bonding with the H-atoms labeled 2, 3 and 4 in Fig. 6.32. At least one of the modes visible in the spectrum belongs to a degenerate representation

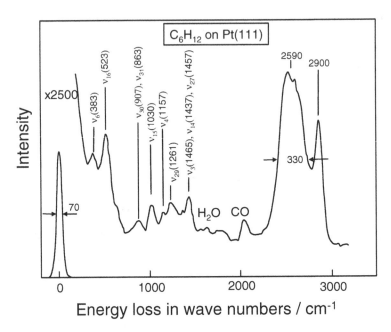

**Fig. 6.31.** Electron energy loss spectrum of cyclohexane ($C_6H_{12}$) adsorbed on Pt(111) representing the vibration spectrum of dipole active modes (after Demuth et al. [6.61]). Most of the modes appear at the same position where the gas phase molecule has them. The broad intense feature with the maximum at 2590 cm$^{-1}$ has no counterpart in the free molecule. It is attributed to the CH stretching vibration of the hydrogen atoms forming a hydrogen bond with surface atoms.

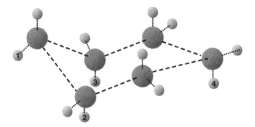

**Fig. 6.32.** The cyclohexane molecule may establish up to three hydrogen bonds with the surface. There are two possibilities; either the hydrogen atoms labeled 1-3 or the atoms labeled 2-4 make the bond with the surface.

(the CH$_2$ twist mode $\nu_{29} = 1261$ cm$^{-1}$). This mode should not possess a dipole moment perpendicular to the surface when the molecular skeleton is oriented parallel, which may be an indication that the molecule would is canted with respect to the surface, thereby realization H-bonding with the atoms 1-3. However, this argument assumes that the carbon skeleton of the cyclohexane molecule remains unperturbed for which there is no proof.

The binding energy of this specific form of hydrogen bonding with surface is much stronger than a van-der-Waals bond, which is consistent with the fact that cyclohexane remains bonded to the surface up to a temperature of 160 K. Annealing to higher temperatures causes dehydrogenation to benzene ($C_6H_6$) via an intermediate having the stoichiometry of $C_6H_9$ [6.62]. Hydrogen bonding of hydrocarbons to surfaces is not confined to cyclohexane, rather it seems to be the standard form of bonding for saturated undissociated hydrocarbons and for unsaturated hydrocarbons when adsorbed at low temperatures [6.61]. In view of this fact and considering that the hydrogen bonding is of outmost importance for all catalytic reactions involving hydrocarbons (hydrogenation, dehydrogenation, hydrogenolysis, isomerization, hydrocracking, and cyclization) it is surprising that little attention has been paid to this form of surface bond.

Unsaturated hydrocarbons may also bond via hydrogen bonding at low temperatures. Transition metal surfaces exposed to unsaturated hydrocarbons at room temperature however either crack the molecule or engage the $\pi$-electrons of the hydrocarbon to form carbon-metal $\sigma$-bonds. A molecule like ethylene ($C_2H_4$) e.g. thereby rehybridizes from $sp^2$ to $sp^3$ type bonding. The carbon-carbon bond order reduces to one. A thoroughly studied system is ethylene on Si(100) [6.63-65]. According to these studies, ethylene binds to the Si-dimers in a di-$\sigma$ configuration with the Si-dimer bond left intact (Fig. 6.33) [6.64].

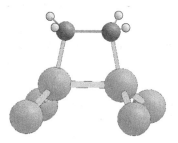

**Fig. 6.33.** Ethylene ($C_2H_4$) bonded to the Si-dimers on the Si(100) surface in a di-$\sigma$ configuration. The state of lowest energy may involve a C-C axis rotated by 11° around the surface normal and the $CH_2$ groups twisted with respect to each other by 27° [6.64].

Acetylene ($C_2H_2$) also rehybridizes upon adsorption to form $\sigma$-bonds with the surface. The frequency of the C-C vibration shifts from 1974 cm$^{-1}$ to about 1200 cm$^{-1}$ [6.66]. The comparison to the frequencies of the C-C stretching vibrations of ethylene and benzene (1623 cm$^{-1}$ and 992 cm$^{-1}$, respectively) indicates that the bond order of the carbon bond changes from three to about 1.5. The symmetry of the acetylene molecule changes along with the rehybridization. The minimum change would involve an upwards bending of the CH-bonds within the plane spanned by the C-C axis and the surface normal to yield $C_{2v}$ symmetry. Electron energy loss spectroscopy on acetylene on Ni(111) [6.66] and Fe(110) [6.67] showed that all modes have a dipole moment perpendicular to the surface which indicates that the molecule/surface complex has no symmetry element. Using X-ray photoelectron diffraction (XPD) the positions of the carbon atoms on Ni(111) were determined to be in the nonequivalent fcc and hcp sites [6.68]. Together with the results from vibration spectroscopy one arrives at the two alternative models displayed in Fig. 6.34. The hydrogen atoms are moved out of the Ni-C-C-Ni symmetry plane to either assume a cis or a trans configuration (Fig. 6.34). The molecule/surface complex contains no symmetry element.

cis                                    trans

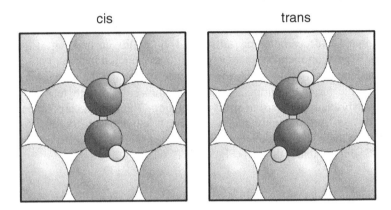

**Fig. 6.34.** Structure models for acetylene on the Ni(111) surface. Neither the cis nor the trans configuration has a symmetry element (point group $C_1$) since the two carbon atoms are in nonequivalent sites.

Benzene is one of the most studied hydrocarbon adsorption system. The NIST structure library lists no less the 15 different solved structures involving benzene on various surfaces of Co, Ni, Pd, Pt, Rh and Ru [6.69]. Among these structures are some very intriguing co-adsorption systems with CO and NO. Early studies of benzene adsorption on Pt(111) and Ni(111) employing electron energy loss spectroscopy found the CH out-of-plane bending mode to have the strongest perpendicular dipole moment [6.70]. It was therefore concluded that the carbon skeleton of the molecule should lay flat on the surface while the C-H bonds may be bend upwards. From the number of $A_1$-type modes (Sect. 7.2.3, Table 7.2) it was concluded furthermore that benzene sits in a site of $C_{3v}$ symmetry with the $\sigma_v$-planes cutting through the C-C bonds. Later structure analysis confirmed both, the flat orientation and the position of the $\sigma_v$-planes. As an example Fig. 6.35 shows the structure of benzene in the ordered $(\sqrt{7} \times \sqrt{7}) R19.1°$ structure on Ni(111)

[6.71] and the co-adsorption structure with CO in a (3×3) unit cell on Pd(111) [6.72].

(a)                                         (b)

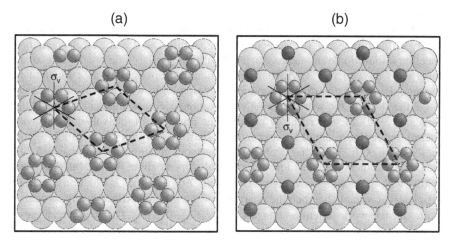

**Fig. 6.35.** Two ordered structures of benzene ($C_6H_6$) on surfaces. **(a)** $(\sqrt{7}\times\sqrt{7})$ R19.1° structure of benzene on Ni(111) [6.71]. The benzene ring lies flat on the surface over the hcp site. The local symmetry is $C_{3v}$ with the $\sigma_v$-planes cutting through the C-C bonds as concluded already in 1978 from vibration spectroscopy [6.70]. **(b)** (3×3) structure of benzene coadsorbed with two CO molecules (dark shaded balls) per unit cell on Pt(111) [6.72]. Here, the benzene rings as well as the CO-molecules are centered on the fcc-site. Dashed lines indicate the surface unit cells.

### 6.4.6 Alkali Metals

The adsorption of alkali metals on surfaces was studied already in the 1920ties. The interest was stimulated by the alkali atom induced reduction of the work function of surfaces, primarily the tungsten surface, and the thereby enhanced thermal emission of electrons[11]. The work function is reduced because alkali atoms have a low ionization threshold and energy is gained by donating the single outer s-electron into the unoccupied states of the substrate. The alkali/surface bond is therefore strongly polar, with the positive end of the dipole moment pointing away from the surface. The reduction of the work function $\Phi_0$ is described by the equation

---

[11] For the younger generation: In those "radio-days" the three terminal device that made radio communication possible was the electron tube with a thermal cathode as an electron emitter, the anode and at least one grid to control the flux of electron toward the anode.

$$\Phi = \Phi_0 - e\frac{p_z}{\varepsilon_0}n_{dip}. \tag{6.50}$$

Here, $p_z$ is the dipole moment of the alkali atom/surface complex, $n_{dip}$ is the area density of the dipoles, $\varepsilon_0$ is the vacuum permeability and $e$ is the electron charge. Since the initial dipole moment associated with the partial ionic bond of the alkali atoms is of the order of 1 $e$Å, the work function drops rapidly as a function of coverage. However, the dipole moment reduces quickly with increasing coverage so that the slope of the work function vs. coverage becomes smaller as soon as the overage exceeds a few percent. We discuss the physics of the alkali-induced work function reduction in detail for a specific system, namely for Li on Mo(110) surfaces. Figure 6.36 shows the decrease of the work function for that case [6.73]. The dipole moment calculated from the initial slope is about 0.5 $e$Å, but decreases rapidly with increasing coverage. At around $\Theta = 0.4$, the 2s-electrons overlap sufficiently to establish a metallic character of the lithium. After passing through a minimum, the work function change levels off at −1.9 eV. The work function of the composite system $\Phi_{comp} = 4.6 - 1.9 \text{ eV} = 2.7 \text{ eV}$ is nearly equal to the work function of lithium ($\Phi_{Li} = 2.9 \text{ eV}$).

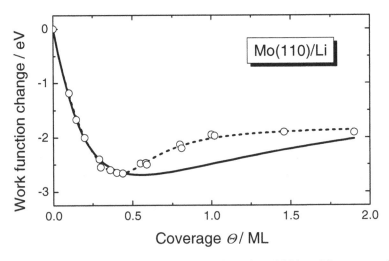

**Fig. 6.36.** Work function change of Mo(110) upon adsorption of lithium. The coverage is in fraction of the number of molybdenum surface atoms. Circles are experimental data [6.73] with the dashed line to guide the eye. The dipole moment drops very rapidly with coverage. At high coverage, the work function of the compound system is essentially the work function of pure lithium. The solid line is a fit to the Topping-model which yields an initial dipole moment of 0.56 $e$Å and an electronic polarizability of $\alpha_e = 10 \text{ Å}^3$.

The decrease of the dipole moment per atom with increasing density of the alkali atoms can be understood as a depolarization effect, at least for very low coverages where the s-wave functions of the alkali atoms do not overlap (*Topping-model* [6.74]). According to that model, the dipole moments of all the alkali atoms produce an electric field at each dipole site that is oppositely oriented to the dipole. If one attributes an electronic polarizability $\alpha_e$ to the alkali/surface complex, this electric field reduces the effective dipole moment associated with the alkali atoms. The electric field generated by the 2D-lattice of dipoles is

$$\mathscr{E} = -p_z \sum_{j \neq i} r_{ij}^{-3} \tag{6.51}$$

The sum extends over all distance $r_{ij}$ from an arbitrary origin. As an exception, (6.51) is written in the cgs-system, because the standard tables of electronic polarizabilities tabulate $\alpha_e$ in cgs-units[12]. With this depolarizing field, the dipole moment reduces to

$$p_z = p_{z0} + \alpha_e(\text{cgs})\mathscr{E} = p_{z0} - \alpha_e(\text{cgs})p_z \sum_{j \neq i} r_{ij}^{-3}$$

$$p_z = p_{z0} / (1 + \alpha_e(\text{cgs}) \sum_{j \neq i} r_{ij}^{-3}) \tag{6.52}$$

For a hexagonal close packed lattice the lattice sum is

$$\sum_{j \neq i} r_{ij}^{-3} = 8.9 n_{\text{dip}}^{3/2} \tag{6.53}$$

The reduction of the work function has then the form

$$\Phi = \Phi_0 - e \frac{p_z}{\varepsilon_0} n_s \Theta / \left(1 + 8.9 \alpha_{el}(\text{cgs}) n_s^{3/2} \Theta^{3/2}\right) \tag{6.54}$$

Here we have replaced the area density of dipoles $n_{\text{dip}}$ by the density of surface atoms $n_s$ and the fractional coverage $\Theta$. The solid line in Fig. 6.36 is a fit of (6.54) to the data points up to $\Theta = 0.5$. From the fit, one obtains the dipole moment and the electronic polarizability of the Li/Mo surface complex as

$$p_z = 0.56 \ e\text{Å} \text{ and } \alpha_e(\text{cgs}) = 10 \ \text{Å}^3 \tag{6.55}$$

---

[12] In the cgs-system $\alpha_e$ has the dimension of a volume that is of the order of the atom volume. In the S.I: system $\alpha_e$(S.I.) has also the dimension of a volume if it is defined by the equation $p = \varepsilon_0 \alpha_e(\text{S.I.})\mathscr{E}$. This $\alpha_e$(S.I.) differs from $\alpha_e$(cgs) by a factor of $4\pi$, so that $\alpha_e$(S.I.) $= 4\pi \alpha_e$(cgs).

The model reproduces the experimental data quite well for small coverages. The Topping model even produces a minimum in the work function change, which, however, lies outside the realm of validity of the model.

In order to understand the alkali-induced change in the work function from the standpoint of quantum mechanics it is useful to look at the spatial extensions of the wave function participating in the bonding. Quantum chemical calculations of the charge density distribution are helpful in that regard. With respect to the discussed example, we refer to charge density calculations performed on a $Mo_{23}$-cluster [6.75]. The (110) surface is represented by a layer of 14 Mo-atoms. The second layer consists of 9 atoms (Fig. 6.37). Figure 6.38 shows the charge density contours of the difference between the Li/Mo surface complex and the Mo-cluster and the Li-atom

$$\Delta\rho = \rho(Mo_{23}Li) - \rho(Mo_{23}) - \rho(Li) \tag{6.56}$$

Dashed and solid contour lines indicate negative and positive charges, respectively. The large extension of the Li2s electron catches ones eye. In the outer sphere, the Li2s charge density is diminished and placed mainly between the Li atom and the Mo surface atoms, but even there the charge spreads over several neighbors and spills out over the edges of the cluster. The reason for this large extension of the bonding charge is that the Li2s electrons, because of their spatial extension cannot establish a bond with the localized Mo4d-orbitals, as positive and negative overlap integrals cancel. The charge transfer from the lithium atom to the molybdenum surface is therefore into the unoccupied sp-states above the Fermi level. The broad spatial extension of the charge density enables the use of electron states right above the Fermi-level. As more Li atoms assemble on the surface the Li atoms are competing for the empty states right above the Fermi level, and the net charge transfer to the surface is reduced. This explains qualitatively the reduced dipole moment per Li-atom for larger coverages. In view of the large lateral extension of the bonding charge density one would expect the reduction of the dipole moment to become effective already for coverages when every second long bridge site is occupied, i.e. for $\Theta = 0.25$. This is about where the Topping model begins to fail (Fig. 6.36).

A further consequence of the large extension of the alkali valence s-electron is that the adsorption of alkali metals is not very site specific. Together with the large repulsive interaction likewise mediated by extended s-electrons, this gives rise to an enormous diversity of surface structures. The NIST library lists over 140 alkali structures for which the crystal structure has been determined [6.69]. These structure determinations merely amount to a very small fraction of the ordered structures that have been observed qualitatively and the solved crystallographic structures only concern the low index surfaces of Ag, Al, Cu, Ni, Pt, Pd, Rh, Ru and Si. A review of the crystallographic surface structures shows that all the standard surface lattices $(2\times2)$, $c(2\times2)$, $c(4\times2)$, $(\sqrt{3}\times\sqrt{3})R30°$ are realized. In agreement with the reasoning above, there is no universally preferred site. On (100) surfaces one finds on top sites as well as hollow sites, on (111) surfaces atop, hcp and fcc site-occupation is realized. For aluminum, surface alkali atoms frequently replace a substrate surface atom and even surface alloy formation has been reported. The (110) surfaces of fcc metals are all relatively unstable with respect to a (2×1) reconstruction. The adsorption of alkali metals can therefore trigger a (2×1) reconstruction even on those fcc metal that possess unreconstructed (110) surfaces when adsorbate-free (Sects. 1.2.1 and 1.2.4, Fig. 1.14).

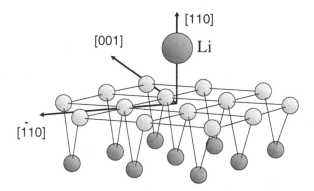

**Fig. 6.37.** $Mo_{23}$-cluster representing the (110) surface. Light and dark shaded balls represent first and second layer Mo-atom, respectively. The long bridge site of the Li-atom has the highest binding energy.

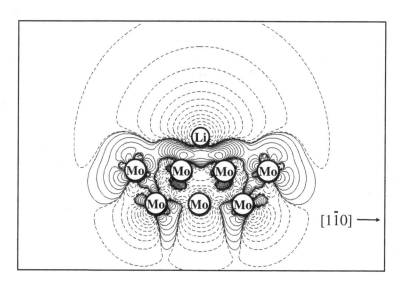

**Fig. 6.38.** Charge density difference $\Delta\rho = \rho(Mo_{23}Li) - \rho(Mo_{23}) - \rho(Li)$ in the (001)-plane (Fig. 6.37). Dashed and solid contour lines indicate negative and positive charges, respectively. The contour lines represent densities $2n^3 \times 10^{-4}$ $e\text{Å}^{-3}$ for $n = -15,...+15$ (after Müller [6.75]).

A very interesting consequence of the low site specificity of the alkali atom bonding is the small activation barrier for diffusion. Diffusion of alkali atoms on metal surfaces is extremely fast: At room temperature local deposits can spread over distances of mm within minutes [6.76, 77].

It should be mentioned finally that alkali metals at surfaces play an important role as *promoters* of catalytic reaction. For a comprehensive treatise of alkali atom adsorption the reader is referred to a volume edited by Bonzel, Bradshaw and Ertl [6.78].

### 6.4.7 Hydrogen

Among the first row elements hydrogen is a special case. Contrary to the alkali metals, the ionization energy is high. The electronegativity is in the range of standard transition metals. The bonding to surfaces is therefore local and covalent. As a molecule, hydrogen does not posses a HOMO or LUMO near the Fermi-level of substrates. Thus, the hydrogen molecule does not chemisorb at surfaces. The van-der-Waals interaction with surfaces is likewise small because of the small polarizability of the hydrogen molecule. On transition metal surfaces, hydrogen dissociates spontaneously without a significant activation barrier. The atom prefers sites of high coordination. The W(100) surface and presumably the other (100) surfaces of bcc-metals are exceptions. The smallness of the hydrogen atom renders it impossible to establish a chemical bond with all four W-atoms in the fourfold hollow site. Here, the bridge site is preferred. The hydrogen atom even draws the two W-atoms closer together, causing a reconstruction (Fig. 1.29). On the standard semiconductor surfaces such as Si, Ge, GaAs etc. the activation barrier for dissociative adsorption is large, the sticking coefficient for the $H_2$ molecule therefore extremely low. The formation of a surface bond to hydrogen atoms requires an exposure to atomic hydrogen.

Because of its small mass the hydrogen atom acts as a quantum particle which acts as a quantum wave packet with respect to diffusive and vibrational motion on surfaces. Figure 6.39 shows the diffusion coefficient of hydrogen versus the reciprocal temperature for W(110) [6.79]. Above 150 K, the diffusion coefficient follows the standard Arrhenius-type behavior. However, the diffusion coefficient levels off to a constant value for temperatures below 150 K. In that temperature regime, the hopping from site to site takes place via a quantum mechanical tunneling process, which is temperature independent.

A further manifestation of the quantum nature of hydrogen motion is the occasionally broadening of the features in the vibration spectrum of hydrogen. An example is hydrogen and deuterium on Ni(110) at monolayer coverage (Fig. 6.40 [6.80]). The vibration features of hydrogen at $1070 \, \text{cm}^{-1}$ and $640 \, \text{cm}^{-1}$ are signifi

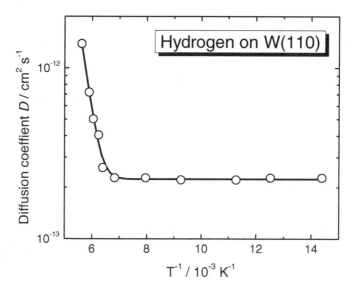

**Fig. 6.39.** Diffusion coefficient of hydrogen on W(110) as a function of the reciprocal temperature (after Auerbach et al. [6.79]). Above 150 K, the diffusion coefficient shows the normal Arrhenius-type behavior. At low temperatures, the diffusion is via a tunneling process from site to site and therefore temperature independent.

cantly broader than the corresponding features of deuterium at $750 \, \text{cm}^{-1}$ and $510 \, \text{cm}^{-1}$. A similar broadening was observed for hydrogen on Mo(110) [6.81]. The reason for the broadening is that the first excited vibration state of hydrogen is delocalized and acquires a band structure. The energy of the vibration state depends on the parallel component of the wave vector $q_\parallel$ [6.82] in the same way as an electron energy depends on the $k$-vector.

The wave vector dependence of the energy of the excited state $\hbar\omega(q_\parallel)$ is not to be confused with a phonon dispersion relation for periodic lattices when the atoms are treated as classical particles. The quantum dispersion here exists even for a single hydrogen atom in a periodic potential. Vibrational excitations, e.g. by inelastic electron scattering, correspond to transitions between band states, e.g. between the ground state and the first excited state. For zero momentum transfer to the electron the transitions between the band states are vertical (Fig. 6.41). The width of the energy loss peaks in Fig. 6.40 therefore results from the quantum dispersion of the excited state. The dispersion vanishes if the excited states are perfectly localized, and the description of the hydrogen atom as a wave packet becomes identical to the description as a classic particle.

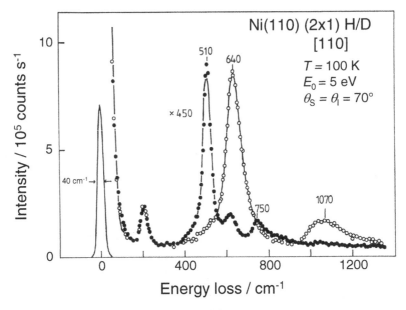

**Fig. 6.40**. Electron energy loss spectrum of the surface vibrations of the (2×1) hydrogen and deuterium covered Ni(110) surface. The vibration features of hydrogen at 640 cm$^{-1}$ and 1070 cm$^{-1}$ are significantly broader than the corresponding features for deuterium due to the quantum nature of the hydrogen atom [6.80].

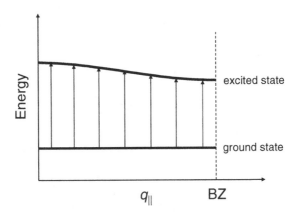

**Fig. 6.41**. Because of the small mass of the hydrogen atom, the excited vibration states of hydrogen may become delocalized quantum states possessing energy bands. Due to the dispersion of the energy bands of the excited states, the vibrational losses acquire a width even at fixed momentum transfer $\Delta q_{\parallel}$. The arrows indicate the possible transition for $\Delta q_{\parallel} = 0$ The dashed line marks the zone boundary.

## 6.4.8 Group IV-VII Atoms

This section considers the surface bonding and the structure of adsorbed oxygen, sulfur, nitrogen and carbon atoms. The surface phases of these elements, in particular on metal surfaces, belong to the most studied surface systems. Although oxygen, sulfur, nitrogen and carbon belong to different rows in the Periodic Table, structures and properties of surface phases of these elements on metals are rather similar. Adsorption phases of oxygen are typically prepared by dissociation of the oxygen molecule. The nitrogen molecule dissociates spontaneously only on very reactive surfaces, e.g. the transition metals on the left side of the Periodic Table. On other surfaces, nitrogen surface layers can be prepared by adsorption of $NH_3$ at low temperature followed by thermal annealing to dissociate $NH_3$ and desorb the less strongly bound hydrogen. Sulfur and carbon layers are conveniently prepared by $H_2S$ and acetylene adsorption, respectively, and the subsequent decomposition of the molecule and hydrogen desorption. To improve the long-range order on the surface the samples are typically annealed to higher temperatures either during or after completion of the adsorption process.

Carbon, nitrogen, oxygen, sulfur, bond to surface atoms via their $p_{x,y,z}$-electrons. The occupation of the p-electron system for these adsorbates ranges from four to six. Correspondingly, these atoms can entertain four, three and two covalent bonds (for carbon, nitrogen and oxygen/sulfur, respectively), which would call for adsorption sites involving the corresponding number of surface atoms. However, on metals, in particular on transition metals, the high density of unoccupied electron states at the Fermi-level enables covalent bonding also with fully occupied orbitals of the adsorbate (cf. the $\sigma$-bond of CO, Sect. 6.4.1). Therefore, all the 2p and 3p-atoms (including fluorine and chlorine) typically assume the sites of highest coordination on metals to maximize their bond energy. Examples are the threefold hollow site on (111) surfaces of fcc-metals, the fourfold hollow sites on the fcc(100) and bcc(100) surfaces, or one of the two equivalent three-atom coordination sites on the bcc(110) surfaces. In most cases, two or more ordered structures exist for different surface coverages. Examples are the p(2×2) and c(2×2) structures of oxygen and sulfur on Ni(100), corresponding to coverages of $\Theta = 0.25$ and $\Theta = 0.5$, respectively, and the p(2×2) and $(\sqrt{3} \times \sqrt{3})R30°$ structure of oxygen on Ni(111) corresponding to $\Theta = 0.25$ and $\Theta = 1/3$, respectively. On the (100) faces, in particular of bcc metals, adsorbed atoms in the fourfold hollow site bond strongly to the substrate atom in the second layer underneath. In the case of nitrogen adsorption on W(100), the nitrogen atom even has its shortest bond distance and therefore its strongest bond with the second layer atom. The surface structure of that system is displayed in Fig. 6.42 [6.83]. As illustrated in the figure, the size of the fourfold hollow site is larger than the size of the nitrogen atom. The site has room for a sphere of 0.87 Å radius while the radius of nitrogen is only about 0.76 Å. On the Ni(100) surface, the fourfold hollow site would have room for a sphere of merely 0.73 Å radius, just not enough to accommodate the nitrogen atoms in the surface plane. While the nitrogen atom can establish a bond also with the second layer nickel atom, the length of that bond remains with 1.99 Å larger than the typical Ni-N bond length. The bond to the Ni-atom in the second layer is therefore under tensile strain; consequently, the bonds to the surface Ni-atom are compressively strained. The Ni-atoms yield to the compressive in-plane strain by engaging in a clockwise and counterclockwise rotation, which causes the p4g reconstruction [6.84].

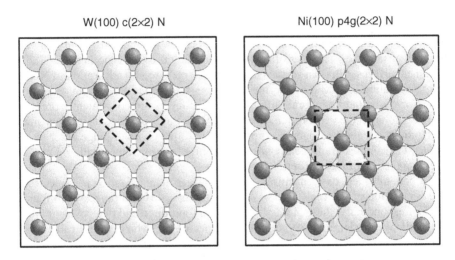

W(100) c(2×2) N          Ni(100) p4g(2×2) N

**Fig. 6.42.** Two surface structures involving nitrogen at 50% coverage [6.83-85]. The dashed lines indicate the unit cells. On the tungsten surface, the nitrogen atom has the strongest bond with the second layer W-atom. The bond distance is 2.13 Å while the bond distance to the atoms in the surface plane is larger 2.275 Å. On the Ni(100) surface, nitrogen causes a strong compressive surface stress in the surface layer, forcing the Ni-atoms to move out of their position by a clockwise and counterclockwise rotation.

The clockwise and counterclockwise rotation of the nickel atoms can be understood as a phonon softening phase transition on the c(2×2) surface. We discuss this issue here in the framework of a simple force field model (see also Sect. 7.1). The displacement pattern of the reconstructed phase corresponds to the $A_2$-phonon on the c(2×2) surface at the $\overline{X}$ -point of the surface Brillouin zone. As this surface phonon mode is entirely localized in the first Ni-layer its frequency is easily expressed in terms of nearest neighbor central force field [6.86]: the atoms are thought to be connected by springs. The force constant of the springs are equivalent to the second derivative of the nearest neighbor potential between the nickel atoms $\varphi''$ (cf. Sect. 7.1.2). For the particular displacement pattern shown in Fig. 6.43, the bonds to the adsorbates do not enter as long as one stays with central forces. To mimic the stress in the surface one assumes that the springs connecting the Ni surface atoms in the surface plane are loaded, hence one assumes the first

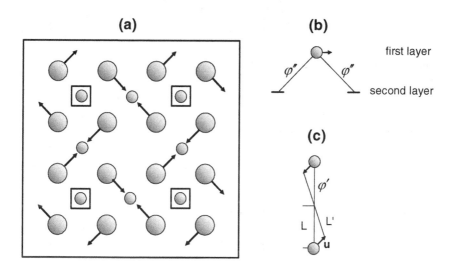

**Fig. 6.43.** (a) Displacement pattern of the $A_2$-phonon on the c(2×2) surface at the $\overline{X}$ -point of the surface Brillouin zone. The small balls represent second layer atoms. The squares mark the position of the carbon/nitrogen atoms. (b) and (c) sketches serving for the calculation of the frequency of the $A_2$-mode in terms of the nearest neighbor central force field (see text).

derivative of the potential between the nickel atoms in the surface plane $\varphi'_{11}$ to be nonzero (Stability of the lattice requires that the first derivative vanishes in the bulk). Such a nonzero $\varphi'_{11}$ corresponds to a macroscopic surface stress $\tau^{(s)}$ (Sect. 4.2.2)

$$\tau^{(s)} = \varphi'_{11} / a_{nn} \qquad (6.57)$$

when $a_{nn}$ is the nearest neighbor distance between the nickel surface atoms. A compressive stress between the nickel surface atoms corresponds to a negative $\varphi'_{11}$. The frequency of the $A_2$ surface phonon is calculated with the help of the sketches in Fig. 6.43. The bonds to the second layer (Fig. 6.43b) contribute an effective spring constant of

$$f_{12} = 2\varphi'' \cos^2 45° = \varphi'' . \qquad (6.58)$$

The springs to the second layer orthogonal to the plane of drawing are not strained to first order and do not contribute. The bonds in the first layer are also not strained to first order (Fig. 6.43c), but now we have assumed a nonzero first derivative of the potential. Expanding the potential $\varphi$ into the displacements $u$ yields

$$\varphi(u) = 2\varphi'u^2 / 4L = \frac{1}{2}\frac{2\varphi'}{a_{nn}}u^2 \tag{6.59}$$

so that this term adds a force constant

$$f_{11} = 2\varphi' / a_{nn} . \tag{6.60}$$

The frequency is therefore given by

$$m_{Ni}\,\omega^2(A_2, \overline{X}) = \frac{1}{2}\varphi'' + 2\varphi_{11}' / a_{nn} . \tag{6.61}$$

The $A_2$-phonon mode becomes soft when

$$\varphi_{11}' / a_{nn} = -\varphi'' / 4 . \tag{6.62}$$

The bulk force constant $\varphi''$ can be calculated from the maximum phonon frequency of bulk nickel to $\varphi'' = 37.9$ N/m. With (6.62) one can estimate the surface to become unstable with respect to the p4g-reconstruction for a surface stress of $-9.4$ N/m [6.87]. Direct measurements of the adsorbate induced surface stress (Sect. 3.3.3) have been carried out for carbon on the Ni(100) surface. Carbon causes the same p4g-reconstruction. It was found that the reconstruction begins when the stress reaches $-6$ N/m in reasonable agreement with the simple force field model (see [6.87] for further details).

The strong bonding and the preference of high coordination sites adsorption of O, S, C, and N may lead to a complete restructuring of the surface [6.88]. An example is the $(2\sqrt{2}\times\sqrt{2})R45°$ structure of oxygen on Cu(100) shown in Fig. 6.43 [6.88]. When oxygen is adsorbed in low doses at room temperature or below, it first adsorbs in the fourfold hollow site [6.89, 90]. At a coverage of $\Theta = 0.34$ the surface undergoes a first order disorder-order phase transition and eventually, after annealing to 500 K, displays a well ordered $(2\sqrt{2}\times\sqrt{2})R45°$ pattern at a coverage of $\Theta = 0.5$. The surface crystallographic structure of this phase was analyzed by LEED [6.88], *Photo Electron Diffraction* (PED) [6.91] and by *Surface Extended X-ray Absorption Fine Structure* (SEXAFS) [6.90]. According to these studies, the surface structure involves a missing row of copper atoms. The oxygen atoms reside merely 0.25 Å above the Cu-surface plane. Each oxygen atom has bonds to four copper atoms with bond distances of 1.92 Å and 2×1.82 Å to the Cu-atoms in the surface plane and 2.15 Å to one Cu-atom in the second plane. This geometry engages all three p-orbitals of oxygen and therefore makes for a particular stable bonding.

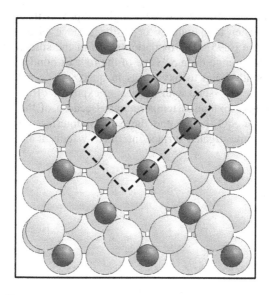

**Fig. 6.44.** The $(2\sqrt{2} \times \sqrt{2})R45°$ structure of oxygen on Cu(100) [6.88, 90, 91]. The dashed line marks the surface unit cell.

Another manifestation of the strong preference of the 2p and 3p atoms for the fourfold hollow site is the restructuring of the Ni(111) surface by carbon, nitrogen and sulfur. For theses adsorbates, the first layer of nickel has the structure of a (100) surface [6.92] with the Ni-atoms rotated in a clockwise and counterclockwise pattern as in the p4g reconstruction of Ni(100), however with a larger unit cell [6.93, 94].

In most cases, the surface phases of oxygen, sulfur, carbon and nitrogen are metastable with respect to the nucleation of clusters of a bulk compound. Extended exposure of nickel to oxygen, e.g., leads to the formation of NiO clusters in the surface. The nucleation of NiO clusters is furthered by defects. Nickel surfaces containing many defects continue to adsorb oxygen after completion of a c(2×2) oxygen surface layer with only a mild drop in the sticking coefficient and form NiO clusters readily, whereas on defect-free surfaces the oxygen uptake saturates with the c(2×2) oxygen layer.

Silicon represents the other extreme with respect to oxygen adsorption. Silicon is so reactive that no particular surface phase with oxygen atoms exists. Oxygen dissociates readily on clean silicon surfaces and the oxygen atoms are inserted into Si-Si bonds to form Si-O-Si bonds as in quartz. A molecular state can also be stabilized at low temperatures [6.95, 96]. No structures with long-range order exist for either the molecular or the dissociated state. Upon adsorption of oxygen at room temperature, a severely disordered silicon oxide grows with a wide range of Si-O-Si bond angles and Si-O distances. At room temperature, this so-called *natural oxide layer* grows up to a thickness of about 30 Å in air. The local order in terms of the range bond lengths and bond angles improves if the exposure to oxygen is made at higher temperatures or if the surface is annealed after room temperature exposure [6.95]. The oxidation states of the silicon atoms after room temperature exposure range from $Si^I$ (one of the four Si-bonds bond to oxygen) to $Si^{IV}$ (all four Si-bonds bond to oxygen as in quartz). The mean oxidation state increases towards the $Si^{IV}$-state with increasing exposure of the silicon surface to oxygen and for higher oxidation temperatures. Because of the technological importance of ultrathin $SiO_2$-layers and the $SiO_2/Si$ interface, this surface system belongs to the most investigated surface/interface systems at all [6.97].

# 7. Vibrational Excitations at Surfaces

Many aspects of the chemistry of adsorbates were revealed through the vibration spectra of the adsorbed species on surfaces, as the vibration frequency spectrum is characteristic for the strength and the type of the bonds. The development of experimental techniques for studying surface vibrations was therefore a major step forward in the understanding of surface chemistry. Of fundamental interest are the vibrational excitations of clean, two-dimensional periodic surfaces, the surface phonons. The general basis for the consideration of vibrational excitations is the Born-Oppenheimer approximation [7.1]. According to this approximation, the electronic eigenstates follow the moving atoms *adiabatically*, which means than the electron energy levels change with the atom positions, but the electrons remain in the same eigenstates. The total energy of the solid as a function of the atom position therefore plays the role of a potential for the atom motion. We begin this section by looking into the consequences of the Born-Oppenheimer approximation for periodic surfaces.

## 7.1 Surface Phonons of Solids

### 7.1.1 General Aspects

The potential $\Phi$ for the atom motion depends on the coordinates $\{r(n)\}$ of all atoms in the solid. The potential can be expanded into a Taylor series around the equilibrium positions denoted as $r_0(n)$.

$$\Phi(\{r(n)\}) = \Phi_0(\{r_0(n)\}) + \sum_{n,m,\alpha,\beta} \Phi(n,\alpha;m,\beta) u_\alpha(n) u_\beta(m)... \qquad (7.1)$$

Here, $u_\alpha(n)$ is the deviation of the atom $n$ from the equilibrium position in the Cartesian direction $\alpha$. The first derivatives vanish, as the expansion is around the global minimum of the potential. If higher order terms are neglected, the equation of motion for a particular atom $n$ in the Cartesian direction $\alpha$ is

$$M(n)\ddot{u}_\alpha(n) + \sum_{m,\beta} \Phi(n,\alpha;m,\beta) u_\beta(m) = 0, \qquad (7.2)$$

in which $M(n)$ is the mass of the atom $n$. The system of equations (7.2) technically couples all atom motions to each other. In reality, the coupling vanishes rapidly for larger distances. For an approximate picture of the vibration spectrum the consideration of nearest and possibly next-nearest neighbors often suffices. For a three-dimensional periodic structure, the system of equations (7.2) is separated into a subsets of $3s$ equations, with $s$ the number of atoms in the unit cell, by the ansatz

$$u_\alpha(n) = u_{0,\alpha}\, e^{-i\left(\omega(q)t - q\cdot r_0(n)\right)} \qquad (7.3)$$

The quantized plane wave solutions of the type (7.3) are the *phonons* of the 3D-solid. A flat surface/interface breaks the 3D-translational symmetry of the solid, which gives rise to solutions of (7.2) that are localized at the surface/interface in the sense that the vibrational amplitude decays in an essentially exponential manner away from the surface/interface. These modes are called *surface (interface) modes*, or *surface (interface) phonons*. A schematic overview over the spectrum of eigenmodes at a surface is shown in Fig. 7.1. The frequencies are displayed as function of the component of the wave vector parallel to the surface $q_\parallel$. In this projection, bulk modes form a continuum because their wave vector perpendicular to the surface remains arbitrary. The continuum of bulk phonons is shown as a shaded area in Fig. 7.1. The frequency of a surface phonon is uniquely determined by $q_\parallel$. Surface phonons have therefore a defined dispersion branch in this graph. They are plotted as solid lines. Genuine surface modes can exist only if no bulk phonons of the same symmetry are present in the same $\omega$, $q_\parallel$ range. In addition to the surface phonons, so-called *surface resonances* exist.

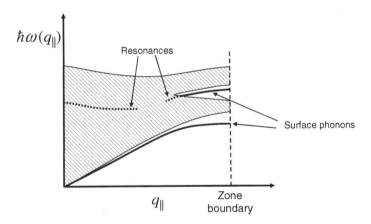

**Fig. 7.1.** Schematic drawing of the spectrum of phonons and resonances vs. the wave vector parallel to the surface $q_\parallel$. The shaded area is the continuum of bulk modes. Solid lines represent surface phonons of the substrate; dotted lines are resonances.

Surface resonances are bulk phonon modes, which have large amplitudes at the surface (dotted lines in Fig. 7.1). One of the surface phonons shown in the figure has an acoustic limit ( $\lim_{q \to 0} \omega(q) = 0$ ). The number of surface phonons with an acoustic limit can range between one and three, depending on the elastic constants, the structure of the substrate, the orientation of the surface, and the direction of $q_{\parallel}$ on the surface (Sect. 7.1.4). Along high symmetry directions, the plane spanned by the surface normal and the direction of $q_{\parallel}$, called *sagittal plane*, may coincide with a mirror plane of the structure. An example would be the [011] direction on a (100) surface. In that case, the surface phonons are even or odd with respect to the sagittal plane. The odd mode is polarized perpendicular to the mirror plane. In other words, the mode is a shear horizontal mode, while the polarization vectors of the even modes lie in the sagittal plane.

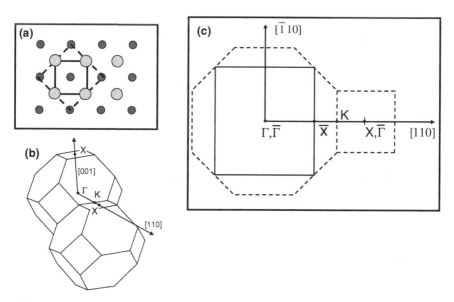

**Fig. 7.2. (a)** Top view on the (001) surface of an fcc-structure (see also Appendix). Large light-grey and small dark-grey circles represent surface and second layer atoms, respectively. The solid line is the surface unit mesh. The dashed line is the projection of the fcc-lattice on the surface. **(b)** Bulk Brillouin zone of the fcc-structure. A few high symmetry points are marked. Starting from the center $\Gamma$, the X-point can be reached in two ways: by moving in the [001] and in the [110] direction. **(c)** *Surface Brillouin Zone* (SBZ) and projected bulk Brillouin-zone are shown as solid and dashed lines, respectively. The X-point of the projected bulk zone coincides with the $\overline{\Gamma}$-point of the adjacent SBZ.

The maximum of $q_\parallel$ is given by the boundary of the *Surface Brillouin Zone* (SBZ) which may differ from the boundaries of the projection of the bulk Brillouin-zone onto the surface plane. This is illustrated in Fig. 7.2 for the (001) surface of the fcc structure (see also Appendix). Along the [110] direction the $\overline{X}$-point of the SBZ is half way to the X-point of the projected bulk zone which therefore coincides with the $\overline{\Gamma}$-point of the adjacent SBZ.

## 7.1.2 Surface Lattice Dynamics

Theoretical studies of surface modes are most conveniently performed on a slab of $N$ layers of unit cells with two equivalent surfaces. We denote the 2D-unit cells by $l_\parallel = (l_1, l_2)$, the layers by $l_z$, and the atoms in the unit cells by the index $\kappa$. A displacement of the atom $\kappa$ in the unit cell $l_\parallel$ of the layer $l_z$ in the direction $\alpha$ is denoted as $u_\alpha(l_\parallel l_z \kappa)$. In this notation the equations of motion become

$$M(l_z\kappa)\ddot{u}_\alpha(l_\parallel l_z\kappa) + \sum_{\beta l_\parallel' l_z' \kappa'} \Phi_{\alpha\beta}(l_\parallel l_z\kappa; l_\parallel' l_z'\kappa')u_\beta(l_\parallel' l_z'\kappa') . \qquad (7.4)$$

It is convenient to introduce mass-normalized amplitudes

$$\xi_\alpha(l_\parallel l_z\kappa) = \left[M(l_z\kappa)\right]^{1/2}u_\alpha(l_\parallel l_z\kappa) . \qquad (7.5)$$

We are interested in solutions with a time dependence of the form $\exp(-i\omega t)$. After inserting this time dependence, (7.4) becomes

$$\omega^2\xi_\alpha(l_\parallel l_z\kappa) - \sum_{\beta l_\parallel' l_z' \kappa'} D_{\alpha\beta}(l_\parallel l_z\kappa; l_\parallel' l_z'\kappa')\xi_\beta(l_\parallel' l_z'\kappa') = 0 , \qquad (7.6)$$

where we have introduced the *dynamical matrix*

$$D_{\alpha\beta}(l_\parallel l_z\kappa; l_\parallel' l_z'\kappa') = \frac{\Phi_{\alpha\beta}(l_\parallel l_z\kappa; l_\parallel' l_z'\kappa')}{\left[M(l_z\kappa)M(l_z'\kappa')\right]^{1/2}} \qquad (7.7)$$

The dynamical matrix is symmetric with respect to the interchange of primed and non-primed symbols (action forces are equal to reaction forces) and depends only on the difference $l_\parallel - l_\parallel'$ because of the translation symmetry. Equation (7.6) admits solutions in the form of plane waves parallel to the slab.

$$\xi_\alpha(l_\parallel l_z\kappa) = e_\alpha(q_\parallel; l_z\kappa)e^{-iq_\parallel \cdot r_0(l_\parallel l_z\kappa)} \qquad (7.8)$$

Inserting (7.8) into (7.6) yields the secular equation for the eigenvectors $e_\alpha(q_\parallel; l_z \kappa)$ and the eigenfrequencies $\omega_s(q_\parallel)$

$$\omega_s^2(q_\parallel)e_\alpha^{(s)}(q_\parallel; l_z \kappa) - \sum_{\beta l_z' \kappa'} d_{\alpha\beta}(q_\parallel; l_z \kappa; l_z' \kappa')e_\beta^{(s)}(q_\parallel; l_z' \kappa') = 0 , \qquad (7.9)$$

in which $d_{\alpha\beta}(q_\parallel; l_z \kappa, l_z \kappa')$ is the Fourier transformed dynamical matrix defined as

$$d_{\alpha\beta}(q_\parallel; l_z \kappa; l_z' \kappa') = \sum_{l_\parallel'} D_{\alpha\beta}(l_\parallel l_z \kappa; l_\parallel' l_z' \kappa')e^{-i q_\parallel \cdot (r_0(l_\parallel l_z \kappa) - r_0(l_\parallel' l_z' \kappa'))} \qquad (7.10)$$

The eigenfrequencies are given by the zeros of the determinant

$$\det\left(\omega_s^2(q_\parallel)\mathbf{1} - d_{\alpha\beta}(q_\parallel; l_z \kappa; l_z' \kappa')\right) = 0 \qquad (7.11)$$

The eigenmodes of an fcc-slab and (100), (111) and (110) surfaces were first studied by Allan, Aldredge and deWette invoking the Lennard Jones potential (6.1) [7.2-4]. While this central force potential is inadequate to describe the force field in a metal quantitatively, the qualitative features of the phonon spectrum and the character of the surface modes are correctly reproduced. Therefore, nearly all subsequent papers followed the notation introduced in the pioneering work of Allen et al. We discuss their results for the (100) surface (Fig. 7.3).
The surface phonon $S_1$ exists throughout the SBZ. However, its character changes from purely shear horizontal polarization, i.e. odd with respect to the sagittal mirror plane, along $\overline{\Gamma}\overline{X}$ to become even with respect to the sagittal mirror plane

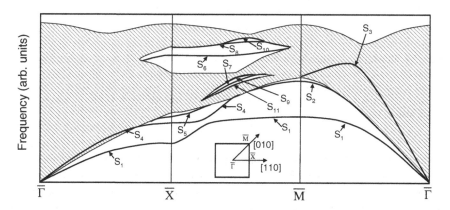

**Fig. 7.3.** Spectrum of modes of a 21-layer fcc-slab with (100) surfaces after Allen et al. [7.3]. The *Surface Brillouin Zone* (SBZ) is shown as an insert.

along $\overline{\Gamma}\overline{M}$ with nearly shear vertical polarization. The modes $S_3$ and $S_4$ are polarized in the sagittal plane along $\overline{\Gamma}\overline{X}$ and $\overline{\Gamma}\overline{M}$. They can exist as surface modes within the bulk band since the surrounding bulk modes are odd with respect to the sagittal plane. $S_2$ and $S_5$ are primarily localized in the second layer. $S_2$ is even along $\overline{\Gamma}\overline{M}$ while $S_5$ is shear horizontal (=odd) along $\overline{\Gamma}\overline{X}$. A prominent mode with a large amplitude at the surface is also the gap mode $S_6$ which is horizontally polarized and localized almost entirely in the first layer at $\overline{X}$. The higher numbered modes have their main amplitude in the second or third layer beneath the surface layer.

The large variety of surface modes with their different degrees of localization in first second and third layers may be bewildering. It is however straightforward to understand the origin of these modes if one considers particular high symmetry points in the SBZ. The discussion is particularly easy for the $\overline{M}$-point of the SBZ if interactions between the atoms are restricted to nearest neighbor central forces. In the bulk of the material one has only one force constant, which is the second derivative of the pair potential between the atoms denoted as $\varphi_b''$. The first derivative vanishes in the bulk, as the crystal is assumed strain-free.

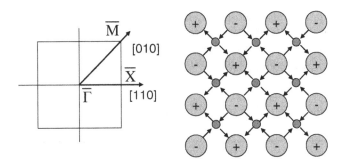

**Fig. 7.4.** Illustration of the $S_1$ surface vibration mode at the $\overline{M}$-point of the SBZ (left panel) on a (100) surface of an fcc (or bcc) structure. The forces on the second layer atoms (small spheres) cancel. To the extent that the force field is restricted to nearest neighbor central forces the mode is completely localized in the surface layer. Corresponding modes exists also in the second layer and in all bulk layers.

To account for differences in the bonding of atoms near the surface the force constants connecting atoms within the surface layer, $\varphi_{11}''$, and between the first and second layer, $\varphi_{12}''$, differ from the bulk force constant. As an example we consider the $S_1$ surface vibration mode at the $\overline{M}$-point of the SBZ. The mode is entirely localized to the first layer because the forces on the second layer atoms (small spheres in Fig. 7.4) cancel. The frequency can therefore be calculated without explicitly solving the secular equation.

$$M\omega_{S_1}^2 (\overline{M}) = 4\varphi_{12}'' \cos^2 45° = 2\varphi_{12}'' \tag{7.12}$$

The second layer and further layer modes of the same type as depicted in Fig. 7.4 are likewise decoupled from the rest of the lattice. Thus, the frequency of the second layer mode, the $S_2$-mode in the notation of Allan et al., is

$$M\omega_{S_2}^2 (\overline{M}) = 2(\varphi_{12}'' + \varphi_b''). \tag{7.13}$$

The maximum frequency bulk mode is likewise easily calculated in this model. It is found, e.g., at the $\overline{M}$-point (cf. Fig. 7.3) and has the polarization vectors along [010]-direction. Every alternate layer vibrates with $180°$ phase shift so that the reduced mass appears in the equation of motion. The mode involves spring forces from eight of the twelve nearest neighbors around each atom. The frequency of the mode is therefore

$$M/2\,\omega_b^2 = 8\varphi_b'' \cos^2 45° = 4\varphi_b''$$
$$M\,\omega_b^2 = 8\varphi_b'' \tag{7.14}$$

If $\varphi_{12}''$ is equal to the bulk force constant $\varphi_b''$, the $S_1$ and $S_2$ mode have their frequencies at $1/2$ and $1/2^{1/2}$ of the maximum bulk frequency $\omega_b$. In the calculations of Allen et al. (Fig. 7.3) the frequency of these modes relative to the bulk modes is a little lower because of the longer range of the Lennard-Jones potential. At the surface, the mode with the same polarization as the bulk mode calculated with (7.14) becomes a genuine surface mode, namely the longitudinal polarized $S_3$-mode at the $\overline{M}$-point. Its frequency is lower than the frequency of the bulk mode because of the missing neighbors. Since the mode is not localized to a single layer, the calculation of its frequency requires the solution of the secular equation (7.11).

### 7.1.3 Surface Stress and the Nearest Neighbor Central Force Model

While the strain in the bulk of the crystal vanishes in the absence of an external pressure, solid surfaces possess a surface stress that typically amounts to a few N/m (Sect. 4.2). If one models the surface stress within the nearest neighbor central force model, the stress corresponds to loaded springs between the surface atoms, i.e. to a non-vanishing first derivative of the pair-potential. For the fcc (100) surface the relation between the first derivative of the potential connecting the surface atoms $\varphi_{11}'$ and the surface stress $\tau_{11}^{(s)}$ is

$$\varphi_{11}' = a_{nn}\tau_{11}^{(s)}, \tag{7.15}$$

in which $a_{nn}$ is the nearest neighbor distance. We study the effect of the stress on the $S_1$-mode at $\overline{M}$. By expanding the potential for vertical displacements of the atoms one obtains the frequency

$$M\omega_{S_1}^2(\overline{M}) = 2\varphi_{12}'' + 4\varphi'/a_{nn}\,. \qquad (7.16)$$

A tensile stress, $\tau_{11}^{(s)} > 0$ leads to an upward shift of the frequency, a compressive stress to a downshift. The effect is the analogue of tuning a string instrument. For very large compressive stresses, the mode would become soft and the lattice would become unstable with respect to a reconstruction pattern that corresponds to the mode depicted in Fig. 7.4. The instability occurs at

$$\tau_{11}^{(s)} \le -\varphi_{12}''/2\,. \qquad (7.17)$$

The issue of stability of surfaces under stress was already discussed in Sect. 6.4.7 in connection with the adsorption of oxygen, carbon and nitrogen atoms in a c(2×2) pattern. There it was shown that another mode, the $A_2$-mode at $\overline{X}$ (6.62), becomes soft and correspondingly the surface would be unstable if

$$\tau_{11}^{(s)} = \varphi_{11}'/a_{nn} \le -\varphi_{12}''/4\,, \qquad (7.18)$$

which is for half the compressive stress than the $S_1$-mode at $\overline{M}$. Because of a selection rule (see Sect. 7.2.3) the $A_2$-mode at $\overline{X}$ cannot be observed by inelastic scattering of particles in the conventional scattering geometry, but the $S_1$-mode

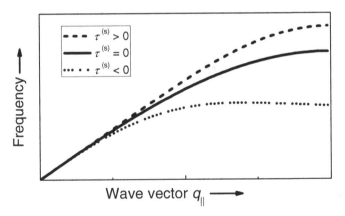

**Fig. 7.5.** Schematic plot of the dispersion of a mostly perpendicularly polarized surface phonon for surfaces under tensile surface stress, without surface stress and with compressive surface stress (dashed, solid, and dotted line, respectively).

can, because the $S_1$-mode is polarized within the sagittal plane and mostly vertical. The dispersion of the $S_1$-mode along the $\overline{\Gamma M}$-direction and of the likewise mostly vertically polarized $S_4$-mode along the $\overline{\Gamma X}$ - direction has been studied for several fcc(100) and some bcc(100) surfaces. The dispersion of these two modes can be taken as a qualitative indicator of the sign and magnitude of the surface stress. The effect is schematically shown in Fig. 7.5. Tensile and compressive surface stresses lead to upwards and downwards shifts of the dispersion curve, respectively. The shift increases with the wave vector $q_{||}$. Dispersion curves showing the signature of strong compressive stress have been observed in ultra thin iron films on Cu(100) [7.5] and for the c(2×2) oxygen covered Ni(100) surface [7.6].

The picture sketched above is appealing and it seems to work qualitatively. It should be said however that the relation between surface stress and the dispersion exists only in the framework of the nearest neighbor model. Rigorous theory does not produce a relation between the macroscopic quantity surface stress and the interatomic forces that determine the dispersion [7.7].

### 7.1.4 Surface Phonons in the Acoustic Limit

The simple force field discussed above provides an easy access to the understanding of the phonon spectrum of fcc-crystals. It even performs not badly on the quantitative side in materials like Cu or Ni, in particular if spring constants at the surface are adjusted to experimental data. However, central forces, regardless how many neighbors are considered, fail in the acoustic limit. It can be shown that central forces establish *Cauchy relations* between the elastic constants. For the cubic structure the Cauchy relation is $c_{12} = c_{44}$. Iridium is the only fcc-material for which that relation approximately fulfilled. All other fcc- and bcc-materials are far off. On the other hand, the sound velocity of acoustic surface phonons can be calculated within the framework of elasticity theory. The mathematics involved is elementary however quite unwieldy. The reader interested in details is referred to the definitive review of Wallis [7.8] and we limit the discussion here to a few qualitative aspects.

We consider first the elastically isotropic case. Since on an elastically isotropic medium the sagittal plane is a mirror plane in all directions, all modes belong either to the even or odd representation. The elastic isotropic medium sustains only a single, even surface wave, which is called the *Rayleigh wave*, named after Lord Rayleigh who calculated the sound velocity of these surface waves already in 1887 [7.9]. For a derivation, see [7.10]. The displacement pattern of the Rayleigh wave is a mixture of longitudinal and transversal motion (Fig. 7.6) as the displacements execute ellipses. Only in a particular depth the displacements vectors are transversal to the direction of propagation. The depth at which that happens depends slightly on the Poisson number $\nu$. The displacement pattern in Fig. 7.6 is calculated for $\nu = 0.3$. Then the displacements are transverse at $z = -0.144\lambda$ when $\lambda$ is the wavelength. The sound velocity of the Rayleigh wave $c_{Rayleigh}$ is

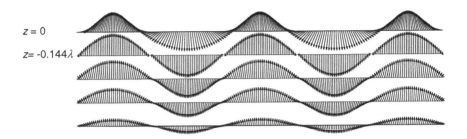

**Fig. 7.6.** Displacement pattern in a Rayleigh wave for a Poisson number $\nu = 0.3$. The displacements are neither longitudinal nor transverse, but closer to the latter. Only in a particular depth $z = -0.144\lambda$, $\lambda$ being the wavelength, the displacements are purely transversal.

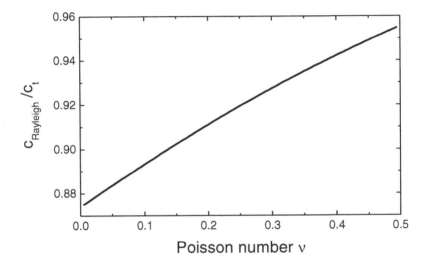

**Fig. 7.7.** Velocity of the Rayleigh wave relative to the sound velocity of the transverse wave as a function of the Poisson number $\nu$.

between 0.874 and 0.955 of the velocity of the transversal sound wave $c_t$, depending on the Poisson number (Sect. 3.3.1). Figure 7.7 shows the ratio $c_{\text{Rayleigh}}/c_t$ as a function of the Poisson number. For anisotropic cubic crystals elastic surface waves may exist in particular directions, however merge with continuum solutions in other directions, depending on the elastic constants. Shear horizontal surface waves (*Love waves*) also exist in certain high symmetry directions and on low index surfaces.

*Surface Acoustic Waves* (SAWs) are important for many technical applications. For example, they are employed in the construction of high frequency filters [7.11]. Figure 7.8 shows the scheme of such a filter. A slab of a piezoelectric material, e.g. LiNbO$_3$, is equipped with interdigital metallic stripes. An electrical impulse on the transmitter imprints a surface wave train on the slab that travels to the receiver where it generates a train of electrical pulses. By proper shaping of the interdigital stripes, a sharp pulse representing a white spectrum is converted into a sinusoidal signal with a nearly Gaussian envelope. The Fourier spectrum of the signal at the receiver is a single frequency (because of the Gaussian envelope) with a bandwidth that is inversely proportional to the number of metallic stripes. Hence, the device acts as a frequency filter.

**Fig. 7.8.** Scheme of a high frequency filter using *Surface Acoustic Waves* (SAWs). The white spectrum of a short pulse imprinted on the device is converted into a harmonic signal of a narrow bandwidth in the receiver on the right.

## 7.1.5 Surface Phonons and Ab-Initio Theory

By virtue of the Born Oppenheimer approximation, phonons are a property of the electronic ground state. Since the eigenvectors of bulk phonons are solely determined by the symmetry at certain high symmetry points of the Brillouin zone, it is relatively straightforward to calculate the frequency of those phonons by ab-initio methods. For example, the longitudinal acoustical mode in an fcc-structure at the X-point (zone boundary in [100]-direction) consists of the motion of rigid next-nearest (100)-planes moving against each other, leaving the center of gravity at rest. The parabolic potential associated with that motion can be calculated as the total energy for three positions of the (100)-sublattices with respect to each other. The calculation requires a doubling of the size of the unit cell. The method is known as the *frozen phonon* method. Modes at other critical points of the Brillouin zone and the elastic constants can be calculated within the same scheme. In order to obtain the phonon frequencies at arbitrary wave vectors the results are mapped onto the parameters of a physical meaningful interatomic potential that serves as an interpolation scheme. The type of potential depends on the nature of the chemical bonds in the crystal. For covalently bonded materials, central forces need be supplemented by angle bending forces. Furthermore, electric dipole fields originating in the polarization of the electron orbitals upon a displacement of the atom position need be considered.

For surface phonons, the situation is more complicated as the frozen phonon method does not work. Even at the high symmetry points of the SBZ, the eigenvectors are not known a priory, except for the few cases discussed in Sect. 7.1.2 where the nearest neighbor model suffices. In all other cases, interlayer force constants must be calculated for at least several layers below the surface, proceeding along the same lines as for frozen phonon calculations [7.12]. This requires a large number of total energy calculations for slabs with huge super-cells, at the corresponding large expense of computer time and power. The results of such effort need still be mapped onto an interpolation scheme in order to obtain the full dispersion curve of surface phonons.

More effective is a method that calculates the elements of the dynamical matrix directly via a perturbation approach within in the *Local Density Approximation* (LDA) [7.13]. The method makes use of the *Hellman-Feyman theorem* according to which the force associated with a variation of a external parameters $\lambda = \{\lambda_i\}$ (e.g. the displacements of atoms from their equilibrium position) is given by the ground-state expectation value of the derivative of the bare external potential acting on the electrons. We illustrate the method for a local potential $V_\lambda(r)$. The derivative of the ground-state energy of the electron system $E_\lambda^{(el)}$ relative to a set parameters $\lambda$ with respect to a particular parameter $\lambda_i$ is

$$\frac{\partial E_\lambda^{(el)}}{\partial \lambda_i} = \int n_\lambda(r) \frac{\partial V_\lambda(r)}{\lambda_i} d^3r , \qquad (7.19)$$

where $n_\lambda(r)$ is the electron density distribution and the integral extends over the entire space where $n_\lambda(r) \neq 0$. Total energy variations with respect to $\lambda_i$ are obtained by integration of (7.19). In order to have the energy variations correct up to second order in $\lambda$ it is necessary to consider the expansion of the integrand to linear order. The dependence of the energy of the electron system on $\lambda = \{\lambda_i\}$ is therefore

$$E_\lambda^{(el)} = E_0^{(el)} + \sum_i \lambda_i \int n_0(r) \frac{\partial V_\lambda(r)}{\partial \lambda_i} d^3r$$
$$+ \frac{1}{2} \sum_{i,j} \lambda_i \lambda_j \int \left\{ \frac{\partial n_\lambda(r)}{\partial \lambda_j} \frac{\partial V_\lambda(r)}{\partial \lambda_i} + n_0(r) \frac{\partial^2 V_\lambda(r)}{\partial \lambda_i \partial \lambda_j} \right\} d^3r \qquad (7.20)$$

The required expansion of the total energy is obtained by adding the corresponding expansion of the interaction energy between the ion cores.

An alternative to full-scale ab-initio theory is the use of semi-empirical potentials that mimic certain features of more rigorous exact theories. The so-called *Embedded Atom Model* (EAM) has been used quite successful [7.14]. In that model, the total energy of a system is the energy gained by "embedding" an atom into the background charge density of all other atoms to which the pair repulsive interaction between the ion cores is added.

$$\phi(\{r_i\}) = \sum_i F_i(\rho_i) + \frac{1}{2} \sum_i \sum_{j \neq i} \phi_{ij}(r_{ij}) \tag{7.21}$$

In this expression, $\rho_i$ is the electron density at the position of an atom $i$ due to the other atoms in the system, $F_i(\rho_i)$ is the energy to embed the atom $i$ into the background density $\rho_i$, and $\phi_{ij}(r_{ij})$ is the core-core pair-repulsion between the atoms $i$ and $j$ separated by the distance $r_{ij}$. A weakness of the method is that the charge density is not calculated self-consistently. Rearrangements of the electron density and changes of the screening properties are not taken into account. The advantage of the method is the high computation speed. The phonon spectra of surfaces with large unit cells, e.g. vicinal surfaces are therefore readily calculated [7.15-18].

### 7.1.6 Kohn Anomalies

Electron-hole pair excitations in metals and phonons cover the same phase space. An electron-hole pair excitation in which the electron is transported from one side of the Fermi-surface to the other changes the momentum of the electron system by about $2k_F$ while requiring only a small amount of energy. Energy and momentum change can match those of phonon states which therefore display an anomaly in the dispersion at $q_c = 2k_F$. These *Kohn-anomalies* [7.19] constitute a breakdown of the Born-Oppenheimer approximation. As the matching condition for momentum and energy is fulfilled only by a few electron states, the effect is very small for bulk phonons. Merely the derivative of the dispersion curve becomes singular. In two dimensions, the effect may cause a noticeable dip in the dispersion curve itself. For a one-dimensional system, finally, the coupling between the electron and phonon excitations is so strong that the lattice becomes instable. This is the so-called *Peierls Transition* [7.20] for one-dimensional metals with a half-filled band. At the surface of a bulk crystal, the surface electrons constitute a two-dimensional electron system (Sect. 8.2). The magnitude of the Kohn-anomaly in the surface phonon dispersion depends on the topology of the two-dimensional Fermi-surface of the surface states and on the strength of the matrix element, which couples electrons and phonons. Figure 7.9 shows schematically a two-dimensional Fermi-surface (i.e. a Fermi-contour) that has extended parallel sections and therefore many possibilities to fulfill energy and momentum conservation in electron phonon scattering. Fermi-surfaces like that are called *nested Fermi-surfaces*.

Fermi-surface nesting and the corresponding strong Kohn-anomalies have been observed for the hydrogen-covered W(110) and Mo(110) surfaces at $q_c \cong 0.9\,\text{Å}^{-1}$ along the [001] direction. The anomalies were discovered in 1992 by Hulpke and Lüdecke using the technique of inelastic scattering of He-atoms [7.21]. The results

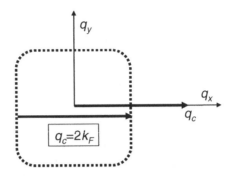

**Fig. 7.9.** Schematic picture of a *nested Fermi-surface* of surface states. The large number of electron-hole pair excitations with the same momentum change causes a pronounced anomaly in the dispersion of the surface phonon at $q_c = 2k_F$.

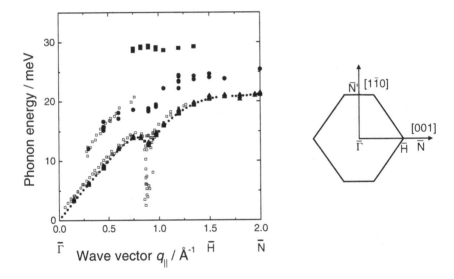

**Fig. 7.10.** Surface phonon dispersion curves on the fully hydrogen covered Mo(110) surface. The SBZ is shown on the *right*. The open squares are the data of Hulpke and Lüdecke [7.22], the filled symbols data of Kröger et al. [7.24]. The same large Kohn-anomaly is also found on the hydrogen covered W(110) surface [7.21, 23].

of Hulpke and Lüdecke actually displayed two branches in the critical region, one with an extremely deep dip, another with a smaller dip. The same two-branched anomaly was observed on the hydrogen covered Mo(110) surface [7.22]. Inelastic electron scattering on the same systems showed later that only the branch with the small dip is a phonon dispersion branch while the very deep dip is due to inelastic scattering of He-atoms from the electronic degrees of freedom [7.23, 24]. The results are shown in Fig. 7.10 for the case of the Mo(110) surface. According to photoemission spectroscopy the electronic transitions causing the anomaly are transitions between hydrogen-induced surface states [7.25, 26]. However, the nesting of the Fermi-surface is not as pronounced as shown in Fig. 7.9. The lengths of nearly parallel sections of the Fermi-surface hardly differ from surfaces that do not show anomalies in the dispersion. A particular large electron-phonon coupling on the hydrogen-covered surfaces must therefore be an important factor.

### 7.1.7 Dielectric Surface Waves

Dielectric surface waves are eigenmodes of dielectric continua for which the dielectric function displays a resonance behavior. They represent a class of surface excitations of its own. Dielectric surface waves exist as phonons and as plasmons, i.e. as excitation of the electron system. In the present context, we are interested in dielectric surface phonons and focus on infrared-active materials with a single infrared-active eigenmode. Examples are the ionic crystals with ZnS, NaCl, or CsCl structure. The dielectric function is given by

$$\varepsilon(\omega) = \varepsilon_\infty + \frac{\omega_0^2(\varepsilon_{st} - \varepsilon_\infty)}{\omega_0^2 - \omega^2 - i\gamma\omega}, \tag{7.22}$$

in which $\varepsilon_{st}$ and $\varepsilon_\infty$ are the static dielectric constant and the dielectric constant at a frequency much larger than the resonance frequency $\omega_0$ and $\gamma$ is a damping constant. The real and imaginary parts of $\varepsilon(\omega)$ are plotted in Fig. 7.11. The bulk of a crystal whose dielectric function is described by (7.22) sustains longitudinal and transverse polarization waves. A longitudinal wave is characterized by the condition

$$\text{div}\,\boldsymbol{P} \neq 0, \quad \text{curl}\,\boldsymbol{P} = 0\,. \tag{7.23}$$

The electric field $\mathscr{E}$ obeys

$$\text{div}\,D = \text{div}\,(\varepsilon_0\varepsilon(\omega)\mathscr{E}) = \text{div}\,(\varepsilon_0\mathscr{E} + \boldsymbol{P}) = \rho\,. \tag{7.24}$$

In the absence of a charge density $\rho$, longitudinal polarization waves exist if

$$\varepsilon(\omega) = 0\,, \tag{7.25}$$

which defines the frequency of the longitudinal polarization wave $\omega_L$ (Fig. 7.11). Transverse polarization waves obey

$$\text{div } \boldsymbol{P} = 0, \ \text{curl } \boldsymbol{P} \neq 0 \qquad (7.26)$$

In the realm of electrostatics, curl $\mathscr{E}$ vanishes identically. The frequency of transverse polarization wave $\omega_T$ is therefore the resonance frequency $\omega_0$ where

$$\varepsilon(\omega) = \infty \ . \qquad (7.27)$$

This exhausts the possibilities for self-sustained electrostatic waves in the bulk.

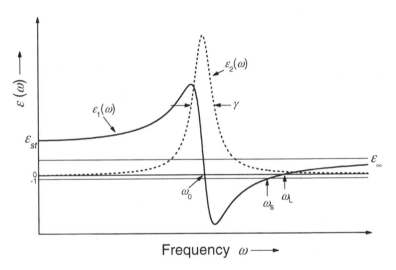

**Fig. 7.11** Real and imaginary part of the dielectric function of an infrared-active material with a single resonance frequency $\omega_0$ (solid and dashed line, respectively).

An additional solution exists at a flat interface between two dielectric half spaces that obeys

$$\text{div } \mathscr{E} = 0, \ \text{curl } \mathscr{E} = 0 \ . \qquad (7.28)$$

Because of (7.28) the electric field is the gradient of a potential $\varphi$ that fulfills the Laplace-equation

$$\Delta\varphi = 0 \qquad (7.29)$$

The solution to (7.29) is a wave that is localized to the interface defined by $z = 0$.

$$\varphi(x, z, t) = \varphi_0 e^{-q|z|} e^{i(qx - \omega t)} \qquad (7.30)$$

The electric field lines of this surface wave are displayed in Fig. 7.12[13].

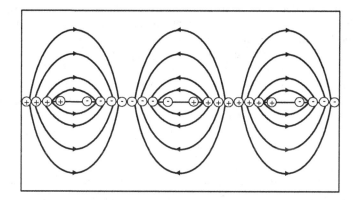

**Fig. 7.12**. Field lines of a dielectric surface wave. Inside the material, the lines mark also the polarization (with opposite direction). The divergence of the polarization at the surface leads to polarization charges as indicated.

The frequency of the wave is defined by the boundary condition that the normal component of the dielectric displacement $D$ must be continuous at the interface $z = 0$.

$$D_z = -\varepsilon_0 \varepsilon^{(-)}(\omega)\frac{\partial \varphi}{\partial z}\bigg|_{z=-0} = -\varepsilon_0 \varepsilon^{(+)}(\omega)\frac{\partial \varphi}{\partial z}\bigg|_{z=+0} \qquad (7.31)$$

This requires that

$$\varepsilon^{(-)}(\omega) = -\varepsilon^{(+)}(\omega) \qquad (7.32)$$

For the special case of a semi-infinite dielectric half space the condition reduces to

$$\varepsilon(\omega) = -1 \qquad (7.33)$$

which is the defining equation for the frequency of the surface wave (Fig. 7.11).

Surface phonons of the type considered above are named *Fuchs-Kliewer surface phonons* after the two researchers who postulated the existence of these phonons [7.28]. Experimentally, these phonons were discovered by their strong

---

[13] An early version of that figure published in 1971 as Fig. 6 of [7.27] has the shape of the field lines incorrect. Despite of that, it has been copied over and over again in textbooks and reviews.

inelastic interaction with electrons reflected from the surface via the electric field that is associated with the surface phonons [7.29] (see also Sect. 7.4.2). Similar modes exist also in thin film systems, where each interface contributes one mode. To satisfy the boundary conditions at the interfaces the solution (7.30) is generalizes to a linear combination of $z$-dependent exponentials that possesses a discontinuity in the first derivative at each interface. Because of the boundary condition, the frequency spectrum becomes $q$-dependent (*coupling dispersion*).

The electrostatic surface waves described above do not interact with light except in special experimental arrangements where they are exposed as frustrated total internal reflections. To understand this, one needs to solve the equation of motion for infrared active phonon modes together with the full Maxwell-equations. The result of such calculation for the semi-infinite half space is displayed in Fig. 7.13. The left panel shows the dielectric function $\varepsilon(\omega)$ vs. frequency for the case of vanishing damping. The right panel displays the various ranges of solutions vs. the reduced wave vector component parallel to the surface $cq_x/\omega_T$. The *light line* $\omega = cq_x$ marks the boundary between radiative and non-radiative solutions. On the vacuum side, light waves can exist only to the left of the light line that is if $q_x < q = \omega/c$. In the ranges $R_1$ and $R'_1$, light waves from the inside are reflected at the boundary and part of the radiation is transmitted into the vacuum. In the range $R_2$, $\varepsilon(\omega)$ is negative which admits only exponentially decaying *evanescent* waves inside the dielectric medium. Hence, one has total reflection of light from the outside (*Reststrahl region*). Solutions in $L_1$ and $L'_1$

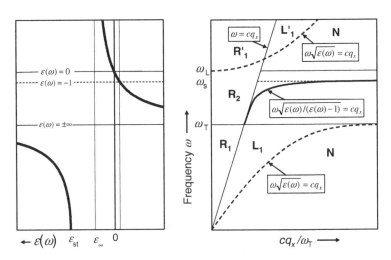

**Fig. 7.13.** *Left panel*: dielectric function $\varepsilon(\omega)$ vs. frequency for the case of vanishing damping. *Right panel*: solutions of the Maxwell equations for the dielectric half space vs. the wave vector component parallel to the surface $q_x$ in reduced units. Radiative and non-radiative solutions exist on the left and right side of the *light line* $\omega = cq_x$, respectively (see text for discussion).

correspond to total internal reflection from the inside. The regions marked by N
sustain no solutions at all. The boundaries to the solutions of total internal reflection are given by the condition $\omega\sqrt{\varepsilon(\omega)} = cq_x$ and are indicated as dashed lines.
The surface wave of the dielectric half space is given by the condition

$$\omega\sqrt{\varepsilon(\omega)/(\varepsilon(\omega)-1)} = cq_x, \qquad (7.34)$$

which reduces to (7.33) for $q_x \gg \omega_T/c$. The coupled solutions of the polarization
wave with the light are called *polaritons*. Since $\omega_T/c$ is of the order of $10^{-5}$ of the
Brillouin zone boundary the polariton dispersion is confined to a narrow region
around the $\overline{\Gamma}$-point.

# 7.2 Adsorbate Modes

## 7.2.1 Dispersion of Adsorbate Modes

The frequencies of the modes of adsorbed atoms depend on their masses and on
the strength of the bond to the substrate. For strongly bound first, second and third
row elements such as hydrogen, carbon, nitrogen, oxygen, sulfur, etc. the frequencies are typically situated above the spectrum of the substrate phonons. The adsorbate modes are therefore genuine surface modes and their eigenvectors are highly
localized to the surface. If the adsorbate atoms form an ordered lattice, the modes
can be expanded into surface phonons and their frequencies are a function of the
wave vector parallel to the surface.

Adsorbate atoms contribute three surface phonon branches for each adsorbate
atom in the surface unit cell. The branches for one adsorbate atom per unit cell are
schematically shown in Fig. 7.14. Frequency ranges and dispersion roughly correspond to the case of Ni(100) covered with a c(2×2) overlayer of oxygen or sulfur.
These atoms reside in the fourfold hollow site. At $\overline{\Gamma}$ and $\overline{X}$ the parallel polarized
modes are degenerate because of the symmetry of the adsorption site. The degen-
eracy is lifted, though marginally in the region between the high symmetry points.
Quite generally, the dispersion is the smaller the larger the gap to the substrate
phonon spectrum is. The reason is that for not too densely packed adsorbate layers
the dispersion arises mostly from the coupling to the substrate motion and not
from direct bonding between the adsorbate atoms. The dispersion branches of
adsorbed hydrogen atoms are practically flat. Hence, the hydrogen atoms vibrate
as if no other hydrogen atoms were present on the surface and the vibration modes
are localized in all three dimensions.

There is a notable exception to the rule "the higher the frequency, the smaller
the dispersion". That is if a vibrational mode bears a large dynamic dipole moment. The dipole moment causes a long range interaction that stiffens the frequency when the amplitudes are in phase, hence at $\overline{\Gamma}$. The effect disappears
quickly for larger wave vectors. To estimate the magnitude of the effect one may

consider a simple model of a 2D-lattice of harmonic oscillators vibrating perpendicular to the surface plane that are coupled via the dipole field $\mathscr{E}^{(\mathrm{dip})}$:

$$m_r \ddot{u}_i + m_r \omega_0^2 u_i = e^* \mathscr{E}_i^{(\mathrm{dip})} .$$ (7.35)

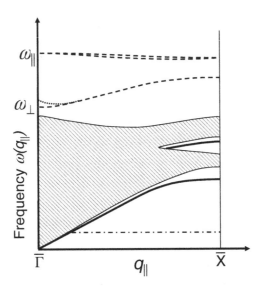

**Fig. 7.14.** Frequency spectrum of an adsorbate covered fcc (100) surface vs. the parallel component of the wave vector $q_\parallel$ in the [110] direction. The shaded area is the range of bulk phonons. Solid lines are dispersion curves of substrate surface phonons. Dashed lines are the dispersion branches of a c(2×2) adsorbate layer for strongly bonded atoms residing in the fourfold hollow site. The two branches of the parallel-polarized modes are degenerate at $\overline{\Gamma}$ and $\overline{X}$ because of symmetry. The dotted line indicates a deviation from the standard dispersion curve near $\overline{\Gamma}$ due to dipole-dipole interactions. Also shown is the nearly dispersion less branch of a vertical motion of an adsorbed layer of rare-gas atoms (dash-dotted line).

Here, $e^*$ is the effective charge associated with the vibration mode and $\omega_0$ is the bare frequency without dipole coupling. The dipole field acting upon the oscillator $i$ is the sum of the fields generated by all other oscillators.

$$\mathscr{E}_i^{(\mathrm{dip})} = -\sum_{j \neq i} p_j |r_i - r_j|^{-3}$$ (7.36)

with

$$p_j = e^* u_j + \alpha_{\mathrm{el}} \mathscr{E}_j^{(\mathrm{dip})}$$ (7.37)

Here, $\alpha_{el}$ is an electronic polarizability. At the $\overline{\Gamma}$-point all oscillators move in-phase. The equation of motion then becomes

$$m_r \ddot{u}_i + m_r \left( \omega_0^2 + \frac{(e^*)^2 \Sigma}{m_r (1 + \alpha_{el} \Sigma)} \right) u_i = 0 \qquad (7.38)$$

where $\Sigma$ stands for the dipole sum (7.36). The resonance frequency is therefore shifted upwards to

$$\omega^2 = \omega_0^2 + \frac{(e^*)^2 \Sigma}{m_r (1 + \alpha_{el} \Sigma)} . \qquad (7.39)$$

For adsorbates that carry a large effective charge like oxygen or carbon monoxide, shift amounts to a few meV.

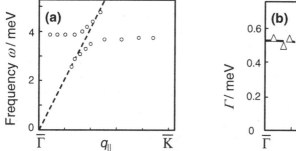 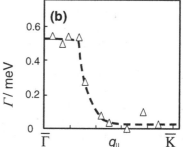

**Fig. 7.15.** (a) Avoided crossing of the vertical polarized mode of a Kr monolayer on Pt(111) (circles) with the Rayleigh wave (dashed line). (b) Line width $\Gamma$ of the mode vs. the wave vector $q_\parallel$. For small $q_\parallel$, the width is large because of radiative damping by anharmonic coupling to the bulk phonon bath (after Hall et al. [7.30]).

Almost free of dispersion are the vertical vibration modes of weakly bound physisorbed atoms whose frequencies for the most part lie well below the substrate phonons (dash-dotted line in Fig. 7.14). In the $\omega(q_\parallel)$ regime where the dispersion branch approaches the Rayleigh wave the adsorbate mode and the Rayleigh wave couple and the dispersion relations make an *avoided crossing*. Crossing is avoided because the Rayleigh wave and the adsorbate mode belong to the same irreducible representations (A' of $C_s$) which has only non-degenerate states, whereas a crossing of the curves would involve one point of twofold degeneracy. Figure 7.15a shows experimental data on the avoided crossing between the Rayleigh wave and Kr-modes on Pt(111) [7.30]. In the region of the avoided crossing, the character of the modes changes from the Rayleigh wave type to a mode that is almost entirely localized on the Kr-atoms, and vice versa. Inside the band of bulk phonons, the Kr-mode is radiatively damped by anharmonic interactions with the thermal bath of bulk phonon modes. This damping is visible as an increased line width (Fig. 7.15b).

## 7.2.2 Localized Modes

The term *localized modes* refers to vibration modes that are not only localized at to surface but also within the surface plane. Typical representatives of such modes are the modes of atoms forming a dilute disordered lattice gas and the internal vibration modes of isolated adsorbed molecules if their frequencies fall outside the range of substrate phonons. As the coverage increases, the vibration modes of the isolated species may couple with each other. The coupling causes a shift in the frequency spectrum of a particular species, which depends on the local environment around the species. The spectral response of the entire disordered layer shows therefore *inhomogeneous broadening*. This inhomogeneous broadening is the equivalent of the spectral width that is covered by the dispersion in ordered layers. If for example the adsorbate layer undergoes an order-disorder transition the width of the inhomogeneous broadening in the disordered phase reflects the magnitude of the dispersion in the ordered layer. In the limit of vanishing dispersion, the vibration modes of an ordered layer may be described either as dispersionless phonons or as localized vibrations. As long as one stays with the harmonic approximation, the distinction is purely semantic. That changes when anharmonic forces are admitted. Consider, e.g., the stretching vibration of an isolated diatomic molecule: Due to the anharmonicity of the potential, the vibration states are not equidistant. For a typical potential the energy differences decrease for higher levels. The excitation energy between the ground state and the first excited state $\hbar\omega_1$ is larger than the difference between the next two levels, and so forth. Consequently, the frequency of the first overtone is less than twice the fundamental. Such shifted overtones are also observed if molecules are adsorbed on a surface [7.31]. The coupling with the substrate opens the possibility for an additional feature in the same spectral range. If the molecules form an ordered lattice on the surface phonons of wave vector $+q$ and $-q$ couple via anharmonic interaction to form a two-phonon bound state [7.32]. This two-phonon bound state is distinct from the overtone frequency of an isolated molecule.

Localized modes of molecules reflect the nature and strength of the internal bonds of the molecules in the adsorbed state, and of the bonds that the molecules make with the surface. The spectroscopy of these modes provides therefore an access to surface chemistry whose importance can be hardly overrated. Several specific examples have already been discussed in Sect. 6 in the context of the surface chemistry of simple molecules. Here, we briefly summarize a few general aspects of surface chemical analysis by vibration spectroscopy.

The question whether a diatomic molecule adsorbs dissociatively or not is answered immediately and conclusively by the presence or absence of the stretching frequency that is characteristic for the molecular bond. This concept was introduced by Propst and Piper in 1967 [7.33] who studied the adsorption of $H_2$, $N_2$, CO and $H_2O$ on the W(100) surface in vacuum using inelastic scattering of electrons. Despite the poor energy resolution at the time (50 meV, equivalent to $\cong 400$ cm$^{-1}$) the authors were able to conclude that all these molecules dissociate upon adsorption on the bare tungsten surface. A more recent example with much better resolution was shown in Fig. 6.29 with the vibration spectrum of dissociated and non-dissociated oxygen molecules. Because of the sharpness of the vibrational features and their wide spectral range, the analytical power of vibration spectroscopy reaches significantly beyond the simple question of dissociation. In Fig. 6.29, we have seen that the frequency of the oxygen-oxygen stretching vibration depends sensitively on the ionicity of the molecule and the strength of the internal bond.

If a molecule adsorbs on the surfaces, its rotational and translational degrees of freedom turn into vibration modes. The frequencies of these modes depend on the adsorption site. The corresponding spectral features can therefore be employed as an indicator of the adsorption site (cf. Fig. 6.27).

Decomposition of complex molecules into fragments and the appearance of reaction intermediates on the surface can be studied as well. An important factor in the chemical analysis of molecules and molecular fragments at surfaces is the existence of characteristic group frequencies, which are salient features of the local environment within a molecule. The vibration frequencies for CH-stretching modes, e.g., are in the wave number range of 2900-3400 cm$^{-1}$. The precise frequency depends on the nature of the bond that the carbon atom makes with the rest of the molecule (sp, sp$^2$, or sp$^3$ hybridization) and to some extent on the symmetry of the mode. The study of the mode spectrum therefore enables the determination of the presence or absence of a particular group of frequencies as well as the analysis of the type of bonding in which the carbon atom is engaged. Figure 7.16 shows a sequence of spectra for methanol on Ni(110) as an example for a study of a decomposition reaction [7.34]. The spectra are obtained by inelastic scattering of electrons. Methanol adsorbed at 110 K displays vibration frequencies of the OH stretching and bending vibrations at 3269 cm$^{-1}$ and 737 cm$^{-1}$, the CH stretching vibrations of the $CH_3$-group at 2821 cm$^{-1}$ and 2948 cm$^{-1}$, the bending vibrations of that group at 1154 cm$^{-1}$ and 1451 cm$^{-1}$ and the very prominent stretching vibration of the single-bonded CO-group in methanol at 1037 cm$^{-1}$. A small amount of carbon monoxide (stretching mode at 2098 cm$^{-1}$) is coadsorbed (The amount is small as the intensity of CO-peaks is very large). After annealing to 180 K for a few minutes, the methanol molecule has lost the hydrogen atom of the OH-group as evidenced by the disappearance of the OH stretching mode. Since the $CH_3$-modes and the CO-group mode remain, the adsorbed species must be a methoxy-group. The methoxy-group decomposes further into adsorbed CO and H after annealing to 325 K (lower panel in Fig. 7.16). The hydrogen vibrations are not visible on the same intensity scale.

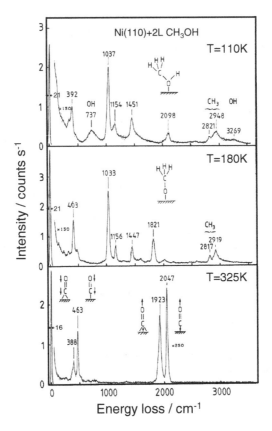

**Fig. 7.16.** Electron energy loss spectra of a Ni(110) surface dosed with 2L of methanol (1Langmuir = $10^{-6}$ Torr s). *Top panel*: after adsorption at 110 K; *center panel*: after annealing to 180 K; *lower panel*: after annealing to 325 K.

Depending on the method used for surface vibration spectroscopy and the nature of the molecular vibration, the minimum surface coverage required for the detection of a species ranges from 1/1000 to 1/10 of a monolayer. This high sensitivity, together with the features described above, makes vibration spectroscopy unrivaled by any other methods concerning the chemical analysis of surface species. The relative ease by which qualitative conclusion can be drawn from an inspection of the vibration spectrum is the strength of the method, and a weakness at the same time! A weakness in so far as not all the qualitative reasoning that has been put forward in the interpretation of vibration spectra have stood up to the test of time. For example, the assignment of adsorption sites of CO and NO molecules via the shifts of the stretching frequency has proved erroneous in several cases. On firm ground however, is the reasoning based on selection rules in connection with the local symmetry of adsorption sites to which we turn now.

### 7.2.3 Selection Rules

Localized modes at surfaces can be classified according to the irreducible representations of the point group of the adsorbate/surface complex. The consideration of points groups, rather than space groups suffices for all but the (few) structures with glide planes. This is because the species are independent of each other if the vibration modes of the adsorbed species are localized in the sense of the preceding section. If adsorbates form an ordered structure and the vibration modes show dispersion, the consideration of merely point group properties suffices for modes at the $\bar{\Gamma}$-point or at other high symmetry points of the SBZ. The possible surface point groups, their irreducible representations and character tables were already introduced in Sect. 1.1.2. Here we are concerned with the classification of vibration modes according to the irreducible representations. In order to focus on the relevant issues, we first consider the selection rules in surface vibration spectroscopy.

A particular selection rule arises in conjunction with inelastic scattering of particles from phonons. As remarked in Sect. 7.1.1, phonons with a $q$-vector in a mirror plane of the substrate separate into even and odd states. In the special case where an atom lies on the mirror plane, the odd phonon mode involves a motion of that atom which is polarized perpendicular to the mirror plane while the two even branches are polarized within the mirror plane. If one considers an inelastic scattering experiment, in which both the incoming particle and the scattered particle have their trajectories in the mirror plane, then the wave functions of the initial and the final state are even with respect to the mirror plane. The intensity of an inelastic scattering event is proportional to the square modulus of the matrix elements between the initial and final state and the phonon state. A matrix element constructed from an odd phonon state and the even initial and final states of the particle must change sign under the mirror operation is therefore identical zero. Hence, only the even modes are detectable in that scattering geometry. The odd modes become visible along the same direction if the mirror plane of the sample is tilted with respect to the scattering plane. Experiments of this type have been carried out only in a few cases because of the required extra degree of freedom of the sample manipulator [7.35, 36].

The most important selection rule in the vibration spectroscopy of adsorbate/surface complexes concerns the modes at the $\bar{\Gamma}$-point. At this point, all species on the surface move in-phase. Modes of this type are excited in infrared spectroscopy and by inelastic electron scattering when energy losses are observed in specular reflection. The interaction with infrared light is via the dipole moment associated with the vibration modes. The same is true for the *dipole scattering mechanism* in inelastic electron scattering, which is active in specular reflection, as will be discussed in Sect. 7.4.2. On a metal surface, the parallel components of dipole moments are perfectly screened by their image dipoles. The interaction is therefore only with those vibrations that bear a dipole moment perpendicular to the surface. This is the so-called *surface selection rule*. For all practical purposes, the rule applies also to semiconductor surfaces. The screening of the parallel component of the dipole moments is proportional to $1/\varepsilon$. The intensity in the spectrum is proportional to $1/\varepsilon^2$. Taking silicon as an example the scattering from vibrations bearing a parallel dipole moment is suppressed by a factor of about 150. Modes bearing a parallel dipole moment need be considered only in special experimental arrangements or on surfaces of wide band-gap insulators with their relatively small dielectric constant of 2-3. In Sect. 7.5.2 we shall see that modes with parallel dipole moments can make themselves visible by a reduction in the infrared surface resistivity of metals.

As shown in Sect. 1.1.2, a molecules/surface complex belongs to either one of the point groups $C_s$, $C_{2v}$, $C_{3v}$, $C_{4v}$, $C_{6v}$, $C_2$, $C_3$, $C_4$, or $C_6$ (or the trivial point group bearing no symmetry element at all). Here, we focus on the most frequent point groups that contain at least one symmetry plane $C_s$, $C_{2v}$, $C_{3v}$, $C_{4v}$, and $C_{6v}$. Table 7.1 shows the character tables of these point groups again, supplemented by two columns. The entries in the penultimate column concern the transformation properties of translation and rotation. The translation along the $z$-axis belongs to the totally symmetric irreducible representations A' ($C_s$-group) and $A_1$ ($C_{nv}$-groups). Only the modes belonging to these representations carry a dipole moment oriented along the $z$-axis. In the language of group theory, the surface selection rule therefore states that only the modes belonging to the total symmetric representation are active in inelastic electron scattering via the dipole scattering or in surface IR-spectroscopy. We note that the IR-selection rule in surface spectroscopy differs from the IR-selection rule for molecules in the gas-phase. In the gas-phase, e.g., IR-activity or Raman-activity are mutually exclusive if molecules have an inversion center. Since surface/molecule complexes are lacking a center of inversion, the rule does not apply there. Even more striking is that the stretching vibration of homonuclear diatomic molecules, which is not infrared-active in the gas-phase, becomes infrared-active on the surface, regardless of the orientation of the molecule. An example was shown in Sect. 6.4.3 with Fig. 6.29, which displays the dipole spectrum of adsorbed oxygen molecules. Even the stretching vibration of the relatively weakly bonded physisorbed oxygen molecule with its bond axis oriented parallel to the surface is clearly discernible in the spectrum. The physical origin of the perpendicular dipole moment associated with the vibration is the change in the charge transfer to the surface with the bond distance in the molecule, which generates a dipole moment that fluctuates along with the vibrational motion.

For the analysis of the vibration spectra of polyatomic molecules, one needs to know the number of modes belonging to the total symmetric representations and to the other representations. These numbers are the entries in the last columns of Table 7.1. The number of modes is expressed in terms of the number of atoms in certain positions with respect to the symmetry elements. The first row of each of the last columns shows the total number of atoms $N$ expressed in terms of the number of non-equivalent atoms on (or off) the symmetry elements. For example, the number of nonequivalent atoms in a general position, i.e. not on any symmetry element is denoted by $m$. The actual number of such atoms in the molecule/surface complex depends on the symmetry elements of the group. For $C_s$ with only one mirror plane, the molecule/surface complex has twice as many atoms as the number $m$ of atoms in non-equivalent positions off the mirror plane. The number of atoms in non-equivalent positions on the $xz$-mirror plane, $m_{xz}$, contributes $m_{xz}$ atoms to the total number of atoms $N$. Hence for the point group $C_s$ the total number of atoms is $N = 2m + m_{xz}$. For the other point groups the entries in the first row of the last columns are constructed accordingly. The number $m_0$ is the number of atoms in non-equivalent positions on the rotation axis.

With the help of that notation, the number of eigenmodes belonging to each representation is easily calculated. We consider again the point group $C_s$ as an example. The three possible displacement vectors of each of the $m$ atoms in a general position with respect to the symmetry element may transform either symmetrically or antisymmetrically with respect to the mirror plane. Hence each atom of type $m$ contributes $3m$ modes to each of the irreducible representation A' and A''. The atoms on the mirror plane have two degrees of freedom within the mirror plane and one perpendicular to the plane and contribute the corresponding number of vibrations to the A' and A'' representations. The numbers of modes for the other groups are derived accordingly. The degenerate modes require extra consideration. The easiest, heuristic way to obtain the number of modes in terms of the atom positions is to subtract the modes of the non-degenerate representations from the total number of modes (which is $3N$) considering that each of the E-modes is doubly degenerate.

With the help of the entries in the last columns, one can calculate the number of modes in each representation, in particular the number of dipole active modes, for a known symmetry of the molecule/surface complex. Moreover, one can establish the type of mode; in the case of a hydrocarbon molecule e.g., whether it belongs to the carbon skeleton or is a mode associated with a hydrogen atom. Vice versa, one can establish the local site symmetry from the number of modes in the various representations, in particular from the number of dipole active modes. An important question in that regard is how the eigenmodes of the molecule, classified according to the irreducible representations of the point group relevant for the gas-phase molecule, break down to the irreducible representations of the point group of the adsorbed state, provided the molecule stays intact. The answer is trivial for diatomic molecules. For a molecule that has many eigenmodes, may be some of them degenerate as e.g. in $C_6H_6$ (point group $D_{6h}$), the breakdown of the modes into the irreducible representations of the surface point group requires a detailed consideration of the various modes, or the employment of the mathematical apparatus of group theory. The results of such analysis are summarized in the so-called *correlation tables*. Table 7.2 and 7.3 are correlation tables for the two important molecular point groups $D_{6h}$ and $D_{2h}$.

The use of these correlation tables is exemplified with the adsorption of benzene on the Ni(111) surface. The maximum possible symmetry on that surface is $C_{3v}$. The analysis of the spectrum of the vibration modes for the ordered $(\sqrt{7} \times \sqrt{7})\,R19.1°$ adsorption phase in connection with the dipole selection rule shows that the carbon ring is oriented parallel to the surface and that the symmetry of the molecular complex on the surface is $C_{3v}$ [7.37]. However, there are two

**Table 7.1.** Character tables of surface point groups as in Table 1.1. The upper left corner denotes the point group. The first column are the irreducible representations, the following columns are the characters of the classes of the group. The penultimate column describes to which irreducible representation the translations along the $x$, $y$ and $z$-axes and the rotations around these axes belong. This classifies the eigenmodes of adsorbates that arise from the translation and rotation degrees of freedom. The first row of the final column gives the total number of atoms $N$ in terms of the number of atoms in specific positions. The following rows contain the number of modes belonging to each representation. The numbers $m_{xz}$, $m_{yz}$, $m_y$ are the non-equivalent atoms on the mirror planes, $m_0$ is the number of non-equivalent atoms on a rotation axis, and $m$ the number of atoms in a general position.

| $C_s$ | $I$ | $\sigma_{xz}$ | | | | $N = 2m + m_{xz}$ |
|---|---|---|---|---|---|---|
| $A'$ | +1 | +1 | | | $z, x, R_y$ | $3m + 2m_{xz}$ |
| $A''$ | +1 | -1 | | | $y, R_x, R_z$ | $3m + m_{xz}$ |

| $C_{2v}$ | $I$ | $C_2$ | $\sigma_{xz}$ | $\sigma_{yz}$ | | $N = 4m + 2m_{xz} + 2m_{yz} + m_0$ |
|---|---|---|---|---|---|---|
| $A_1$ | +1 | +1 | +1 | +1 | $z$ | $3m + 2m_{xz} + 2m_{yz} + m_0$ |
| $A_2$ | +1 | +1 | -1 | -1 | $R_z$ | $3m + m_{xz} + m_{yz}$ |
| $B_1$ | +1 | -1 | +1 | -1 | $x, R_y$ | $3m + 2m_{xz} + m_{yz} + m_0$ |
| $B_2$ | +1 | -1 | -1 | +1 | $y, R_x$ | $3m + m_{xz} + 2m_{yz} + m_0$ |

| $C_{3v}$ | $I$ | $C_3$ | $\sigma$ | | | $N = 6m + 3m_v + m_0$ |
|---|---|---|---|---|---|---|
| $A_1$ | +1 | +1 | +1 | | $z$ | $3m + 2m_v + m_0$ |
| $A_2$ | +1 | +1 | -1 | | $R_z$ | $3m + m_v$ |
| $E$ | +2 | -1 | 0 | | $x, y, R_x, R_y$ | $6m + 3m_v + m_0$ |

| $C_{4v}$ | $I$ | $C_4$ | $C_4^2$ | $\sigma_v$ | $\sigma_d$ | | $N = 8m + 4m_v + 4m_d + m_0$ |
|---|---|---|---|---|---|---|---|
| $A_1$ | +1 | +1 | +1 | +1 | +1 | $z$ | $3m + 2m_v + 2m_d + m_0$ |
| $A_2$ | +1 | +1 | +1 | -1 | -1 | $R_z$ | $3m + m_v + m_d$ |
| $B_1$ | +1 | -1 | +1 | +1 | -1 | | $3m + 2m_v - m_d$ |
| $B_2$ | +1 | -1 | +1 | -1 | +1 | | $3m + m_v + 2m_d$ |
| $E$ | +2 | 0 | -2 | 0 | 0 | $x, y, R_x, R_y$ | $6m + 3m_v + 3m_d + m_0$ |

| $C_{6v}$ | $I$ | $C_6$ | $C_6^2$ | $C_6^3$ | $\sigma_v$ | $\sigma_d$ | | $N = 12m + 6m_v + 6m_d + m_0$ |
|---|---|---|---|---|---|---|---|---|
| $A_1$ | +1 | +1 | +1 | +1 | +1 | +1 | $z$ | $3m + 2m_v + 2m_d + m_0$ |
| $A_2$ | +1 | +1 | +1 | +1 | -1 | -1 | $R_z$ | $3m + m_v + m_d$ |
| $B_1$ | +1 | -1 | +1 | -1 | +1 | -1 | | $3m + 2m_v + m_d$ |
| $B_2$ | +1 | -1 | +1 | -1 | -1 | +1 | | $3m + m_v + 2m_d$ |
| $E_1$ | +2 | +1 | -1 | -2 | 0 | 0 | $x, y, R_x, R_y$ | $6m + 3m_v + 3m_d + m_0$ |
| $E_2$ | +2 | -1 | -1 | +2 | 0 | 0 | | $6m + 3m_v + 3m_d$ |

**Table 7.2.** Correlation table for $D_{6h}$. The table shows to which irreducible representation particular modes of a molecule of $D_{6h}$ symmetry (e.g. $C_6H_6$) belongs when the symmetry is broken at the surface. Thin solid lines are symmetry planes. The black dots symbolize atoms when the molecule is oriented with its plane perpendicular to the surface.

| $D_{6h}$ | $C_{6v}$ | $C_{6v}$ | $C_{3v}$ | $C_{3v}$ | $C_s$ | $C_s$ | $C_{2v}$ | $C_s$ | $C_s$ | $C_{2v}$ | $C_s$ | $C_s$ |
|---|---|---|---|---|---|---|---|---|---|---|---|---|
| $A_{1g}$ | $A_1$ | $A_1$ | $A_1$ | $A_1$ | $A'$ | $A'$ | $A_1$ | $A'$ | $A'$ | $A_1$ | $A'$ | $A'$ |
| $A_{1u}$ | $A_2$ | $A_2$ | $A_2$ | $A_2$ | $A''$ | $A''$ | $A_2$ | $A''$ | $A''$ | $A_2$ | $A''$ | $A''$ |
| $A_{2g}$ | $A_2$ | $A_2$ | $A_2$ | $A_2$ | $A''$ | $A''$ | $A_2$ | $A'$ | $A''$ | $B_1$ | $A'$ | $A''$ |
| $A_{2u}$ | $A_1$ | $A_1$ | $A_1$ | $A_1$ | $A'$ | $A'$ | $A_1$ | $A''$ | $A'$ | $B_2$ | $A''$ | $A'$ |
| $B_{1g}$ | $B_2$ | $B_1$ | $A_2$ | $A_1$ | $A'$ | $A''$ | $B_2$ | $A''$ | $A'$ | $B_2$ | $A''$ | $A'$ |
| $B_{1u}$ | $B_1$ | $B_2$ | $A_1$ | $A_2$ | $A''$ | $A'$ | $B_1$ | $A'$ | $A''$ | $B_1$ | $A'$ | $A''$ |
| $B_{2g}$ | $B_1$ | $B_2$ | $A_1$ | $A_2$ | $A''$ | $A'$ | $B_1$ | $A'$ | $A''$ | $B_2$ | $A''$ | $A'$ |
| $B_{2u}$ | $B_2$ | $B_1$ | $A_2$ | $A_1$ | $A'$ | $A''$ | $B_2$ | $A''$ | $A'$ | $B_1$ | $A'$ | $A''$ |
| $E_{1g}$ | $E$ | $E$ | $E$ | $E$ | $A''+A'$ | $A''+A'$ | $B_1+B_2$ | $A''$ | $A''$ | $A_2+B_2$ | $A''$ | $A''$ |
| $E_{1u}$ | $E$ | $E$ | $E$ | $E$ | $A''+A'$ | $A'+A''$ | $B_2+B_1$ | $A'$ | $A'$ | $A_1+B_1$ | $A'$ | $A'$ |
| $E_{2g}$ | $E$ | $E$ | $E$ | $E$ | $A'+A'$ | $A'+A'$ | $A_2+A_1$ | $A'$ | $A'$ | $A_1+B_1$ | $A'$ | $A'$ |
| $E_{2u}$ | $E$ | $E$ | $E$ | $E$ | $A''+A'$ | $A''+A'$ | $A_2+A_1$ | $A''$ | $A''$ | $A_2+B_2$ | $A''$ | $A'$ |

**Table 7.3.** Correlation table for $D_{2h}$. The table shows to which irreducible representation particular modes of a molecule of $D_{2h}$ symmetry (e.g. $C_2H_4$) belongs when the symmetry is broken at the surface. Thin solid lines are symmetry planes. The black dots symbolize atoms when the molecule is oriented with its plane perpendicular to the surface.

| $D_{2h}$ | $C_{2v}$ | $C_s$ | $C_s$ | $C_{2v}$ | $C_s$ | $C_s$ | $C_{2v}$ | $C_s$ | $C_s$ |
|---|---|---|---|---|---|---|---|---|---|
| $A_g$   | $A_1$ | $A'$  | $A'$  | $A_1$ | $A'$  | $A'$  | $A_1$ | $A'$  | $A'$  |
| $A_u$   | $A_2$ | $A''$ | $A''$ | $A_2$ | $A''$ | $A''$ | $A_2$ | $A''$ | $A''$ |
| $B_{1g}$ | $A_2$ | $A''$ | $A''$ | $B_1$ | $A'$  | $A'$  | $B_1$ | $A'$  | $A'$  |
| $B_{1u}$ | $A_1$ | $A'$  | $A'$  | $B_2$ | $A''$ | $A''$ | $B_2$ | $A'$  | $A'$  |
| $B_{2g}$ | $B_1$ | $A''$ | $A'$  | $A_2$ | $A''$ | $A''$ | $B_2$ | $A''$ | $A''$ |
| $B_{2u}$ | $B_2$ | $A'$  | $A''$ | $A_1$ | $A'$  | $A'$  | $B_1$ | $A''$ | $A''$ |
| $B_{3g}$ | $B_1$ | $A'$  | $A''$ | $B_2$ | $A''$ | $A''$ | $A_2$ | $A''$ | $A'$  |
| $B_{3u}$ | $B_2$ | $A''$ | $A'$  | $B_1$ | $A'$  | $A''$ | $A_1$ | $A'$  | $A'$  |

types of $C_{3v}$ symmetries. One has the $\sigma_v$-planes running through the carbon atoms; the other has the $\sigma_v$-planes cutting through the carbon-carbon bonds. Different molecular modes belong to the $A_1$ total symmetric representation in the two cases. For the former the $B_{1u}$ and $B_{2g}$-modes assume $A_1$-character, for the latter the $B_{2u}$ and $B_{1g}$-modes become $A_1$-modes (Table 7.2). Experiments have shown that the $B_{2u}$ mode of the carbon skeleton (1310 cm$^{-1}$ and 1420 cm$^{-1}$ in the gas-phase and the adsorbed state, respectively) assumes $A_1$-character, which means that the $\sigma_v$-planes are cutting through carbon-carbon bonds [7.37]. About 20 years later, this orientation was confirmed by structure determination (see Sect. 6.4.5; Fig. 6.35).

# 7.3 Inelastic Scattering of Helium Atoms

## 7.3.1 Experiment

This section pays tribute to an experimental technique that has been instrumental in the discovery and in the study of the dispersion of surface phonons. The technique flourished in the 80ties and 90ties of the last century, largely because of instrumental improvements originating in the group of J. P. Toennies [7.38]. It is however almost forgotten by now, partly because science has moved on to other issues of interest, partly because the complexity of the technique prevented its commercialization. Fig. 7.17 illustrates the principle of the generation of a He-beam by a supersonic expansion from a nozzle. He-gas is released from a high-pressure cell into a vacuum chamber through a small orifice of 5-10 μm diameter. The entrance opening of the trumpet-shaped skimmer has a diameter of 0.2-0.4 mm. The skimmer is placed at a distance of about 8 mm from the orifice of the high-pressure cell. The combination of nozzle and skimmer defines the He-beam both with respect to angular and energy resolution. The majority of He-atoms does not move into the desired beam direction and cannot pass the orifice of the skimmer. These atoms are pumped off by a roughening pump that can handle large gas throughputs. In order to reduce the ambient He-pressure in the sample chamber the He-beam passes a further sequence of individually pumped chambers separated by diaphragms with small orifices. In each of these chambers the He-atoms, which do not pass the orifice of the next diaphragm are pumped off. Eventually the ambient pressure in the sample chamber is only due to the He-atoms of the beam. The angular width of the beam is about 0.2°. The energy distribution is also small, for the following reason: Inside the high-pressure cell, the atoms have a thermal velocity distribution. Because of the high starting pressure, the density of atoms in the beam is very high in the initial stages of the expansion, right after the orifice. He-atoms traveling along the beam direction and possessing a different tangential speed undergo scattering events with each other. Except for the unlikely events of head-on collisions, the scattered He-atoms have trajectories that make a larger angle with the beam direction. These atoms can therefore not pass the orifice of the skimmer to enter the next stage of differential pumping. Only those He-

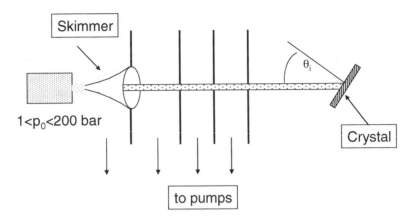

**Fig. 7.17.** Schematic illustration of the production of a He-beam with a narrow energy distribution.

atoms survive in the beam, which have nearly the same velocity. The width of the energy distribution depends on the initial pressure. The larger the pressure, the higher the density of He-atoms in the beam and the smaller is the resulting energy distribution. The width of the energy distribution is furthermore proportional to the temperature of the high-pressure gas cell. The production of undesired He-clusters set an upper limit to the pressure and a lower limit to the temperature. Energy widths of 0.2 meV and 0.08 meV have been achieved for gas cells operating at liquid nitrogen and liquid helium temperature, respectively.

The energy of the beam can be calculated via simple thermodynamic reasoning. During the expansion process, energy must be conserved. In the limit that the velocity perpendicular to the beam direction is vanishingly small, the kinetic energy per atom in the beam is equal to the enthalpy per atom $h_0$ in the high-pressure cell,

$$E = \frac{mv_\parallel^2}{2} = h_0 = c_p T_0 = \frac{5}{3}c_v T_0 = \frac{5}{2}k_B T_0. \tag{7.40}$$

Here, $c_p$ and $c_v$ are the specific heats per atom at constant pressure and volume, respectively, and $T_0$ is the temperature of the cell. A liquid-nitrogen cooled nozzle should therefore produce a beam of 16.6 meV energy. Technically, the energy is a little larger ($\cong 18$ meV).

The energy spectrum of the scattered He-atoms is determined using the time of flight method. A rotating blade that has a narrow slit chops the He-beam into packages. The scattered individual He-atoms are detected by a quadrupole mass spectrometer that is situated at a distance of the order of 1 m from the sample. The time of flight is measured by gating the quadrupole signal with the chopper blade. The duty cycle can be improved to 50% by using choppers that have a pseudorandom distribution of slit widths in combination with mathematical matrix inversion techniques to recover the desired time-of-flight signal.

Inelastic scattering events from surface phonons obey energy conservation and momentum conservation parallel to the surface. The phonon wave vector $q_{\parallel}$ is therefore

$$q_{\parallel} = \frac{\sqrt{2m}}{\hbar} \left( \sqrt{E} \, \sin \theta_1 - \sqrt{E \pm \hbar\omega} \, \sin \theta_S \right) \qquad (7.41)$$

where $\theta_1$ and $\theta_S$ are the angles of the incident and scattered beam with respect to the surface normal, $\hbar\omega$ is the quantum energy of the phonon and $E$ the energy of the He-beam before the scattering event. The plus and minus signs stand for energy gain and energy loss events, respectively. The crossings of the *scan curve* (7.41) with the dispersion curve of a surface phonon mark possible scattering events. Fig. 7.18 shows possible events for a 20 meV beam and the approximate surface phonon dispersion curve $\hbar\omega(q_{\parallel})$ on Ni(100) along the [110] direction. Positive and negative values of the energy correspond to phonon annihilation and creation, respectively.

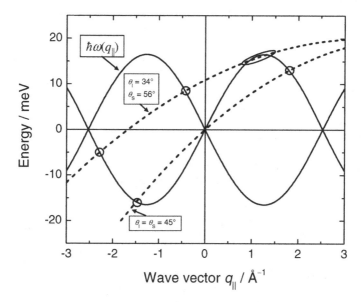

**Fig. 7.18.** Scan curves (dashed line) for a 20 meV He-beam, 90° scattering angle and a phonon dispersion relation (solid line) that resembles the Rayleigh-phonon on Ni(100). The intersections of the scan and dispersion curves mark possible scattering events. Positive and negative values of the energy correspond to phonon annihilation and creation events, respectively.

The intensity of phonon annihilation and creation processes are proportional to $n$ and $n+1$, respectively where $n$ is the occupation number according to Bose statistics,

$$n = \left(e^{\hbar\omega/k_{B}T} - 1\right)^{-1}. \tag{7.42}$$

The intensities of phonon annihilation and creation events are related by the Boltzmann factor

$$\frac{I_{an}}{I_{cr}} = \frac{n}{n+1} = e^{-\hbar\omega/k_{B}T}. \tag{7.43}$$

The experimental set-up typically fixes the scattering angle between the incident and scattered beam. Here the angle is assumed to 90°. The momentum transfer to the surface is varied by rotating the sample. Fig. 7.18 shows two scan curves, one with $\theta_{I} = \theta_{S} = 45°$ and a second one with $\theta_{I} = 34°$ and $\theta_{S} = 56°$. Both scan curves intersect the phonon dispersion curve several times. The maximum number of intersections is four. A time of flight spectrum therefore displays up to four peaks that could result from the same phonon. The phonon dispersion curve of a particular branch is obtained by varying the sample orientation in small increments, by marking all data points in the $\omega(q_{\parallel})$ space, and by running continuous curves through the data points thus obtained. The scan curve for $\theta_{I} = 34°$ and $\theta_{S} = 56°$ runs nearly parallel to the dispersion curve for a while. Considering the finite energy and momentum resolution, this means that scattering is possible in a large phase space. Correspondingly, the peaks from those events posses a high intensity and are broad in energy.

### 7.3.2 Theoretical Background

The simples approach to inelastic scattering of particles is the Born-approximation. In that approximation the scattering cross section is

$$\frac{d\sigma}{d\hbar\omega d\Omega} = \left(\frac{m}{2\pi\hbar^{2}}\right)^{2}\frac{k'}{k}\left|\left\langle\varepsilon\left|\int d^{3}r\, e^{-i(k'-k)\cdot r}V(r)\right|\alpha\right\rangle\right|^{2}\delta(E'-E-\hbar\omega). \tag{7.44}$$

Here $m$ is the mass of the particle, $k$ and $E$ denote wave vector and energy of the particle with the primed quantities referring to the scattered particle, $V(r)$ is the total scattering potential, and $\langle\varepsilon|$ and $|\alpha\rangle$ are the initial and final phonon states. The delta function ensures energy conservation. For scattering from a lattice of localized potentials around each atom, the dynamic part of the atom position vectors can be expanded into phonon annihilation and creation operators and the scattering cross section for one-phonon scattering events is readily calculated.

The Born approximation is valid for the scattering of fast particles from weak, but localized potentials, so that the particles are scattered only once and the interaction times are short. Because of the short interaction times, the scattered particles see the atoms at their instantaneous positions. The inelastic events result from the interference of the waves scattered by atoms in different spatiotemporal positions. Inelastic scattering of neutrons, e.g., is well described in the Born-approximation. However, He-atom scattering is more complicated as the atoms do not enter the crystal at all and are therefore not scattered by local potentials of individual atoms. The trajectories of He-atoms are turned around outside the surface were the electron density is very low, about $10^{-5} e / a_{\text{Bohr}}^3$ (cf. Fig. 6.1). Hence, He-atoms necessarily interact with more than one surface atom at a time. For a qualitative discussion of the inelastic scattered intensity the mechanics of Born scattering can be kept if the atoms are described by the classical phase $\eta$ [7.39, 40]

$$\eta = \frac{1}{\hbar} \mathscr{L} \tag{7.45}$$

with $\mathscr{L}$ the Lagrange function. For small displacements of the atom positions $u_n$ around their equilibrium positions $r_n$ the Lagrange function $\mathscr{L}$ can be expanded as

$$\mathscr{L}(r(t)) = \mathscr{L}_0 + \sum_n \frac{\partial \mathscr{L}(r(t))}{\partial r_n} \cdot u_n(t) = \mathscr{L}_0 + \sum_n F_n(r(t)) \cdot u_n(t) \tag{7.46}$$

where $F_n(r(t))$ is the force that the atom $n$ imposed on the scattered particle at its position $r(t)$. Inelastic scattering from phonons is due to the second term. In the semi-classical picture the amplitude $A$ of the scattered wave is

$$A \propto e^{-i(\eta_0 + \eta_{\text{ph}})} \cong e^{-i\eta_0} (1 - i\eta_{\text{ph}} + \ldots) . \tag{7.47}$$

Here $\eta_{\text{ph}}$ is the phonon-induced total phase shift of the particle in the scattering event which is

$$\hbar \eta_{\text{ph}} = \sum_n \int_{-\infty}^{+\infty} dt\, F_n(r(t)) \cdot u_n(t) . \tag{7.48}$$

The second term of the expansion (7.47) gives rise to one-phonon scattering events. The cross section for one-phonon events is proportional to the square modulus of the scattering amplitude, whence

$$\frac{d\sigma_{1ph}}{d\hbar\omega d\Omega} \propto |A|^2 \propto \left|\sum_n \int_{-\infty}^{+\infty} dt\, \boldsymbol{F}_n(\boldsymbol{r}(t))\cdot\boldsymbol{u}_n(t)\right|^2 . \tag{7.49}$$

As the He-atoms are scattered only from surface atoms we need to consider merely displacement of these atoms. For simplicity, we consider a vertically polarized phonon in a primitive surface unit cell

$$u_{z,n}(t) = u_{z,n}^0\, e^{-i(\omega t - \boldsymbol{q}_\| \cdot \boldsymbol{r}_{n\|})} . \tag{7.50}$$

Here, $\boldsymbol{r}_{n\|}$ is the two-dimensional vector pointing to the atom $\boldsymbol{n}$ in the surface and $\boldsymbol{q}_\|$ is the phonon wave vector as before. Inserting (7.50) into (7.49) yields the one-phonon cross section as proportional to the squared modulus of Fourier-transformed forces on the scattered atom.

$$\frac{d\sigma_{1ph}}{d\hbar\omega d\Omega} \propto \left|F_z(\boldsymbol{q}_\|,\omega)\right|^2 \tag{7.51}$$

In order to discuss qualitative aspects of the frequency dependence of the cross section we consider a simple case, the (inelastic) reflection from a flat surface at perpendicular incidence. The potential $V(z)$ has an exponentially repulsive part and an attractive part from the van-der-Waals forces (Sect. 6.1).

$$V(z) = V_0 e^{-\alpha z} - c(z - z_w)^{-3} \tag{7.52}$$

The Fourier transform of the force acting on the scattered He-atom is thus

$$\frac{d\sigma_{1ph}}{d\hbar\omega d\Omega} \propto \left|\int_{-\infty}^{+\infty} e^{-i\omega t} dt \left\{\alpha V_0 e^{-\alpha z(t)} - 3c(z(t) - z_w)^{-4}\right\}\right|^2 \tag{7.53}$$

in which $z(t)$ is obtained from the integration of

$$\dot{z} = \sqrt{\frac{2}{m}[E - V(z(t))]} \tag{7.54}$$

Fig. 7.19 shows the result for the frequency dependence of the cross section for a particular set of parameters that was calculated by Harris and Liebsch for the Cu(110) surface ($V_0 = 12$ eV, c = 1.52 eV $a_B^3$, $z_w = 0.461 a_B$, $\alpha = 1.39 a_{Bohr}^{-1}$, with the Bohr radius $a_B = 0.529$ Å$^{-1}$ [7.41]).

The decay of the intensity for higher frequencies as shown in Fig. 7.19 was first discussed by Beeby and is therefore called the *Beeby-effect* [7.42] or *slow colli-*

*sion effect.* The critical frequency beyond which the intensity falls off exponentially is given by the condition

$$\omega_c = v_{He} / \Lambda \tag{7.55}$$

in which $v_{He}$ is the velocity of the He-atom and $\Lambda$ is the characteristic range of the potential. The velocity of a 20 meV He-beam is about 980 m/s. In absence of the attractive part the range of the potential can be estimated as $\Lambda = 1/\alpha$. The resulting critical frequency of $v_c = 4$ THz is in approximate agreement with the solid line in Fig. 7.19. The critical frequency increases if the van-der-Waals attraction is included since the potential has a longer range in that case. The slight increase of the cross section above the value at $v = 0$ is because the He-atom picks up some speed in the vicinity of the surface due to the van-der-Waals attraction.

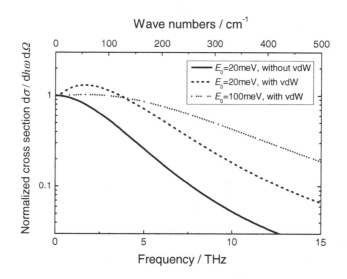

**Fig. 7.19.** Frequency dependence of the one-phonon cross section for He-atom scattering (*Beeby-effect*) according to (7.53). The solid and dashed lines are for beam energies of 20 meV without and with the van-der-Waals interaction, respectively. The dotted line is for beam energies of 100 meV with van-der-Waals interaction. The parameters $V_0 = 12$ eV, $c = 1.52$ eV$a_B^3$, $z_w = 0.461 a_B$, and $\alpha = 1.39 a_B^{-1}$, with $a_B = 0.529$ Å$^{-1}$ are as calculated by Harris and Liebsch for the Cu(110) surface [7.41].

Because of the Beeby-effect, the higher vibration modes of chemisorbed atoms and the internal vibrations of molecules cannot be observed in He-scattering. According to (7.55) the use of higher beam energies should help. Unfortunately, for higher beam energies the spectrum becomes completely dominated by multiphonon events. This is very nicely illustrated in Fig. 7.20 with spectra obtained on a monolayer of xenon adsorbed on Cu(100). The strongest contribution to the inelastic spectrum is from the dispersionless vertical motion of the xenon atoms. Multiphonon events therefore appear only at multiples of the single-phonon energy. For a beam energy of 8.1 meV (7.20a) only single phonon scattering events are visible. At $E_i = 16.3$ meV double and triple phonon loss and gain events appear. At $E_i = 36.3$ meV the spectrum is completely dominated by multiphonon scattering. The symbols "L" and "RW" stand for a longitudinal surface wave and the Rayleigh wave, respectively (after Gumhalter [7.43], see also [7.44]).

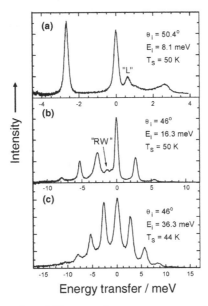

**Fig. 7.20.** Inelastic scattering of He-atoms from a xenon covered Cu(100) surface for three different beam energies (a) $E_i = 8.1$, (b) 16.3, and (c) 36.3 meV. Momentum transfer is along the [100]-direction. The prominent loss and gain events are due to the dispersionless vertical motion of the xenon atoms. The peaks labeled "L" and "RW" correspond to a longitudinal surface wave and the Rayleigh wave. The total scattering angle was 95.8° in all cases (after [7.43, 44]).

Quantitative statements about the dependence of the cross section on the wave vector are less straightforward as it is problematic to calculate the phonon-induced corrugation of the potential far outside the surface. Qualitatively, one expects the phonon-induced corrugation at large distances to be much less if neighboring atoms move in anti-phase compared to the case where they move in-phase. The intensity should therefore decrease as one approaches the boundary of the SBZ. The effect is known as the *Armand effect* after G. Armand who discussed this topic first [7.45]. The decay of the intensity is particularly prominent on metal surfaces because of their smooth charge density outside the surface. Theoretical analysis predicts the decay to be [7.39, 45]

$$\frac{\mathrm{d}\sigma_{1\mathrm{ph}}}{\mathrm{d}\hbar\omega\,\mathrm{d}\Omega} \propto e^{-q_\parallel^2/q_c^2}. \tag{7.56}$$

where $q_c$ is a parameter. Experimental results confirm the trend. For Ag(111) the decay that can be fit to $q_c = 0.62$ Å$^{-1}$ so that the intensity drops by an order of magnitude at the zone boundary [7.46, 47]. For a recent review on the theory as well as on certain experimental aspects the reader is referred to an article of B. Gumhalter [7.43].

# 7.4 Inelastic Scattering of Electrons

## 7.4.1 Experiment

Inelastic scattering of electrons or *Electron Energy Loss Spectroscopy* (EELS) is the most versatile technique for the investigation of vibrational modes on surfaces in vacuum. The energy resolution of present-day spectrometers is as high as 1 meV. All vibration modes from the low frequency vibrations of physisorbed rare-gas atoms to the internal vibrations of molecules are accessible in a single scan. Furthermore, one can probe phonons in the entire Brillouin zone. By varying the electron energy, different scattering mechanisms can be employed, leading to different selection rules for the inelastic scattering. In this first subsection, we discuss some of the salient properties of electron spectrometers that are used for inelastic electron scattering.

The width of the energy distribution of electrons from thermal, field emission or photoemission cathodes is at least 200 meV. Energy selectors are required to make electrons useful for inelastic scattering from phonons. A second energy selector is required for the analysis of the scattered electrons. The combined resolution of the electron *monochromator* and *analyzer* must be in the low meV range. Electrostatic deflector type selectors, as opposed to magnetic deflectors are used exclusively, because of the difficulty to shield magnetic fields effectively. Figure 7.21 displays a typical experimental set-up of an electron spectrometer. It comprises the electron emission system, two monochromators, lenses to image the exit slit of the monochromator onto the sample and further onto the analyzer entrance slit, two analyzers and a channeltron electron multiplier. In the interest of an optimum intensity of the signal, perfect as possible images of each aperture onto the next one are required. Just as in light optics, the phase space is conserved in the process of imaging. Because of the $C_{2v}$ symmetry of the optics around the beam direction, phase space conservation applies to the plane of drawing in Fig. 7.21 and perpendicular to the plane of drawing separately.

$$s \sin\alpha\sqrt{E} = const.$$
$$h \sin\beta\sqrt{E} = const. \tag{7.57}$$

Here, $E$ is the beam energy, $s$ and $h$ are the width and heights of the slit apertures, and $\alpha$ and $\beta$ are the angular apertures in the plane and perpendicular to the plane in Fig. 7.21. The energy range of the electrons to be scattered at the sample is between a few and a few hundred eV. As the quantum energies of vibrational motions are a few hundred meV at the most, the energies of the inelastically scattered electrons are about the same as the energy of the incident electrons. Together with the condition of phase space conservation, this calls for an electron optics that is identical in the monochromator and the analyzer part, in reverse order. A small allowance may have to be made to match the electron optics of the monochromator to the tendency of high intensity beams to diverge because of the repulsive interaction between the electrons.

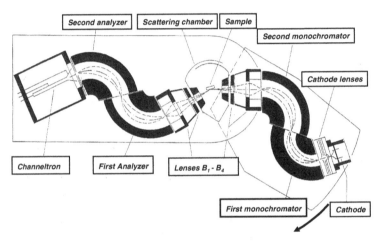

**Fig. 7.21.** Electron spectrometer for inelastic scattering of electrons from surfaces.

Classical energy dispersive deflectors are the spherical deflector, which ideally has stigmatic focusing for a total deflection angle of 180° and the cylindrical deflector that focuses in the plane perpendicular to the cylinder axis at 127°. Metallic plates at both ends with slit apertures to achieve energy selections reduce the focusing angles in both cases whereby the spherical deflector looses its perfect stigmatic focusing. For both devices, the basic equation that describes the image of the entrance slit at the exit-slit position is

$$y_{\text{exit}} = -y_{\text{entrance}} + c_{\text{D}} \, r_0 \, \delta E \, / \, E_{\text{p}} - c_{\alpha\alpha} \, r_0 \, \alpha^2 - c_{\beta\beta} r_0 \beta^2 \, . \tag{7.58}$$

Here, $r_0$ is the radius of the center path of the electron and $y_{\text{exit}}$ and $y_{\text{entrance}}$ are the deviations of the electron trajectory from the radius $r_0$ at the entrance and exit slits, respectively. $E_{\text{p}}$ is the nominal pass energy for electrons traveling tangential to the radius $r_0$, $\delta E$ is the deviation from that energy, $c_{\text{D}}$ is a constant describing the energy dispersion, $\alpha$ is the angle in the dispersion plane that the trajectory makes with the tangential path at the entrance slit, $\beta$ is the corresponding angle perpendicular to the dispersion plane, and $c_{\alpha\alpha}$ and $c_{\beta\beta}$ are the second order angular aberrations in the dispersion plane and perpendicular to it, respectively. The first order term in $\alpha$ vanishes, as the devices possess first order focusing. The first order term in $\beta$ vanishes for the spherical deflector because of symmetry. The constants $c_{\alpha\alpha}$, $c_{\beta\beta}$ and $c_D$ are 4/3, 1 and 1.0 for the cylindrical deflector, and 1, 0 and 2.0 for the ideal spherical deflector.

With the improvements in computing power, bundles of electron trajectories are calculated within a short time even on a PC. This has opened the possibility to design free-form deflectors whose properties can be optimized with respect performance and ease of construction. The deflector that proved best in performance for the purpose of electron energy loss spectroscopy is illustrated in Fig. 7.22.

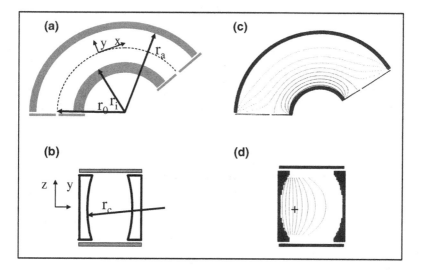

**Fig. 7.22.** Free-form electrostatic deflector that features stigmatic focusing, equipotential metal apertures and a correction of the angular aberration in the dispersion plane [7.48]. **(a)** Cross section in the dispersion plane showing the deflector plates and the entrance and exit apertures. **(b)** Cross section perpendicular to the electron path showing the concavely shaped deflector plates. **(c, d)** As **(a)** and **(b)**, yet with the dashed equipotential lines obtained from the solution of the Laplace-equation for the device.

The cross section in the dispersion plane shows deflector plates that are curved as for a cylindrical deflector (7.22a). Orthogonal to the dispersion plane the deflector plates are concavely shaped (7.22b) with a radius of curvature $r_c$. Since the radial field is zero right after the entrance slit and right before the exit slit, electrons do not travel along the equipotential line that connects entrance and exit slit (7.22c),

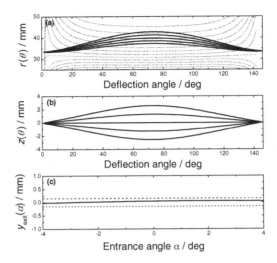

**Fig. 7.23.** Electron trajectories in a free-form deflector of the type shown in Fig. 7.22. Parameter are the radial position of the entrance slit $r_0 = 33.5$ mm, radii of inner and outer deflection plate $r_i = 20.5$ mm and $r_0 = 60$ mm, respectively, and $r_c = 100$ mm. **(a)** Radial position $r(\theta)$ versus the deflection angle $\theta$ for the entrance angles $\alpha = -4°, -2°, 0°, 2°$ and $4°$. **(b)** Position perpendicular to the dispersion plane $z(\theta)$ for the entrance angles $\beta = -4°, -2°, 0°, 2°$ and $4°$. **(c)** Deviation of the radial position from $r_0$ at the exit slit as a function of the entrance angle $\alpha$.

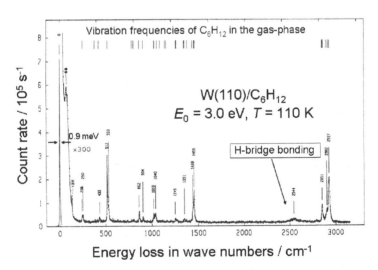

**Fig. 7.24.** Electron energy loss spectrum of cyclohexane on W(110) showing the performance of advanced spectrometers. In addition to the gas-phase modes, the spectrum shows the shifted CH-stretching mode of the hydrogen atoms pointing towards the surface (Sect. 6.4.5) and surface induced splittings of the modes at 250 and 520 cm$^{-1}$.

but rather travel up to larger radii until midway of the path. This is illustrated in the upper panel Fig. 7.23. The concave curvatures in combination with the top and bottom cover plates ensure the stigmatic focusing while minimizing simultaneously the angular aberrations. The angular aberration in the dispersion plane can be reduced to zero, by making $r_c \cong 1.3r_0$. However, the sum of the angular aberration coefficients $c_{\alpha\alpha}+c_{\beta\beta}$ stays constant [7.48]. In total, the device behaves nearly as an ideal sphere without fringe fields from the entrance and exit aperture. Unlike the ideal spherical deflector, the device is very robust with respect to high current loads. This is owed to the fact that the device has active stigmatic focusing whereas the ideal sphere has stigmatic focusing only because of the spherical symmetry. Figure 7.24 shows an example of a high-resolution spectrum obtained with the spectrometer featuring the free-form deflectors.

## 7.4.2 Theory of Inelastic Electron Scattering

In inelastic scattering of electrons one distinguishes three types of mechanisms, scattering from long-range electric fields (*"dipole scattering"*), scattering from the short-range atomic potentials (*"impact scattering"*), and scattering via short lived resonances with molecular orbitals (*"resonance scattering"*). The mechanisms are illustrated in Fig. 7.25. Each mechanism has its own characteristic features, with respect to angular distribution of the scattered electrons, the dependence of the cross section on the electron energy, the selection rules that apply, and each mechanism requires a theory of its own. Resonance scattering, e.g. occurs in weakly bound molecules at particular electron energies. The electron is captured for a short time in a molecular orbit. After some time the molecule undergoes a Frank-Condon transition from the ionic state into the ground state and the electron is re-emitted. The trajectory of the emitted electron is determined entirely by the symmetry of the molecular orbit. As the final state of the molecule may be vibrationally excited, the energy of the emerging electron is correspondingly lower (for a review see [7.49]). Since resonance scattering is less important for surface vibration spectroscopy as an analytical tool, we focus on dipole and impact scattering in the following.

### *Dipole scattering - the dielectric theory of electron solid interaction*

We consider the interaction of electrons with dipolar electric fields of elementary excitations at surfaces. The theory is not confined to vibration spectroscopy. It applies equally to other elementary excitations such as electronic transitions and plasma excitation in the limit of small momentum transfer. The theory of dipole scattering can be approached from three different viewpoints, each one having its own virtues and shortcomings.

One possibility is to consider the Hamiltonian of the free electron and treat the dipole fields associated with the elementary excitations as a perturbation in the spirit of the Born-approximation. The advantage of this approach is that it delivers the scattering kinematics, i.e. energy and momentum conservation. The disadvantage is that one has to focus on a particular type of elementary excitation. The second approach considers the Hamiltonian of the excitation and treats the electron at as external perturbation. Because of the smallness of the momentum change involved in inelastic scattering from long-range dipole fields the electron trajectory can assumed to be unperturbed by the inelastic event (cf. Fig. 7.25a). This treatment delivers the multiple losses straightforwardly. The disadvantage is that energy and momentum conservation have to be introduced ad hoc.

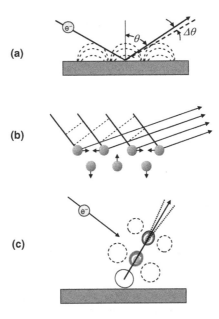

**Fig. 7.25.** Illustration of the three different scattering mechanisms, which are operative at surfaces. Small-angle dipole scattering (**a**), wide-angle impact scattering (**b**), and resonance scattering (**c**).

The third possibility is to consider the inelastic interaction as a classical energy loss of a charged particle reflected from a surface. The advantage of the latter formulation is that it is independent of the specific elementary excitation. Solids and adsorbed species on solids are represented by their complex dielectric functions $\varepsilon(\omega)$ and their complex dynamic polarizabilities $\alpha(\omega)$, respectively. In their specific realms of applicability, all three approaches lead to the same final expression for the inelastically scattered intensity. We pursue the third path in the following, usually known as the *dielectric theory* of inelastic electron scattering. As a matter of convenience, we do the derivation in Gaussian units. The theory was originally developed by Geiger [7.50] and Raether [7.51] for inelastic scattering of high-energy electrons in transmission geometry [7.52]. The low energy reflection case was first studied by Lucas and Sunjic [7.53] using the second approach. The quantum theory of scattering for the reflection geometry was developed by Evans and Mills [7.54].

In classical electrodynamics the total energy dissipation is

$$W = \frac{1}{4\pi} \int dt \int d^3 r \, \mathscr{E}(r,t) \cdot \dot{D}(r,t) \tag{7.59}$$

where $\mathscr{E}(r,t)$ and $\dot{D}(r,t)$ are the electric field and the time derivative of the displacement, respectively. The integral extends over the entire space. By writing the energy loss that way, we have restricted our considerations to local and instantaneous dielectric responses. This approximation suffices in most cases of interest. We remark however that the surface response of metals due to electron-hole pair excitations is not properly described within the framework of local dielectric response [7.55]. The total energy dissipation can also be expressed in terms of the probability $P(q_\parallel, \omega)$ for an energy loss of energy $\hbar\omega$ and momentum $\hbar q_\parallel$,

$$W = \int_{\hbar\omega>0} d(\hbar\omega) \, dq_\parallel \, \hbar\omega \, P(q_\parallel, \omega) . \tag{7.60}$$

For surface losses, one may employ the two-dimensional Fourier expansion

$$\mathscr{E}(r,t) = \int d\omega \, dq_\parallel \, e^{-i\omega t} e^{iq_\parallel \cdot r_\parallel} e^{-q_\parallel |z|} \mathscr{E}(q_\parallel, z, \omega) . \tag{7.61}$$

Here the position vector is $r = (r_\parallel, z)$, and the surface plane is at $z = 0$. The expansion may be considered as an expansion into dielectric surface waves (Sect. 7.1.7, (7.30)). Contrary to (7.60), the integral extends over negative and positive $\omega$. An analogous expansion holds for the displacement $D(r,t)$. The Fourier-components of $D$ and $\mathscr{E}$ are related to each other by the dielectric function $\varepsilon(\omega) = \varepsilon_1(\omega) + i\varepsilon_2(\omega)$, which we assume to be scalar for the moment. A possible dependence on $q_\parallel$ can be neglected as the scattering process is near $q_\parallel = 0$, but still to the far-right of the light line (Fig. 7.13). By inserting (7.61) and the corresponding equation for the displacement $D(r,t)$ into (7.59) and by considering that the integrals

$$\int_{-\infty}^{\infty} dx \, e^{i(q-q')x} = 2\pi\delta(q-q')$$

$$\int_{-\infty}^{\infty} dt \, e^{i(\omega-\omega')t} = 2\pi\delta(\omega-\omega') \tag{7.62}$$

are representations of delta-function, one obtains the expansion for the energy loss

$$W = 2\pi^2 \int d\omega dq_{\parallel} \, \omega \varepsilon_2(\omega) \int_{z \leq 0} dz \left| \mathscr{E}_{int}(q_{\parallel}, z, \omega) \right|^2 e^{-2q_{\parallel}|z|}. \tag{7.63}$$

As the imaginary part of $\varepsilon(\omega)$ is zero in vacuum, only the electric field inside the material $\mathscr{E}_{int}(q_{\parallel}, z, \omega)$ matters and the integral extends only over the half-space $z \leq 0$. The integral over $q_{\parallel}$ and $\omega$ has the form of (7.60). The integral over $z$ is therefore the loss probability $P(q_{\parallel}, \omega)$. The internal electric field is generated by the electron while it is approaching the surface from the outside and after it is reflected from the surface potential barrier or diffracted by the near-surface atoms. The electron also spends a short time inside the solid before it emerges in the vacuum again. For low energy electrons with their small penetration depth, this part may be neglected. The internal electric field therefore stems from a charge outside the solid. If the solid is a semi-infinite dielectric half-space in $z \leq 0$, the relation between the internal field $\mathscr{E}_{int}(q_{\parallel}, z, \omega)$ and an external field $\mathscr{E}_{ext}(q_{\parallel}, z, \omega)$ is

$$\mathscr{E}_{int}(q_{\parallel}, z, \omega) = \frac{2\mathscr{E}_{ext}(q_{\parallel}, z, \omega)}{\varepsilon(\omega) + 1}. \tag{7.64}$$

The loss probability is therefore proportional to

$$P(q_{\parallel}, \omega) \propto \frac{\varepsilon_2(\omega)}{\left| \varepsilon(\omega) + 1 \right|^2} = \mathrm{Im} \frac{-1}{\varepsilon(\omega) + 1}. \tag{7.65}$$

The loss probability has a pole at the frequency of the dielectric surface wave (Sect. 7.1.7). As an example, we consider the dielectric function of an infrared active material in the case of small damping $\gamma$ (7.18).

$$\varepsilon(\omega) = \varepsilon_\infty + \frac{\omega_0^2(\varepsilon_{st} - \varepsilon_\infty)}{\omega_0^2 - \omega^2 - i\gamma\omega} \tag{7.66}$$

Then

$$P(q_{\parallel}, \omega) \propto \lim_{\gamma \to 0} \mathrm{Im} \frac{-1}{\varepsilon(\omega) + 1} = \frac{\varepsilon_{st} - \varepsilon_\infty}{(1 + \varepsilon_{st})(1 + \varepsilon_\infty)} \frac{\pi\omega_s}{4} \delta(\omega - \omega_s) \tag{7.67}$$

with

$$\omega_s = \omega_0 \left( \frac{1 + \varepsilon_{stat}}{1 + \varepsilon_\infty} \right)^{1/2} \tag{7.68}$$

the frequency of the Fuchs-Kliewer surface phonon. We thus learn that surface phonons are excited by electrons while the electrons are outside the solid; longitudinal bulk phonons are excited by electrons inside the solid since the screening factor is $1/\varepsilon(\omega)$ in that case. The same holds for plasmon excitations.

We now consider the case of a thin dielectric layer of thickness $d$ on a non-absorbing substrate. The substrate/layer interface is at $z = 0$. The dielectric functions of layer and substrate are denoted as $\varepsilon_s(\omega)$ and $\varepsilon_b$, respectively. It is assumed that $|\varepsilon_s(\omega)q_\| d| \ll \varepsilon_b$. The potential $\varphi$ originating in a charged particle outside the solid is as if the layer did not exist. The potential is given by the potential of the bare charge and the image charge inside the solid

$$\varphi(\mathbf{r}) = \frac{e}{r} - \frac{\varepsilon_b - 1}{\varepsilon_b + 1}\frac{e}{r'} \tag{7.69}$$

where $\mathbf{r}$ and $\mathbf{r}'$ are the vectors pointing from the charge and its image to a particular point (Fig. 7.26).

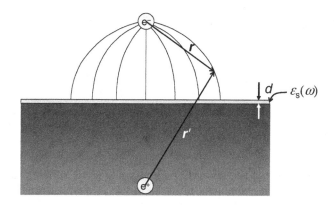

**Fig. 7.26.** Electric field lines of a point charge near a surface of a dielectric material with a dielectric permeability $\varepsilon_b \gg 1$. The dielectric active surface layer is assumed thin so that the field is not affected by the presence of the layer.

To calculate the dielectric losses of the active layer we make use of the fact that the parallel component of the electric field and the normal component of the displacement are continuous at the interface to the substrate and remain constant in the thin active layer. We rewrite the scalar product $\mathscr{E} \cdot \dot{\mathbf{D}}$ under the integral (7.59) as

$$\mathscr{E} \cdot \dot{\mathbf{D}} = \varepsilon_{s\|}\mathscr{E}_\| \cdot \dot{\mathscr{E}}_\| + \frac{1}{\varepsilon_{s\perp}}D_\perp \dot{D}_\perp \tag{7.70}$$

and thereby admit a tensorial dielectric function $\varepsilon_s$ inside the active layer. The parallel component of the electric field $\mathscr{E}_\parallel$ is

$$\mathscr{E}_\parallel\Big|_{z=0} = -e\nabla_\parallel\left(\frac{e}{r} + \frac{1-\varepsilon_b}{1+\varepsilon_b}\frac{e}{r'}\right)\Bigg|_{z=0} = -\frac{2}{1+\varepsilon_b}\nabla_\parallel\frac{e}{r}\Big|_{z=0} = \frac{2}{1+\varepsilon_b}\mathscr{E}_{bare\parallel} \quad (7.71)$$

where $\mathscr{E}_{bare}$ is the field of the electron charge at the interface $z=0$ without the image charge. Note that $\nabla_\parallel r^{-1} = +\nabla_\parallel r'^{-1}$ but $\nabla_z r^{-1} = -\nabla_z r'^{-1}$ at $z=0$. The vertical component of the displacement $D_\perp$ is

$$D_\perp\Big|_{z=0} = -e\varepsilon_b\nabla_z\left(\frac{e}{r} + \frac{1-\varepsilon_b}{1+\varepsilon_b}\frac{e}{r'}\right)\Bigg|_{z=0}$$

$$= \frac{2\varepsilon_b}{1+\varepsilon_b}\nabla_z\frac{e}{r}\Big|_{z=0} = \frac{2\varepsilon_b}{1+\varepsilon_b}\mathscr{E}_{bare\perp} \ . \qquad (7.72)$$

The total energy loss expressed in terms of the Fourier-components thereby becomes

$$W = 8\pi^2 d\int\!\omega d\omega dq_\parallel\left[\frac{\mathrm{Im}\,\varepsilon_{s\parallel}(\omega)}{(\varepsilon_b+1)^2}\left|\mathscr{E}_{bare\parallel}(q_\parallel, z=0, \omega)\right|^2\right.$$

$$\left. -\frac{\varepsilon_b^2}{(\varepsilon_b+1)^2}\mathrm{Im}\frac{1}{\varepsilon_{s\perp}(\omega)}\left|\mathscr{E}_{bare\perp}(q_\parallel, z=0, \omega)\right|^2\right]. \qquad (7.73)$$

For substrates with $|\varepsilon_b|>>1$, hence in particular for metal surfaces, only the term with the perpendicular component of the field survives which constitutes the surface selection rule. The dielectric active surface layer is frequently a monolayer of molecules with dipole active vibration modes. In that case

$$d\,\mathrm{Im}\frac{1}{\varepsilon_{s\perp}(\omega)} \cong 4\pi n_s\mathrm{Im}\,\alpha_{s\perp}(\omega) \qquad (7.74)$$

where $\alpha_{s\perp}(\omega)$ is the dynamic polarizability of the molecule and $n_s$ is the surface density. In order to express the integrand of (7.73) (which is $P(q_\parallel, \omega)$) in terms of the scattering parameters we need the Fourier-transform of the potential $e/r(t)$,

$$\frac{e}{r} = \frac{e}{2\pi}\int dq_\parallel\frac{1}{q_\parallel}e^{-q_\parallel|z|}e^{-iq_\parallel\cdot r_\parallel} \ . \qquad (7.75)$$

The gradient of the potential, the electric field, has therefore the expansion

$$-\nabla \frac{e}{r} = \frac{e}{2\pi} \int d\mathbf{q}_\| \, (i e_\|, 1) e^{-q_\||z|} e^{-i\mathbf{q}_\| \cdot \mathbf{r}_\|} \tag{7.76}$$

in which $(i e_\|, 1)$ is a three dimensional vector with the perpendicular component 1 and $e_\|$ the unit vector in the surface plane.

We now study the probability for a phonon excitation for an electron that is reflected at the surface at the time $t = 0$. In that case

$$\mathbf{r}(t) = \mathbf{r}_\| + v_\| t e_\| + v_\perp |t| e_\perp \qquad -\infty < t < \infty , \tag{7.77}$$

where $v_\|$ and $v_\perp$ are the electron velocities parallel and perpendicular to the surface, respectively. The vector $\mathbf{r}_\|$ has its origin in the point of reflection and points to an arbitrary position on the surface. The Fourier-integral of the electric field is

$$\begin{aligned}
\mathscr{E}_{\text{bare}}(\mathbf{r}, t) &= -\nabla \frac{e}{\left| \mathbf{r}_\| + v_\| t e_\| + v_\perp |t| e_\perp \right|} \\
&= \frac{1}{2\pi} \int d\mathbf{q}_\| (i e_\|, 1) e^{-i(\mathbf{q}_\| \cdot (\mathbf{r}_\| + \mathbf{v}_\| t))} e^{-q_\|(|z| + v_\perp |t|)} .
\end{aligned} \tag{7.78}$$

The Fourier integral of the kernel with respect to time is

$$e^{-i\mathbf{q}_\| \cdot \mathbf{v}_\| t} e^{-q_\| v_\perp |t|} = \frac{1}{2\pi} \int d\omega e^{-i\omega t} \frac{2 q_\| v_\perp}{(\omega - \mathbf{q}_\| \cdot \mathbf{v}_\|)^2 + q_\|^2 v_\perp^2} . \tag{7.79}$$

The Fourier-components of the electric field are therefore

$$\mathscr{E}_{\text{bare}}(\mathbf{q}_\|, \omega) = (i e_\|, 1) \frac{e}{(2\pi)^2} \frac{2 q_\| v_\perp}{(\omega - \mathbf{q}_\| \cdot \mathbf{v}_\|)^2 + q_\|^2 v_\perp^2} . \tag{7.80}$$

By inserting this equation into (7.73) and comparing with (7.60) we obtain the probability for the reflected electron to suffer an energy loss of $\hbar\omega$ with a wave vector $\mathbf{q}_\|$.

$$P(\mathbf{q}_\|, \omega) = \frac{16 e^2 n_s}{\hbar^2 \pi} \frac{q_\|^2 v_\perp^2}{((\omega - \mathbf{q}_\| \cdot \mathbf{v}_\|)^2 + q_\|^2 v_\perp^2)^2} \operatorname{Im} \alpha_{s\perp}(\omega) . \tag{7.81}$$

The probability has a sharp resonance at the *surfing condition*

$$v_\parallel = \omega / q_\parallel . \tag{7.82}$$

The resonance is infinitely sharp (is this approximation) for grazing incidence. Because of wave vector conservation, the surfing condition entails that the electrons are scattered into a small angular cone around all the directions of elastic scattering, around the specular reflected beam as well as around the diffracted beams. In either case, the intensity of the inelastically scattered electrons is proportional to the elastic intensity. In practical applications of (7.81), one is interested in the inelastically scattered intensity observed in a small angular aperture. The aperture angles of spectrometers are small because of angular aberrations in the deflectors and in the lens systems. We denote the angular deviations from the specular beam in the scattering plane and perpendicular to it as $\alpha$ and $\beta$, respectively. From momentum conservation one obtains (cf. (7.41))

$$q_{\parallel x} \cong -\frac{\sqrt{2mE_I}}{\hbar}\left(\alpha \cos\theta_I + \vartheta_E \sin\theta_I\right)$$

$$q_{\parallel y} \cong \frac{\sqrt{2mE_I}}{\hbar}\beta \tag{7.83}$$

$$\text{with } \vartheta_E = \hbar\omega/2E_I, \alpha, \beta \ll 1$$

where $E_I$ is the energy of the incident electron. The condition $\vartheta_E \ll 1$ is consistent with the initial assumption that the trajectory of the electron should remain essentially unaffected by the inelastic scattering process. The element in $q_\parallel$-space is converted into the angular space by

$$dq_\parallel = k_I^2 \cos\theta_I \, d\alpha \, d\beta \tag{7.84}$$

where $k_I$ is the wave vector of the incident electron. We introduce furthermore reduced angles as $\hat{\alpha} = \alpha / \vartheta_E, \hat{\beta} = \beta / \vartheta_E$. With these notations, one obtains the relative intensity of inelastic events as

$$\frac{dI_{inel}}{I_{el} \, d\hbar\omega} = \frac{8 n_s \, \mathrm{Im}\,\alpha_{s\perp}(\omega)}{\pi a_B \, E_I \cos\theta_I}[1 + n(\omega)]\int_{-\hat{\alpha}_c}^{\hat{\alpha}_c} d\hat{\alpha} \int_{-\hat{\beta}_c}^{\hat{\beta}_c} d\hat{\beta}\frac{(\hat{\alpha}\cos\theta_I + \sin\theta_I)^2 + \hat{\beta}^2}{(1 + \hat{\alpha}^2 + \hat{\beta}^2)^2}. \tag{7.85}$$

Here, $\hat{\alpha}_c$ and $\hat{\beta}_c$ are the aperture angles of the spectrometer and $a_B$ is the Bohr-radius. To account for the quantum statistics (not reproduced in the classical derivation) we have added the term $[1+n(\omega)]$ in which $n(\omega)$ is the boson occupation number of a harmonic oscillator of frequency $\omega$. For energy gains, the factor $[1+n(\omega)]$ is replaced by $n(\omega)$.

The form (7.85) is suitable for numerical calculations of the intensity for a given oscillator strength of the vibration, once the angular apertures of the spectrometer are known. For an estimate of the intensity it is useful to integrate (7.85) under the assumption of a circular aperture. With the transformation into polar coordinates $\hat{\alpha} = \hat{\vartheta}\cos\varphi$ $\hat{\beta} = \hat{\vartheta}\sin\varphi$ one obtains after some algebra

$$\frac{dI_{\text{inel}}}{I_{\text{el}}\, d\hbar\omega} = \frac{4\,n_s\, \text{Im}\,\alpha_{s\perp}(\omega)}{a_B\, E_I \cos\theta_I}[1+n(\omega)]$$
$$\times\left[\left(\sin^2\theta_I - 2\cos^2\theta_I\right)\frac{\hat{\vartheta}_c^2}{1+\hat{\vartheta}_c^2} + \left(1+\cos^2\theta_I\right)\ln(1+\hat{\vartheta}_c^2)\right]. \tag{7.86}$$

In the limit $\hat{\vartheta}_c \ll 1$ that is typical for energy losses due to innermolecular vibrations the intensity is

$$\frac{dI_{\text{inel}}}{I_{\text{el}}\, d\hbar\omega} = \frac{32\,E_I}{a_B\,(\hbar\omega)^2}\frac{\sin^2\theta_I}{\cos\theta_I}[1+n(\omega)]\vartheta_c^2\, n_s\, \text{Im}\,\alpha_\perp(\omega)\ . \tag{7.87}$$

The intensity of dipolar excitations, vibrational as well as electronic excitations, therefore falls off for higher frequencies. This explains why electronic transitions, e.g. interband transitions are very weak features in an energy loss spectrum. The frequency dependence of the intensity is in remarkable contrast to infrared absorption spectroscopy (Sect. 7.5). It is occasionally useful to express $\text{Im}\,\alpha_{s\perp}(\omega)$ in terms of an effective charge $e_{\text{eff}}$ associated with a mode. The dielectric function of a medium with a density $n$ of harmonic oscillators in Gaussian units has the form (7.18)

$$\varepsilon(\omega) = \varepsilon_\infty + \frac{4\pi e_{\text{eff}}^2\, n/m_{\text{red}}}{\omega_0^2 - \omega^2 - i\gamma\omega} \tag{7.88}$$

in which $m_{\text{red}}$ is the reduced mass. For a dilute layer $\varepsilon_\infty = 1$. In the SI-system, the factor $4\pi$ is to be replaced by $1/\varepsilon_0$. Using the identity

$$\lim_{\gamma\to 0} \text{Im}\,\frac{1}{z-i\gamma} = \pi\delta(z) \tag{7.89}$$

one obtains with (7.74)

$$\text{Im}\,\alpha_{s\perp}(\omega) = \frac{\pi}{2}\frac{n_s\, e_{\text{eff}}^2}{\varepsilon_\infty^2\, m_{\text{red}}\,\omega_R}\delta(\omega_R - \omega) \quad\text{with}\quad \omega_R^2 = \omega_0^2 + 4\pi n e_{\text{eff}}^2/m_{\text{red}}\ . \tag{7.90}$$

### Impact scattering

The Born approximation (7.44) may again serve as a starting point for a treatment of the impact scattering. Inserting initial and final states of the solid expanded into phonon creation and annihilation operators yields the momentum conservation law for the components parallel to the surface and an expression for the inelastic cross section in terms of the scattering potential. The salient features of this so-called *kinematic scattering theory* are more easily derived by considering the scattering of a classical wave from an arrangement of point scatterers (see e.g. [7.56], Chapt. 4). With reference to the notation introduced in Sect. 7.1.2, one can write the scattering amplitude as

$$f = f_0 e^{-i\omega_0 t} \sum_{l_\parallel, l_z, \kappa} e^{-iK \cdot r(l_\parallel l_z \kappa, t)} \tag{7.91}$$

in which $K$ is the difference between the wave vector of the incident wave $k_I$ and the scattered wave $k_S$. The time dependent position vectors of the scattering centers $r(l_\parallel l_z \kappa, t)$ can be expanded for small phonon amplitudes $u(l_\parallel l_z \kappa, t)$

$$r(l_\parallel l_z \kappa, t) = r_0(l_\parallel l_z \kappa) + u_0(q_\parallel l_z \kappa) e^{\pm i(q_\parallel \cdot r_0(l_\parallel l_z \kappa) - \omega(q_\parallel)t)} . \tag{7.92}$$

With this expansion the scattering amplitude becomes

$$f = f_0 e^{-i\omega_0 t} \delta(K_\parallel) \sum_{l_z, \kappa} e^{-iK \cdot r(0 l_z \kappa)}$$
$$+ f_0 e^{-i(\omega_0 \pm \omega(q_\parallel))t} \delta(K_\parallel \mp q_\parallel) \sum_{l_z, \kappa} K \cdot u_0(q_\parallel l_z \kappa) e^{-iK \cdot r(0 l_z \kappa)} \tag{7.93}$$

in which the second term describes the classical analogue of inelastic scattering with energy and momentum conservation. The $\delta$-functions result from the summation over $l_\parallel$. The scalar product of $K$ and $u_0$ vanishes if the displacements are perpendicular to the scattering plane, or more generally phrased, if the mode is odd with respect to the scattering plane. This selection rule was discussed already in Sect. 7.3.2. The energy dependence of the inelastically scattered intensity is given by the product $(K \cdot u_0)^2 \propto E_I$ and the energy dependence of the atomic form factor $f_0(E_I)$. Neglecting the latter, the intensity of phonon scattering should be roughly proportional to the energy of the incident electron beam, i.e. a smooth function of the energy. The kinematic model would furthermore predict that the intensity of a phonon that is polarized parallel to the surface should be much less than the intensity of a perpendicular polarized phonon, since the vertical component of $K$ is much larger than the parallel component.

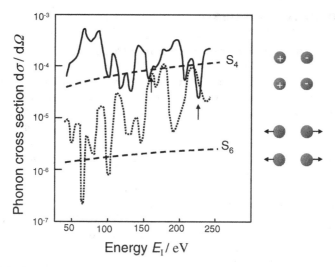

**Fig. 7.27.** Calculated cross section for inelastic scattering from the $S_4$ and $S_6$-phonon at $\overline{X}$ on the Ni(100) surface as function of the energy of the incident electron beam (solid and dashed lines respectively). The polar angle of detection is fixed to $\theta_S = 65°$. The angle of the incident beam is adjusted to keep the momentum transfer at $\overline{X}$. The arrows point to energies at which the intensity of the $S_6$-phonon exceeds the intensity of the $S_4$-phonon (after Xu et al. [7.57]). The dashed lines mark the intensities according to the kinematic theory.

The kinematic scattering theory described above assumes that electrons are scattered only once whereas in reality low energy electrons scatter many times elastically before they emerge in a diffracted beam. A proper treatment of inelastic phonon scattering requires the consideration of these the multiple elastic scattering events. The *dynamic theory of scattering from phonons* has been worked out by Tong et al. in 1980 [7.58, 59]. With multiple elastic scattering included, the cross sections become a strongly oscillating function of energy (cf. Sect. 1.1.1). This is illustrated in Fig. 7.27 for the $S_4$ and $S_6$-phonon on the Ni(100) at the $\overline{X}$-point of the SBZ (Sect. 7.1.2, Fig. 7.3). The kinematic model predicts that the intensity of the $S_6$-phonon should be almost two orders of magnitude lower than the intensity of the $S_4$-phonon. This is also the overall trend according to the dynamic scattering theory. However, the strong oscillations effect that the intensity of the $S_6$-phonon exceeds the intensity of the $S_4$-phonon at certain energies (arrows in Fig. 7.27). Experiments have confirmed the predictions of the theory [7.57]. From the standpoint of an experimentalist, the strong oscillations in the cross section are an advantage as they enable the observation of modes polarized in the surface plane. Furthermore, one can distinguish phonons that have only a very small difference in energy, below the resolution of the spectrometer, by their different energy dependence of the cross section. This greatly expands the possibilities of inelastic electron scattering for the investigation of surface phonon dispersion.

## 7.5 Optical Techniques

### 7.5.1 Reflection Absorption Infrared Spectroscopy

In 1966 R. G. Greenler laid the theoretical foundations of *Reflection Absorption InfraRed Spectroscopy* (RAIRS) [7.60]. He showed that the infrared absorption of a thin layer deposited on a metal surface observed in reflection geometry is greatly enhanced at grazing incidence. The reflectivity of the metal is near unity and the infrared absorption of a particular eigenmode of an adsorbed molecule should appear as a dip in the reflectivity, hence the name *Reflection Absorption Spectroscopy*. Since the electric field is perpendicular on a metal surface, only modes with a perpendicular dipole moment, the totally symmetric modes are visible. For the same reason the absorption occurs only in the reflectivity of p-polarized light.

The angular dependence of the change in the reflectivity due to surface absorption can be calculated by applying the Fresnel-boundary conditions of conventional optics to a three-layer system consisting of vacuum, the adsorbate layer and the metal substrate [7.60, 61]. The calculation is somewhat cumbersome and does not reveal the physical nature of the enhancement. We therefore follow a different route that considers the power absorption $\dot{W}$ in the adsorbate layer that has the area $A$ and the thickness $d$ (cf. 7.59, 7.73).

$$\dot{W}_{\text{abs}} = \frac{Ad}{2} E_{\text{s}\perp}^2 \, \omega \, \text{Im} \frac{-1}{\varepsilon_\perp(\omega)} \, . \tag{7.94}$$

Here, $E_{\text{s}\perp}$ is the perpendicular component of the amplitude of the electric field at the surface, which varies periodically in time. The factor 1/2 results from the integration over one period. The power in the infrared beam is

$$\dot{W}_{\text{in}} = \frac{1}{2} A_0 c E_{0,\text{p}}^2 \tag{7.95}$$

where $E_{0,\text{p}}$ is the electric field amplitude of the p-polarized light and $A_0$ is the cross section of the beam. There is no power absorption for s-polarized light since the electric field is zero at a metal surface for s-polarized light. The change in reflectivity $\Delta R_\text{p}$ is the ratio of the absorbed power to the incident power. For an ideal metal the field at the surface is

$$E_{\text{s}\perp} = 2 E_{0,\text{p}} \sin \theta \tag{7.96}$$

and the cross section of the beam and the absorbing surface area are related by

$$A = A_0 / \cos \theta \, . \tag{7.97}$$

The change in reflectivity is thus

$$\Delta R_p = \frac{8\pi}{\lambda} \frac{\sin^2\theta}{\cos\theta} d \, \mathrm{Im} \frac{-1}{\varepsilon_\perp(\omega)} \cong \frac{32\pi^2}{\lambda} \frac{\sin^2\theta}{\cos\theta} n_s \, \mathrm{Im}\,\alpha_\perp(\omega). \qquad (7.98)$$

The approximation is again for a low surface concentration of adsorbates $n_s$. For non-ideal metals with a finite dielectric constant $\varepsilon_b$ the electric field eventually drops to zero at grazing incidence. The correction factor that accounts for this drop-off is

$$F = \left( 1 + \frac{1}{|\varepsilon_b|^2} \frac{\sin^2\theta}{\cos^4\theta} \right)^{-1}. \qquad (7.99)$$

Because of this factor, the change in the reflectivity passes through a maximum that lies at about 85°-88° for metal surfaces.

Equation (7.98) is remarkable in several ways. Firstly, we see that the enhancement of the reflectivity is mostly due to the increase in the surface area. Secondly, we realize that RAIRS probes the same optical properties of the surface as EELS in the dipole scattering regime (cf. 7.87). An important difference is in the spectral sensitivity. EELS is significantly more sensitive in the low frequency range because of the factor $\omega^2$ that appears in the denominator of (7.87).

Reflection-absorption infrared spectroscopy was introduced as an experimental technique in 1970 by the group of J. Pritchard [7.62, 63] and was quickly adopted by several groups thereafter. The signal to noise in the spectra was greatly improved with the advent of *Fourier-Transform InfraRed* (FTIR) spectroscopy. Figure 7.28 shows an experimental set-up as used in our lab in connection with other UHV-studies [7.64]. Both the spectrometer and the detector chamber are evacuated to avoid absorption from the IR-bands of water vapor. The sample sits in a narrow tube on top of the main UHV-chamber, which houses several other surface science techniques. The spectrometer is a Michelson-interferometer. The intensity at the detector as a function of the position $x$ of the movable mirror is

$$dI(\omega, x) = \left| \mathscr{E}_1(\omega) + \mathscr{E}_2(\omega) \right|^2 d\omega \qquad (7.100)$$

with

$$\mathscr{E}_1 = \sqrt{I_0 R(\omega)} \sin\omega t \text{ and } \mathscr{E}_2 = \sqrt{I_0 R(\omega)} \sin(\omega t + 2\omega x/c) \qquad (7.101)$$

where $R(\omega)$ is the reflectivity of the sample and $I_0$ the intensity of the beam. The detector has approximately a white spectral response function; hence, it detects the integral over the frequency, which is the Fourier transform of the spectral function of interest,

$$I(x) = const + I_0 \int_0^\infty R(\omega)\cos(2\omega x/c)\,d\omega . \qquad (7.102)$$

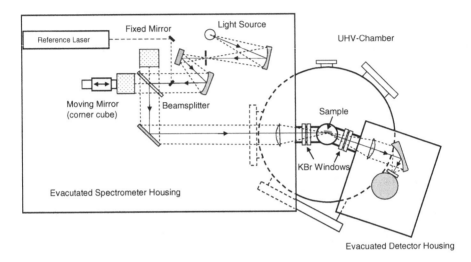

**Fig. 7.28.** An experimental set-up for Fourier-transform reflection-absorption infrared spectroscopy at a single crystal surface in UHV.

Figure 7.29 shows the reflection-absorption spectrum of CO adsorbed on Cu(111) at 90 K. The spectrum represents the ratio of the spectral response of the clean surface and the CO covered surface. The total measuring time for the small spectral range was only 60 s. The resolution of the FTIR-technique is significantly better than the natural line width of adsorbed CO. The line width of adsorbed species is broadened because of anharmonic coupling to other low-frequency modes. In the case of CO, the large dipole moment associated with the vibration gives rise to electron/hole pair excitations, causing an additional broadening.

An advantage of infrared spectroscopy is that it can be applied to surfaces in contact with a gas phase, as long as the density of the gas phase is not so high that it blocks the IR-beam in the spectral ranges of interest. Adsorbed species and gas phase species are distinguished by taking the difference of the spectral response for s- and p-polarized light. For infrared spectroscopy at the surface/electrolyte interface, the *Attenuated Total Reflection* (ATR) technique in the geometry introduced by Kretschmann [7.65] has proved useful [7.66]. Fig. 7.30 displays schematically the experimental set-up. The IR-beam is internally total reflected at the surface of a prism made of Si, Ge, or ZnS. A very thin metal film is evaporated onto the prism. The vibration modes of molecules that are adsorbed on the film

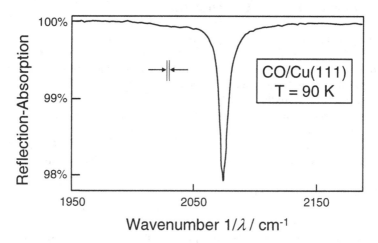

**Fig. 7.29.** Infrared absorption spectrum of CO on Cu(111). The spectral resolution is significantly better than the line width.

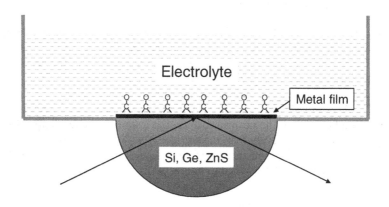

**Fig. 7.30.** Experimental set-up for *Attenuated Total Reflection* (ATR) spectroscopy at the metal/electrolyte interface in the Kretschmann geometry [7.65].

from the electrolyte extract energy from the IR-beam because they reside in the field of the evanescent wave at the surface of the prism. The infrared signal obtained from adsorbed species in the ATR-geometry may be enhanced when the films are rough because of the enhancement of the electric field at the apices of the surface.

## 7.5.2 Beyond the Surface Selection Rule

Conventional IR-sources and detectors are not very effective in the far infrared. RAIRS is therefore mainly used in the frequency range above 700-1000 cm$^{-1}$. Around 1990 researchers began to explore the use of synchrotron radiation as a far-infrared light source in order to extend the spectral range of IR-spectroscopy into the realm of low frequency vibrations. Hirschmugl et al. were indeed able to find a weak absorption line of the metal carbon stretching vibration of CO on Cu(100) using synchrotron radiation [7.67]. The great surprise of that study however was a strong spectral signature of the frustrated rotation of CO, mode that is dipole forbidden according to the surface selection rule, since its dipole moment lies parallel to the surface. The spectral feature of that mode was an anti-absorption line with a Fano-type line shape (Fig. 7.31). A similar spectral feature was actually observed earlier for hydrogen adsorbed on W(100) [7.68], however misinterpreted as an overtone of a parallel vibration.

This section considers the essential physics of the effect as first described by Persson [7.69, 70]. For a proper quantitative treatment of the change in the reflectivity, one would have to resort to the formalism non-local optics. However, the Fresnel-equations of local optics suffice for a parameterized qualitative description. An important clue to the origin of the effect is the experimental observation of a continuous frequency dependent reduction in the reflectivity with increasing coverage of CO observed by Hirschmugl et al. [7.67]. The reflectivity of a metal in the infrared deviates from R = 1 because of the finite conductivity. The reduction in the reflectivity can therefore be attributed to a reduction of the conductivity in the near-surface region by CO adsorption. In fact, the reduction of the conductivity of thin films upon adsorption is a well-studied phenomenon [7.71] (see also Sect. 8.4.1). The anti-absorption line in Fig. 7.31 represents an increase of the electronic conductivity in the near-surface region at the resonance frequency of a mode that is polarized parallel to the surface. We cast this statement into a mathematical form by first considering the equation of motion for an ensemble of $N$ adsorbed atoms of mass $M$ parallel to the surface,

$$NM\ddot{u} + NM\omega_0^2 u + NM\eta(\dot{u} - \dot{x}) = 0 \ . \tag{7.103}$$

Here, $u$ is the amplitude of vibration and $x$ is the coordinate of uniform motion of the metal electron gas parallel to the surface. The third term in (7.103) describes a friction between the electron gas and the adsorbate that is proportional to the relative speed of motion $\dot{u} - \dot{x}$. The friction term is the classical analogue to the mechanism of electron/hole pair damping of vibrational modes. To see this, one solves the Fourier transform of (7.103) for a vanishing motion of the electron gas. The term $\tau_{\text{eh}} = 1/\eta$ then plays the role of an energy relaxation time. The corresponding equation of motion of the electron gas must contain the same friction term with reverse sign, hence

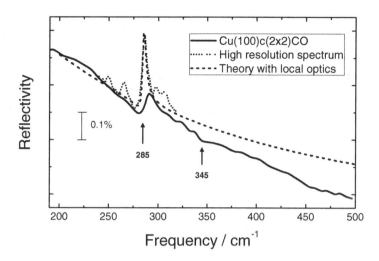

**Fig. 7.31.** Infrared reflectivity of a CO covered Cu(100) surface at 100 K relative to the clean surface after Hirschmugl et al. [7.67]. In addition to the anti-absorption line at 285 cm$^{-1}$ the spectrum (solid line) shows a very weak signature of a conventional absorption line at 345 cm$^{-1}$ (barely significant in this particular spectrum). The dotted line is the same spectrum with high resolution in a narrow range. The dashed line is calculated from a friction model in the framework of local optics.

$$N_e m \ddot{x} + N_e m \frac{1}{\tau} \dot{x} - NM\eta(\dot{u} - \dot{x}) = -N_e e \mathscr{E} \qquad (7.104)$$

where $m$ and $N_e$ are the electron mass and the number of electrons in a thin layer below the surface. As we are interested in the principal effect, we can leave the thickness of that layer, and hence the value of $N_e$, open for the moment. The relaxation time $\tau$ is related to the dc-conductivity $\sigma$ via

$$\sigma = \frac{ne^2}{m}\tau \qquad (7.105)$$

where $n$ is the electron concentration. The additional friction term in (7.104) due to the relative motion of adsorbate at electron gas vanishes at the resonance frequency $\omega_0$ of the adsorbate. The model should therefore produce an increase in the reflectivity at $\omega_0$. The coupled equations (7.103) and (7.104) are solved by introducing the Fourier-transformed amplitudes $x(\omega)$, $u(\omega)$, $P(\omega)$, $\mathscr{E}(\omega)$,

$$\left\{-\omega^2 - i\omega(\frac{1}{\tau} + \tilde{\eta}) + \frac{\omega^2\eta\tilde{\eta}}{\omega_0^2 - \omega^2 - i\omega\eta}\right\}x(\omega) = -\frac{e}{m}\mathscr{E}(\omega) \qquad (7.106)$$

where $\tilde{\eta} = \eta NM / N_e m$. With the definition of the polarization $P(\omega)$

$$P(\omega) = -enx(\omega) = \varepsilon_0(\varepsilon(\omega) - 1)\mathscr{E}(\omega) \qquad (7.107)$$

one obtains the complex dielectric function $\varepsilon(\omega)$ of the electron gas near the surface as

$$\varepsilon(\omega) = 1 - \frac{\omega_p^2}{\omega^2 + i\omega/\tau + i\omega\dfrac{\tilde{\eta}(\omega_0^2 - \omega^2)}{\omega_0^2 - \omega^2 - i\eta\omega}} \qquad (7.108)$$

in which $\omega_p$ is the plasmon frequency

$$\omega_p^2 = \frac{e^2 n}{m\varepsilon_0}. \qquad (7.109)$$

In the absence of friction ($\eta = 0$), equation (7.108) reduces to the dielectric function of a free electron gas. From standard optics, we take the reflectivity of p-polarized light,

$$R(\omega) = \left|\frac{\varepsilon(\omega)\cos\theta - \sqrt{\varepsilon(\omega) - \sin^2\theta}}{\varepsilon(\omega)\cos\theta + \sqrt{\varepsilon(\omega) - \sin^2\theta}}\right|^2. \qquad (7.110)$$

The dashed line in Fig. 7.31 shows a numerical solution of (7.110) with $\omega_0 = 285\ \text{cm}^{-1}$, $\eta = 4\ \text{cm}^{-1}$, $\tilde{\eta} = 100\ \text{cm}^{-1}$. The plasma frequency $\omega_p$ and the relaxation time $\tau$ are calculated from the electron density of copper and the conductivity at 100 K, respectively. The agreement with the experiment is quite pleasing, although one should mention that continuous part of the reduction of the reflectivity, in particular its dependence on the adsorbate concentration, is not properly represented in local optics. In summary, one can state that IR-reflection absorption spectroscopy is also sensitive to modes which carry a dipole moment parallel to the surface.

### 7.5.3 Special Optical Techniques

This section briefly addresses two further techniques of surface vibration spectroscopy that are used in certain niches, *Surface Enhanced Raman Spectroscopy* (SERS) and *Sum Frequency Generation* (SFG) or *Vibration Sum Frequency Spectroscopy* (VSFS). For details the reader is referred to special monographs [7.72-75].

#### Surface enhanced Raman effect

The surface enhanced Raman Effect was discovered in 1974 by Fleischmann et al. on electrolytic roughened silver electrodes covered with pyridine. From their experimental results, it could be inferred that the cross section for Raman scattering was enhanced by factor of the order of $10^{10}$-$10^{12}$ over the cross section of molecules dissolved in a liquid. The huge enhancement factor created a storm of activity and many a controversy as to the possible origin of the effect. Since the very large enhancements were observed only on rough, preferably silver surfaces, it was proposed that the enhancement should be due to the enhancement of the electromagnetic field on rough surfaces. An elementary access to the problem is obtained by considering the enhancement of the field near the surface of a homogenously polarized sphere in an external electric field $\mathscr{E}_0$. The tangential and radial components of the electric field at any point at a distance $R$ from the center of the sphere of radius $r$ is

$$\mathscr{E}_r = \mathscr{E}_0\left(1+\frac{\varepsilon-1}{\varepsilon+2}\frac{2R^3}{r^3}\right)\cos\theta, \quad \mathscr{E}_\theta = \mathscr{E}_0\left(1-\frac{\varepsilon-1}{\varepsilon+2}\frac{2R^3}{r^3}\right)\sin\theta. \quad (7.111)$$

The real part of the denominator vanishes at the frequency where $\varepsilon_1(\omega) = -2$, which is the plasma resonance of a sphere with homogeneous polarization of the sphere. This plasma resonance occurs at wavelength $\lambda = 355$ nm and $\lambda = 490$ nm for silver and gold respectively. Its excitation requires that the diameter of the sphere is small compared to the wavelength $\lambda$. The radial field $\mathscr{E}_r$ is particular large. At resonance and at the point $R = r$, $\theta = 0$, the modulus of the field is

$$|\mathscr{E}_r| = 6\mathscr{E}_0/\varepsilon_2(\omega_p) \quad (7.112)$$

The intensity of the primary light beam at the source of the Raman scattering is proportional to the square of the field. Since Raman scattering originates from the light emitted from an induced dipole moment, the radiation from the fluctuating dipole is also proportional to the square of the field. In total, the intensity of the Raman signal is proportional to the fourth power of the electric field. Silver has a particular small damping at the plasmon resonance [7.76]. From (7.112) one calculates a Raman enhancement factor of about $10^6$.

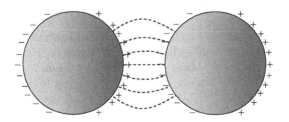

**Fig. 7.32.** The electric field between two homogeneously polarized spheres (at plasmon resonance) is larger than on the surface of a single sphere when the polarization is along the axis connecting the centers of the spheres. For 10 nm Ag spheres and a gap of 1 nm the field is enhanced by about a factor of 30 compared to the field on the surface of a single sphere and about a factor of nearly 1000 compared to the field in vacuum [7.77].

Even stronger enhancements are calculated in the gap between two spheres if the applied field is along the axis connecting the centers of the spheres [7.77]. The effect can be qualitatively understood by considering the surface charges on homogeneously polarized spheres (Fig .7.32). The field enhancement at midpoint between the two Ag-spheres of 10 nm radius can be almost a factor of 1000, when the gap between the spheres is 1 nm and the frequency of the light matches the plasmon resonance of the spheres. This renders a Raman enhancement factor of almost $10^{12}$! With a judicious choice of molecules one may profit furthermore from resonant electronic excitations of the molecule or charge transfer resonances with the silver surface, so that enhancement factors of even $10^{14}$ can be achieved, permitting Raman spectroscopy on a single molecule. The molecule sending the signal is not individually addressed however, contrary to inelastic tunneling spectroscopy.

The variation of the cross section by many orders of magnitude and the low cross section on perfectly flat surfaces makes the Raman Effect less suitable for studies of adsorbate phases on single crystal surfaces. After much initial excitement, the field of SERS has therefore become nearly dormant until lately when it revived in the emerging field of Nanoscience. The ability to make nanowires and nanospheres in a controlled way has offered novel applications for Raman spectroscopy [7.78]. In connection with certain standardized nanostructured collectors, Raman spectroscopy is one of the most sensitive quantitative methods for trace analysis!

## Vibration sum frequency spectroscopy

Nonlinear optics is concerned with the susceptibility of matter in high electric fields at which the susceptibility becomes a function of the electric field. Within the framework of local optics we assume that the polarization $P(r,t)$ at the position $r$ and at time $t$ depends only on the field $\mathscr{E}(r,t)$ at the same position and time. The Taylor-expansion of the polarization in terms of the field is then

$$P_\alpha(r,t) = \varepsilon_0 \left\{ \sum_\beta \chi_{\alpha\beta}^{(1)} \mathscr{E}_\beta(r,t) + \sum_{\beta\gamma} \chi_{\alpha\beta\gamma}^{(2)} \mathscr{E}_\beta(r,t)\mathscr{E}_\gamma(r,t) + ... \right\} \qquad (7.113)$$

where $\alpha$, $\beta$ and $\gamma$ denote the components in Cartesian coordinates, $\chi_{\alpha\beta}^{(1)}$ is the conventional susceptibility tensor, and $\chi_{\alpha\beta\gamma}^{(2)}$ is the second-order susceptibility tensor. By introducing the Fourier-transformed components $\mathscr{E}(\omega)$ and $P(\omega)$ it is easily seen that the second order susceptibility creates polarizations at the sum and difference of the frequencies of the electric fields. For practical applications in spectroscopy, only the sum-frequency is of interest as the signal of the difference frequency is buried in the fluorescence spectrum.

$$P_\alpha(\omega = \omega_1 + \omega_2) = \varepsilon_0 \left\{ \sum_{\beta\gamma} \chi_{\alpha\beta\gamma}^{(2)}(\omega = \omega_1 + \omega_2) \mathscr{E}_\beta(\omega_1)\mathscr{E}_\gamma(\omega_2) \right\} \quad (7.114)$$

The intensity of the beam emerging at $\omega = \omega_1 + \omega_2$ is proportional to the squared modulus of the polarization $|P(\omega)|^2$ and hence proportional to the product of the intensity of both beams at $\omega_1$ and $\omega_2$. The surface sensitivity stems from the fact that $\chi_{\alpha\beta\gamma}^{(2)}$ is a third rank tensor that vanishes identically in the bulk of centrosymmetric media. Only the signal from a surface or an interface between two such media survives, as the symmetry is broken there. This makes vibrational sum frequency spectroscopy (VSFS) a potent tool for studies of solid/liquid and solid/solid interfaces. For the solid/vapor and solid/liquid interfaces of the same material, e.g. for the ice/water vapor and the ice/liquid water interface, VSFS is the only available method of vibration spectroscopy. In conventional UHV-surface physics, the enormous experimental effort in VSFS does not pay off unless one uses the ability of VSFS to perform time resolved experiments with picosecond and even femtosecond time resolution, e.g. in the form of pump-probe experiments. Such experiments have opened new opportunities to study state selected chemistry at surfaces.

In a typical VSFS-experiment, one of the two frequencies is in the range of visible light while the other one lies in the infrared regime of the surface vibration modes of interested. The susceptibility, and therefore the polarization and the amplitude of the sum frequency signal passes through a resonance if either one frequency matches a resonance frequency of the system. If one is interested in resonances in the infrared regime, the frequency of the infrared beam must be varied to match the frequencies of vibration modes.

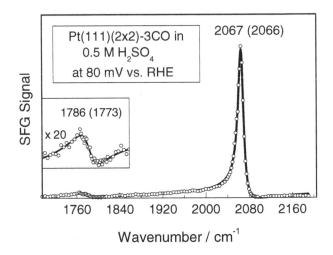

**Fig. 7.33.** Sum frequency spectrum at the Pt(111)/electrolyte interface. The spectrum shows two features due to carbon monoxide adsorbed on the surface in an atop site (2067 cm$^{-1}$) and in the threefold hollow site (1786 cm$^{-1}$). The wave numbers quoted in brackets are taken from an infrared study [7.79]. The solid curve represents a fit with a frequency independent phase shift between the resonant and the non-resonant contribution to the total susceptibility (courtesy of W. Daum, see also [7.80]).

The susceptibility $\chi^{(2)}_{\alpha\beta\gamma}$ can be split into a non-resonant contribution $\chi^{(2,\mathrm{NR})}_{\alpha\beta\gamma}$ and a resonant contribution

$$\chi^{(2)}_{\alpha\beta\gamma} = \chi^{(2,\mathrm{NR})}_{\alpha\beta\gamma} + \sum_i \frac{A^{(i)}_{\alpha\beta\gamma}}{\omega_2 - \omega_i + i\Gamma_i} \qquad (7.115)$$

where $\omega_i$ are the frequencies of the vibration modes. Both $\chi^{(2,\mathrm{NR})}_{\alpha\beta\gamma}$ and $A^{(i)}_{\alpha\beta\gamma}$ are complex quantities. The intensity of the scattered light is proportional to squared modulus of $\chi^{(2)}_{\alpha\beta\gamma}$. Because of a possible phase shift between $\chi^{(2,\mathrm{NR})}_{\alpha\beta\gamma}$ and $A^{(i)}_{\alpha\beta\gamma}$ the shape of the resulting resonance differs from a Lorenzian line. Weak resonant signals merely modulate the background with a Fano-type line shape. We illustrate the line shape with the example of CO on the Pt(111) surface in 0.5 M H$_2$SO$_4$ at 80 mV vs. RHE in Fig. 7.33 [7.80]. At this potential CO that is dissolved in the aqueous electrolyte adsorbs on Pt(111) in a (2×2) structure [7.79]. The (2×2) unit cell contains three CO-molecules, one in the atop-position and two in the three-fold site. The line shape depends on the various parameters in the susceptibility (7.115) but also on the frequency dependence of Fresnel-coefficients. As these parameters are not known a-priori, the precise values of the resonance frequency cannot be determined from the VSFS experiment in particular when the signal is as weak as for the three-fold site. The frequencies noted in Fig. 7.33 are taken from a fit to the experimental data. The numbers in brackets are the resonance frequencies in infrared absorption on the same system [7.79].

## 7.6 Tunneling Spectroscopy

Of all vibration spectroscopies, tunneling spectroscopy on a single molecule individually addressed by the STM-tip undoubted has the largest intellectual appeal. Tunneling spectroscopy of molecular vibrations as-such existed before the invention of the scanning tunneling microscope (see e.g. [7.81]). The spectroscopy was performed in extended tunnel junctions between two metals that were interfaced by an insulating layer and a thin layer of the molecular species of interest. In most cases, the tunnel junctions were made from oxidized aluminum films with a spray-coated layer of molecules covered by a lead top electrode. Although the principle feasibility of single molecule vibration spectroscopy with an STM-tip could be inferred from these experiments the considerable technical difficulties of single molecule vibration spectroscopy could be overcome only 17 years after the invention of the STM [7.82, 83].

We consider the basics of the tunneling process with the help of the scheme in Fig. 7.34a. The number of channels for elastic tunneling from one metal to another one increases with the voltage $V$ between the metals. The tunneling current is therefore roughly proportional to the applied voltage $V$. If $eV$ exceeds the quantum energy $\hbar\omega$ of a vibration mode, additional channels for inelastic tunneling open up and the current rises more steeply. This is schematically illustrated in Fig. 7.35a. The probability of tunneling between two states is proportional to the occupation of the initial state and proportional to the probability that the final state is not occupied. Assuming that the matrix element is constant, the inelastic current is

$$I_{\text{inel}} \propto \int d\varepsilon_i d\varepsilon_f\, f(\varepsilon_i, T)(1 - f(\varepsilon_f + eV, T))\delta(\varepsilon_i - \varepsilon_f - \hbar\omega). \quad (7.116)$$

Here $f(\varepsilon, T)$ is the Fermi-function. Note, that $V < 0$ for the case shown in Fig. 7.34a. In a real experiment, the onset of inelastic tunneling cannot be discerned from the energy dependence of the elastic tunneling current caused by the energy dependence of the matrix element for tunneling and the energy dependence of the density of initial and final states. One therefore identifies the threshold for inelastic tunneling at $\hbar\omega/e$ in the second derivative of the current with respect to

**(a)**                                    **(b)**

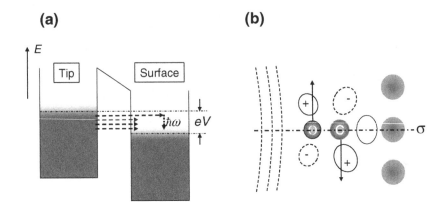

**Fig. 7.34. (a)** Illustration of elastic and inelastic tunneling processes from occupied states of the tip to the unoccupied states of the substrate (negative voltage $V$ on tip). The inelastic process opens additional channels for tunneling, causing a rise in the tunneling current for a voltage beyond a threshold voltage $V_{\text{thr}} \geq \hbar\omega/e$, modulo the Fermi-distribution. **(b)** The matrix element between the symmetric tip-state, the vibrational motion and the electronic state of the molecule must be totally symmetric. If the relevant electronic state is antisymmetric with respect to a symmetry plane $\sigma$ then the vibration state must also be antisymmetric for a non-vanishing matrix element.

the voltage where the threshold shows as a peak (Fig. 7.35c). Experimentally the second derivative is obtained as the second harmonic current-response to a sinusoidal modulation of the voltage. The width of the peak is determined by the width of the Fermi-distribution. The full width at half maximum is $\Delta E_{1/2} = 3.5 k_B T$. Sufficient resolution for vibration spectroscopy requires cooling to liquid-He temperatures ($\Delta E_{1/2}(4 \text{ K}) \cong 1.2 \text{meV}$). The practical achievable resolution is less since the temperature of the microscope is typically in the range of 10 K, higher than the temperature of the liquid-He cooled shroud surrounding the microscope. Once the microscope is cooled down to 4 K by contact with the liquid-He bath, the thermal contact is broken. The microscope is then suspended on soft springs inside the surrounding shields. The required electric wiring to the microscope and the heat dissipated in the operation of the microscope causes the temperature to rise above the bath temperature of 4 K.

Figure 7.36 shows the first published single-molecule tunneling spectra of CO-vibrations [7.84]. Different CO-isotopes are employed to distinguish vibration modes from structure in the background. Two CO-modes are discerned, a strong signal from the hindered rotation (around 36 meV) and a weak signal from the CO-stretching mode (around 256 meV). The fact that the signal from the stretching mode is so weak came as a surprise at the time. Before the experiments of

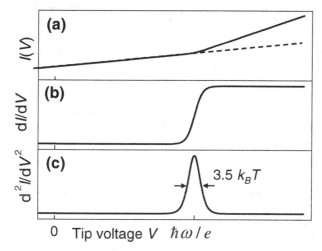

**Fig. 7.35.** The principle of tunneling spectroscopy illustrated with **(a)** the tunneling current $I(V)$, **(b)** its first derivative and **(c)** its second derivative. The dashed line in the upper panel shows schematically the tunneling current as a function of the tip voltage with reference to the Fermi-level of the sample. If the tip voltage exceeds $\hbar\omega/e$ with $\hbar\omega$ the eigenfrequency of a vibrational mode, a channel for additional inelastic tunneling opens and the current increases more steeply than for pure elastic tunneling. Center and lower panel show the derivative and the second derivative of the tunneling current, respectively. The intrinsic half width of the peak in the second derivative is $\Delta E_{1/2} = 3.5 k_B T$.

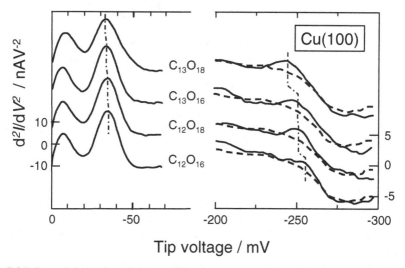

**Fig. 7.36.** Second derivative of the tunneling current vs. the tip voltage for CO on Cu(100) at 8 K with the tip place directly over the CO-molecule (after Lauhon and Ho [7.84]). The second derivative spectra were obtained by modulating the voltage with 7 mV ac-voltage

and by detection of the second harmonic of the tunneling current. Sampling time was 8h. The dashed lines are reference spectra on the Cu-surface.

Lauhon and Ho, inelastic tunneling from adsorbate modes was believed to be mediated via the interaction with the dipole field, much the same way as in inelastic electron scattering [7.81]. If that were the case, the stretching mode should yield the strongest signal and the hindered rotation should be silent as this mode is dipole forbidden. The prejudice that the CO-stretching mode should yield a strong signal may in fact have prevented an earlier success in inelastic tunneling spectroscopy of single molecule vibrations. The experiments of Lauhon and Ho demonstrated experimentally that the dipole field of a vibration mode is irrelevant, a fact that is corroborated by current theory of the inelastic tunneling process [7.85].

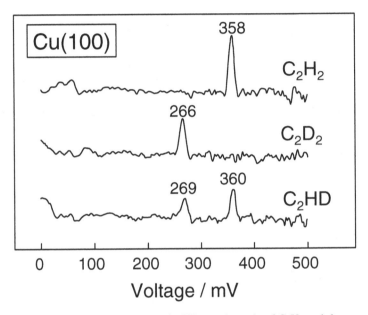

**Fig. 7.37.** Tunnel spectra of the asymmetric CH-stretch mode of $C_2H_2$ and the corresponding partially and totally deuterated molecule at 8 K. Total scan time was about 30 min for each spectrum (after Stipe et al. [7.83]).

The tunneling process involves the overlap of the extended s-states originating in the apex of the tip [7.86] and the orbitals of the molecule that contribute to the density of states near the Fermi-level. For carbon monoxide, these are the partially occupied $2\pi^*$-orbitals (Fig. 7.34b). The matrix element for inelastic tunneling is constructed from the initial and final electron states and the deformation potential induced by the normal mode of the molecule. When the tip is centered on the molecule, the s-states of the tip are symmetric with respect to the $\sigma$-plane of the molecule in $C_{2v}$-symmetry. The final state, the $2\pi^*$-orbital, is antisymmetric ($B_1$ or $B_2$-representation, depending on the definition of the $\sigma$-plane). As the matrix element (because it is an observable entity) must belong to the totally symmetric representation, the normal coordinate of the vibration mode must belong to same B-representation as the final state. Hence, the selection rule is just opposite to the selection rule for inelastic electron scattering. The small interaction with the CO-stretching mode arises from the less effective tunneling into symmetric states.

The peculiar selection rule was first formulated by Lorente et al. [7.85, 87]. The selection rule depends on the symmetry of the molecular electronic state that is involved in the tunneling process. If that state were symmetric, then the symmetric vibration should appear with the highest intensity. However, this would be an untypical situation, at least for small molecules. A molecule resides in a high symmetry site because the bonding to the substrate atoms involves symmetric molecular orbitals. The energy levels of the electrons engaged in that bond are necessarily far below the Fermi-level. These electrons do not participate in the tunneling process.

Tunneling spectroscopy is not confined to CO-molecules. The CH-stretching vibrations of hydrocarbons yield even stronger signals because of the larger amplitude of hydrogen atoms. Fig. 7.37 shows a the vibration spectra of single molecules of acetylene on Cu(100) and its deuterated counterparts. The deuterium modes appear with smaller intensity since the mean square amplitude is smaller by $2^{-1/2}$. According to the selection rule, the observed frequency should correspond to the asymmetric CH-stretch mode of acetylene. Theoretical calculations have shown that the intensity of the symmetric stretch should be roughly an order of magnitude lower [7.87].

Because of the considerable experimental effort, the required low temperatures, the moderate resolution and the long sampling time, tunneling spectroscopy is not going to replace the conventional methods of vibration spectroscopy. However, as a tool to identify individual molecules and to monitor (tip induced) chemical reaction between molecules it serves a useful purpose.

# 8. Electronic Properties

## 8.1 Surface Plasmons

### 8.1.1 Surface Plasmons in the Continuum Limit

Surface plasmons in the continuum limit are dielectric eigenmodes of a half-space of a free electron gas, which exists at the frequency where the real part of the dielectric function equals minus one.

$$\varepsilon_1(\omega) = -1 \qquad (8.1)$$

As discussed in Sect. 7.1.7, the electrostatic potential describing the plasmon is a solution of the Laplace equation and has the form

$$\varphi(x,z,t) = \varphi_0 e^{-q|z|} \sin(qx - \omega t) \qquad (8.2)$$

where $x$ is the coordinate parallel to the surface and $z$ perpendicular to it. The $x$ and $z$-components of the electric field vector are

$$\mathscr{E}_x = q\varphi_0 e^{-q|z|} \cos(qx - \omega t)$$
$$\mathscr{E}_z = -q \operatorname{sgn}(z)\varphi_0 e^{-q|z|} \sin(qx - \omega t) \ . \qquad (8.3)$$

Inside the solid, the electric field is accompanied by a polarization wave of the same form with $\boldsymbol{P}(x,z,t) = -\varepsilon_0 \mathscr{E}(x,z,t)$. The divergence of the normal component of the polarization at the surface is equivalent to a surface charge density wave with a charge density $\rho_P$ that is a delta-function in $z$.

$$\rho_P \propto \delta(z)\sin(qx - \omega t) \qquad (8.4)$$

The frequency of the surface plasmon is obtained by inserting the dielectric function of the free electron gas into (8.1),

$$\varepsilon(\omega) = 1 - \frac{\omega_p^2}{\omega^2 + i\varepsilon_0 \omega \omega_p^2 / \sigma(\omega)} . \tag{8.5}$$

Here, $\sigma(\omega)$ is the conductivity and $\omega_p$ is the plasma frequency which is related to the electron density $n$ via

$$\omega_p^2 = ne^2 / m\varepsilon_0 \tag{8.6}$$

with $m$ the electron mass. In the absence of damping ($\sigma(\omega) = \infty$), $\omega_p$ is the frequency of the bulk plasmon. This bulk plasmon is a longitudinal wave that exists at $\varepsilon(\omega) = 0$. For the free electron gas the frequency of the surface plasmon is

$$\omega_s = \omega_p / \sqrt{2} . \tag{8.7}$$

Table 8.1 shows that this relation is well fulfilled for free-electron like metals. A notable exception is silver. For silver, the plasmon frequency is downshifted by a d-band excitation, which makes the slope of $\varepsilon(\omega)$ very steep, so that bulk and surface plasmon come close together.

**Table 8.1.** Frequencies of bulk and surface plasmons for metals that have well-defined plasmon excitations.

| Metal | $\hbar\omega_p$/eV | $\hbar\omega_s$/eV | Ratio | Metal | $\hbar\omega_p$/eV | $\hbar\omega_s$/eV | Ratio |
|-------|------|------|-------|-------|------|------|-------|
| Li | 7.12[8.1] | 4.28[8.2] | 0.60 | Cs | 2.9[8.3] | 1.99[8.4] | 0.69 |
| Na | 5.72[8.3] | 3.99[8.4] | 0.70 | Mg | 10.4[8.5] | 7.38[8.2] | 0.71 |
| K | 3.72[8.3] | 2.73[8.4] | 0.73 | Al | 15.1[8.1] | 10.3[8.4] | 0.68 |
| Rb | 3.41[8.3] | 2.46[8.3] | 0.72 | Ag | 3.74[8.6] | 3.68[8.7] | 0.98 |

As discussed for phonons in Sect. 7.1.7 with Fig. 7.13, surface plasmons do not interact with light on a flat surface since they exist only to the right of the light line. Because of the electric field (8.3), they do interact strongly with electrons reflected or diffracted from the surface, giving rise to intense energy losses. The shape of the loss function is given by $-\text{Im}(\varepsilon(\omega) + 1)^{-1}$ (7.65). The width of the surface plasmon peak is given by the imaginary part of the dielectric function $\varepsilon_2(\omega)$. For free electron metals, $\varepsilon_2$ is small in the relevant frequency range. The plasmon loss is therefore a sharp feature that dominates the loss spectrum. Energy losses of electrons reflected from a surface due surface plasmon excitations were first studied by Powell and Swan on aluminum in 1959 [8.8]. The authors could prove their surface character by the sensitivity of the observed frequency to contamination. The frequency shifts because on a contaminated surface the number "1" in (8.1) is to be replaced by an effective dielectric constant of the contaminant. Since the dielectric constant is larger than one, the contamination shifts the surface plasmon frequency downwards.

For insulators, semiconductors and transition metals the loss function displays a broad spectral distribution, but the integrated intensity is about the same as for free electron metals because of a sum-rule, which is related to the venerable *f*-sum rule in optics. Before high intensity UV-light sources became available with the synchrotrons, inelastic scattering of electrons was used to determine the loss function and thereby the optical constants in the UV-range via a Kramers-Kronig analysis [8.9]. Plasmon losses and the corresponding features on non free-electron like materials are also observed in the spectra of electrons that emerge from the solid due to a photoemission process or an Auger-process.

### 8.1.2 Surface Plasmon Dispersion and Multipole Excitations

Inelastic scattering of electrons by surface plasmons is dominated by the contributions of small wave vectors $q_\parallel$ because of the surfing condition (7.82). As long as one focuses on the intense loss spectrum observed near specular reflection and on high electron energies, the dispersion of surface plasmons plays no role. On the other hand, surface plasmons at $q_\parallel \cong 0$ do not carry information on surface properties. For example, the surface plasmon frequency at $q_\parallel \cong 0$ is independent of the crystallographic structure of the surface. For larger wave vectors $q_\parallel$, the surface plasmon is more localized to the surface and its frequency becomes sensitive to the charge density distribution at the surface, in other words, the surface plasmon shows dispersion. The dispersion of surface plasmons for the alkali metals K, Na and Cs was investigated by Tsuei et al. [8.4, 10]. The result for Na is shown in Fig. 8.1. The initial reduction of the frequency can be understood qualitatively as follows: At the frequency of the surface plasmon, the centroids of the induced charge density $\delta n(z,\omega)$ and the induced potential $\varphi(q_\parallel,z)$ lie outside the jellium edge in the region where the electron density is low (Fig. 8.2). The induced potential spreads over a smaller $z$-range for larger $q_\parallel$. The potential sees a lower mean electron density with increasing wave vector $q_\parallel$ and the plasmon frequency therefore becomes smaller with increasing $q_\parallel$. Within the *Random Phase Approximation* (RPA), the frequency shift can be expressed quantitatively in terms of the position of the centroid of the induced charge density $d(\omega)$ [8.10]

$$\omega_s(q_\parallel) = \omega_s(0)(1 - q_\parallel d(\omega_s)/2...) \tag{8.8}$$

where $d(\omega)$ is defined as

$$d(\omega) = \int dz\, z\, \delta n(z,\omega) / \int dz\, \delta n(z,\omega) . \tag{8.9}$$

The polynomial fit to the dispersion curve in Fig. 8.1 yields $d(\omega_s) = 0.77\,\text{Å}$. The negative slope of the dispersion curve can also be derived in continuum theory if a smooth rather than an abrupt change of the electron density at the surface is invoked [8.11].

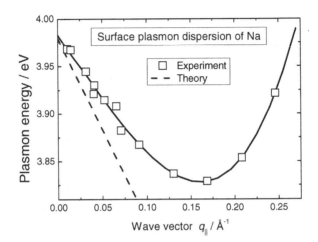

**Fig. 8.1.** Dispersion of the surface plasmon of sodium. The surface was prepared by evaporation on an Al(111) surface. The dashed line is the dispersion according to a first order perturbation theory. The solid line is a fit to a fourth order polynomial (After Tsuei et al. [8.10]).

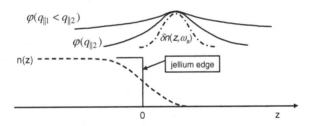

**Fig. 8.2.** At the frequency of the surface plasmon, the centroid of the induced charge density $\delta n(z,\omega)$ lies outside the jellium edge. For larger wave vectors, the potential sees a smaller electron density. The frequency of the surface plasmon becomes smaller with increasing $q_{\parallel}$ (After Tsuei et al. [8.10]).

The centroid of the induced surface charge shifts towards the interior of the jellium edge when the frequency approaches the bulk plasmon frequency, so that $d < 0$. For the special case of silver, the surface plasmon frequency is close to the bulk plasmon frequency. The initial dispersion is therefore positive for silver [8.7].

Another experimental manifestation of the dynamic surface response of the free electron gas is the *multipole surface plasmon*. The name multipole plasmon refers to the fact that the induced surface charge density has a multipole character rather than the monopole character of the normal surface plasmon. The multipole plasmon causes a strong feature in the off-specular energy loss spectrum of electrons scattered from the surface of alkali metals [8.12] (Fig. 8.3). For further details on the dynamic response of free electron metals the reader is referred to the book of A. Liebsch [8.13].

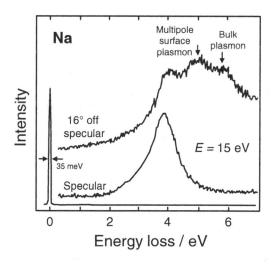

**Fig. 8.3.** Electron energy loss spectrum of 15 eV electrons from a polycrystalline sodium films (after Tsuei et al [8.12]). The energy loss in specular reflection is the surface plasmon excitation. At 16° off-specular the multipole surface plasmon and bulk plasmon contribute to the spectrum with about the same intensity. Note that specular reflection at $\theta_S \cong 60°$ corresponds to a larger $q_{\|}$-value than the off specular spectrum due to the magnitude of the energy loss in comparison to the impact energy (cf. 7.37).

## 8.2 Electron States at Surfaces

### 8.2.1 General Issues

#### *A historical remark*

Electronic surface states were first postulated by Tamm based on a crude one-dimensional model [8.14]. The solid was represented as a one-dimensional potential well with a periodic potential in the form of positive delta-functions. The general solutions for that potential are plane waves. Shockley considered a more realistic, yet still one-dimensional potential and Bloch-wave solutions [8.15]. Shockley showed that the surface states found by Tamm were due to an incompleteness of the potential at the surface, i.e. due to a surface potential that was altered compared to the bulk. In the later literature, the distinction between surface states caused by a changed potential vs. surface states caused by the finiteness of the lattice was frequently made by referring to "Tamm-states" and "Shockley-states", respectively. Some authors also refer to Tamm-states and Shockley-states in the sense that the former should be localized, tight-binding type surface states and the latter free-electron type surface states, a distinction that has little foundation in the original work of Tamm and Shockley.

Early experimental evidence for the existence of surface states came from the electric properties of space charge layers at semiconductor surfaces. As discussed in Sect. 3.2.2, a high density of surface states pins the Fermi-level at the surface to a fixed position with respect to the valence and conduction band edges. Depending on the doping, the pinning gives rise to depletion, accumulation or inversion layers. The measurement of the surface conductivity and the changes in the surface conductivity with an externally applied electric field normal to the surface are therefore traditional methods to probe for the existence of surface states and for the dependence of their density of states and energy level and on surface structure and deposited adlayers. Surface conductivity probes only for surface states inside the valence band gap of semiconductors. The investigation of surface states below the valence band edge and above the conduction band edge as well as of surface states in metals requires spectroscopic techniques. Today we possess a rich arsenal of such techniques supplemented by powerful theoretical methods from which a complete picture of surface state bands has emerged. Simultaneously the interest has expanded into related areas such as quantum size effects on surfaces, the effect of the electronic structure on lateral interactions between adatoms and other defects, and the stability of nanostructures on surfaces. In the following, we first consider the general theoretical concepts of surface states on 2D-periodic surfaces. We proceed with experimental methods to probe for such surface states and discuss examples of typical surfaces. Issues related to a lateral confinement of surface states are discussed in Sect. 8.3.

### Surface states and bulk states

The electronic eigenstates of a three-dimensional solid are Bloch-waves classified by an index for the electron band $i$ and by the three-dimensional (3D) $k$-vector.

$$\left| i, k \right\rangle = u_{i,k}(r) e^{-ik \cdot r} \tag{8.10}$$

The function $u_{i,k}(r)$ has lattice periodicity. For the core levels, the lattice periodic functions $u_{i,k}(r)$ are completely localized atomic orbitals and the energy levels become independent of $k$. In the valence band regime that is of prime interest here, the energy levels form continuous bands. In order to count the number of possible wave-vectors one subdivides the infinite solid into subsections of $N$ unit cells, where $N$ is a large number. Complete 3D-translational symmetry is preserved by requiring that the all properties of the solid repeat after a supercell which contains $N$ unit cells. These *periodic boundary conditions* require that

$$\left| i,k \right\rangle = u_{i,k}(r)\mathrm{e}^{-ik\cdot r} = u_{i,k}(r)\mathrm{e}^{-ik\cdot(r+r_N)}. \tag{8.11}$$

The phase factor in the exponent must be a multiple of $2\pi$. Without loss of generality we can describe the $k$-vector in cartesian coordinates (rather that in crystal base vector coordinates), hence one has

$$k_x, k_y, k_z = \frac{2\pi}{L_{x,y,z}} n_{x,y,z} \tag{8.12}$$

where $L_{x,y,z}$ are the dimensions of the supercell and $n_{x,y,z}$ are integer numbers. The maximum and minimum values are such that $k$ remains within the first Brillouin zone. Because of the large extension of the solid, the $k$-vectors form a quasi-continuum.

If the solid has a form of a slab with surface perpendicular to the $z$-axis, the solutions become standing waves with respect to $k_z$. As long as the slab is thick so that it contains many atom layers along the $z$-axis, the modified boundary condition with respect to $k_z$ has no effect on the eigenstates. For very thin layers the eigenstates with different $k_z$-vectors have distinctively different energies. This can lead to interesting *quantum size effects* in the electronic structure and physical properties of slabs to which we devote a separate section.

The introduction of surfaces also gives rise to new eigenstates for which the wave function is localized at the surface, the so-called *surface states*. There are two possible categories of surface states: Those that are localized entirely in the first layer and those whose localization is owed to an imaginary part of $k_z$. The magnitude of the imaginary part depends on the relative position of the energy with respect to the bulk states. Typical completely localized surface states are those that are associated with the chemical bonds of adsorbates or those with the dangling bonds on covalently bonded crystals (cf. Sect. 1.2.3, Fig. 1.24; Sect. 3.2.2). These states can exist inside the energy range of bulk states. Strictly speaking, theses states are surface resonances, as the wave function does have a continuation into extended bulk states. However, the energy of these states would be just at the same value if the slab were only a few monolayers thick. The extended surface states with a complex $k_z$ on the other hand, exist only in the forbidden gaps of the bulk states (of the same irreducible representation). The *intrinsic surface states* on metal surfaces are typical examples. Surface states of either type are completely characterized by a 2D-wave vector $k_\parallel$ and a band index

$$\left| i,k_\parallel \right\rangle = u_{i,k_\parallel}(r)\mathrm{e}^{-ik_\parallel\cdot r} \tag{8.13}$$

and the energy spectrum is described by $E(k_\parallel)$. As for bulk states, core levels can, but need not be characterized by a $k_\parallel$-vector, as their energy is independent of $k_\parallel$.

Before we present tutorial examples of surface states that exist on metals and semiconductors, we need to educate ourselves about the experimental methods to probe for electronic surface states. Unlike to other surface properties, experimental studies on electronic surface and bulk states are dominated by a single technique, which is *photoemission electron spectroscopy*. Our present experimental knowledge of the band structure of bulk solids and their surfaces is almost completely based on that technique.

## 8.2.2 Probing Occupied States – Photoelectron Spectroscopy

### Basic elements

Photoelectron spectroscopy is a technique that probes bulk electron states as well as surface states. The relative weight of both depends on the kinetic energy $E_{kin}$ of the photoemitted electron. In Sect. 2.2.2, we have considered the mean free path of electrons $\lambda$, which is also the information depth of electrons with a characteristic energy (Fig. 2.16). The information depth has its minimum of about $\lambda \cong 5$ Å for electron energies around 50 eV. Even at this minimum, the photoemitted electrons carry information of several atom layers and therefore on bulk as well as on surface states. The relative weight of bulk states in the spectrum of photoemitted electrons increases for energies below and above 50 eV.

The photocurrent is proportional to the photon absorption within the information depth and proportional to the coupling of the wave function of the excited electron to the wave functions of electrons emerging from the solid. A full theory of the photoemission process requires the consideration of the non-local optics at surfaces. However, simplifications can be made in the photon energy range that is of interest in practical photoemission spectroscopy. Surface and bulk states in the valence band regime are probed with ultraviolet/soft X-ray light for which the quantum energy is between 20 eV and 2 keV. The wavelength of that light ranges between 620 and 124 Å. All atoms within the information depth are therefore subjected to electromagnetic radiation with nearly the same phase. An alternative way of expressing this fact is that the light contributes nearly no momentum and the transitions between the initial and final electron states are vertical. The photon energy is furthermore outside the regime of collective excitations of the electron gas and outside the regime of extremely strong absorption. The modifications of the electromagnetic field at the surface may then be disregarded. In that case, the absorption of electromagnetic radiation and therefore the intensity of the photo-emission current is proportional to the square of the matrix element of the momentum vector operator $p$ with the initial and final state. The vector operator $p$ has the same orientation as the electric field of the UV-light. Here we are interested in the current carried by electrons of a particular kinetic energy $E_{kin}$ into a particular direction given by the wave vector $k_\parallel^{(el)}$.

$$I_{\text{ph}}(k_\parallel^{(\text{el})}, E_{\text{kin}}) \propto \sum_{i,j} \sum_{k_z} \left| \left\langle i, k_\parallel, k_z \mid p \mid j, k_\parallel, k_z \right\rangle \right|^2$$

$$\times \delta(E_{\text{kin}} + E_{\text{vac}} - E^{(i)}(k_\parallel, k_z) - h\nu)\,\delta(k_\parallel - k_\parallel^{(\text{el})}) \tag{8.14}$$

Here, $i$, $j$ denote the initial and final bands, $k_\parallel$, $k_z$ are the components of the wave vector in the initial and the final state, and $E^{(i)}$, $E_{\text{vac}}$ are the energies of the initial state and the vacuum level. The matrix element is an integral over the unit surface cell and over the information depth along the $z$-coordinate. The photoemitted electron therefore carries the information on the energy of the initial state by virtue of the energy conservation

$$E_{\text{kin}} = -E_{\text{vac}} + E^{(i)}(k_\parallel, k_z) + h\nu . \tag{8.15}$$

The parallel component of the $k$-vector is also conserved, as the electron wave function inside has to phase-match to the wave function outside for periodic surfaces (Fig. 8.4). The $k_\parallel$-vector of surface states is therefore fully determined by the $k_\parallel$-vector of the photoemitted electron. The perpendicular component of the $k$-vector of bulk states remains unknown, but can be established by special techniques (see e.g. Sect. 4.2 of [8.16]).

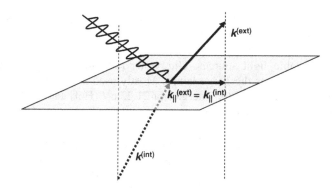

**Fig. 8.4.** Illustration of the wave vector conservation in photoemission spectroscopy. The determination of the perpendicular component of the wave vector of bulk states requires special techniques that may involve more than a single experiment.

It is well known that a free electron cannot absorb a photon since energy and momentum conservation cannot be fulfilled simultaneously. It is therefore essential for the photoemission process that the electron be bound by a potential prior to the photo-excitation process. This can be shown in a formal way by considering the photoemission matrix element in real space. Using the identity

$$[\boldsymbol{p}_i, H] = -i\hbar \frac{\partial}{\partial x_i} V(\boldsymbol{r}) \tag{8.16}$$

with $H$ the unperturbed one-electron Hamilton operator and $V(\boldsymbol{r})$ the one-electron potential one may rewrite the real space photoemission matrix element as

$$\int d\boldsymbol{r}\, \psi_\beta^*(\boldsymbol{r}) p_i \psi_\alpha(\boldsymbol{r}) = -i\hbar (E_\alpha - E_\beta)^{-1} \int d\boldsymbol{r}\, \psi_\beta^*(\boldsymbol{r}) \frac{\partial V(\boldsymbol{r})}{\partial x_i} \psi_\alpha(\boldsymbol{r}). \tag{8.17}$$

Here, the index $i$ denotes a Cartesian component and $\alpha$ and $\beta$ arbitrary electron states. The matrix element (8.17) is proportional to the gradient of the potential. For surfaces one may distinguish two contributions to the matrix element: one from the ion core potentials and a second one from the perpendicular gradient of the potential at the surface. This so-called surface photoemission dominates the photoemission process at low photon energies for p-polarized light. As the process occurs at the surface potential barrier, the excitation probability is completely independent of the electron wave vector perpendicular to the surface. At high photon energies and for electrons excited from deeper energy levels the steep gradient of the ion core potential dominates the photoemission process.

### Light sources

Depending on the photon energy of the light sources, one distinguishes between *Ultraviolet Photoemission Spectroscopy* (UPS) and *X-ray Photoemission Spectroscopy* (XPS). The distinction between these two regimes is according to the traditionally available light sources. Ultraviolet radiation is provided by open, differentially pumped gas discharge lamps. Mostly used are the sharp and intense spectral lines of helium at photon energies of 21.22 eV (HeI) and 40.82 eV (HeII). Less common but occasionally also used are neon and argon at 16.85/16.67 eV (NeI), 26.9 eV (NeII), 11.83 eV (ArI) and 13.3/13.48 eV (ArII). Standard laboratory sources for X-rays are the Mg-$K_{\alpha 1,2}$ and Al-$K_{\alpha 1,2}$ emission lines at 1253.6 eV and 1486.6 eV, respectively. The traditional sources lost some of their importance when synchrotron sources for the entire spectral range between 10 eV and several keV became available. The advantage of synchrotron sources, in particular of undulator beam lines is that they combine high photon fluxes with tunability in a wide spectral range. By tuning the wavelength and the energy selector for the photoemitted electrons simultaneously, one can e.g. optimize the cross section for photoemission from particular electron states. Synchrotron light is polarized, linear-horizontal in the synchrotron plane, left-circular and right-circular below and above the plane of the ring, respectively (for electrons traveling clockwise in the ring, as seen from a position above the ring). The polarization offers additional possibilities. By using selection rules, the symmetry of electron orbitals can be determined with linear polarized light, and with circular polarized light, one can distinguish between spin-up and spin-down states via the effect of magnetic circular dichroism. Figure 8.5 displays the experimental set-up at one of the undulator beam lines (beam line 7) of the *Advanced Light Source* (ALS) at Berkeley, USA [8.17]. This beam line has been used in many surface studies of the last decade. With three different gratings, the monochromator covers the photon energy range between 60 and 1200 eV. The resolving power of the monochromator $E/\Delta E$ is about 8000. The energy resolved photon flux is between $10^{12}$ and $10^{13}$ s$^{-1}$.

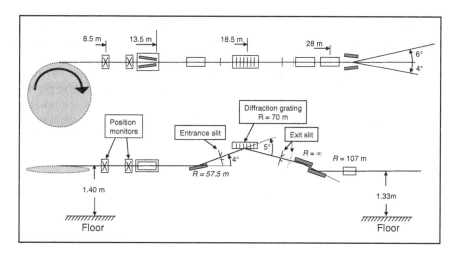

**Fig. 8.5.** The set-up of a the undulator beam line 7 at the *Advanced Light Source* (ALS) at Berkeley, USA that is commissioned to extended UV-light and soft X-rays. The synchrotron is not drawn to scale. (After Warwick et al. [8.17])

As far as UV-light is concerned these figures of merit are not much better than those of the HeI-lamp ($\cong 10^{12}$ photons per second, $\Delta E = 3$ meV, hence $E/\Delta E = 7000$). Hence, for experiments in which the photon energy need not be varied, polarization is not required and where the photon energy of 21.2 eV suffices the HeI discharge lamp is still competitive in performance, and orders of magnitude lower in operational cost. For example, for the investigation of the band structure of electronic surface states in the valence band regime, discharge lamps do well. Most experimental data on the surface state dispersion have been obtained that way.

## Energy analysis

The typical experimental arrangement for photoelectron spectroscopy of the valence band structure consists of a combination of an energy analyzer and a lens system (Fig. 8.6). The sample is mounted on a goniometer that defines the polar and azimuthal angle of electron emission with respect to the surface orientation. Energy analysis is performed mostly with hemispherical electrostatic deflectors (Sect. 7.4.1). When equipped with corrections for the fringe fields at the entrance and exit apertures by a suitable grading of the aperture potential (Fig. 8.6), energy resolutions down to 1 meV can be realized. To keep the resolution constant during energy scans the analyzer operates at constant pass energy while a lens system provides the energy retardation or acceleration. The lens system also defines the angular aperture. The combination of exit slit and electron multiplier shown in Fig. 8.6 is occasionally replaced by a position sensitive detector for the parallel detection of electrons of different energy. Different attempts have been made to combine energy analysis with a simultaneous display of the emission angles on a position sensitive detector. The price for the multiplex gain is usually a lower energy resolution and image distortions in the angular pattern. The most advanced spectrometers of SCIENTA keep both effects well under control.

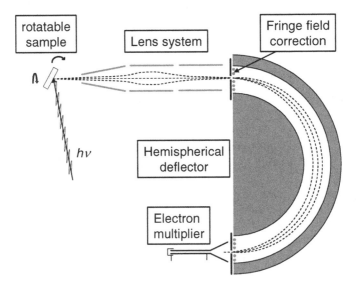

**Fig. 8.6.** Typical experimental set-up for photoemission spectroscopy with a hemispherical analyzer for energy selection and a simple electrostatic lens system. The angle between direction of the incident light and the spectrometer is fixed. The emission angle is varied by rotating the sample with respect to the polar and azimuthal axis.

### Selection rules in UPS

The photoemission process obeys certain selection rules that follow from the matrix element in (8.14). Consider for example a surface with a mirror plane. The electron eigenstates belong to either the odd or the even representation. If one looks for electrons with trajectories lying within the mirror plane, the final state belongs to the even representation. The matrix element is then nonzero for s-polarized light and even electron states and for p-polarized light and odd electron states. Hence, by orienting the sample with respect to the polarization plane of the synchrotron light and by observing electrons emitted in the mirror plane one can immediately determine the symmetry of the initial state. This symmetry analysis does not require synchrotron light. Linear polarized light can also be obtained from gas discharge lamps at the price of a loss in intensity by reflecting the light at the Brewster angle from a pair of gold mirrors.

For atomic orbitals with vector character, the p-states, one can even go a step further and determine the orientation of the orbitals in space from the intensity profile with reference to the polarization of light. Since the momentum vector operator in the matrix elements is oriented as the electric field $\mathscr{E}^{(s)}$ of the UV-light at the surface, the photoemission intensity from a $p_z$-orbital is proportional to $|\mathscr{E}_z^{(s)}|^2$, from a $p_x$-orbital proportional to $|\mathscr{E}_x^{(s)}|^2$ and from a $p_y$-orbital proportional to $|\mathscr{E}_y^{(s)}|^2$. A calculation of the electric field components as a function of the polarization and the angle of incidence requires the complex reflection coefficients for the UV-light. However, the orientation of the p-orbitals is easily determined just from the angular orientation of the polarization with respect to the surface coordinate system for which certain components of the field vanish.

The flux of electrons that are photoemitted from localized orbitals also shows an interesting dependence on the energy of the final state, which follows directly from the matrix element. If, for example, the initial state is an s-orbital the intensity goes through a minimum when the wavelength of the electron in the final state matches approximately the spatial extension of the initial state since positive and negative contributions to the matrix element cancel. Such minima have been discussed first by J. W. Cooper [8.18] and are therefore named *Cooper minima*.

## 8.2.3 Probing Unoccupied States

### Constant initial state photoemission

Several techniques have been developed to probe unoccupied electron states. Information on the states above the vacuum level can be obtained from photoemission experiments that are carried out in a specific form, namely by a simultaneous variation of the photon energy and the kinetic energy window of the analyzer. Thereby the initial state remains constant and the intensity of the photocurrent reflects the transition probability and thus the density of the final states. The principle of this *constant initial state spectroscopy* is illustrated in Fig. 8.7a.

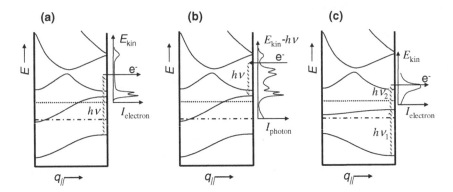

**Fig. 8.7.** Probing unoccupied (surface) states: **(a)** In *constant initial state photoemission* the kinetic energy and the photon energy are varied simultaneously. The photocurrent $I(q_\parallel)$ increases when the density of final states for a particular $q_\parallel$ is high. **(b)** In *isochromat spectroscopy (inverse photoemission)*, the photon yield for a particular $h\nu$ is observed as a function of the kinetic energy of incoming electrons. The yield $Y_{ph}(q_\parallel)$ increases when the density of final states for a particular $q_\parallel$ is high. **(c)** In *two-photon photoemission*, the electron yield is observed as a function of the kinetic energy of the electron while the energies of the two photons are kept constant. The electron yield peaks at a kinetic energy that corresponds to an emission from the pumped, but thermally unoccupied state.

### UV-isochromat spectroscopy – inverse photoemission

In 1977, V. Dose introduced *UV-isochromat spectroscopy* as a technique for probing unoccupied surface states [8.19]. The technique became popular later under the name *inverse photoemission*. The surface of a solid is subjected to an (intense) beam of electrons of defined energy at a defined angle of incidence (mostly vertical). The electrons enter the solid in one of the higher energy bands above the vacuum level and drop into one of the lower bands under emission of a photon (Fig. 8.7b). Since the quantum efficiency of that process is low, very sensitive photon detectors covering a wide solid angle are required. Dose employed a Geiger-counter for that purpose (Fig. 8.8). The Geiger counter is filled with iodine and equipped with a $SrF_2$ window [8.20]. The combination of the photoionization threshold of iodine and the transmission cut-off endows the counter with an energy window of 0.4 eV. The sample is mounted on a two-axis goniometer to define the $k_\parallel$-vector of the incident beam and thereby the $k_\parallel$-vector of the final state. The energy resolution can be improved when UV-monochromators are used instead of the Geiger-counter. Because of the low photon count rates parallel processing of the different photon energies is required.

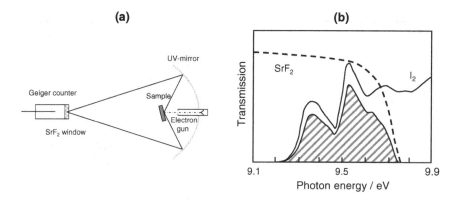

**Fig. 8.8.** (a) Experimental set-up for inverse photoemission spectroscopy. To define the $k_\parallel$-vector of the incident beam the sample is placed on a two-axis goniometer. Photons are collected in a wide solid angle and focused on a Geiger-counter. (b) The Geiger-counter is filled with iodine, which has an ionization threshold of 9.23 eV (solid line). Together with the spectral cut-off of the $SrF_2$-window at 9.75 eV (dashed line), the photon energy window (shaded area) has a spectral range of 0.4 eV (after Goldmann et al. [8.20]).

### Two-Photon PhotoEmission

*Two-Photon PhotoEmission* (2PPE) is a pump-probe technique for thermally un-occupied states (Fig. 8.7c). A first high-power laser pulse fills the state temporarily and a second pulse brings the electron from the intermediate state into the vacuum as a photoemitted electron. The method requires short laser pulses in the femtosecond regime and sufficiently long life times of the intermediate state. The technique has been employed to probe the so-called *image potential states* (Sect. 8.2.5) as these states couple very little to the substrate electronic state and therefore enjoy a long lifetime. The experiment probes these states by looking for the intensity of the photoemitted electrons as a function of their kinetic energy while the photon energies are kept fixed (for convenience). The electron intensity peaks at the kinetic energy that corresponds to the photoemission of the intermediate state with the probe photon $h\nu_2$. An experimental arrangement after Shumay et al. [8.21] is shown in Fig. 8.9a. An argon ion laser feeds a Ti-sapphire laser, which produces 70 fs pulses of IR-light at 790 nm wavelength. The pulse is tripled in frequency and brought to the sample after proper shaping with the help of two quartz prisms to pump electrons into empty states. The split-off 790 nm pulse runs through a variable delay line and probes the population in states between the vacuum level and $-1.57$ eV below the vacuum level (790 nm corresponds to $h\nu_2 = 1.57$ eV). Figure 8.9b shows the 2PPE intensity as a function of the kinetic energy of photoemitted electrons and the delay time between the pump and probe pulses. The maxima correspond to image potential states on the Cu(100) surface. The states labeled $n = 1, 2, 3$ have different lifetimes (for details see Sect. 8.2.5).

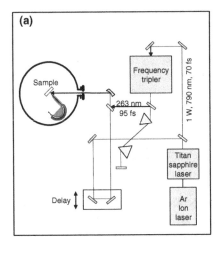
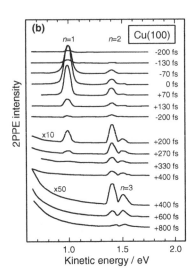

**Fig. 8.9. (a)** Experimental arrangement for *Two Photon PhotoEmission* (2PPE). The sample is subjected to a 263 nm, 95 fs pump laser pulse followed by a delayed 790 nm, 70 fs probe pulse. The prisms compensate the group velocity dispersion arising from the UHV-window. **(b)** 2PPE intensity as a function of the kinetic energy of photoemitted electrons and the delay time between the pump and probe pulses. The maxima correspond to image potential states on the Cu(100) surface (after Shumay et al. [8.21], see Sect. 8.2.5).

## 8.2.4 Surface States on Semiconductors

### Silicon and germanium surfaces

This section considers surface state bands on clean semiconductor surfaces that arise from the *dangling bonds* at surfaces (Sect. 1.2.3) and from the *back-bonds* of the surface atoms. In their seminal paper of 1996, Allen and Gobeli [8.22] measured the band bending at the surface of silicon as a function of bulk doping. The authors found that the Fermi-level remains pinned up to very high $n$- and $p$-type doping levels and showed thereby that on cleaved Si(111) surfaces the dangling bonds contribute about one occupied and one unoccupied surface state per atom to the density of states in the band gap. Spectroscopic evidence for a large density of surface states and for a gap of about 0.3 eV between the occupied and unoccupied band of surface states came from optical experiments (1971 [8.23]) and electron energy loss spectroscopy (1975 [8.24]). The $E(k)$-dependence of these dangling bond states and those on other surfaces was mapped out later with the help of photoemission spectroscopy and with band structure calculations. Research in this area was largely driven by technological developments, both with respect to issues and materials, since surface states on semiconductors have a significant effect on the electronic properties of surfaces/interfaces (Sect. 3.2.2). Presently, the band structure of surface states is known for a large number of materials and surface orientations. Here, we focus on a few pedagogical examples and begin with Si(100) and Ge(100).

The surface structure, the surface Brillouin zone (SBZ), the structure of the symmetric and asymmetric dimers and the electronic states for Si and Ge(100)(2×1) surfaces are displayed in Fig. 8.10. The calculation of the band structures by Krüger and Pollmann [8.25] shows that the symmetric dimer model leads to a metallic character of the surface state band, as the Fermi-level cuts across the two bands. This metallic character is in accordance with the qualitative reasoning that the dangling bond orbitals on the two atoms forming the symmetric dimer should be degenerate when the overlap between neighboring dimers is neglected. The formation of asymmetric dimers can be understood as a Jahn-Teller distortion, which leads to a state of lower electronic energy as the degeneracy is lifted by the distortion and the occupied band shifts down below the Fermi-energy. This lifting of the degeneracy is reflected in the larger gap between the surface state bands (lower panels of Fig. 8.10). The lower occupied bands correspond to the electronic states of the upper atoms in the dimers.

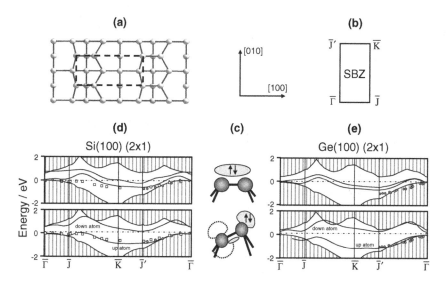

**Fig. 8.10.** Dangling bond surface states on Si(100) (2×1) and Ge(100) (2×1) surfaces: **(a)** the surface unit cell; **(b)** the surface Brillouin zone; **(c)** models for the symmetric and asymmetric dimer; **(d)** and **(e)** the surface state bands for the symmetric and asymmetric dimer models for Si and Ge, respectively (after Krüger and Pollmann [8.25]). The dash-dotted lines represent the Fermi-levels. The surface state bands are metallic and insulating for the symmetric and asymmetric dimer respectively, as expected for a Jahn-Teller distortion (Sect. 1.2.3). Experimental data from [8.26-28] are plotted as squares and circles for Si and as circles for Ge, respectively.

The experimental data displayed in Fig. 8.10 were obtained from angle resolved photoemission. As shown in Sect. 1.3.1 (Fig. 1.34) (100) surfaces of Si and Ge in general consist of two domains with mutual orthogonal orientations of the dimers. The domains are separated by monolayer high steps. Angle resolved photoemission spectroscopy on such a surface would produce a mixture of electronic states of the two domains except for the [010] direction which is common to both domains. Fortunately, single domain (100) surfaces can be prepared from vicinal surfaces that form an angle of about 4° with the (100) plane. On such surfaces the step height doubles after annealing so that only a single reconstruction domain exists with the dimer rows perpendicular to the step orientation [8.29]. Experiments agree well only with the asymmetric dimer model, supporting the existence of a Jahn-Teller distortion of the dimers on Si and Ge surfaces. In that sense, photoemission spectroscopy provides information on an element of the surface structure. This reasoning became important insofar as room temperature tunneling microscopy images display the dimers as symmetric (Fig. 1.35) while low temperature STM images show the dimers as asymmetric [8.30]. STM observations alone would therefore indicate, rather suggestively, a phase transition from asymmetric to symmetric dimers. In the light of the photoemission results, the STM observations have to be interpreted differently: at room temperature, the dimers flip back and forth between the two asymmetric configurations so that the slow STM sees only the mean dimer structure.

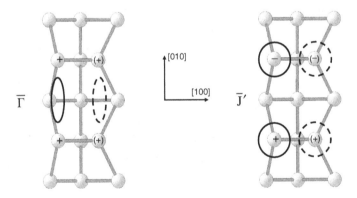

**Fig. 8.11.** The phases of the dangling bond orbitals at $\overline{\Gamma}$ and $\overline{J}'$. At $\overline{\Gamma}$ the orbitals of equivalent atoms are in phase, which leads to a high charge density between the dimers for the occupied state, which is symbolizes as a nominally bonding orbital by the solid ellipse. The equivalent orbital arising from the sp$^2$ bonded, lower dimer atom is unoccupied (dashed ellipse). At the $\overline{J}'$ point, the maximum charge density sits on the upper surface atom in the region of lower potential. The energy of the electron is therefore lower at $\overline{J}'$.

We note from Fig. 8.10 that the dispersion of the surface states along the $\overline{\Gamma}\overline{J}$ - and the $\overline{\Gamma}\overline{J}'$-direction is rather different. Along the $\overline{\Gamma}\overline{J}$-direction, the band is almost flat because the overlap of the surface states concerns fourth nearest neighbor atoms and is therefore marginal. The overlap between surface states of adjacent dimers concerns second neighbor atoms along $\overline{\Gamma}\overline{J}'$. The dispersion is therefore larger. One can even understand the sign of the dispersion. At the $\overline{\Gamma}$-point, the orbitals of two adjacent dimers along the [010] direction overlap in phase, i.e. constructively, causing the charge density to have its maximum between the dimers (Fig. 8.11). This is a region of high potential. At $\overline{J}'$, the orbitals have a node between the dimers. The maximum charge density is therefore on the dimer and therefore in the region of low potential. The electron energy must therefore be lower at $\overline{J}'$. Overall, the dispersion is weak because of the small overlap.

### (110) surfaces of zincblende structures

The surface states arising from the dangling bonds on the (110) surfaces of zincblende (ZnS) structures (the III-V compounds, some II-VI compounds and cubic SiC) are similar to those of the (100) surfaces of diamond structure elements. At first this may be surprising as there is no dimer reconstruction on the (110) surfaces of ZnS structures. Rather, "anions" (group VI, V-atoms) and "cations" (group II, III-atoms) form a zigzag chain along the [001] direction. Nevertheless, each surface atom contributes one surface state. Figure 8.12 shows the unit cell in real space (a) and in reciprocal space (b) together with side view of the ideally terminated and the relaxed surface ((c) and (d)) for GaAs(110). On the relaxed surface, the zigzag-chains between Ga and As-atoms are tilted, so that the surface As-atoms reside about 0.7 Å above the surface Ga-atoms. This is in accordance with the reasoning that the orbital character at the surface reflects the number of electrons on the atoms more closely than in the bulk since there is no need to form $sp^3$-hybrides in order to build a 3D-structure. The three electrons on the Ga-atoms form planar $sp^2$-bonds; the five electrons on the As-atom (electron configuration $4s^2 4p^3$) do not hybridize and the half-filled p-orbitals, ideally at angles of 90° with each other, make p-bonds with the neighboring atoms. The relaxation of the surface in the form of a tilt is thus a natural consequence of the number of electrons on the atoms. The local electronic configuration is very similar to the case of the dimers on the (100) surface of Si and Ge (Fig. 8.10). The $sp^2$-configuration of the occupied orbitals on the Ga-atoms leaves the surface state, the $p_z$-state of the Ga-atom empty. The surface state on the As-atom arises from the unused "backs" of the p-orbitals. Because of the bonding of the As-atoms to the substrate, all p-orbitals are filled with two electrons. Thus, the surface state is also filled. This qualitative reasoning is nicely confirmed by calculations of the surface band structure. Figure 8.12e and 8.12f show the surface band structure for the truncated bulk surface and the relaxed surface. The two bands associated with the dangling bond states are well separated in both cases, however more so for the relaxed case. The band arising (mostly) from the gallium dangling bond states is unoccupied (dashed line) while the band arising from the As-states is occupied.

Again we have the similarity to the dimers on Si(100) and Ge(100) (Fig. 8.10) with a notable difference of practical importance: The separation between the empty and full surface state bands is larger for GaAs. Clean and defect-free GaAs surfaces possess no surface states in the conduction band gap that would pin the Fermi-level. Surfaces prepared by cleaving may contain steps however. These steps contribute gap states of a concentration, which may be large enough to pin the Fermi-level up to moderate doping levels.

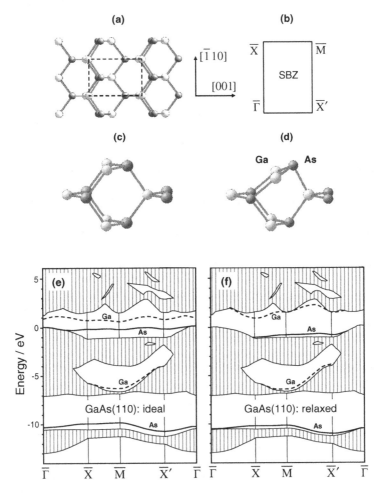

**Fig. 8.12.** Surface state bands on the ideally terminated and the relaxed GaAs(110) surface: **(a)** the surface unit cell; **(b)** the surface Brillouin zone; **(c)** and **(d)** models for the ideal surface (truncated bulk) and the relaxed surface; **(e)** and **(f)** calculated band structures. Bands arising from the Ga and As surface atoms are shown as dashed and solid lines respectively. The shaded areas are the bulk bands projected onto the SBZ (after Sabisch et al. [8.31]).

The absence of surface states with a concentration that would correspond to the concentration of surface atoms was demonstrated by electron energy loss spectroscopy (EELS) [8.32]. The spectrum displayed in Fig. 8.13 merely shows the onset of the interband transitions at 1.4 eV, while equivalent spectra of Si-surfaces show strong absorption due to surface states [8.24]. The absence of surface states in the band gap is also very nicely demonstrated in scanning tunneling spectroscopy (STS). The solid line in Fig. 8.13 is the normalized derivative of the tunneling current with respect to the tunnel voltage $(dI/dV)/(I/V)$ versus the tunnel voltage [8.33]. This quantity is roughly proportional to the density of states of the sample. If the sample is at negative potential with respect to the tip then electron from the occupied states, i.e. the conduction band and possibly occupied surface states tunnel into the tip. For reversed bias, electrons tunnel from the tip into the conduction band and into empty surface states. The fact that the tunnel spectrum shows defined onsets of the current, separated by 1.4 eV proves that there are no surface states of significant concentration in the gap. The small hump in the STS spectrum is due to tunneling into depleted bulk dopant states, as was demonstrated by STS on differently doped samples [8.33].

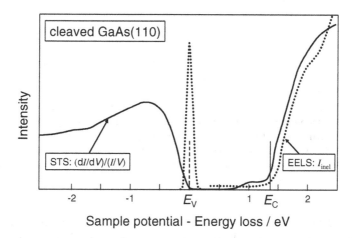

**Fig. 8.13.** Scanning tunneling spectrum (STS) and electron energy loss spectrum (EELS) of the cleaved GaAs(110) surface (solid and dotted lines, respectively, after Feenstra [8.33] and Froitzheim et al. [8.32]). The elastic peak of scattered electrons is placed at the zero of the STS to make the techniques comparable. Both spectra show the absence of surface states in the 1.4 eV direct gap between the valence band and the conduction band edges.

From the orbital picture of the surface states, one would assume that electron tunneling is mostly into (out of) the occupied (unoccupied) surface states since their orbitals represent the outmost electronic states. That this is indeed so was nicely demonstrated by STM images obtained at negative and positive bias by Feenstra et al. [8.34]. Figure 8.14 shows STM images of the cleaved (110) surface of *p*-doped

GaAs obtained at +1.9 V and −1.9 V. At a sample voltage of +1.9 V with respect
to the Fermi-level of the tunneling tip, electrons tunnel into the empty surface
states that are localized on the Ga-atoms. Only these atoms are visible at this bias
as bright spots. When the bias is reversed the electrons tunnel out of the surface
states localized on the As-atoms so that now these atoms appear as bright spots.
The localization of surface states on atoms thus opens the door to chemically specific STM imaging. As seen from the band structure in Fig. 8.12 the surface state
bands merge with the bulk states in parts of the Brillouin zone. The localized density of states on the surface atoms has therefore contributions from electron states,
which are true surface states and from states that decay into bulk electronic states,
i.e. from surface resonances (see also [8.35]).

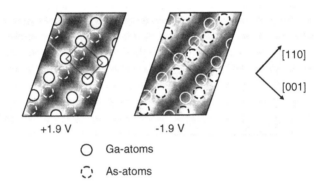

+1.9 V            -1.9 V

○    Ga-atoms

◌    As-atoms

**Fig. 8.14.** STM images of the cleaved GaAs(110) surface at two different voltages (after
Feenstra et al. [8.34]). The sample was *p*-doped. At a sample voltage of +1.9 V with respect to the Fermi-level of the tip electrons tunnel into the empty surface states that are
localized on the Ga-atoms. These atoms appear as bright spots. With reversed bias the
electrons tunnel out of the surface states localized on the As-atoms so that these atoms
appear as bright spots.

Of considerable technical importance are the interface states in semiconductor
heterojunctions and semiconductor/metal junctions as they have a decisive influence on the electrical properties of the semiconductor space charge layers at the
interface (Sect. 3.2.2). Naively one might expect that the dangling bond states on a
semiconductor surface should disappear when a metal is deposited on the surface.
In that case, the band bending of the semiconductor should solely depend on the
work function of the metal and the electron affinity (Sect. 3.2.2). However, experiment as well as theory shows that the metal can induce particular interface
states, which have energies inside the conduction band gap of the semiconductor.
These states are called *Metal Induced Gap States* (MIGS) (see e.g. [8.36]).

Further electronic surface states exist in band gaps that are deeply immersed
either in the conduction or in valence bands. Fig. 8.12 shows several surface bands
that have split off from the lower energy valence bands. These surface states are

associated with the bonds of the Ga and As surface atoms to the next layer below. They have therefore been named *back-bond surface states*. As their energy levels are far away from the Fermi-level, they play no role in electric properties of surfaces. First experimental evidence of back-bond surface states came from inelastic electron scattering on Si(111) (7×7) and Si(100) surfaces [8.37]. The position of these states within the valence bands was obtained a little later by photoemission spectroscopy [8.38].

### 8.2.5 Surface States on Metals

For a long time surface states on metals have been studied entirely because of their intellectual appeal rather than for any importance for other surface phenomena, not to speak of practical relevance. This picture has changed quite dramatically in recent years. We now understand that surfaces states, in particular those around the Fermi-level, have a decisive influence on nucleation as well on the formation and the stability of nanostructures via *quantum size effects*. The investigation of surface states on metals has therefore experienced a considerable revival. Quantum size effects associated with surface states will be considered in Sect. 8.3. This section is devoted to the electronic properties of surface states on metals as such. We begin with a particular kind of surface states that was already briefly addressed in Sect. 8.2.3 in the context of experimental probes for unoccupied states.

### Image potential states

Image potential states are special unoccupied electronic surface states that exist between the vacuum level and the Fermi-level. As their name says, they result from the image potential (Sect. 3.2.1). Sufficiently far away from the surface, the image potential has the form (3.21)

$$\phi(z) = -\frac{1}{4}\frac{e^2}{4\pi\varepsilon_0}\frac{1}{|z - z_0|} \tag{8.18}$$

where $z_0$ is the position of the image plane. The image plane is approximately one Bohr radius outside the jellium edge. The image potential also exists on semiconductor and insulator surfaces. The potential is then (cf. 7.65)

$$\phi(z) = -\frac{1}{4}\frac{\varepsilon-1}{\varepsilon+1}\frac{e^2}{4\pi\varepsilon_0}\frac{1}{|z - z_0|} \tag{8.19}$$

where $\varepsilon$ is the static dielectric constant. A prerequisite for the existence of an electronic state in the image potential as a localized surface state, however, is that there are no bulk states at the same energy and wave vector $k_\parallel$. The material must have an energy gap around and below the vacuum level. The (100) surfaces of fcc metals have an energy gap in the required energy range and for a large $k_\parallel$-range. They are therefore prototype surfaces for image potential states. The eigenstates of the $z^{-1}$-potential (8.18) are the same as for the hydrogen atom with the quantum number for the orbital momentum $l = 0$. Parallel to the surface the eigenstates are free-electron solutions. The energy levels are therefore

$$E_n = E_{\text{vac}} - \frac{13.55\,\text{eV}}{16\,n^2} + \frac{\hbar^2 k_\parallel^2}{2m} = E_{\text{vac}} - 0.85\,\text{eV}/n^2 + \frac{\hbar^2 k_\parallel^2}{2m}. \qquad (8.20)$$

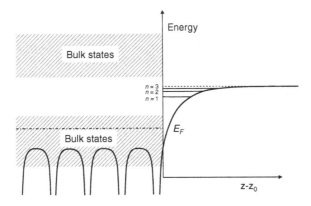

**Fig. 8.15.** Image potential states between the vacuum level and the Fermi-level. Their existence e.g. at $k_\parallel = 0$ requires an energy gap in the bulk states at $k_\parallel = 0$, so that the wave function is totally reflected from the surface. The energy levels of image potential states are fixed to the vacuum level.

Note that the energy scale always refers to the vacuum level for image potential states, not to the Fermi-level. If the work function of a surface is modified by adsorption, the difference between the Fermi-energy and the energy levels of the image potential states shifts according to the work function change. On a real surface, the energy levels for a pure $z^{-1}$-potential are modified since the wave function penetrates into the bulk region as an evanescent wave. Echenique and Pendry [8.39] have shown that the eigenstates are well described by adding a constant term $a = 0.21$ to the quantum number $n$ so that

$$E_n = E_{\text{vac}} - 0.85\,\text{eV}/(n+a)^2 + \frac{\hbar^2 k_\parallel^2}{2m}. \qquad (8.21)$$

Table 8.2 lists the eigenvalues for $k_\parallel = 0$ according to (8.20) and (8.21) together with experimental values from two-photon photoemission spectroscopy [8.21]

(Fig. 8.9). Two-photon photoemission spectroscopy probes the states at $k_\parallel = 0$ since the photoemitted electrons are observed at normal emission and photons carry nearly zero momentum.

**Table 8.2.** Energy of image potential states in eV below the vacuum level according to the hydrogen-model (8.20), the modified hydrogen model (8.21) [8.39] and experimental data obtained for Cu(100) [8.21].

| $n$ | $E_n = -0.85n^{-2}$ | $E_n = -0.85(n+0.21)^{-2}$ | $E_n(\text{exp})$ |
|---|---|---|---|
| 3 | -0.09 | -0.08 | -0.06 |
| 2 | -0.21 | -0.17 | -0.17 |
| 1 | -0.85 | -0.58 | -0.58 |

### Nearly free electron surface states

The sp-metals have surface states associated with the gaps in the bulk bands. Surface states also derive from the localized d-electrons. For the coinage metals Cu, Ag and Au, the bands of d-surface states are well below the Fermi-level while the surface state bands arising from the sp-states have their energy levels around the Fermi-energy. The latter surface states are of particular interest for several reasons. Firstly, the occupation of these surface states depends significantly on parameters such as temperature and strain, and furthermore on the presence or absence of adsorbed species. As the wave functions are partially localized outside the surface (defined by the jellium edge or the boundary of the valence d-electrons) the surface states are sensitive to the structure and morphology of the surface (for a review see [8.40]). Defects such as steps, islands or adatoms act as potential barriers for the wave functions. These defects can be of the same or a different material. Because of the confinement by defect barriers, the eigenstates are standing waves. Their energy levels shift upwards, the more, the smaller the area of confinement is, as known from the particle-in-a-box-model (Sect. 3.1.1). As the energy of the surface state bands are around the Fermi-energy, the energy levels are partially filled, partially empty. The Fermi-vector of these surface states is rather small on the (111) surfaces ($k_{fs} = 0.08$ Å$^{-1}$ for Ag(111) and $k_{fs} = 0.22$ Å$^{-1}$ for Cu(111)). The Fermi-vector cut-off gives rise to pronounced and long-range Friedel-oscillations (Sect. 3.2.1, Fig. 3.5), and to periodic oscillations of the total electronic energy with the distance between the potential barriers (*quantum size effect*).

The wave functions of the surface states being localized outside the surface can be probed by tunneling microscopy and tunneling spectroscopy and the potential barriers can be constructed by manipulating atoms with the STM-tip. Thus, individual adatoms and small clusters together with the STM-tip make a neat construction kit to build two-dimensional quantum systems with the surfaces of coinage metals as the playground.

Partially occupied surface state bands exist on all three low-index surfaces of the coinage metals. On the (100) surface they are localized in $k$-space around the $\overline{\text{X}}$ - point of the SBZ (Fig. 7.2). On the (110) surface they exist around $\overline{\text{Y}}$ (the zone boundary in [100] direction. On the (111) surfaces the surface state band centers at $\overline{\Gamma}$. These surface states are therefore of particular interest for quantum size effects in the nm-range and we focus on these surfaces in the following.

Early photoemission work on metal surface states was performed by the group of Neddermeyer (see e.g. [8.41]). Figure 8.16 shows a set of photoemission data by Kevan [8.42] with improved resolution permitting the measurement of the dispersion curve. The lowest energy of the surface state band is at the $\overline{\Gamma}$-point, corresponding to electron emission in normal direction. The dispersion of the band is directly seen in the set of spectra displayed in Fig. 8.16a as a function of the polar emission angle in the $\overline{\Gamma}\overline{\text{M}}$ (or $\overline{\Sigma}$) direction. The doublet structure of the photoemission peak arises from the doublet character of the ArI line. The resulting dispersion curve is shown in Fig. 8.16b together with the bulk bands depicted as shaded areas. The same dispersion curve is found for NeI light (16.8 eV) assuring the surface character of the photoemission peaks. The dispersion curve is well described by a parabola underlining the free-electron character of the surface state band.

Figure 8.16b shows that the (111) surface has a gap around the $\overline{\Gamma}$-point in the energy range of the surface state band. Consequently, the electron hole left behind after the photoemission process has a very long lifetime, in particular at the $\overline{\Gamma}$-

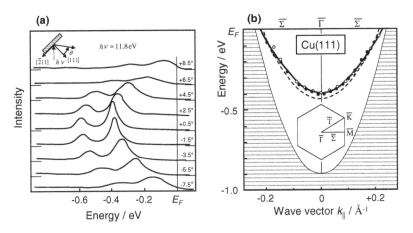

**Fig. 8.16.** (a) Ultraviolet photoemission spectra from a surface state on Cu(111) obtained with ArI-light. The second peak is because of the Ar-doublet. (b) Dispersion of the surface state: data points obtained with $h\nu = 11.8$ eV (ArI) and $h\nu = 16.8$ eV (NeI) are shown as open and solid circles, respectively (after Kevan [8.42]). The dashed line is the dispersion curve at 30 K as obtained more recently in high-resolution experiments by Reinert et al. [8.43]. The insert shows the SBZ.

point and at low temperatures. Hence, the intrinsic line width of the photoemission peak is quite small on well-ordered surfaces. The surface state is therefore an excellent benchmark for the quality of the experimental equipment. Figure 8.17a shows the improvements over the years as assembled by Reinert et al. [8.43] for the example of the particular narrow photoemission line of the $\overline{\Gamma}$-state on Ag(111).

Figure 8.17b shows the photoemission peaks of the $\overline{\Gamma}$-states for all coinage metals. Table 8.3 finally summarizes the energies of the $\overline{\Gamma}$-states with respect to the Fermi-level, the effective masses and the Fermi-vectors. On the Au(111) surface, the surface state band shows a noticeable spin-orbit splitting so that there are two Fermi-vectors corresponding to the two bands [8.43]. The lifetime of the electron hole as well as the energy of the surface state depends critically on the structural order. Earlier results on lifetime and energy differ from the values quoted in Fig. 8.17b and in Table 8.3. The differences between various authors are particularly large for Ag(111). There, the energy of the surface state band depends significantly on the surface lattice constant and therefore on the temperature. The surface state band shifts upwards with increasing interatomic distances and hence with rising temperature. For temperatures larger than about 450 K the $\overline{\Gamma}$-point is pushed above the Fermi-level [8.44]. The surface state is also shifted upwards in laterally strained Ag(111) films. A strain of merely $\varepsilon = +0.5\%$ suffices to push the $\overline{\Gamma}$-point above the Fermi-level [8.45].

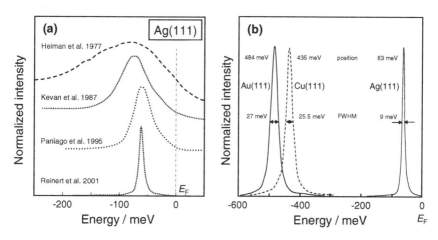

**Fig. 8.17.** (a) Photoemission spectra from the Ag(111) surface state for normal electron emission demonstrate the progress in resolution (after Reinert et al. [8.43] with data from [8.41, 44, 46]. (b) High resolution photoemission spectra of Au(111), Ag(111) and Cu(111) at 30 K for normal electron emission (after Reinert et al. [8.43]). The instrumental resolution is 3.5 meV so that the measured *Full Width at Half Maximum* (FWHM) for the most part represents the intrinsic line width due to finite lifetime broadening.

**Table 8.3.** Parameters of the nearly free electron surface state band on the (111) surfaces of noble metals at 30 K [8.43]. The parameters differ not insignificantly from earlier results. The difference is partly due to the temperature dependence of the energy levels, partly due to an incorrect positioning of the Fermi-level in earlier work and partly due to the improved resolution in the work of Reinert et al..

| Property | Cu(111) | Ag(111) | Au(111) |
|---|---|---|---|
| $E_F - E_{min}$ / eV | 0.435 | 0.063 | 0.484 |
| $m_{eff}$ / $m_0$ | 0.412 | 0.397 | 0.255 |
| $k_F$ / Å$^{-1}$ | 0.215 | 0.080 | 0.167/0.192 |

Free-electron surface states are also seen in scanning tunneling spectroscopy (STS). The differential tunnel conductance $dI/dV$ is roughly proportional to the combined local density of states of the tip and the surface under the tip. Assuming that the density of state of the tip is about constant near the Fermi-level then $dI/dV$ curves should reflect the density of surface states. The density of states of a parabolic band of two-dimensional surface states is

$$D(E) = \frac{2\pi m_{eff}}{h^2} \Theta(E - E_{min}) \tag{8.22}$$

in which $\Theta(E-E_{min})$ is the step-function with the onset at the minimum of the band $E_{min}$. The tunnel conductance should therefore be a step function. Figure 8.18 shows

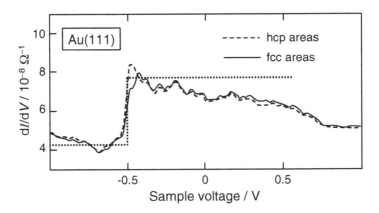

**Fig. 8.18.** Differential tunneling conductance $dI/dV$ on Au(111) surfaces (after Chen et al. [8.47]). The conductivity shows a sharp onset at the minimum of the surface state band. The dotted line represents the density of surface states. The $dI/dV$-curves are slightly different for the hcp and fcc areas of the reconstructed surface (Sect. 1.2.1, Fig. 1.12).

that this is approximately the case [8.47]. Unlike photoemission spectroscopy, STS probes the occupied as well as the unoccupied states with the same conductance. Only the current reverses from sample-to-tip at negative sample voltages to tip-to-sample at positive voltages. Interestingly, STS shows a small difference in the conductance on the hcp and the fcc areas of the reconstructed Au(111) surface (Fig. 1.12).

### 8.2.6 Band Structure of Adsorbates

A special type of surface states arises from the valence electrons of adsorbed species on surfaces. The dispersion of these surface state bands reflects the lateral coupling of the valence orbitals. As for the surface states on clean surfaces, the adsorbate valence states may couple to the bulk bands of the substrate if the substrate has electronic energy levels in the same $E(k_\parallel)$-space. Strictly speaking, the states are then resonances. The adsorbate valence electrons nevertheless retain a localized character even when they are situated within a bulk band. The surface band structure of adsorbates has been investigated for many common adsorbate systems. Most of the work has been performed in the early days of angle resolved photoemission spectroscopy with modest resolution and less advanced equipment than available nowadays. Since qualitative as well as quantitative aspects of the valence band structure of adsorbates are well understood by now research in this area has become less fashionable. Within the framework of a textbook, it is nevertheless useful to discuss the matter briefly, as it proves to be tutorial. We choose a particular model case for this purpose, the c(2×2) overlayers of chalcogenides on fcc(100) and focus on the c(2×2) oxygen layer on Ni(100). The qualitative picture is similar for other chalcogenides and other transition metal substrates.

The valence band arises from the oxygen 2p-electrons. Following A. Liebsch, we consider first the band structure of the bare c(2×2) oxygen layer without a substrate, which is shown in Fig. 8.19a. The band structure is easily understood in terms of symmetry properties and the lateral bonding between the atoms in the spirit of the tight binding model. We begin the discussion with the $\overline{\Gamma}$-point. The valence bands of the $p_x$- and $p_y$-states are degenerate there. Their energy is nearly the same as for the $p_z$-band since the overlap between the orbitals of the nearest neighbor atoms is small in both cases (Fig. 8.20). The overlap is antibonding for the $p_x$- and $p_y$-states and bonding for the $p_z$-state. Hence, the energy of the latter band is a little lower. As one moves towards $\overline{M}$, all (weak) nearest neighbor bonds become antibonding for the $p_z$-orbitals. The energy of the $p_z$-band therefore rises along the $\overline{\Gamma} \overline{M}$ direction (Fig. 8.19a). The same happens with the $p_z$-band along the $\overline{\Gamma} \overline{X}$ direction, however, the increase is only half of the increase along $\overline{\Gamma} \overline{M}$ (Fig. 8.19a) since one half of the nearest neighbor bonds are antibonding and one half is bonding at $\overline{M}$ (Fig. 8.20). The $p_z$-bands belong to the even representation along $\overline{\Gamma} \overline{M}$ and $\overline{\Gamma} \overline{X}$. The $p_x$- and $p_y$-bands are also degenerate at $\overline{M}$.

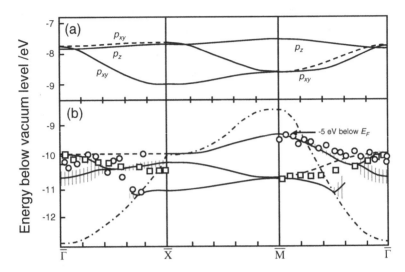

**Fig. 8.19.** Electronic band structure of the c(2×2) oxygen layer on Ni(100); **(a)** the band structure of a bare oxygen layer; **(b)** with the Ni-substrate (after Liebsch [8.48]). The lines are the oxygen 2p-bands. The odd states along high symmetry directions are designated by dashed lines. The Ni sp-bulk states lie above the dash-dotted line in **(b)**. Oxygen valence band states in that region couple to bulk states and may lose their character as surface states. The perpendicular shading indicates the broadening of the oxygen states due to coupling to bulk states. Experimental data of Kilcoyne et al. are plotted as circles and squares for the even and odd modes, respectively [8.49]. The absolute energy scale is fitted to the theory. The arrow in the lower right panel marks the experimental $-5$ eV level below the Fermi-energy.

Between $\overline{\Gamma}$ and $\overline{M}$ the bands separate into odd and even bands of slightly different energy. At the $\overline{M}$-point the nearest neighbor orbitals are bonding with a stronger overlap than for the $p_z$-states (Fig. 8.20). The energy of the $p_x$ and $p_y$-bands therefore disperses downwards towards $\overline{M}$ and the dispersion is larger than for the $p_z$-band (Fig. 8.19a). The even and odd $p_x$, $p_y$-bands split significantly along the $\overline{\Gamma}\,\overline{X}$-direction. Note that the $p_x$- and $p_y$-orbitals at the $\overline{\Gamma}$ are drawn such as to meet the symmetry required along the $\overline{\Gamma}\,\overline{M}$ and $\overline{\Gamma}\,\overline{X}$-directions. The odd $p_x$-$p_y$ band disperses mildly in upwards direction as half of the bonds assume a small antibonding character at $\overline{X}$ (Fig. 8.20). The even $p_x+p_y$ orbitals form strong bonds between two nearest neighbors at $\overline{X}$. The bands therefore disperses downwards and the energy at $\overline{X}$ is the lowest in the entire 2D-band structure (Fig. 8.19a).

The simple valence band structure of the oxygen layer remains essentially unaltered by the presence of the Ni-substrate if the valence bands are outside the bulk electronic states of nickel. That is the case between $\overline{M}$ and $\overline{X}$. The dash-dotted

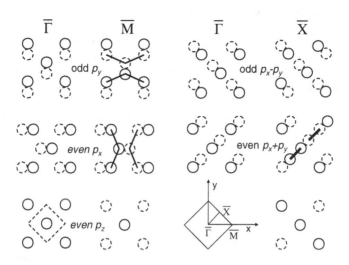

**Fig. 8.20**. Orbital character of the 2p-states of oxygen in a c(2×2) layer at the various points in the Brillouin zone (see insert at the bottom). At $\overline{\Gamma}$ the $p_x$- and $p_y$-states are degenerate. The same holds at $\overline{M}$. There the $p_x$- and $p_y$-orbitals overlap constructively. The formation of bonds lowers the energy of the $p_x$- and $p_y$-bands at $\overline{M}$. Strong lateral bonding also occurs for the even $p_x+p_y$ band at $\overline{X}$ (see text for further discussion).

line in Fig. 8.19b marks the lower boundary of Ni-bulk states. The Ni-bands are of sp-type in that energy range, as is apparent from the large dispersion of the boundary. The d-states lie above the oxygen valence-band region. Along the $\overline{\Gamma}\,\overline{M}$ and $\overline{\Gamma}\,\overline{X}$-directions the oxygen bands overlap and partially hybridize with the Ni-sp-bands. Starting from the $\overline{M}$ and $\overline{X}$ points, the even $p_x$, $p_y$-bands make an *avoided crossing* with the Ni-states and therefore disperse downwards instead of upwards. The odd $p_y$-band does not couple with the even Ni-bulk states in that region and remains unaffected therefore. The oxygen valence bands loose their localized character inside the Ni-bulk bands. They become broad resonances (indicated by the vertical lining in Fig. 8.19) or disappear completely. Figure 8.19 includes experimental data of Kilcoyne et al. [8.49]. The absolute energies are shifted to match the theory. For reference: the experimental energy of the highest oxygen band at $\overline{M}$ is 5 eV below the Fermi-level. The experiment distinguishes between even and odd bands. Circles and squares, respectively mark the corresponding data points. The experimental data roughly trace the theoretical bands, although there is some scattering in the data which may be partly due to the intrinsic width of the energy levels, partly due to the, by present standards low resolution.

## 8.2.7 Core Level Spectroscopy

Core level spectroscopies are frequently employed to determine the concentration of impurities on surfaces. This aspect was already discussed in Sect. 2.2.2. In this application, angular resolution is neither required nor appropriate. On the contrary, analyzers with large acceptance angles such as the cylindrical mirror analyzer are preferred to achieve high sensitivity (Fig. 2.17). Although the core levels are not engaged in the chemical bond, the kinetic energy of electrons photoemitted from a core level depends on the local environment of the atom. For a reference scale, which reflects solely the properties of the solid one defines the *binding energy* of an electron $E_b$ as the difference between the photon energy and the kinetic energy of the electron.

$$E_b = h\nu - E_{kin} = E(N-1) - E(N) \qquad (8.23)$$

Energy conservation requires that the so-defined binding energy equals the difference between the total energy of the solid with $N$ electrons minus the energy for the system with $N$-1 electrons. Shifts of the binding energy due to changes in the local environment of an atom that originate in $E(N)$ are called *initial state effects*, shifts originating in $E(N-1)$ are denoted as *final state effects*. Initial state effects are caused by a change in the local potential. If e.g. the atom in question engages in polar bonds that remove part of the valence charge from the atom then the electrostatic potential at the atom is more positive, hence, the total energy of the electron system of that atom is lower and the binding energy $E_b$ increases. The binding energies of core electrons therefore reflect the charge transfer from atoms to their bonding partners. This charge transfer is often expressed in terms of the electronegativity difference between the bonding partners. The correlation between the shift in $E_b$ and the charge transfer is well established and unambiguous for molecules. Since the binding energies of core electrons reflect the nature of a chemical bond of the atom, the shifts are called *chemical shifts*. The relation between the shifts in $E_b$ and the electronegativity difference to the bonding partners is unique only if the coordination number does not change. Surface atoms have a lower coordination than bulk atoms. The electrostatic potential resulting from neighboring ion cores as seen by a core electron of a surface atom is less attractive than for bulk atoms. Core electrons of surface atoms have therefore a smaller binding energy than core electrons of bulk atoms. The effect is even more pronounced for surface atoms in sites of particular low coordination, e.g. at steps. The shifts in the binding energy for surface atoms are referred to as *surface core-level shifts*. Surface core-level shifts and chemical shifts are of comparable magnitude.

The energy of the final state of the solid after the photoemission process depends on the screening of the core hole by the electron system. The more effective the screening process, the lower the energy of the final $(N-1)$-state and thus the lower the binding energy $E_b$. For the transition metals a simple argument can be brought forward as to how the screening should be affected by the lower coordination at the surface. Screening involves a relocation of electronic charge from the surrounding atoms to the atom with the core hole. This relocation requires the filling of empty states. Filling the states right above the Fermi-level requires the least energy. Because of the lower orbital overlap at the surface, the width of the valence d-electron bands of transition metals is smaller at the surface, the density of states around the Fermi-level therefore higher. Thus, core holes in surface atoms are more effectively screened and the binding energy of electrons is reduced by this effect. Depending on the type of surface and the material, the screening contribution to the total surface core-level shift is of the order of 25%. For free-electron metals and the coinage metals screening is provided by the sp-electrons. According to the jellium model the density of sp-electrons is lower at the surface leading to less effective screening (Fig. 3.5) and thus to a reduction of the shift caused by initial state effects. The overall surface core-level shift remains negative, however, i.e. towards smaller binding energy.

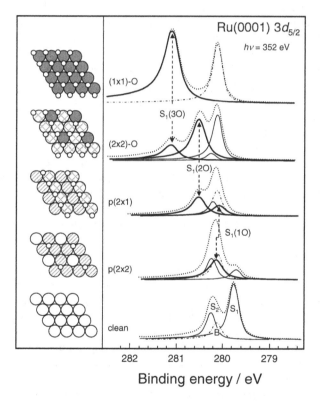

**Fig. 8.21** Core level spectra on the clean and oxygen covered Ru(0001) surfaces. Dotted lines are the experimental spectra. Experimental data are fitted to various contributions. Dash-dotted lines and thin solid lines are contributions from bulk atoms (B), first layer ($S_1$) and second layer ($S_2$) atoms of the clean surface, respectively. The fat solid lines mark the contributions from surface atoms with bonding to one, two or three oxygen atoms (After Lizzit et al. [8.50]).

Core-level shifts on clean as well as on adsorbate-covered surfaces have been observed since the early days of X-ray photoemission spectroscopy. The use of high-resolution monochromators on synchrotron beam lines in combination with high-resolution electron spectrometers has enabled very detailed studies of the core level shifts as a function of the local environment of a surface atom. Figure 8.21 shows a nice example concerning the photoelectron spectrum of $3d_{5/2}$-electrons on Ru(0001) [8.50]. Ruthenium is a 4d-transition metal, hence the 3d-electrons are core electrons. The clean surface shows two surface peaks. One originating from the first layer surface atom ($S_1$) is shifted by about $\Delta E_{SCLS} \cong -0.37$ eV towards lower binding energies, hence in the expected direction. The binding energy of the second layer atom is shifted upwards by $\Delta E_{SCLS} = +0.125$ eV. Upon oxygen adsorption, the binding energy of the surface core electron increases and the amount of that increase depends on the number of oxygen atoms bonding to the particular surface atom. On the p(2×2) surface three out of four surface atoms bond to an oxygen atom. Electrons from these surface atoms have a binding energy close to the bulk value. The intensity of the $S_1$-peak is reduced. In the p(2×1) structure the surface atoms bond to either one or two oxygen atoms. The $S_1$-peak of the clean surface has disappeared. The spectrum displays two surface peaks $S_1(1O)$ and $S_1(2O)$ with core-level shifts $\Delta E_{CLS} \cong 0$ and $\Delta E_{CLS} = +0.4$ eV. At even higher coverages the peak $S_3(3O)$ appears at $\Delta E_{CLS} = 0.96$ eV which corresponds to surface atoms bonding to three oxygen atoms.

The intensities of the various contributions to the experimental spectra are not proportional to the number of atoms from which the peaks originate. This has to do with diffraction of the photoelectron. The photoemitted electron undergoes a multiple scattering process in much the same way as an external electron is scattered from the substrate in the LEED process (Sect. 1.1.1), only that the source of the electron is now a particular atom. The multiple diffraction process carries the information about the local environment of the source atom. The local structure can therefore be analyzed by comparing experimental energy and angle distributions of the intensities to a multiple scattering calculation. Contrary to LEED, no ordered structures are required. *Photo Electron Diffraction* (PED) is therefore a powerful tool for the analysis of the structure of adsorbed species and their adsorption sites on surfaces (For a review see e.g. [8.51]). Some structure elements can even be ascertained without invoking multiple scattering calculations. The scattering of electrons with a kinetic energy above about 400 eV is strongly peaked into the forward direction. For example, the angular distribution of the C1s photoelectrons of carbon monoxide peaks along the axis of the CO-molecule. The orientation of the molecular axis with respect to the surface is therefore obtained directly from the angular distribution (*search light effect*). This analysis even works when the CO is embedded into an adsorbate layer of hydrocarbons, as the carbon atoms in hydrocarbons possess a different binding energy.

# 8.3 Quantum Size Effects

## 8.3.1 Thin Films

The confinement of the electron wave function within the two surfaces of a thin film can cause oscillations in the density of states at a particular energy, e.g. the Fermi energy. The oscillations in the density of states affect the total energy of the film and quite generally all properties of the film that depend on the density of electrons at the Fermi-level: electrical and thermal conductivity, the magnetic susceptibility and most importantly the magnetic coupling across non-magnetic thin films (Sect. 9.8.3). The latter effect has gained considerable technical and economical importance in the *Giant Magneto Resistance* (GMR) effect, which is discussed in Chapt. 9.

Quantum size effects in thin films were first observed as oscillations in the current in tunneling spectra of Al/Al$_2$O$_3$/Pb and Mg/MgO/Pb sandwiches, in which the Pb-films were about 250 Å thick [8.52, 53]. Later Jonker and Park reported on quantum oscillations in the sample current of Ag and Cu covered W(110) substrates subjected to a normal incidence low energy electron beam [8.54]. Investigations on other thin-film systems followed, including Cu(100) films deposited on Ni(100) surfaces [8.55]. Quantum size effects in photoelectron spectroscopy were first predicted by Loly and Pendry for Ag-films on Pd(100) [8.56] and soon thereafter discovered in experiments on Ag-films on Si(111) by Wachs et al. [8.57]. Using elastic scattering of He-atoms Hinch et al. reported in 1989 that certain film thicknesses are avoided in the epitaxial growth of lead on Cu(111) by double layers growth. The effect was (more convincingly) confirmed in STM-studies of Ortero et al. [8.58]. The stabilization of at times rather sophisticated nanostructures by quantum size effects has also been reported (see e.g. [8.59]). All these various effects are specific manifestations or consequences of the confinement of the electron wave function.

This section considers the one-dimensional confinement of bulk electron states in thin films deposited on a substrate. On one side, the electron wave function is confined by potential barrier to the vacuum. There is also a barrier on the substrate side if the substrate has a band gap. The band gap may be an absolute one as in semiconductors or a partial gap for electron states with a particular wave vector parallel to the film orientation. The simplest model for electron confinement is the one-dimensional quantum well with infinite walls. The electron wave functions are then standing waves that obey the condition (Sect. 3.1.1)

$$2k_\perp D = 2\pi n \tag{8.24}$$

in which $k_\perp$ is the wave vector perpendicular to the film of thickness $D$ and $n$ is an integer. The condition (8.24) can be interpreted as a phase matching condition stating that the total phase shift for a wave traveling back and forth between the boundaries must be an integer multiple of $2\pi$. Walls of a finite height have a complex amplitude-reflection coefficient $r$, which introduces an additional phase shift so that the phase matching condition becomes

$$2k_\perp D + \varphi_A + \varphi_B = 2\pi n \tag{8.25}$$

where $\varphi_A$ and $\varphi_B$ are the phase shifts at the two boundaries. We consider first the total number of states $N(E_m)$ up to a certain energy level $E_m$ for a one-dimensional quantum well with infinite barriers. When the width of the quantum well $D$ is expanded, the energy of a particular state denoted by the quantum number $n$ drops below the energy $E_m$ and the number of electron states increases by 2 (2, because of spin degeneracy). Upon further expansion, the number of states stays constant until the next state fall below the energy $E_m$, and so forth. The density $N(E_m)$ is therefore a staircase function with the period $2k_\perp D$ (Fig. 8.22). The analytical equation that describes the staircase function is

$$N(E_m) = \frac{2}{\pi} \arctan\left[\sin(2k_\perp D)/(1 - \cos(2k_\perp D))\right] + 2k_\perp D / \pi$$
$$\equiv -\frac{2}{\pi} \mathrm{Im}\left[\ln(1 - e^{i2k_\perp D})\right] + 2k_\perp D / \pi \ . \tag{8.26}$$

The first term in (8.26) is a sawtooth function with sudden jumps from $-1$ to $+1$ at $2k_\perp D = 2\pi n$ and a smooth decay from $+1$ to $-1$ when the argument $2k_\perp D$ varies by 1. The second term, linear in $2k_\perp D$, converts the sawtooth into a staircase. The analytical function (8.26), valid for infinite walls with reflection coefficients $r_A = r_B = 1$, can be extended analytically to the case $r_A$, $r_B < 1$. We can guess how this extension must look since we know that the steps must vanish when the reflection coefficients of the walls approach zero. A function that has this analytical property and has also the correct limit for $r_A = r_B = 1$ is

$$N(E_m) = \frac{2}{\pi} \arctan\left[r_A r_B \sin(2k_\perp D)/(1 - r_A r_B \cos(2k_\perp D))\right] + 2k_\perp D / \pi$$
$$\equiv -\frac{2}{\pi} \mathrm{Im}\left[\ln(1 - r_A r_B e^{i2k_\perp D})\right] + 2k_\perp D / \pi \ . \tag{8.27}$$

This is indeed the correct functional form for arbitrary $r_A$, $r_B$ [8.60]. Figure 8.22 displays the function for several values of

$$R = |r_A||r_B| \ . \tag{8.28}$$

The total phase shift $\varphi_{AB} = \varphi_A + \varphi_B$ after one round of reflections at the two interfaces is taken as zero. Otherwise, the $x$-axis would be shifted by $\varphi_{AB}$. For small values of $R$, $N(E_m)$ approaches the linear increase with $2k_\perp D$. The oscillating part of $N(E_m)$ assumes a sinusoidal shape in that case.

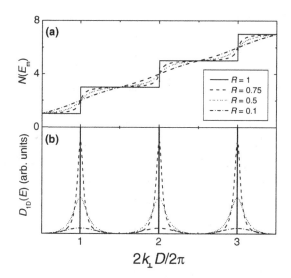

**Fig. 8.22.** (a) The total number of states $N(E_m)$ up to a particular energy $E_m(k_\perp)$ in a one-dimensional quantum well. The combined phase shift at the interfaces $\varphi_{AB}$ is set to zero. (b) One-dimensional density of states $D_{1D}(E)$ without the factor from the dispersion $\partial k_\perp / \partial E$.

The intensity of photoelectrons in normal emission is proportional to the one-dimensional density of states $D_{1D}(E)$ where $E$ is the binding energy of the electron. Taking the derivative of (8.27) one obtains for the one-dimensional density of states

$$D_{1D}(E) = \frac{\partial N(E)}{\partial E}$$
$$= \frac{4D}{\pi} \left\{ \frac{1}{2} + \frac{R\cos\left[2k_\perp(E)D + \varphi_{AB}\right] - R^2}{1 + R^2 - 2R\cos\left[2k_\perp(E)D + \varphi_{AB}\right]} \right\} \frac{\partial k_\perp(E)}{\partial E} . \tag{8.29}$$

The function represents a series of peaks with a periodicity $2k_\perp D$ and an amplitude factor $\partial k_\perp / \partial E$. The density of states $D_{1D}(E)$ without the factor $\partial k_\perp / \partial E$ is plotted in Fig. 8.22b. The function is a series of peaks located at $2k_\perp D = 2\pi n$. The peaks are $\delta$-functions for $R = 1$, and become the broader the smaller $R$ is. The peak structure is due to the denominator in (8.29) which is the same as for the transmitted light intensity in a Fabry-Perot interferometer. It should be pointed out however that the density of states concerns the stationary solutions of the wave function, not the intensity of a transmitted beam.

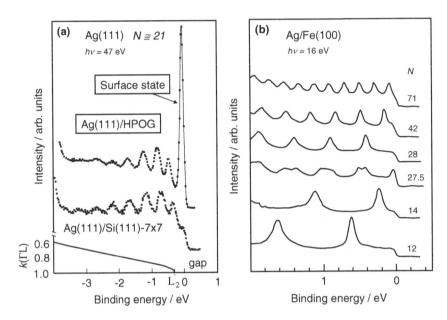

**Fig. 8.23.** Spectra of photoemitted electrons normal to the surface for **(a)** $N = 21$ monolayers of Ag(111) on *Highly Ordered Pyrolytic Graphite* (HPOG) and the Si(111)-7×7 surface [8.61] and **(b)** for various layer thicknesses of Ag(100) on Fe(100) [8.62]. Sample temperatures were $T = 130$ K and $T = 100$ K in **(a)** and **(b)**, respectively.

The oscillations in the one-dimensional density of states are seen in photoemission spectra with normal electron emission. Figure 8.23 shows examples of such spectra. Figure 8.23a displays photoemission spectra of (111) oriented silver films deposited on *Highly Ordered Pyrolytic Graphite* (HOPG) and Si(111)-7×7 as reported by Neuhold and Horn [8.61]. The $E(k_\perp)$ relation of silver along the [111] direction is shown at the bottom. In the energy range shown the electrons have sp-character. The d-bands of silver are 4.5 eV below the Fermi-level. For both substrates, the photoemission spectrum shows a series of oscillations whose intensities reflect the derivative $\partial k_\perp / \partial E$ for higher energies (8.29) and the lower reflectivity of the interface barrier in the lower energy range. Figure 8.23a is an interesting comment on the surface states discussed in Sect. 8.2.5. On the HOPG substrate, the surface state is visible as a strong peak right below the Fermi level. On Si(111)-7×7 the silver film is strained by 0.95%. This strain is enough to push the surface state above the Fermi level so that it becomes invisible. Figure 8.23b shows the oscillations in the photoemission spectrum for various thicknesses of Ag(100) films deposited on Fe(100) whiskers [8.62]. The perfect layer-by-layer growth and the low defects concentration permits observation of quantum oscillations beyond film thicknesses of 100 monolayers. The effect of not perfectly flat surfaces becomes apparent in the spectrum for nominally $N = 27.5$ monolayers where the peaks are split into doublet corresponding to $N = 26$ and $N = 28$ layers. These doublets are seen only because the surface consists of large areas covered with 26 and 28 layers. Less perfectly grown films show much broader and weaker oscillations up the point where oscillations disappear completely. Quantum oscillations in photoemission were therefore discovered relatively late when scientists had learned how to grow well-ordered films. In the original work of Wachs et al. [8.57] which likewise concerned Ag on Si(111) the oscillations were much weaker than in the later work of Neuhold and Horn (Fig. 8.23a). The widths of quantum well peaks depend also on the temperature. With increasing temperature, the peaks broaden due to electron-phonon coupling. The spectra in Fig. 8.23 were recorded at low temperatures.

### 8.3.2 Oscillations in the Total Energy of Thin Films

According to Fig. 8.22, the total number of electrons in the quantum well up to the Fermi energy is an oscillating function of the film thickness. This causes characteristic oscillations in the energetics of thin films. To calculate the thermodynamic relevant energy one has to observe that electrons that are squeezed out of the quantum well end up in bulk states of the substrate at the Fermi-level. The thermodynamic energy to consider is therefore the Gibbs free energy of electrons

$$G = \int_0^{E_F} (E - E_F) D(E) \mathrm{d}E .$$  (8.30)

The lower integration limit is the bottom of the quantum well. Integrating (8.30) by parts yields

$$G = - \int_0^{E_F} N(E) \mathrm{d}E$$  (8.31)

in which $N(E)$ is the total number of electrons up the energy $E$ (8.27). We are interested in the oscillations only. The oscillating part of the energy is

$$\Delta G = \frac{2}{\pi} \operatorname{Im} \int_0^{E_F} \ln\left[1 - r_A r_B \exp(2ik_\perp D)\right] \mathrm{d}E .$$  (8.32)

In three dimensions, the oscillations are obtained by integrating over the parallel wave vectors.

$$\Delta G_{3D} = \frac{1}{4\pi^2} \int \Delta G \, dk_{\parallel} = \frac{1}{2\pi^3} \, \mathrm{Im} \int dk_{\parallel} \int_0^{E_F} \ln\left[1 - r_A r_B \exp(2ik_{\perp} D)\right] dE \tag{8.33}$$

Because of the upper limit, this function oscillates with the period $2k_{F\perp}D$ where $k_{F\perp}$ is the Fermi-vector perpendicular to the film plane. The actual position of the minima and maxima and the precise oscillation period depends on the band structure $E(k)$, and the energy dependence of the modulus of the total reflection coefficient $R$ and of the total phase shift $\varphi_{AB}$. If $\varphi_{AB}$ is zero, then the minima and maxima occur at

$$
\begin{aligned}
2k_{F\perp}D\big|_{min} &= (2n+1)\pi \\
2k_{F\perp}D\big|_{max} &= 2n\pi \ .
\end{aligned}
\tag{8.34}
$$

Figure 8.24 displays the oscillations in $\Delta G$ for a quantum well system in one-dimension (8.32).

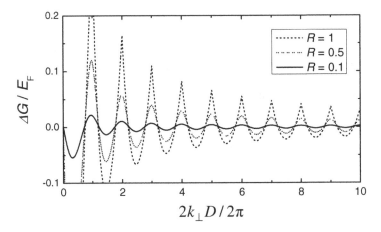

**Fig. 8.24.** Oscillations in the Gibbs free energy of a one-dimensional quantum well for reflection coefficients $R = 0.1, 0.5, 1$ and zero phase total shift $\varphi_{AB}$.

In three dimensions, the oscillations are less pronounced. Nevertheless, they amount to several tens of meV per surface atom. The energy is large enough to make films unstable that have thicknesses corresponding to maxima in $\Delta G$. Such films decompose into films that have thicknesses above and below the critical thickness. A particular nice example is the bifurcation that occurs with the well-ordered Ag(100) layers on Fe(100) [8.63]. The pronounced quantum oscillations in the density of states were already shown in Fig. 8.23. Figure 8.25 shows normal emission angle photoemission spectra of six monolayer (6 ML) films and 3 ML films deposited at low temperatures. In both cases, the spectrum is dominated by a single sharp peak in the density of states. The peak position is a clear indication of the particular thicknesses. Upon annealing the 6 ML film decomposes into areas with 5 ML and 7 ML thickness (Fig. 8.25a). The spectrum at the bottom was obtained after cooling again to low temperatures. The 3 ML film decomposes into a 2 ML and 4 ML film during annealing (Fig. 8.25b). One of the nice features of photoemission spectroscopy is that in combination with the band structure the energy dependence of reflectivity and phase shift can be obtained from the experiment by fitting the calculated 1D-density of states to the spectrum. Using this method, Luh et al. were able to calculate the energy gain in the bifurcation process to 20 meV and 40 meV per atom area for the 6 ML and 3 ML films, respectively [8.63]. One might wonder whether these relatively small energies comparable with $k_B T$ can shift the balance completely to the bifurcated state. However, for the equilibrium structure it is not the energy per atom area which counts but the total energy of an extended films. The kinetics of the process is driven by the energy gain for individual atoms. There, even a tiny energy gradient causes a net mass transport towards equilibrium (see Chaps. 10 and 11).

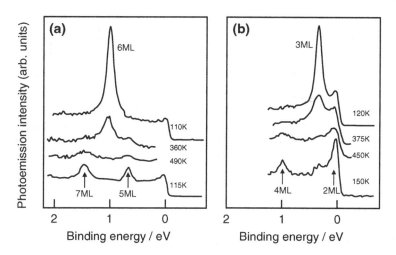

**Fig. 8.25.** Normal emission angle photoemission spectra of Ag(100) films on Fe(100) whiskers showing the bifurcation of 6 ML **(a)** and 3 ML **(b)** films into 5/7 ML and 2/4 ML films, respectively, upon annealing (after Luh et al. [8.63]).

*Magic thicknesses* of thin films or *magic heights* of islands caused by quantum interferences have been reported also for other system. A well-studied example concerns Pb-islands on Cu(111) surfaces [8.58]. Figure 8.26a shows the height distribution of islands after room temperature growth and a brief annealing to 400 K. Clearly, certain heights like 5, 9, 12, 13, 14, and 18-19 are "forbidden", while heights of 6, 8, 11, 17, and 20 are strongly preferred. The preferred island heights fulfill the condition (8.34) for a minimum in the energy while the totally excluded island height of 9 ML corresponds to a maximum (Fig. 8.26b). The phase is calculated with a mean *nesting vector* of the Fermi surface (Sect. 7.1.6, Fig. 7.9) of 0.625 $\text{Å}^{-1}$ and a zero phase shift $\varphi_{AB}$ for each round trip in the multiple reflection.

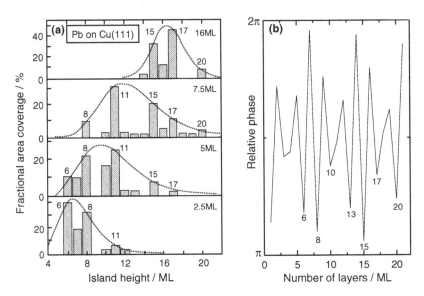

**Fig. 8.26.** (a) Histograms of the areas occupied by Pb(111) islands vs. their height in monolayers (ML) for mean coverages 2.5ML, 5ML, 7.5ML and 16ML (from bottom to top). (b) Relative phase with respect to the condition of a maximum or minimum in the energy (8.34).

### 8.3.3 Confinement of Surface States by Defects

Defects on surfaces represent potential barriers for surface states that are localized in the outmost surface layer (Sect. 8.2.5). The scattering of the electron wave function results in a standing wave pattern around the defects. As described in the previous section, this causes oscillations in the local density of states that in turn can be probed by scanning tunneling spectroscopy. Images of the tunneling conductivity $dI/dV$ or the normalized tunneling conductivity $(dI/dV)/(I/V)$ represent approximately the local density of states at a particular energy. Occupied (unoccupied) states are probed for negative (positive) sample bias with respect to the tip. Particular instructive are the standing wave patterns inside closed boundaries.

Figure 8.27 shows the standing wave pattern inside a *quantum corral* of Fe-atoms on a Cu(111) surface as published by M. F. Crommie, C. P. Lutz, and D. M. Eigler in 1993 [8.64, 65]. In the same year Y. Hasegawa and P. Avouris reported on the corresponding pattern near monatomic steps on the Au(111) surface [8.64, 65]. The dispersion of the surface state can be obtained from the standing wave pattern between two parallel steps [8.66].

**Fig. 8.27.** Perspective view on the standing wave pattern of surface states on Cu(111) inside a *quantum corral* that is formed by iron atoms. The Fe-atoms are placed into their position using the STM-tip as a manipulator (after Crommie et al. [8.64, 65]).

The confinement of the wave function between the potential barriers represented by parallel steps has interesting consequences on the energetics of steps. We consider these consequences for a vicinal surface with regularly spaced steps separated by a distance denoted as $L_\perp$. For simplicity we assume that the steps are non-penetrating, i.e. lines of infinitely high potential. Then, the surface state wave function has nodes at the steps and the wave vector perpendicular to the step direction is quantized.

$$k_\perp = \frac{\pi}{L_\perp} n \quad \text{with} \ n = 1,2,3... \tag{8.35}$$

The energy dispersion $E(k)$ consists of a series of one-dimensional bands. The bottom of each band is a function of the quantum number $n$

$$E(k) = \frac{\hbar^2}{2m^*}(k_\parallel^2 + \frac{\pi^2}{L_\perp^2 n^2}) = \frac{\hbar^2}{2m^*} k_\parallel^2 + E_{n0} \tag{8.36}$$

The total density of states is a sum over one-dimensional density of states $D_n(E)$

$$D(E) = \sum_{n=1} D_n(E) \tag{8.37}$$

with

$$D_n(E) = \frac{2L_{\parallel}}{\pi} \sqrt{\frac{2m^*}{\hbar^2}} \left( E - \frac{\hbar^2}{2m^*} \frac{\pi^2 n^2}{L_{\parallel}^2} \right)^{-1/2} \tag{8.38}$$

where $L_{\parallel}$ is the length of the steps. The factor of two in the nominator takes care of the assumed spin degeneracy of the bands.

To calculate the consequences of the confinement for the projected surface tension $\gamma_p$ (Sect. 4.3.1, eq. 4.45) we rewrite (8.30) for the present case.

$$\gamma_{p,el} = G / L_{\parallel} L_{\perp} = \frac{1}{L_{\parallel} L_{\perp}} \int_{E_{n0}}^{E_F} (E - E_F) \sum_{n=1} D_n(E) \, dE \tag{8.39}$$

After integration one obtains

$$\gamma_{p,el} = -\frac{8k_F E_F}{3\pi L_{\perp}} \sum_{n=1}^{n_{max}} (1 - \frac{\pi^2}{L_{\perp}^2 k_F^2} n^2)^{3/2} \tag{8.39}$$

where $n_{max}$ is the maximum $n$ that still leads to a positive value of the expression in brackets. The summation can be carried out numerically. However, it is instructive to consider also the continuum approximation of the sum.

$$\sum_{n=1}^{n_{max}} (1 - a^2 n^2)^{3/2} \cong a^{-1} \int_{a/2}^{1} (1 - x^2)^{3/2} \, dx$$

$$= a \left( \frac{1}{4} x(1 - x^2)^{3/2} + \frac{3}{8} x(1 - x^2) + \frac{3}{8} \arcsin x \right) \Big|_{a/2}^{1} \tag{8.40}$$

$$\cong \frac{3\pi}{16a} - \frac{1}{2}$$

A continuum approximation to $\gamma_{p,el}$ is therefore

$$\gamma_{p,el}^{(cont)} = -\frac{1}{2\pi} k_F^2 E_F + \frac{4k_F E_F}{3\pi h} \tan\theta . \tag{8.41}$$

We have replaced $L_\perp$ in the denominator by $\tan\theta = h/L_\perp$ with $h$ the step height to bring the electronic contribution into the form of an expansion of the projected surface tension in powers of $p = \tan\theta = h/L_\perp$ (Sect. 4.3.1, eq. (4.46)). For further discussion we consider the surface tension multiplied by the area of one atom on the surface $\Omega_s$ and subtract the value at $p = 0$. Figure 8.28 shows the numerical solution of (8.39) and the continuum approximation for surface states on Cu(111) (Table 8.3). The exact solution (dashed line) oscillates around the continuum solution (solid line). The oscillations become larger for large $p$.

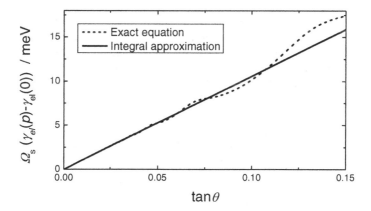

**Fig. 8.28.** Contribution of the electron confinement energy to the surface tension of vicinal surfaces vs. $\tan\theta \equiv p = h/L_\perp$. Plotted is the surface tension is multiplied by the area of one atom on the surface $\Omega_s$ with the value for $p = 0$ subtracted. For large $\tan\theta$, the exact solution (dashed line) oscillates around the approximate continuum solution (solid line).

According to Fig. 8.28 the surface tension of a stepped surface should be up to 17 meV per atom higher than for a flat surface. An interesting question is whether this extra energy is to be attributed to the step energy or a step/step interaction. Intuitively one might be inclined to consider the confinement energy as a step/step interaction[14] Intuition leads astray however. As shown in Sect. 4.3.1, the first term in the expansion of $\gamma_p$ with respect to $\tan\theta = h/L_\perp$ is the step line tension $\beta$ divided by the step height $h$.

$$\gamma_p = \gamma_0 + \frac{\beta}{h}\tan\theta + \ldots \tag{8.42}$$

---

[14] This was erroneuosly claimed by N. Garcia and P. A. Serena [8.67]. The authors considered merely the upshift of the ground state not the integral over the 1D-dispersion.

Hence, the second term in (8.41) is an electronic contribution to the step line tension and not to the step/step interaction. The magnitude of this contribution is

$$\beta_{el}a_{\parallel} = \frac{4}{3\pi} a_{\parallel} k_F E_F \tag{8.43}$$

where $a_{\parallel}$ is the diameter of an atom so that $\beta_{el}a_{\parallel}$ is the step line tension for the length of one atom. For the Cu(111) surface (Table 8.3), the electronic contribution to the step line tension would amount to 100 meV for perfectly reflecting steps on vicinal surfaces. The experimental value obtained from the fluctuations of large islands, which has no contribution from the electron confinement, is 270 meV [8.68]. The step energy on vicinal surfaces should therefore be higher by a considerable amount. The calculation overestimates the effect as in reality the steps are not perfectly reflecting. Sanchez et al. estimated intensity reflection coefficients on Cu(111) vicinals to $R = 0.3$-$0.4$ [8.69]. Bürgi et al. made very careful measurements of the reflection coefficient as a function of energy on individual terraces on Ag(111) and found reflection coefficients of $R = 0.25$ and $0.6$ at $E_F$ for ascending and descending steps, respectively. The energy dependence of the reflectivity followed the expected trend to smaller reflection coefficients for higher energies. Reflection coefficients smaller than $R = 1$ reduce the confinement energy and thereby the electronic contribution to the step line tension (cf. Fig. 8.24).

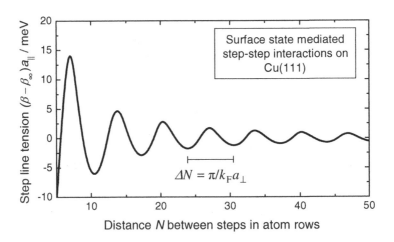

**Fig. 8.29.** Difference in the step line tension of vicinal surfaces to the step line tension at large step distances vs. the step distance. Perfectly reflecting steps are assumed. The oscillations correspond to alternating attractive and repulsive step/step interactions and represent another manifestation of quantum size effects on the energetics of surface structures.

The oscillations of the exact solution (dashed line in Fig. 8.28) can be viewed as oscillations of step line tension with the step distance, or as an oscillating step/step interaction. Figure 8.29 shows the oscillations in the step energy as a function of the distance. The distance $L_\perp$ is denoted in units of atom rows $N = L_\perp / a_\perp$ with $a_\perp$ the length unit of one atom row. The oscillation period is determined by the condition (8.34).

$$\Delta N = \pi / k_F a_\perp = 6.6 \qquad (8.44)$$

As the oscillations in the energy are larger than the elastic repulsive energies between steps, preferred orientations should exist for surfaces vicinal to Cu(111). For example, a vicinal surface with $\tan \theta = 0.07$ should phase-separate into vicinals with $\tan \theta \cong 0.05$ and $\tan \theta \cong 0.09$.

### 8.3.4 Oscillatory Interactions between Adatoms

Oscillatory step/step interactions result from the boundary condition that the surface state wave function be zero at the steps. For the same reason the interaction between other defects should be repulsive and oscillating if the defects represent areas of a high potential for the surface state electrons. The oscillation period is $2k_F D$ with $D$ now the distance between the defects or the distances between shells of high potential around the defect sites. For one-dimensional defects, one expects the oscillatory decay to be faster than for line defects. Long-range oscillatory interactions between surface defects mediated by substrate electrons were predicted by Grimley [8.70] and Einstein and Schrieffer [8.71] many years before any experimental evidence existed. Lau and Kohn showed in 1978 that the interactions caused by a partly filled band of surface states decay as $D^{-2}$ and that the interactions oscillate with a period $2k_F D$ [8.72]. Hyldgaard and Persson [8.73] proposed that the interaction should obey the simple analytical form

$$\Delta G = -E_F \left( \frac{2\sin(\delta_F)}{\pi} \right)^2 \frac{\sin(2k_F D + 2\delta_F)}{(k_F D)^2}. \qquad (8.45)$$

Experimental evidence for the oscillatory interactions between adatoms was first reported in 2000. Repp et al. studied the pair distribution of Cu atoms on Cu(111) surfaces at low temperatures using STM [8.74]. Figure 8.30 shows the Cu-atoms as white 0.4 Å high protrusions and a monolayer high step to an upper terrace. The atoms were deposited at a sample temperature of 15 K. At this temperature, the Cu-atoms are mobile. The STM image was obtained at 9 K where the Cu atoms are immobile enough to produce defined images. Contrary to what one would expect for metals, the Cu-atoms do not coalesce into densely packed islands but rather form local hexagonal arrangements with a closest distance of about 12.5 Å.

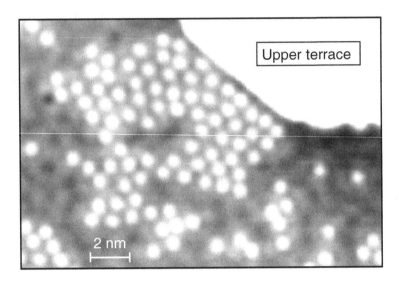

**Fig. 8.30.** STM image of Cu adatoms (white circles) on Cu(111) surfaces at 9 K. The pattern between the adatoms evolves from surface state standing waves. The adatoms do not coalesce but rather form an approximately hexagonal lattice with a lattice constant of 12.5 Å. The adatoms also keep a distance to the ascending step that forms the boundary to the next higher terrace. Both results have considerable consequences for nucleation and growth (Courtesy of Jascha Repp, [8.74]).

This indicates the existence of an attractive interaction with a potential minimum at 12.5 Å and a barrier for nucleation into islands. The interaction potential is determined from the probability to find a Cu atom at a distance $D$ divided by the probability to find an empty surface site at that distance within a given frame size. The result is plotted in Fig. 8.31. The potential has minima at distances $D_{min} = 12.5$ Å, 27 Å, 41.5 Å and 71 Å. The solid line is a fit of (8.45) to the data with an arbitrary pre-factor, a phase shift $\delta_F$ of $0.38\pi$ and the Fermi wave vector of 0.215 Å$^{-1}$. The long-range part of the interaction potential is well reproduced. The model does not include contributions from elastic and electrostatic dipolar repulsions (Sect. 3.4.2) and attractive chemical interactions at short distances.

The surface state mediated oscillatory interaction and in particular, the repulsive interaction at short distances has considerable consequences for nucleation and growth. Without the repulsive interaction, a critical nucleus for layer growth is formed on metal surfaces whenever two atoms meet. The concentration of critical nuclei for growth is then given by the ratio of the flux and the diffusion coefficient (Sect. 11.1). A surface state mediated repulsive interaction and the activation barrier for nuclei formation that goes along with it, reduces the concentration of nuclei, mimicking a surface possessing a smaller diffusion coefficient.

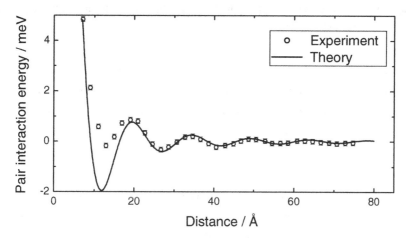

**Fig. 8.31.** Pair interaction energy between Cu-adatoms on Cu(111) surfaces obtained from the pair distance frequency in STM images [8.74]. The solid line is a fit to (8.44) with an arbitrary pre-factor and a phase shift of $0.38\pi$.

# 8.4 Electronic Transport

As one of the traditional topics of thin film physics, electron transport in systems with restricted dimensions has become an issue of fundamental and technological importance in the nanosciences. The following section concentrates on those effects that are quantitatively important and/or concern the transition between classical conduction and ballistic quantum conduction. For the quantitatively small (although theoretically very interesting) effects that go under the name *weak localization* the reader is referred to the tutorial review of G. Bergmann [8.75].

## 8.4.1 Conduction in Thin Films – the Effect of Adsorption

We first reconsider the simple model that was discussed in Sect. 7.5.2 in the context of infrared reflection-absorption spectroscopy. The model concerns the friction between adsorbed particles moving parallel to the surface and a Drude electron gas and yields a simple equation for the change in the resistivity of thin films due to absorption. The magnitude of the effect on the resistivity is related to the infrared absorption properties of modes vibrating parallel to the surface. In equation (7.106) the Fourier component of the displacement of the electrons in $x$-direction was expressed in terms of the electric field along the $x$-axis $\mathscr{E}_x$. In the present context, we are interested in the small frequency limit of that equation. With the current density being

$$j_x = -en\dot{x} = \sigma \mathcal{E}_x, \tag{8.45}$$

with $n$ the electron concentration and $\sigma$ the conductivity, one obtains for the conductivity

$$\sigma = \frac{e^2 n}{m(\tau^{-1} + \tilde{\eta})}. \tag{8.46}$$

The result differs from the conventional Drude conductivity by the friction term $\tilde{\eta} = \eta NM / N_e m$, with $M$ and $m$ the adsorbate and the electron mass, respectively. The ratio of the number of adsorbed particles $N$ and the number of electrons $N_e$ in the film of thickness $D$ can be replaced by

$$\frac{N}{N_e} = \frac{n_s}{n D} \tag{8.47}$$

in which $n_s$ is the area density of the adsorbate atoms. The resistivity of the film $\rho = \sigma^{-1}$ therefore becomes

$$\rho = \rho_0 (1 + \frac{\eta M \tau}{m n} \frac{n_s}{D}) \tag{8.48}$$

where $\rho_0$ is the bulk resistivity. The additional term in (8.48) scales inversely proportional to the film thickness $D$ and proportional to the density of adsorbates $n_s$. The crucial quantity that determines the dependence of the film resistivity on the density is the friction coefficient $\eta$. Its value depends strongly on the type of bonding of the adsorbate with the surface and so do various approaches to calculate $\eta$ from theory [8.76]. The simple model considered above (likewise introduced by Bo Persson) links experimental data in infrared spectroscopy with dc-resistance on thin films. For example for CO on Cu(100), the infrared spectrum in Fig. 7.31 could be fitted with a friction coefficient $\eta = 1.2 \times 10^{11}$ s$^{-1}$. With that number, the relaxation time $\tau = 2.5 \times 10^{-14}$ s$^{-1}$, and the electron density $n = 8.47 \times 10^{22}$ cm$^{-3}$ one would calculate the change in the resistivity for CO on Cu(100) at 100 K to

$$D \frac{\partial \rho}{\partial n_s} = 2300 \ \mu\Omega \ \text{cm} \ \text{Å}^3. \tag{8.49}$$

Experimental measurements of the film resistivity are not available for that particular system, but similar systems show adsorbate induced resistivities of about the same magnitude (Table 8.4).

**Table 8.4.** Initial slope of the product of the resistivity $\rho$ and the film thickness $D$ with respect to the concentration of adsorbates. The values were originally assembled by Persson [8.77] from experimental results published by (a) [8.78] and (b) [8.79]. The values shown in this table are revised. With the exception of (b) the data refer to polycrystalline films.

| System | H/Ni[a] | CO/Ni[a] | CO/Cu[a] | Ag/Ag[b] |
|---|---|---|---|---|
| $D\partial\rho/\partial n_s$ /$\mu\Omega$ cm Å$^3$ | 850 | 2000 | 600 | 1800 |

Measurements of the resistivity are performed on films deposited on glassy substrates. Depending on the annealing temperature such films have a texture to show preferentially (111) surfaces or even quite well ordered (111) surfaces. However, most of the experiments were performed at a time when methods of thin film characterization were less well developed. Preferential adsorption at surfaces with a particular orientation and at defects may therefore obscure the result. An exception is the deposition of silver atoms on an a well ordered Ag(111) film at 10 K (Fig. 8.32). At this low temperature, the Ag atoms are immobile on the time scale of the measurement. The atoms therefore stay where they have hit the surface at random positions. These randomly arranged Ag atoms cause a relatively large change in the resistivity. The solid line in Fig. 8.32 is a fit to

$$D \, \Delta\rho = D\Delta\rho_{max} (1-e^{-C n_s})$$  (8.50)

where $C$ is

$$C = \frac{\partial\rho}{\partial n_s} \frac{1}{\Delta\rho_{max}}.$$  (8.51)

The value in the last column of Table 8.4 was obtained from that fit. The effect on the resistivity disappears for deposition at 350 K because at that temperature silver atoms agglomerate on the surface in large islands of monolayer height.

Changes in the resistivity of thin films can also be observed upon adsorption of ions from an electrolyte. Figure 8.33 shows an example [8.80]. A 20 nm silver film was deposited on a clean MgO surface on which silver grows pseudomorphic with (100) orientation. The resistive stripes had a length of 6 mm and a width of 1.8 mm. The base resistance without adsorption was 12 $\Omega$. The observed increase in the resistance as function of the electrode potential is caused by the isothermal adsorption of I$^-$, Br$^-$, and Cl$^-$. The small hysteresis results from the limited diffusion in the electrolyte in combination with the speed of the potential variation.

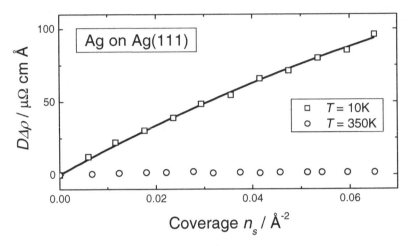

**Fig. 8.32.** Change in the resistance multiplied by the thickness $D = 20$ nm for a silver film vs. coverage with Ag adatoms for two temperatures (after D. Schumacher [8.79]). The resistance does not change if silver is deposited at 350 K since the atoms coalesce into large islands. At low temperatures, the Ag atoms stay as single atoms in random positions as deposited. The friction with the electron gas causes an increase in the film resistivity.

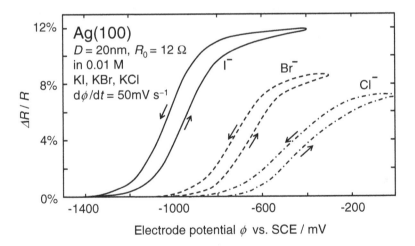

**Fig. 8.33.** Relative change in the resistance of a thin film Ag(100) electrode as function of the electrode potential in 0.01 M KI, KBr, and KCl, respectively (after Winkes et al. [8.80]). Iodine, bromine and chlorine adsorb "*specifically*" on silver. The change in the resistivity follows the isothermal absorption curve (cf. Sect. 6.2.5). The hysteresis results from the limited diffusion in the dilute electrolyte.

The maximum coverage corresponds to a c(2×2) layer at 50% coverage [8.81]. From the dimensions of the sample and the base resistance one calculates values for $D\partial\rho/\partial n_s$ of 720 μΩ cm Å$^3$, 550 μΩ cm Å$^3$, and 430 μΩ cm Å$^3$ for Γ, Br$^-$, and Cl$^-$, respectively. These numbers are of the same order as for chemisorbed species in vacuum.

## 8.4.2 Conduction in Thin Films - Solution of the Boltzmann Equation

Already in 1938, Fuchs calculated the resistivity of thin metal films based on the solution of the Boltzmann transport equation in two-dimensions [8.82]. The material was reviewed (with some corrections made) by Sondheimer in 1952 [8.83]. The theory is therefore referred to as the Fuchs-Sondheimer theory. The corresponding solution for the resistivity of a thin circular wire was reported by Dingle in 1950 [8.84].

The microscopic reason for the increase in the resistivity of thin films is the reduction of the average mean free path of electrons. Electrons, which are accelerated by the external field, may loose their increased momentum in scattering processes when the electron trajectories hit the surface. The effect on the mean free path vanishes if the momentum parallel to the field direction is conserved (mirror reflection). The effect of the scattering is the largest if the scattering is completely random. It is useful to consider primarily these extreme cases and describe the scattering process by a parameter $p$, the probability for mirror reflection with momentum conservation. Since even on a clean and perfectly flat surface there is a finite probability for non momentum-conserving scattering events, the scattering from a perfect surface corresponds to a parameter $p$ that is slightly smaller than one. A rough surface corresponds to $p \ll 1$ and causes a higher resistivity of the film. There is in fact plenty of experimental evidence for an increased resistivity of films when their surfaces are rough [8.78]. The effect of adsorbates on the resistivity may also be considered in terms of a "roughness" since adsorbates represent surface centers for non momentum-conserving scattering.

In order to describe the resistivity of thin films as a function of the mirror reflection coefficient $p$ and the film thickness Fuchs and Sondheimer solved the Boltzmann transport equation for the thin film case. The stationary linearized Boltzmann equation in the relaxation time approximation is (see e.g. [8.85])

$$v(k)\cdot\nabla_r f(k,r,T) - \frac{e}{\hbar}\mathscr{E}\cdot\nabla_k f(k,r,T) = -\frac{f(k,r,T)-f_0(k,r,T)}{\tau(k)}. \quad (8.52)$$

Here, $f(k, r, T)$ is the probability for electrons to possess (under the influence of an electric field $\mathscr{E}$) a vector $k$ at the macroscopic position $r$ and at temperature $T$ and $v(k) = \hbar^{-1}\partial E(k)/\partial k$ is the group velocity. It is important to realize that the Boltzmann equation describes the transport in a semi-continuum approximation which is valid only if the mean free path is large compared to the lattice constant. In other words, the spatial variation of the distribution function must be negligibly small on the atomic length scale and $r$ must be understood as a macroscopic position vector. For a bulk system, the distribution function does not depend on $r$. The solution of the Boltzmann equation is then straightforward if one assumes a $k$-independent relaxation time $\tau$. The distribution function $f(k, T)$ deviates from the Fermi equilibrium distribution $f_0(k, T)$ by a small amount $f_1(k, T)$ because of the electric field $\mathscr{E}$. We assume that the field is in $x$-direction and obtain

$$f(k) = f_0(k) + f_1(k) = f_0(k) + \frac{e}{\hbar}\tau\mathscr{E}_x\frac{\partial f_0}{\partial k_x} . \tag{8.53}$$

The electric current density $j$ is

$$j = -\frac{e}{8\pi^3}\int v(k)f(k)\,\mathrm{d}^3k . \tag{8.54}$$

The integral over the $k$-space converts into an integral over the Fermi-surface, which is easily calculated for a free electron gas metal with a spherical Fermi surface to yield the Drude conductivity

$$j = \sigma\mathscr{E} = \frac{e^2\tau n}{m^*}\mathscr{E} , \tag{8.55}$$

in which $m^*$ is the effective mass and $n$ is the carrier concentration. We note that the electron concentration $n$ here appears incidentally and not as a result of an ansatz as in (8.45) (see [8.85]).

If one admits surfaces that scatter the conduction electrons randomly, the deviation of the distribution function from equilibrium due to the electric field is smaller at the surface than in the bulk; hence, the distribution function depends on the position with respect to the surface. The non-vanishing gradient of $f$ with respect to $r$ in (8.52) complicates the solution of the Boltzmann equation considerably. We introduce surfaces in the $xy$-plane at $z = 0$ and at $z = D$ and assume perfect random scattering at the surfaces for the moment. At the surface $z = 0$ the deviation of the distribution function from equilibrium $f_1(k, z=0)$ must vanish for $k_z > 0$. Vice versa, $f_1(k, z=D)$ must vanish for $k_z < 0$. The distribution function therefore splits into two functions, one for $k_z > 0$ and another one for $k_z < 0$. To be able to calculate an analytical expression for the current density one must introduce the simplification of spherical Fermi-surface so that the $k$-vector can be replaced by the group velocity $\hbar k = mv$ and $\mathrm{d}f/\mathrm{d}k_x = (\hbar/m)(\mathrm{d}f/\mathrm{d}v_x)$. The Boltzmann equation (8.52) is then

$$\frac{\partial f_1}{\partial z} + \frac{f_1}{\tau v_z} = \frac{e\mathscr{E}_x}{mv_z}\frac{\partial f_0}{\partial v_x} \tag{8.56}$$

The solutions that fulfill the boundary conditions are

$$f_1^+(v,z) = \frac{e\tau\mathscr{E}_x}{m}\frac{\partial f_0}{\partial v_x}\{1-\exp(-z/\tau v_z)\} \qquad v_z > 0$$

$$f_1^-(v,z) = \frac{e\tau\mathscr{E}_x}{m}\frac{\partial f_0}{\partial v_x}\{1-\exp((a-z)/\tau v_z)\} \quad v_z < 0$$

(8.57)

The distribution function $f_1^+(v,z)$ is plotted in Fig. 8.34.

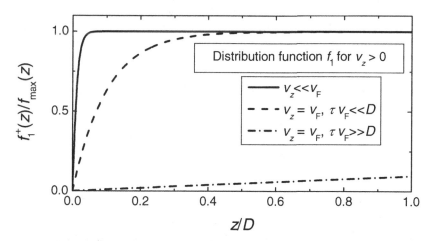

**Fig. 8.34.** Boltzmann distribution function $f_1^+$ for $v_z > 0$ in a thin film of thickness $D$ vs. the distance $z$. See text for discussion.

If the electron moves parallel to the film plane ($v_z \ll v_F$) the distribution function is at its maximum in almost the entire film save for a small range near $z = 0$, which shrinks to zero width as $v_z$ approaches zero (solid line). For electrons traveling perpendicular to the film ($v_z = v_F$) the limiting distribution function depends on the ratio of the mean free path $\Lambda = \tau v_F$ to the thickness. For $\Lambda \ll D$, $f_1^+(v,z)$ still approaches its maximum value within the film. If $\Lambda \gg D$, then $f_1^+(v,z)$ stays small in the entire film. For the calculation of the current density and the conductivity one introduces polar coordinates ($v, \theta, \varphi$) in the $v$-space such that $v_z = v\cos\theta$. The current density then becomes

$$j(z) = -\frac{2e^2 m^2 \mathscr{E}_x}{h^3} \int_0^\infty dv \int_0^{2\pi} d\varphi \, \tau v^3 \frac{\partial f_0}{\partial v} \cos^2 \varphi$$

$$\times \left[ \int_0^{\pi/2} \sin^3\theta \left\{ 1 - \exp\left( -\frac{z}{\tau v \cos\theta} \right) \right\} d\theta \right. \tag{8.58}$$

$$\left. + \int_{\pi/2}^{\pi} \sin^3\theta \left\{ 1 - \exp\left( -\frac{D-z}{\tau v \cos\theta} \right) \right\} d\theta \right] .$$

The current density depends now on the position $z$ because the distribution function does. The total current density is obtained by integration over the film thickness. Relatively simple analytical solutions of the rather beastly integral (8.58) exist in the limit of small and large thicknesses [8.83]. These solutions are

$$\frac{\rho}{\rho_0} = 1 + \frac{3}{8}\frac{\Lambda}{D}(1-p) \quad D/\Lambda \gg 1 \tag{8.59}$$

$$\frac{\rho}{\rho_0} = 1 + \frac{4}{3}\frac{1-p}{1+p}\frac{\Lambda}{D\ln(\Lambda/D)} \quad D/\Lambda \ll 1. \tag{8.60}$$

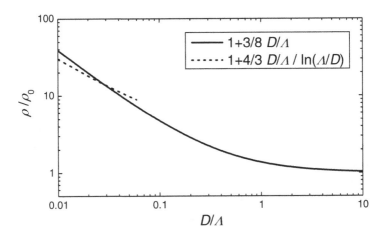

**Fig. 8.35.** Ratio of thin film resistivity to bulk resistivity vs. the ratio of the mean free path $\Lambda$ to the thickness $D$ for $p = 0$ according to (8.59) and (8.60) (solid and dashed line, respectively).

We have admitted the possibility of partial randomness of the surface scattering $(1-p)$ in (8.59) and (8.60). The case $D/\Lambda \gg 1$ has the same functional dependence on the film thickness $D$ as derived for the adsorbate-induced resistivity (8.48) although the physical picture is different here. In applications of current interest, the other limit $D/\Lambda \ll 1$ is more important. Given experimental uncertainties, the function valid for the alternative limit $D/\Lambda \gg 1$ (8.59) describes experimental data quite well also for $D/\Lambda \ll 1$ (Fig. 8.35).

### 8.4.3 Conduction in Space Charge Layers

Electrical conduction in the space charge layers at surfaces of semiconductors or at the interfaces of semiconductor-heterostructures differs from metallic conduction insofar as the carrier density varies as a function of the position $z$ from the interface (Sect. 3.2.2). Because of the strong variation of the carrier concentration, it is not meaningful to define a conductivity in the space charge layer but rather a conductance with reference to the bulk.

$$\Delta\sigma = e(\mu_{ns}\Delta N + \mu_{ps}\Delta P) \qquad (8.61)$$

Here, $\mu_{ns}$ and $\mu_{ps}$ are the mobilities of electrons and holes in the space charge layer, respectively and $\Delta N$ and $\Delta P$ are the excess concentrations of electrons and holes defined as

$$\Delta N = \int_0^\infty [n(z) - n_b]\,dz \qquad \Delta P = \int_0^\infty [p(z) - p_b]\,dz \qquad (8.62)$$

with $n_b$ and $p_b$ the bulk concentrations of electrons and holes, respectively. Because of the integration over the $z$-axis, the surface conductance has the dimension $\Omega^{-1}$. Figure 8.36 shows the surface conductance of a slightly $p$-doped silicon sample versus the surface voltage $\phi_s$. For negative $\phi_s$ (corresponding to upwards bending of the bands, see panel (a)), one has a positive excess concentration of holes $\Delta P$. By definition, the surface conductance is zero for a flat band situation (panel (b)) and can even become negative when the Fermi-level at the surface is farther apart from either the conduction or the valence band than in the bulk (panel (c)). For positive $\phi_s$, the excess concentration of electrons rises exponentially with $\phi_s$, and so does the surface conductance. For very positive or negative surface voltages, in particular for higher doping levels the space charge layer becomes very thin (Sect. 3.2.2). The confinement of the charge carriers into the thin layer causes additional scattering processes with the surface, leading to a reduced mobility compared to the bulk. For a full theoretical treatment of that reduction, one would have to consider the solution of the Boltzmann transport equation in the space charge layer with the additional complication of a strong electric field $\mathscr{E}_z$ perpendicular to the direction of carrier transport. A remarkably simple argument how this reduction should depend on the electric field $\mathscr{E}_z$ was brought forward by Schrieffer [8.86].

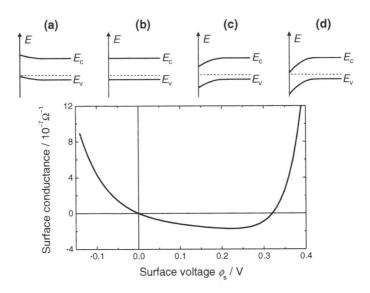

**Fig. 8.36.** Surface conductance of a slightly $p$-doped silicon sample as a function of the surface voltage (after W. Mönch [8.87]). The corresponding band bending is sketched in the panels (**a-d**) at the top of the graph.

Here, Schrieffer's reasoning is presented in a slightly modified form. We consider the mobility in non-degenerate accumulation layers (panel (a) in Fig. 8.36) and in non-degenerate inversion layers (panel (d) in Fig. 8.36). A fraction (1-p) of the charge carriers is assumed to scatter randomly at the surface. The mean velocity of those carriers leaving the surface after a scattering event equals the mean velocity in one direction in a Boltzmann gas

$$\langle |v_x| \rangle = \sqrt{k_B T / 2\pi m^*} \tag{8.63}$$

where $m^*$ is the effective mass of the charge carrier. The electric field in the space charge layer drives the charge carrier back to the surface. The mean time after that process is completed is

$$\tau = \frac{2m^*}{e \mathscr{E}_z} \langle |v_x| \rangle = \frac{\sqrt{2k_B T m^* / \pi}}{e \mathscr{E}_z} . \tag{8.64}$$

The mean time for a random surface scattering is therefore $\tau_{s0} = \tau/(1-p)$ .

According to the Matthiesen rule, the scattering from phonons and the surface scattering are independent. Noting that $\mu_b = e\tau_b/m^*$ the ratio of surface to bulk mobility becomes

$$\frac{\mu_s}{\mu_b} = \frac{1}{1 + \tau_b/\tau_{s0}} = \left(1 + \mu_b \frac{(1-p)\sigma_{sc}}{\varepsilon\varepsilon_0\sqrt{2k_BT/(\pi m^*)}}\right)^{-1} \tag{8.65}$$

where we have replaced the electric field at the surface by the area density of the total charge in the space charge layer $\sigma_{sc} = \varepsilon\varepsilon_0\mathscr{E}_z$.

The reduction of the mobility in space charge layers and the general trend to lower mobility for increasing charge density has been observed many times (For a review on the earlier literature see e.g. [8.88]). Figure 8.37 shows examples of experimental data on non-degenerate inversion layers in mildly $p$-doped silicon. The solid line is a fit to the Schrieffer model for $(1-p) \approx 0.05$. For further details on the conduction in space charge layers, the reader might consult the tutorial volumes of W. Mönch and H. Lüth [8.87, 89].

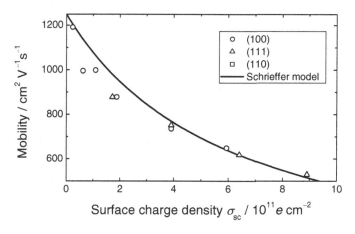

**Fig. 8.37.** Mobility in non-degenerate n-inversion channels of $p$-doped silicon (100), (111) and (110) surfaces (after Sah et al. [8.90]). The doping level was $N_A = 2.2\times10^{14}$ cm$^{-2}$. The solid line is a fit with the Schrieffer model for $(1-p) \approx 0.05$.

### 8.4.4 From Nanowires to Quantum Conduction

The problem of electronic conduction in long and thin wires can also be treated within the framework of the Boltzmann transport equation. As in the case of thin films, analytic expressions are available in the limits that the diameter $D$ of the wire is much larger or much smaller than the mean free path $\Lambda$ [8.83, 84].

$$\frac{\rho}{\rho_0} = 1 + \frac{3}{4}\frac{\Lambda}{D}(1-p) \quad D/\Lambda \gg 1 \tag{8.66}$$

$$\frac{\rho}{\rho_0} = 1 + \frac{1-p}{1+p}\frac{\Lambda}{D} \quad D/\Lambda \ll 1 \tag{8.67}$$

To bring these equations into perspective of the dimensions of technical electric wiring we recall that the mean free path for copper is about 40 nm at room temperature and increases, to the order of millimeters at 4 K, depending on the quality of the sample. These dimensions are in the range of present device technology. Electron transport in systems with restricted dimensions is therefore of fundamental as well as of technological interest, and has become a field of intense activity. By using special preparation techniques, wires with diameters in the nm range can be prepared as single crystals. For example, on 4° vicinal surfaces of Si(100) the steps form doublets and these doublets bunch together upon Ag-deposition. Along these bunches, crystalline silver nanowires grow with a width of 200-1000 nm, heights of about 150 nm and lengths up to 100 μm [8.91].

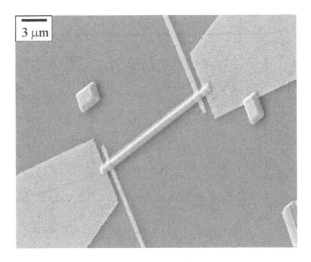

**Fig. 8.38.** Scanning electron microscopy image of a single crystal silver nanowire with contact leads for a four-probe resistance measurement (courtesy of M. Hartmann and G. Dumpich).

Figure 8.38 shows an example of such a nanowire with contact leads for a four-probe resistance measurement. These nanowires are ideal objects to study the effect of surfaces on electric conduction while remaining in the limits of classic transport. The dotted line in Fig. 8.39 shows the measured resistance of a nanowire like the one displayed in Fig. 8.38. Its length was determined to

$L = 16.5$ μm. The temperature dependence of the resistance follows closely the temperature dependence of pure bulk silver material (dashed line). The effective cross section of the nanowires $A_{eff}$ can therefore be calculated from the temperature dependent part of the resistance and the resistivity of pure bulk silver. To calculate the contribution from surface scattering we replace the temperature dependent mean free path $\Lambda$ in (8.67) by the resistivity $\rho_0(T)$

$$\Lambda(T) = \frac{m * v_F}{e^2 n \rho_0(T)} \tag{8.68}$$

where $v_F$ is the Fermi velocity, $m*$ the effective mass and $n$ the electron concentration. If one further replaces the diameter $D$ by the square root of the effective area $A_{eff}^{1/2}$ the resistance $R(T)$ becomes

$$R(T) = \rho(T)\frac{L}{A_{eff}} = \left(\rho_0(T) + \frac{1-p}{1+p}\frac{m v_F}{e^2 n}\frac{1}{\sqrt{A_{eff}}}\right)\frac{L}{A_{eff}}. \tag{8.69}$$

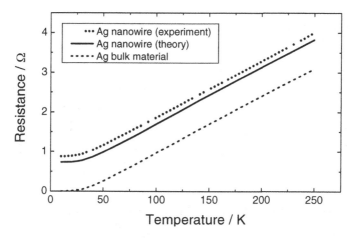

**Fig. 8.39.** Experimental resistance of a nanowire as function of temperature (dotted line) [8.92]. The dashed line represents the resistance of a thick wire made from pure silver whose cross section is matched to scale with the temperature dependent part of the resistance of the nanowires. The solid line is calculated from (8.69).

The solid line in Fig. 8.39 is calculated from (8.69) under the assumption of perfect random scattering from the surfaces ($p = 0$) and agrees quite well with the experiment. The assumption of random surface scattering is meaningful as the silver wires have a rough surface and as they are covered by a carbonaceous layer of residues from the manufacturing and contacting process.

If the length of the nanowire between the two contact plates (Fig. 8.38) becomes smaller, the diffusive transport (Fig. 8.40a) blends into the ballistic electron transport from one reservoir into another. Electrons moving along the direction of the wire ($z$-direction) pass the nanowire without a scattering event (Fig. 8.40b). Electrons, which enter the wire at some larger angle with respect to the axis of the wire, may still lose their momentum in the forward direction by scattering from the surfaces. The possibility of such electron trajectories however diminishes if the energy levels become discrete with respect to $k_x$ and $k_y$ because of the lateral confinement of the wave function. Only $k_z$ remains as a continuous variable and electrons can acquire momentum continuously only along the $z$-direction. This is the regime of ballistic quantum transport. The number of conductance channels in the $z$-direction depends on the cross section of the wire. For simplicity, we discuss the case of a rectangular shaped wire with dimensions $D_x$ and $D_y$. As the surface boundary condition, we assume the electron wave function is zero at the surface (infinitely high potential wall). We assume further a spherical Fermi surface.

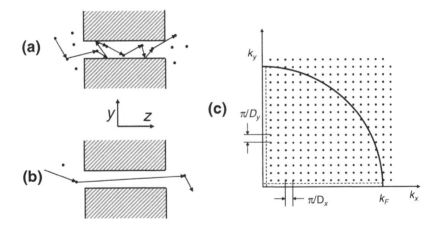

**Fig. 8.40.** Illustration of (a) diffusive and (b) ballistic transport from one reservoir on the left side to another reservoir on the right. (c) The dots indicate the allowed $k$-vectors. The solid line is the circular cross section through the Fermi-surface, which is assumed to be spherical. The number of 1D channels participating in ballistic quantum transport is given by the number of dots inside the circle.

The allowed $k$-vectors are marked in Fig. 8.40c as dots in the $k_x$, $k_y$ plane. Note that the $k_x = 0$ and $k_y = 0$ axes are excluded, as the wave function would be identical to zero for either $k_x = 0$ or $k_y = 0$. With the help of Fig. 8.40c, the number of allowed states $N$ in the continuum limit is easily calculated as the ratio of the area of the Fermi quarter circle to the area occupied by the allowed states.

$$N = \frac{D_x D_y}{\pi^2} \left\{ \frac{1}{4} \pi k_F^2 - \frac{1}{2} \frac{\pi}{D_x} k_F - \frac{1}{2} \frac{\pi}{D_y} k_F + \frac{1}{4} \frac{\pi}{D_x} \frac{\pi}{D_y} \right\}$$

$$= \frac{k_F^2}{4\pi} D_x D_y - \frac{k_F}{2\pi} (D_x + D_y) + \frac{1}{4} \, . \tag{8.70}$$

The product $D_x D_y$ in the leading first term of (8.70) is the cross section area. In the limit of very thick wires, the number of conduction channels depends therefore only on the cross section area, not on the shape. The remaining two terms are first order corrections to the asymptotic result and do depend on the shape. For small dimensions, $N$ becomes a discrete function of $D_x$ and $D_y$. The minimum cross section is given by the condition that a single channel must fit into the wire. From Fig. 8.40c we see that for a rectangular wire both dimensions $D_x$ and $D_y$ must exceed

$$D_{x,y} > \frac{\pi\sqrt{2}}{k_F} \, . \tag{8.71}$$

Since for metals the minimum dimension $D_{x,y}$ approximately amounts to the diameter of an atom, ballistic quantum transport through a single metal atom or a chain of metal atoms should be possible and was indeed observed. Before we turn to the experimental side, we derive the conductance of a single channel. The current in one channel is (cf. 8.54)

$$I = e \int_{E_F^{(L)}}^{E_F^{(R)}} D^{(1)}(E) v(E) \, dE \tag{8.72}$$

where $D^{(1)}(E)$ is the one-dimension density of states of the spin-degenerate free electron states

$$D^{(1)}(E) = \frac{1}{\pi \, \partial E / \partial k_z} \, , \tag{8.73}$$

and $v(E)$ is the group velocity in the one-dimensional band

$$v(E) = \frac{1}{\hbar} \frac{\partial E}{\partial k_z} \, . \tag{8.74}$$

$E_F^{(L)}$ and $E_F^{(R)}$ are the Fermi levels in the bulk reservoirs at the left and right side of the channel so that

$$V = (E_F^{(R)} - E_F^{(L)})/e \qquad (8.75)$$

is the applied voltage. The current is therefore simply

$$I = \frac{2e^2}{h}V . \qquad (8.76)$$

Hence, the conductance of each channel is

$$G_0 = \frac{2e^2}{h} = 7.74809 \times 10^{-5}\ \Omega^{-1} = 1/(129064\ \Omega). \qquad (8.77)$$

The total conductance is the product of $G_0$ and the number of channels $N$. Experimental manifestations of the conduction quantum came first from the quantum Hall effect, there as quantum steps of $e^2/h$ (instead of $2e^2/h$) since the spin degeneracy is lifted in high magnetic fields. The high precision by which these quantum jumps can be measured permitted the definition of the resistance unit $\Omega$ in terms of universal constants. Quantized conduction in the sense considered here is readily observed in so-called *break-junctions* where a mechanical contact between two electrodes is broken. The conductivity immediately before the contact breaks is quantized in units of $G_0$. While quantum conduction through atom size wires is a very active field of research with a sophisticated methodology of its own (for a review see [8.93]) quantum jumps in the conductance can even be observed in conventional mechanical relays. Figure 8.41 shows the quantum jumps during contact breaking in a relay with AuCo electrodes [8.94].

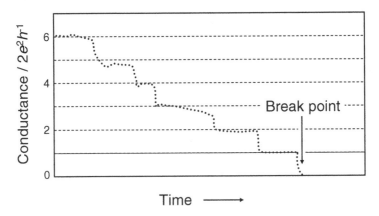

**Fig. 8.41.** Conductance jumps in a conventional mechanical relay with AuCo electrodes (after Hansen et al. [8.94])

These quantum jumps could have been discovered with the technology that was available around 1930 or so. It is amusing to speculate how quantum physics might have developed if the discovery were made at the time!

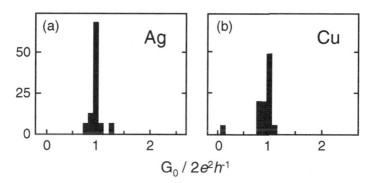

**Fig. 8.42.** Probability distribution for finding a conductivity $G_0$ during repeated attempts to contact a single silver or copper atom on a Ag(111) or Cu(111) surface with a tunneling tip (after Limot et al. [8.95]).

We have argued that a metal wire, which has the conductance of $G_0$, should have the diameter of about one atom. From observations like in Fig. 8.41 and in other specially designed break-junctions one would therefore conclude that the last contact before fracture is made by a single atom or a short chain of single atoms. That the conductance of a single atom is indeed $G_0$ can be proven with the help of an STM. Limot et al. have placed the STM tip directly over Ag and Cu adatoms on the Ag(111) and Cu(111) surfaces and have measured the tunneling current as a function of distance until contact was made. The conductance was nearly always equal to $G_0$ or at least very close to this value as shown in Fig. 8.42.

# 9. Magnetism

## 9.1 Magnetism of Bulk Solids

Magnetic phenomena have fascinated humanity for millennia. Scientific research on magnetic properties of materials began with the development of electrification in the late 19th century. Ever since then, research on magnetism has been an important part of Material Science. The understanding of magnetism as a collective quantum property is one of the finest achievements of Solid State Physics. The diversity and complexity of magnetic phenomena as well as the range of applications was further enlarged with the development of Surface and Thin Film Physics, and more recently with the Nanosciences. Currently, research is spurred by technological demands for larger and faster data storage devices. Present day computer performance were not feasible without the specific use of magnetic properties of interfaces, thin films and nanostructures, be it in sensors employing the *Giant Magneto Resistance* (GMR) effect or in high-density magnetic storage media. The finite dimensions, the crystallography, the specific composition and the presence of interfaces in thin film systems and in nanostructures have a pronounced and occasionally unexpected effect on the magnetic properties. This chapter is devoted to these aspects. As a preparation, however we need to familiarize ourselves with some basic properties of bulk magnetism.

### 9.1.1 General Issues

The possibility to store information in the orientation of magnetic moments in small structures is based on the existence of a magnetic anisotropy, which keeps the magnetization parallel or antiparallel to a particular orientation, thereby providing the physical realization of binary digital information. For bulk materials, the source of anisotropy is the magneto crystalline anisotropy. The spin orientation is coupled to the orientation of the orbitals in the solid, and thereby to the crystal orientation by the spin-orbit coupling. Because of the spin-orbit coupling, the energy stored in the magnetization of a ferromagnetic crystal depends on the orientation of the magnetization relative to the crystal axes. The energies associated with the magneto crystalline anisotropy are relatively small. Nevertheless, without the magnetic anisotropy even the simplest macroscopic manifestation of magnetism, the permanent magnet, would not exist!

The orientation with the lowest energy is called the *easy axis* or easy orientation. A ferromagnetic crystal in thermal equilibrium has its magnetization along the easy axis. For hcp cobalt, e.g. this is the hexagonal c-axis. The magnetization will therefore be either parallel to the c-axis. In order to minimize the energy of the external field, the magnetization breaks up into domains of parallel and antiparallel orientations. If the crystal is annealed to above the Curie temperature and cooled down again, the number and size of parallel and antiparallel domains will be about equal, so that little magnetization is noticed from the outside. In an external magnetic field oriented along the direction of the c-axis, the magnetization rises quickly with an applied field by moving the boundaries between the domains and the magnetization saturates at comparably small fields, since the external field need not work against the magnetocrystalline anisotropy (Fig. 9.1). Merely the domain walls have to move to let one type of domain grow at the expense of the other (insert in Fig. 9.1).

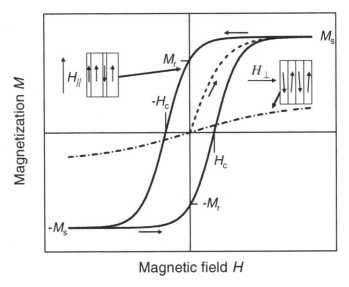

**Fig. 9.1.** Schematic plot of the magnetization $M$ of a ferromagnet with a single easy axis versus the magnetic field $H$. Dashed line: initial magnetization starting from zero magnetization, $H$ parallel to easy axis. Solid lines: magnetization hysteresis, $H$ parallel to easy axis. Dash-dotted line: $H$ perpendicular to easy axis.

Once a single domain or nearly a single domain crystal is obtained, the domain structure may remain with a preponderance of the once achieved orientation even when the external field is removed. The remaining magnetization is called the *remanence* $M_r$. To bring the magnetization of the sample back to zero one needs to reverse the external field up to a value $-H_c$ that is called the *coercive field*. With increasing magnetic field in the reverse direction, the magnetization eventually saturates again, now in the reverse direction. In total one runs through a hysteresis loop. The loop is symmetric if the system is symmetric with respect to the $H_{\parallel}$-axis.

A magnetization perpendicular to the c-axis in response to an external field requires the energy to turn the magnetization out of the easy direction (insert Fig. 9.1) The magnetization therefore rises much less with increasing external field in that case, and there is no remanent magnetization of the sample as a whole. As soon as the external field is removed the magnetization in the internal domains snaps back to be parallel or antiparallel to the c-axis.

The magnitude of the remanence $M_r$ depends on how easy domain walls can move. Perfect crystals and certain alloys (*μ-metal, permalloy, ferrites*) have a small remanence. Defects tend to pin domain walls and make the crystal magnetically harder. In some technical applications, e.g. in transformer cores, one wants no remanence to keep the energy loss low. For data storage on the other hand, one likes a remanence near saturation.

### 9.1.2 Magnetic Anisotropy of Various Crystal Structures

The crystalline magnetic anisotropy is specific to the crystal structure. To express the change of the energy per volume with the orientation of the magnetization quantitatively one expands the energy in terms of the directional cosines with the crystal axes and keeps the lowest and second lowest non-vanishing terms. It is customary to write the directional cosines with the axes i as $\alpha_i$. Only even orders have nonzero coefficients because of time reversal symmetry; Inversion of time reverses the orientation of the spin and the energy is invariant with respect to time reversal. The expansion is furthermore written in such a way that the identity

$$\sum_i \alpha_i^2 \equiv 1 \tag{9.1}$$

is obeyed. For a cubic crystal the expansion up to fourth order would formally be

$$u_{\text{cub}} = u_2(\alpha_1^2 + \alpha_2^2 + \alpha_3^2) + u_4(\alpha_1^4 + \alpha_2^4 + \alpha_3^4) \tag{9.2}$$

The first term is constant and can therefore be omitted; using (9.1) the second term can be rearranged to

$$u_{\text{cub}} = -2u_4(\alpha_1^2\alpha_2^2 + \alpha_1^2\alpha_3^2 + \alpha_2^2\alpha_3^2) \tag{9.3}$$

One sees from this result that there was no need to include mixed terms of the type $\alpha_1\alpha_2$ into the formal expansion (9.2), as they are automatically included. With the fourth order expansion (9.3) the ratio of the energies along the space diagonal and the area diagonal are fixed to 4/3. To allow for a deviation from that ratio one needs to include a sixth order term. The formally simplest way to do that is by writing the anisotropy as

$$u_{\text{cub}} = K_1(\alpha_1^2\alpha_2^2 + \alpha_1^2\alpha_3^2 + \alpha_2^2\alpha_3^2) + K_2\alpha_1^2\alpha_2^2\alpha_3^2 \tag{9.4}$$

in which $K_1$ and $K_2$ are the anisotropy constants. For iron $K_1 = 4.2\times10^4\,\text{Jm}^{-3}$ and $K_2 = 1.5\times10^4\,\text{Jm}^{-3}$ at room temperature[15]. The easy directions are therefore $\langle 100\rangle$, the hard directions $\langle 111\rangle$. The difference in energy between an easy and a hard direction is $1.5\times10^4\,\text{Jm}^{-3}$. For nickel the constant $K_1$ is negative ($K_1 = -5.7\times10^3\,\text{Jm}^{-3}$, $K_2 = -2.3\times10^3\,\text{Jm}^{-3}$), which makes $\langle 111\rangle$ the easy direction.

For hexagonal crystals, the second order term in the expansion does not vanish. When the hexagonal axis is denoted by the index 3, the expansion up to forth order is formally

$$u_{\text{hex}} = u_{2\parallel}\alpha_3^2 + u_{2\perp}(\alpha_1^2 + \alpha_2^2) + u_4\alpha_3^4. \tag{9.5}$$

There is no fourth order term in $\alpha_1$ and $\alpha_2$ because of the hexagonal symmetry. The second term is equal to $u_{2\perp}(1-\alpha_3^2)$ and can be integrated into the first term. The anisotropy can therefore be expressed in terms of a single angle, the angle $\theta$ with the c-axis

$$u_{\text{hex}} = u_2\alpha_3^2 + u_4\alpha_3^4 = -K_2\cos^2\theta - K_4\cos^4\theta \tag{9.6}$$

in which $K_2$ and $K_4$ are the anisotropy constants. The negative signs are chosen to be in keeping with the standard notation. For cobalt, the constants are $K_2 = 4.1\times10^5\,\text{Jm}^{-3}$ and $K_4 = 1.0\times10^5\,\text{Jm}^{-3}$. The c-axis is therefore the easy axis. Changing the orientation of the magnetization from the easy to the hard direction costs an energy of $5\times10^5\,\text{Jm}^{-3}$, which is 30 times more than for iron. As cobalt has a smaller saturation magnetization than iron (1400 G vs. 1700 G at room temperature, for magnetization $M$: 1 G = 1000 A/m), the large energy difference is entirely due to fact that the second order effects do not vanish as for cubic symmetry.

Thin films frequently grow pseudomorphic with the substrate up to a critical thickness $t_c$ (Sect. 4.2.5). Pseudomorphic films of cubic materials grown on a (100) and (111) surfaces become tetragonal and hexagonal, respectively. Furthermore, the alloys MnAl, CoPt and FePt that have a large magnetocrystalline anisotropy belong to the tetragonal class. It is therefore useful to consider also tetragonal

---

[15] As elsewhere in this volume we use SI units and write equations accordingly. It is unfortunate that the SI-system introduces the magnetization $M$ differently than the electric polarization $P$: It is $D = \varepsilon_0 E + P$, but $B = \mu_0(H+M)$. The awkward consequence is that in the SI system $H$ and $M$ have the same dimensions, but the conversion factors from Gaussian units are different: For $H_{\text{Gaussian}}$ into $H_{\text{SI}}$ it is 1 Oe = 1 G $\Rightarrow$ $1000/4\pi$ Am$^{-1}$; for $M_{\text{Gaussian}}$ into $M_{\text{SI}}$ it is 1 emu cm$^{-3}$= 1 G $\Rightarrow$ 1000 A m$^{-1}$. For the conversion between SI units and Gaussian units an article of Arrot is quite useful [9.1].

crystals. The expansion of the magnetic energy density in terms of the directional cosines for a tetragonal material can be written as

$$u_{\text{tetr}} = u_{2\parallel}\alpha_3^2 + u_{2\perp}(\alpha_1^2 + \alpha_2^2) + u_{4\parallel}\alpha_3^4 + u_{4\perp}(\alpha_1^4 + \alpha_2^4). \tag{9.7}$$

The angular dependence of the second order term is again described already by the first term, since $\sum_i \alpha_i^2$. It is customary to introduce the angles with the tetragonal axis and with one of the basal axes as $\theta$ and $\varphi$, so that $\alpha_3 = \cos\theta$, $\alpha_1 = \sin\theta\cos\varphi$, and $\alpha_2 = \sin\theta\sin\varphi$. After some algebra (9.7) acquires the form

$$u_{\text{tetr}} = -K_2\cos^2\theta - \frac{1}{2}K_{4\perp}\cos^4\theta - \frac{1}{2}K_{4\parallel}\frac{1}{4}(3 + \cos 4\varphi)\sin^4\theta. \tag{9.8}$$

We have again introduced the anisotropy constants such as to be in keeping with the conventional notation in the literature.

$$K_2 = u_{2\perp} - u_{2\parallel}, \; K_{4\parallel} = -2u_{4\parallel}, \; K_{4\perp} = -2u_{4\perp}. \tag{9.9}$$

Since the tetragonal distortions in pseudomorphic films of cubic materials may be rather small, it is useful to be able to relate the anisotropy constants for the tetragonal system to those of the cubic system in the limit of vanishing tetragonal distortion. It is not at all obvious from (9.8) how to do that. However, the task is easily performed by using the generic expansions (9.2) and (9.7). If the tetragonal distortion vanishes one has $u_2 = u_{2\parallel} = u_{2\perp}$ and $u_4 = u_{4\parallel} = u_{4\perp}$. Using the transformation from (9.2) to (9.3) one obtains

$$K_{4\parallel} = K_{4\perp} = K_1(\text{cubic}). \tag{9.10}$$

An access to the in general larger second order term $K_2$ for a slightly tetragonally distorted cubic crystal is obtained by considering the effect of *magnetostriction*. Magnetostriction in the classical understanding describes the change in the dimensions of the sample upon reorientation of the magnetization. In the context here, it is useful to consider energy density as a function of the elastic strain tensor $\varepsilon_{ij}$ (Sect. 3.3.1) and the orientation of the magnetization. For a cubic material the *magnetoelastic energy density* $u_{\text{me}}$ is [9.2]

$$\begin{aligned} u_{\text{me,cub}} = {} & B_1(\varepsilon_{11}\alpha_1^2 + \varepsilon_{22}\alpha_2^2 + \varepsilon_{33}\alpha_3^2) \\ & + 2B_2(\varepsilon_{12}\alpha_1\alpha_2 + \varepsilon_{23}\alpha_2\alpha_3 + \varepsilon_{13}\alpha_1\alpha_3). \end{aligned} \tag{9.11}$$

In the most general case, the magnetoelastic coupling coefficients $B$ are a fourth rank tensor $B_{ijkl}$. A fourth rank tensor has three independent components for cubic systems, which reduce to two in this case because of the normalization condition (9.1). In terms of the tensor components the coefficients $B_1$ and $B_2$ in (9.11) are

$$B_1 = B_{1111} - B_{1122}, \quad B_2 = 2B_{2323} . \tag{9.12}$$

The magnetoelastic energy for the hexagonal system is [9.2]

$$
\begin{aligned}
u_{\text{me,hex}} = {}& B_1(\varepsilon_{11}\alpha_1^2 + 4\varepsilon_{12}\alpha_1\alpha_2 + \varepsilon_{22}\alpha_2^2) + B_2\varepsilon_{33}(1-\alpha_3^2) \\
& B_3(\varepsilon_{11} + \varepsilon_{22})(1-\alpha_3^2) + 2B_4(\varepsilon_{23}\alpha_2\alpha_3 + \varepsilon_{13}\alpha_1\alpha_3)
\end{aligned}
\tag{9.13}
$$

with

$$
\begin{aligned}
B_1 &= B_{1111} - B_{1122}, \quad B_2 = B_{3311} - B_{3333} \\
B_3 &= B_{1122} - B_{1133}, \quad B_4 = 2B_{1313} .
\end{aligned}
\tag{9.14}
$$

The values of $B_1$ and $B_2$ for Fe, Ni, fcc-Co and hcp-Co are tabulated in Table 9.1. As an example, we consider the energy density of a (100) oriented film of a cubic material, which experiences an isotropic strain in the basal $x_1$, $x_2$-plane and is free to expand in the $x_3$-direction. The condition that the film is stress free in the $x_3$-direction relates the strain $\varepsilon_{33}$ to $\varepsilon_{11} = \varepsilon_{22}$

$$\tau_{33} = c_{11}\varepsilon_{33} + 2c_{12}\varepsilon_{11} = 0 . \tag{9.15}$$

**Table 9.1.** Magneto elastic coupling constants in MJm$^{-3}$ for Fe, Ni and fcc-Co (after Sander [9.3]).

| Material | $B_1$ | $B_2$ | $B_3$ | $B_4$ |
|----------|-------|-------|-------|-------|
| bcc Fe | −3.43 | 7.83 | | |
| fcc Ni | 9.38 | 10 | | |
| fcc Co | −9.2 | 9.38 | | |
| hcp Co | −9.1 | −29 | 29.2 | 29.4 |

The energy density (9.11) becomes thereby

$$
\begin{aligned}
u_{\text{me,cub}} &= B_1(\varepsilon_{11}(\alpha_1^2 + \alpha_2^2) + \varepsilon_{33}\alpha_3^2) \\
&= B_1\varepsilon_{11}\left(1 - (1 + 2c_{12}/c_{33})\alpha_3^2\right).
\end{aligned}
\tag{9.16}
$$

The second order anisotropy constant $K_2$ for the tetragonal distorted cubic lattice is therefore

$$K_2 = B_1 \varepsilon_{11} (1 + 2c_{12} / c_{33}) \,.$$ (9.17)

An intensively studied system is Ni(100) epitaxially grown on Cu(100). Up to about 10 monolayers of nickel grow pseudomorphic. The strain $\varepsilon_{11}$ is therefore +2.5%. The nickel film expands in the basal plane and therefore contracts along the surface normal. With $B_1$ taken from Table 9.1 one obtains

$$K_2 = 5.35 \times 10^5 \ \mathrm{Jm}^{-3} \,.$$ (9.18)

Note that because of the definition of $K_2$ (9.10) this means that the easy axis is along the surface normal. The calculated value for $K_2$ is in good agreement with the experimental value determined for thin epitaxial nickel films on Cu(100) ($K_2 = 4.4 \times 10^5 \ \mathrm{Jm}^{-3}$ [9.4]). It is quite remarkable that the relatively small tetragonal distortion of 2.5% causes an anisotropy that is two orders of magnitude larger than the anisotropy of bulk nickel. The easy axis is therefore entirely determined by the strain. Because of the positive sign of $K_2$ the energy is minimal for $\theta = 0$, that is for a perpendicular orientation of the magnetization. As a warning, it should be mentioned that the real system Ni/Cu(100) is more complicated because other sources of anisotropy to which we attend later.

## 9.2 Magnetism of Surfaces and Thin Film Systems

The finite dimensions, the crystallography, the specific composition and the presence of interfaces in thin film systems and in nanostructures have pronounced and occasionally unexpected effects on the magnetic properties. To study these effects a large variety of experimental tools have been developed which we describe in the following on an elementary level.

### 9.2.1 Experimental Methods

The simplest question one might ask concerning surface and thin film magnetism is that of the magnitude of the magnetization. As simple as the question is, it is not so easily addressed experimentally as genuinely surface sensitive methods are difficult to calibrate for quantitative measurements. Vice versa, the classical methods to determine the magnitude of magnetic moments in a bulk material such as *Ferro Magnetic Resonance* (FMR), and the *Torsion Oscillation Magnetometry* (TOM) are not particularly sensitive and not surface specific. However, with considerable experimental effort these methods have been made sensitive enough to probe the magnetization in thin and ultra-thin films [9.5, 6]. Both FMR and TOM integrate over the entire volume of the material. The magnetization of a thin film can be measured only when grown on a non-magnetic material.

A relatively straightforward access to hysteresis loops and therefore to the orientation of the easy axis is provided by the *Magneto-Optic Kerr Effect* (MOKE). This experiment measures the change of the polarization state of a light beam upon reflection from a magnetic material. One distinguishes three types of Kerr-effects: the longitudinal, the polar and the transverse Kerr-effect. The longitudinal effect turns s-polarized into elliptic polarized light upon reflection when the magnetization is parallel to the surface and in the scattering plane of the light. The polar effect turns likewise linear polarized light in elliptically polarized light, but for a magnetization perpendicular to the surface. The transverse effect changes the orientation of the ellipse for elliptically polarized light when the magnetization is in the surface plane and perpendicular to the scattering plane. In all cases, the change in the ellipticity is proportional to the magnetization. MOKE is easily integrated into a UHV-system for thin film and surface analysis. All it takes are two windows in the UHV-system at appropriate positions and means to subject the sample to a magnetic field. Conventional optical elements can be used as the entire optical setup remains outside the vacuum chamber. Because of the easy integration and because of its high sensitivity, MOKE has become a workhorse for experiments in thin film magnetism. Just as FMR and TOM, MOKE measures integral magnetic properties. As the magneto-optical constants of thin films are not known, only relative values of magnetizations are obtained. In current research on periodic magnetic nanostructures the magneto-optic Kerr effect in diffracted beams has become a valuable tool [9.7]

Experimental techniques that use electrons have genuine surface sensitivity. The probing depth depends on the energy of the electrons that are employed (Sect. 2.2.2). Several techniques that use electrons have been developed. One is the measurement of the asymmetry in the diffraction of spin-polarized electrons. The technique has been named *Spin Polarized Low Energy Electron Diffraction* (SPLEED). In this experiment, electrons from a source of spin-polarized electrons are diffracted from a surface of a magnetic material. Measured is the asymmetry of the exchange scattering

$$A_{ex} = \frac{1}{P_0} \frac{I_{\uparrow\uparrow} - I_{\uparrow\downarrow}}{I_{\uparrow\uparrow} + I_{\uparrow\downarrow}} \tag{9.19}$$

Here $P_0$ is the polarization of the incident beam and $I_{\uparrow\uparrow}$ and $I_{\uparrow\downarrow}$ are the intensities of the non spin-flip and spin-flip events. Measuring $A_{ex}$ requires sources for spin polarized electrons and the spin analysis of the diffracted electrons. The latter is performed either with the *Mott detector* or with a LEED-detector. Both detectors work with the asymmetry of the cross section for scattering parallel or anti-parallel to the vector $S \times k$, with $S$ the spin orientation and $k$ the k-vector of the electron. Responsible for this asymmetry is the spin-orbit coupling. The LEED detector determines the asymmetry with the backwards diffracted beams for an incident beam at normal incidence. The asymmetry of the intensity

$$A = \frac{I(hk) - I(\overline{h}\overline{k})}{I(hk) + I(\overline{h}\overline{k})} \qquad (9.20)$$

for a given electron energy is proportional to the spin polarization of the incident beam.

The standard source for spin-polarized electrons is the GaAs photocathode [9.8] or lately the strained layer $GaAs_{0.95}P_{.05}$ photocathode [9.9]. Due to spin-orbit splitting of the valence band states, electrons photo-excited into the conduction band by circular polarized light are spin polarized. By treating the (110) surface of the GaAs-crystal with cesium and oxygen a negative electron affinity can be achieved so that electrons in the conduction band have an energy above the vacuum level and are therefore photo-emitted. The polarization of the photo-emitted electrons varies between 40% for the standard GaAs cathode and 75% for the strained layer $GaAs_{0.95}P_{.05}$ cathode. The polarization direction is along the surface normal.

GaAs sources are also used in other types of experiments with spin-polarized electrons such as *Spin Polarized Low Energy Electron Microscopy* (SPLEEM) and *Spin Polarized Electron Energy Loss Spectroscopy* (SPEELS). SPLEEM images the surface in the light of a particular diffracted electron beam. Magnetization contrast can be obtained for all orientations if the spin of the primary electrons are appropriately rotated with respect to the beam direction by electromagnetic fields. The lateral spatial resolution of LEEM is a few nm and the vertical resolution suffices to see monatomic steps. The very complicated, expensive equipment and the difficulty to operate a LEEM system have prevented its widespread use so far. The same can be said about SPEELS. For the investigation of the magnetic excitation spectrum of thin films, no alternative to SPEELS exist. Special applications to spin waves are discussed in Sect. 9.9.

A technique that combines high spatial resolution of about 5-10 nm presently with the complete determination of the polarization vector is *Scanning Electron Microscopy with Polarization Analysis* (SEMPA) [9.10]. This instrument is based on conventional *Scanning Electron Microscopy* (SEM) equipped with the standard non-polarizing electron cathode. The information on the magnetization of the sample lies in the polarization of the secondary electrons. A LEED diffraction detector with a pair of channel electron multipliers detect the scattering asymmetry (9.20) in the plane set up by the incident electron beam and the surface normal. A second pair of multipliers measures the asymmetry orthogonal to this plane and therefore in the surface plane. By rotating the surface around the incident electron beam, contrast is achieved for all orientations of the magnetization. With proper calibration specific to the system investigated, the method is quantitative in terms of the orientation and magnitude of the magnetization.

The availability of synchrotron sources for high-intensity tunable wavelength X-rays has enabled a new type of magnetic spectromicroscopy based on the *Magnetic Circular X-ray Dichroism* (MCXD) [9.11]. Magnetic circular dichroism is the dependence of the absorption cross section on the helicity of the light. For the 3d-transition metals e.g. the $L_{2,3}$ edge absorption involves transitions from the $2p_{3/2}$ and $2p_{1/2}$ state into the empty 3d-states above the Fermi level (Fig. 2.15). For the ferromagnets Fe, Co and Ni, the densities of empty spin up and spin down states are different, and so are the absorption cross sections for left- and right-handed circular polarized light. The sign of the difference depends on the orientation of the magnetization. The difference in the X-ray absorption leads to a locally different emission of secondary electrons. Hence, the intensity of these secondary electrons reflects the orientation of the magnetization in the sample. An image of the secondary electron emission therefore bears magnetic domain contrast. The image is element specific as the X-ray absorption has a sharp threshold at the L-edge and decays thereafter and the photon energy for L-edge absorption is element specific. Imaging of the secondary electrons is performed with the *Photo Emission Electron Microscope* (PEEM). In this instrument, the secondary electrons are accelerated in a cathode lens to 10-20 KeV, pass one or two projective lenses, and form a magnified image of the surface on a channel plate electron multiplier. The optics is identical to the viewing optics in LEEM. The ultimate lateral resolution of PEEM is 5nm and 20 nm for instruments featuring magnetic and electrostatic lenses, respectively. In combination with X-ray absorption dichroism the working resolution is of the order of 100 nm. Attempts are underway to improve the resolution by adding elements that filter the energy to reduce chromatic aberrations and to correct for angular aberrations. By employing the *Magnetic Linear X-ray Dichroism* (MLXD) domain contrast is obtained also for antiferromagnets [9.12].

The ultimate lateral resolution is provided by the *Spin Polarized Scanning Tunneling Microscope* (SP-STM) [9.13, 14]. The instrument is based on the sensitivity of the tunnel current to the relative orientation of the majority spins in tip and substrate. The effect is known as *Tunnel Magneto Resistance* (TMR). One way to achieve magnetic contrast is to coat a conventional tungsten tip with a few monolayers of a ferromagnetic material such as Fe, Gd, GdFe [9.13]. The coverage should be limited to a few layers to avoid high magnetic field that might switch the domain orientation in the substrate. Using this technique domain walls with a width of 1.1 nm have been resolved on samples from hexagonal cobalt. An approach alternative to coating is the use of magnetically soft materials for the tip such as the metallic glass CoFeSiB [9.14]. By ac-modulation of the magnetization with the help of a coil wrapped around the tip, magnetic and topological contrast is separated.

Lateral resolutions of about 30 nm are achieved in *Magnetic Force Microscopy* (MFM) [9.15]. As the image contrast is caused by magnetic forces, the contrast does not reflect the polarization of the spin state in the solid. MFM is therefore less suitable for fundamental studies. Nevertheless, the instrument has its place in imaging magnetic contrast on a routine basis because it is a low effort instrument, robust and easy to handle.

## 9.2.2 Magnetic Anisotropy in Thin Film Systems

For thin film systems and nanostructures, further sources of anisotropy exist. Firstly, there is an anisotropy associated with the shape of the film or the shape of the nanostructure that results from the depolarization field. Secondly, the surface and the interface to the substrate is a source of anisotropy. If the interface is between a ferromagnetic film and an antiferromagnetic substrate, the magnetization of the ferromagnetic film may be fixed by the exchange coupling across the material boundary, which leads to an *exchange bias* for the magnetization of the film.

Magnetic nanostructures are frequently single domain magnets. For thin films, the domain size is large compared to the thickness. In those cases, a *shape asymmetry* arises from the asymmetry of the depolarization field[16]. Consider for example a homogenously magnetized thin film. If the film is magnetized perpendicular to the film plane then the boundaries give rise to a depolarizing field $H_d = -M$ (in S.I. units). If the film is polarized parallel to the plane then there is no depolarizing field. If the direction of the magnetization is turned from parallel to perpendicular, the perpendicular component builds up gradually. To calculate the energy associated with the depolarization effect one has to integrate the differential form of the magnetostatic energy density $u$

$$du = -\mu_0 M_\perp dH_\perp .$$
(9.21)

With $\theta$ the angle between the orientation of the magnetization and the axis perpendicular to the film one obtains after integration

$$u = \frac{1}{2}\mu_0 M^2 \cos^2 \theta .$$
(9.22)

Here, $\mu_0$ is the vacuum permeability (=$4\pi\times10^{-7}$ Vs/Am). In most papers on thin film magnetism, cgs units are used. In these units, (9.22) assumes the form $u = 2\pi M^2 \cos^2 \theta$. For cobalt the depolarization energy is $1.23\times10^6$ Jm$^{-3}$ at room temperature, about a factor of two larger than the crystalline anisotropy energy of $5.1\times10^5$ Jm$^{-3}$. A Ni(100) film would have a magnetocrystalline anisotropy energy of $-2\times10^3$ Jm$^{-3}$ in favor of the [111] direction. The depolarization energy is $1.5\times10^5$ Jm$^{-3}$. The easy axis should therefore be always in the surface plane, if it were not for the two other sources of anisotropy, the strain as discussed above and the surface and interface anisotropy.

The source of surface and interface anisotropy is the modified spin orbit coupling or the exchange coupling at the surface or the interface [9.16]. The energy associated with that asymmetry is proportional to the film area, not the volume. If

---

[16] The magnetic depolarization field is the analog of the electrostatic depolarization field. In the electrostatic case, the field is caused by the charge density of the surface arising from the termination of the polarization.

one still considers the energy per volume, as is conventionally done, then the surface and interface contributions to the anisotropy are proportional to the inverse of the thickness. We consider the technical simple case of a strained cubic film with (100) surfaces and the rotation of the magnetization from the perpendicular orientation to the parallel orientation in the [100] zone. The anisotropy energy is then

$$u = \left\{ \frac{1}{2} \mu_0 M^2 - B_1 \varepsilon_{11} (1 + 2 c_{12} / c_{11}) - \frac{K_s + K_{int}}{t} \right\} \cos^2 \theta$$

$$+ K_1 \cos^2 \theta \sin^2 \theta \ . \tag{9.23}$$

Here, $K_s$ and $K_{int}$ are constants describing the surface and interface anisotropy and $t$ is the thickness of the film. A positive (negative) value means that the interface and surface anisotropy favor the perpendicular (parallel) orientation. Equation (9.23) offers a wealth of different scenarios, in particular since the anisotropy constants as well as the magnetization are temperature dependent. We consider a few simple cases.

### Case I: $K_s + K_{int} > 0$, $K_1 > 0$, $\varepsilon_{11} = 0$; the thickness t is varied.

For very small thickness, $u$ is minimal for $\theta = 0°$, the perpendicular orientation of the magnetization. Above a critical thickness $t_c$

$$t_c = 2(K_s + K_{int}) / \mu_0 M^2 \tag{9.24}$$

the magnetization flips into the parallel direction. This is the frequently occurring, classical situation.

### Case II: $K_s + K_{int} < 0$, $K_1 > 0$, $B_1 \varepsilon_{11} > 0$; the thickness t is varied.

For very small thickness, $u$ is minimal for $\theta = 90$, the parallel orientation. Beyond a critical thickness $t_c$

$$t_c = 2(K_s + K_{int}) / (\mu_0 M^2 - 2 B_1 \varepsilon_{11} (1 + 2 c_{12} / c_{11})) \tag{9.25}$$

the magnetization flips into the perpendicular direction. This is basically the mechanism that causes a reorientation of the magnetization between one and two monolayer thick iron films on W(110) from parallel to perpendicular [9.17]. For a quantitative analysis, the (110) orientation of the iron film as well as higher order contributions of the strain to the magnetic anisotropy have to be taken into account [9.3]. The sharp reorientation transition between one and two monolayers can be exploited to make an interesting structure of nanowires. On stepped W(110) surfaces the iron film grows nucleationless from the steps. For a coverage slightly above one monolayer, the surface organizes in wires of double layers with

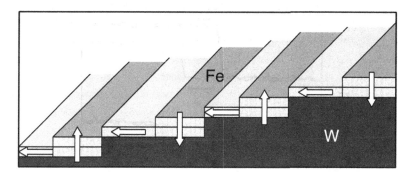

**Fig. 9.2.** Monolayer and by-layer stripes of Fe(110) on W(110) prepared via epitaxial growth [9.17]. The magnetic orientation changes between perpendicular and parallel.

perpendicular magnetization, separated by areas with parallel magnetization (Fig. 9.2). To minimize the field energy the perpendicular magnetized double layerwires are alternatively polarized up and down. If one has a domain boundary in the wire so that the magnetization changes from up to down, then the entire staircase follows. The magnetic stripes have been observed directly by spin dependent scanning tunneling microscopy [9.13].

### Case III: $K_s + K_{int} < 0$, $K_1 < 0$, $\varepsilon_{11} > 0$; the temperature is varied.

For a particular range of constants (9.23) has a minimum for a finite angle $\theta_{min}$ is given by

$$\sin^2 \theta_{min} = 0.5 + \left\{ \begin{aligned} & B_1 \varepsilon_{11}(1 + 2c_{12}/c_{11}) \\ & + \left(K_s(T) + K_{int}(T)\right)/t - \frac{1}{2}\mu_0 M^2(T) \end{aligned} \right\} / 2K_1(T) . \quad (9.26)$$

For tutorial purposes, we neglect the temperature dependence of the interfacial anisotropy constants and of the magnetoelastic constant $B_1$ and merely consider the temperature dependence of the magnetization. The magnetization is large for low temperatures. We assume that the term in the bracket is negative so that the complete second term in (9.26) is positive (Note $K_1 < 0$!). The angle $\theta_{min}$ is larger than 45° and may be close to or exactly 90° depending on the magnitude of the second term. As the magnetization decreases with temperature, the second term can pass through zero. Now the $K_1$-term makes the $\theta = 45°$-orientation the preferred one. At high temperature finally, the second term may become negative and the magnetization is parallel to the film. In total, one can have a continuous change in the orientation of the magnetization from $\theta = 90°$ to 0°.

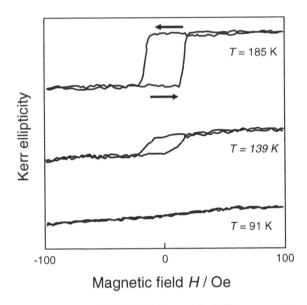

**Fig. 9.3.** Kerr ellipticity of a 8 ML Ni(100) film on Cu(100) measured by the polar Kerr effect. The magnetic field is oriented perpendicular to the plane (after Farle et al.[9.5]). At low temperatures, the magnetization is in-plane. At higher temperatures, the magnetization rotates out of the film plane causing the characteristic hysteresis (Fig. 9.1).

Such a continuous variation occurs for Ni(100) films on Cu(100). Figure 9.3 shows the results of an experiment on an 8.2 monolayer thick Ni(100) film pseudomorphic with the Cu(100) substrate [9.5] using the *Magneto-Optic Kerr Effect* (MOKE). The results shown in Fig. 9.3 were obtained with the magnetic field applied perpendicular to the film-plane. As discussed in Sect. 9.1 and illustrated with Fig. 9.1, the absence of a hysteresis implies that the magnetic field is perpendicular to the easy axis. For the Ni(100) film, this is case at low temperatures (Fig. 9.3); the magnetization lies therefore in the film plane at this temperature. With increasing temperature, the magnetization rotates out of plane. For the 8 ML film, perpendicular orientation of the magnetization is reached at 185 K.

Another case of practical importance is that of hexagonal cobalt layers growing with the c-axis perpendicular to the surface. The energy density up to second order in $\cos^2\theta$ is

$$u = \left\{ \begin{matrix} \frac{1}{2}\mu_0 M^2 - K_2 - (B_1 + 2B_3 - 2B_2 c_{13}/c_{33})\varepsilon_{11} \\ -\frac{K_s + K_{\text{int}}}{t} \end{matrix} \right\} \cos^2\theta . \qquad (9.27)$$

An experimental example is the growth of hexagonal cobalt on an Au(111) film on top of a W(110) substrate. Up to four monolayers of cobalt, the interface anisotropy in combination with the uniaxial anisotropy keeps the easy axis perpendicular to the film. Between four and five monolayers the influence of the interface anisotropy becomes small enough so that the easy axis flips into the film plane [9.18, 19].

### 9.2.3 Curie Temperature of Low Dimensional Systems

The anisotropy has also has a strong influence on the Curie temperature $T_c$ of thin films, in particular for ultra-thin films of a few monolayers thickness. The reason is that according to the Mermin-Wagner theorem [9.20] a single magnetic monolayer has no long-range order at finite temperature, unless there is some magnetic anisotropy. The effect of the anisotropy on the Curie temperature of ultra-thin films has been studied theoretically for the Heisenberg model. The Hamiltonian for this model is

$$H = -\frac{1}{2} J \sum_{i,\delta} S^{(i)} \cdot S^{(i+\delta)} - \frac{1}{2} K \sum_j (S_\perp^{(j)})^2 \qquad (9.28)$$

The innocent looking Hamiltonian is in fact rather complex as it is non-linear in the spin operators $S$. The indices $i$ and $\delta$ in the first sum run over all spins in the system and their nearest neighbors. $J$ is the isotropic exchange coupling constant. The second term introduces anisotropy in the first layer with the anisotropy constant $K$ and a summation over all spins $j$ in that first layer. The anisotropy is of the same type as the bulk crystalline anisotropy, i.e. proportional to $S^2$, although here also a polar anisotropy is conceivable, if the substrate is a magnetized ferromagnet or an antiferromagnet (*exchange bias*).

Solutions of (9.28) were studied by Erickson and Mills using the Monte Carlo method [9.21]. The result for a monolayer is shown in Fig. 9.4 as squares. For this particular case an analytical solution exists from renormalization group theory, which reads [9.21, 22]

$$T_c(2D)/T_c(3D) = 2/\ln(\pi^2 J / K) \qquad (9.29)$$

This equation is plotted as the solid line in Fig. 9.4. The fact that very small anisotropies $K/J = 0.01$ suffice to bring the Curie temperature of the monolayer up to 30% of the bulk value makes the much celebrated Mermin-Wagner Theorem a somewhat artificial one.

Erickson and Mills investigated solutions to (9.28) for film thicknesses up to six monolayers assuming $K/J = 0.1$. The result is shown in Fig. 9.5 as circles. The bulk value of the Curie temperature is approached rather quickly with increasing thickness. It is instructive to compare the exact result of Erickson and Mills to the

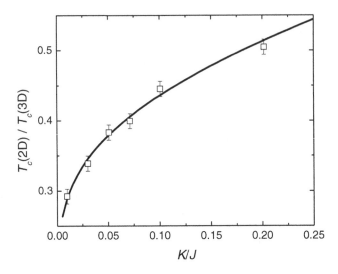

**Fig. 9.4.** Ratio of the Curie temperature of a square monolayer to the Curie temperature of a simple cubic lattice in the Heisenberg model. Even an extremely small anisotropy of $K/J = 0.01$ pushes the Curie temperature of the 2D-system from zero to 30% of the bulk value.

mean-field model. In the mean-field model, the exchange interaction with the neighboring spins in (9.28) is replaced by a coupling to the mean spin orientation. For a bulk system the temperature dependence of the magnetization is then easily calculated as (see e.g. [9.23])

$$\frac{M(T)}{M_{\mathrm{s}}} = \tanh\left(\frac{T_{\mathrm{c}}}{T}\frac{M(T)}{M_{\mathrm{s}}}\right). \tag{9.30}$$

$M(T)$ is the temperature dependent magnetization, $M_{\mathrm{s}}$ is the saturation magnetization and $T_{\mathrm{c}}$ is the Curie temperature given by

$$T_{\mathrm{c}} = \nu J / 4k_{\mathrm{B}} \tag{9.31}$$

with $\nu$ the number of nearest neighbors and $k_{\mathrm{B}}$ the Boltzmann constant. The same two equations hold for a monolayer, so that the Curie temperature of a monolayer is reduced merely according to the reduced number of nearest neighbors. A (100) layer of an fcc crystal has four instead of twelve nearest neighbors; a (111) layer six instead of twelve, so that $T_c$ reduces to one third and one half of the bulk Curie temperature, respectively. This is in gross disagreement with the Mermin-Wagner theorem and due to the total neglect of fluctuations by the mean field model. However, for real systems which have always some anisotropy the mean field model is useful as it serves as an interpolation scheme and provides an easy access to those physical properties that are less affected by fluctuations.

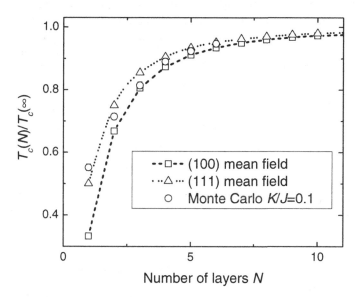

**Fig. 9.5.** Ratio of the Curie temperature for an $N$-layer film to the bulk Curie temperature as a function of the number of monolayers $N$. Circles are Monte Carlo solutions of the Heisenberg model with a surface anisotropy of $K/J = 0.1$ [9.21]. The triangles and squares are mean-field solutions of the Ising-model with nearest neighbor coupling for (111) and (100) oriented films of fcc- crystals, respectively.

Within the mean field model, the Curie temperature as well as the temperature dependence of the magnetization is easily calculated by a self-consistent solution of the mean field coupling in and between layers. With the reduced magnetization and temperature

$$m(T) = M(T)/M_s, \, t = T/T_c \qquad (9.32)$$

one has a set of equations for the $N$-layer system.

$$
\begin{aligned}
m_1(t) &= \tanh\!\big(t(\nu_\perp m_2(t) + \nu_\| m_1(t))/12\big) \\
m_i(t) &= \tanh\!\big(t(\nu_\perp m_{i-1}(t) + \nu_\| m_i(t) + \nu_\perp m_{i+1}(t))/12\big) \qquad (9.33) \\
m_N(t) &= \tanh\!\big(t(\nu_\perp m_{N-1}(t) + \nu_\| m_N(t))/12\big)
\end{aligned}
$$

Here, $\nu_\|$ and $\nu_\perp$ are the number of nearest neighbors within a layer and between two layers, respectively. The set of equation is solved by starting from an arbitrary initial magnetization, e.g. $m_i \equiv 1$ on the right hand side, solve for the resulting new $m_i(t)$, insert this set into the tanh-function, and repeat the process until convergence is achieved. Figure 9.5 shows the Curie-temperatures as a function of the number of layers for the (100) and (111) orientation of an fcc film. The mean field model can easily be adapted to have an arbitrary Curie temperature of the monolayer film, by introducing an effective number of nearest neighbors or a different exchange coupling constant $J$ in the surface layer. The model then serves as an interpolation scheme.

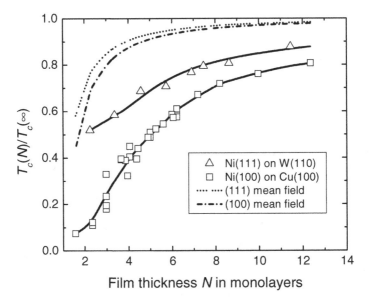

**Fig. 9.6.** Compilation of experimental data on the Curie temperatures for Ni(111) and Ni(100) films in relation to the bulk Curie temperature (after Baberschke [9.4]). The solid lines are to guide the eye. The dotted and dash-dotted lines are the mean fileld results from Fig. 9.5. The large deviation shows that the itinerant ferromagnet is not well described by models with localized spins.

Experimental studies on the Curie temperature of thin films concern mostly 3d-transition metals. A compilation of data for nickel films is shown in Fig. 9.6 [9.4]. The data on the Ni(100) surface were obtained by different groups and by using different techniques. Whereas the experimental results for $T_c$ agree quite well among each other, they fall far below the prediction of the Heisenberg model, regardless of the special approach to solve the Hamiltonian (9.28). In particular, the difference for thicker layers is quite striking. It constitutes a qualitative failure of the Heisenberg model for a proper description of nickel films. Nickel, along with the two other 3d-transition metals Co and Fe, is a so-called *itinerant ferromagnet*. The ferromagnetism is associated with delocalized spin-up and spin-down electrons that occupy different electron bands. As for the Heisenberg model, one can devise a simple mean field model which relates the Curie-temperature to the average magnetic moment and the exchange splitting between the spin-up and spin-down band at zero temperature (see e.g. [9.23]). The mean field model fails on the quantitative side as it overestimates the Curie temperature by far. This is again because of the neglect of fluctuations, in particular of nonlinear spin wave excitations. The mean field model predicts furthermore that the exchange splitting should vanish at $T_c$. Experiment as well as theory has shown that the exchange splitting persists above $T_c$ within a correlation length of about 25 Å [9.24-26]. One might therefore speculate that the Curie temperature is affected when the system dimensions are of the order of the spin correlation length. Quantum size effects (Sect. 9.4.3) may also play a role.

### 9.2.4 Temperature Dependence of the Magnetization

The temperature dependence of the magnetization in thin film systems has been studied quite extensively. Aside from the Curie temperature itself, the critical behavior near the Curie temperature has attracted a substantial interest. Close to the Curie-temperature $T_c$ the temperature dependence of the magnetization is described by

$$M(T) \propto \left(1 - T/T_c\right)^{\beta_1},  \tag{9.35}$$

in which $\beta_1$ is the critical exponent. Many attempts have been made to determine critical exponents for particular thin film systems and surfaces and to compare them to theory. A considerable number of experimental techniques were developed for this purpose. Despite this entire effort very little can be said in terms of definitive and general statements. This section discusses some of the reasons for this failure.

One reason is that various experimental techniques measure the magnetism with a different spatial resolution and probing depth (Sect. 9.2). Some techniques integrate over the entire film area and depth. Techniques that employ spin polarized electrons measure the magnetism in the near-surface region weighted by the not too well known electron escape depth (Sect. 2.2.2) or by some completely unknown spin persistence length. For surface and thin films systems, the magnetization is different in each layer, as we shall see, and each layer appears to posses a different critical exponent near the Curie temperature. Experiments that probe the magnetization with different depth resolution are therefore bound to give different results.

A second reason is that the magnetic properties depend very critically on the number of layers, on the surface anisotropy, on the strain, on the surface structure and on the concentration of defects. It is nearly impossible to have a full characterization of the systems investigated and establish consistency among different groups. To aggravate the situation the true critical exponent is revealed only very close to the Curie temperature where the magnetization is low and even minute inhomogeneities of the system have a dramatic effect on the magnetic moment. Not infrequently, the effect of inhomogeneities is all too obvious from experimental data when the magnetization as a function of temperature displays a tail above $T_c$ rather than dropping to zero with a critical exponent. Extracting critical exponents further away from the Curie-temperature from such data is a fruitless exercise as the critical exponent is only defined in the limit $T \rightarrow T_c$.

The simple mean field model introduced in the previous chapter is very suitable to illustrate the problems mentioned above. We consider first the magnetization of a thick film of 50 (100) fcc layers in Fig. 9.7. The results of the simulation are plotted versus a reduced temperature

$$T_{red} = 1 - T / T_c . \tag{9.34}$$

A critical behavior therefore corresponds to a straight line in the double-logarithmic plot of Fig. 9.7 and the slope is the critical exponent $\beta_1$. The bulk layer in the center of the film (solid line) has the critical exponent of (about) 0.5, as follows analytically from an expansion of (9.30) around $T_c$. A single layer has the same critical exponent, yet a lower $T_c$ because of the lower number of nearest neighbors (9.31). Thus the surface layer by itself would posses a Curie temperature $T_{c,surf} = T_{c,bulk} / 3$. However, because of the coupling to the bulk, the Curie temperature of the surface layer is dragged up to the Curie temperature of the entire slab, which for a 50 layer slab is practically equal to the Curie temperature of the 3D crystal. The temperature dependence of the magnetization in the surface layer and the layers beneath is a consequence of this dragging. It is not possible to determine meaningful critical exponents from experimental data between $T_{red} = 0.1$ and 0.01 because the deeper layers first follow the magnetization of the bulk and then gradually approach the slope of the surface layer. This change in slope occurs closer to $T_c$ the deeper the layer is. Three sets of experimental data prove the point. The open and crossed squares represent the spin asymmetry of electrons diffracted from Ni(100) and Ni(110) surfaces, respectively [9.27]. The asymmetry is assumed proportional to the magnetization. The circles represent the spin polarization of 10 eV secondary electrons as measured by Abraham and Hopster [9.28] for a Ni(110) surface. Again, the spin polarization is believed to be proportional to the magnetization. All three data sets are matched to each other at high $T_{red}$ to eliminate the unknown proportionality factors. While the data obtained from spin polarized electron diffraction agree for the two surface orientations, they fall below the data obtained from the spin polarization of secondary electrons, at least in an intermediate range of $T_{red}$. The difference must be attributed to the larger mean information depth of 10 eV secondary electrons. This interpretation is corroborated by the fact that Alvarado et al. found a systematic trend to lower apparent critical exponents for lower electron energies. Both experiments for the most part cover a temperature range where the true critical behavior is not reached.

Figure 9.8 shows the results for the magnetization in the surface layer for film thicknesses varying between $N = 3$ and $N = 50$ monolayers. The apparent critical exponents vary between nearly 1.0 for the three-layer slab and 0.5 for a thick slab. Experimental data obtained with an experimental technique sensitive to the magnetization of the surface layer will therefore show a strong layer dependence of the apparent critical exponents.

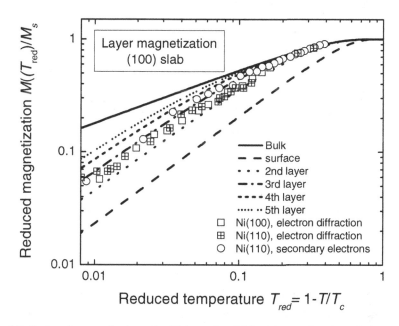

**Fig. 9.7.** Reduced magnetization of a 50 layer fcc (100) slab as a function of the reduced temperature $T_{red} = 1 - T/T_c$ according to the mean field model. The apparent critical exponent varies between 0.5 for the bulk layer (solid line) to 1 for the surface layer (dashed line). Squares and circles represent experimental data of Alvarado et al. [9.27] and Abraham and Hopster [9.28], respectively.

Figure 9.9 represents the same simulation, but now the mean magnetization of the layer is shown. All curves stay close to the bulk curve. The apparent critical exponents vary between 0.5 and 0.56. The largest deviation is for the 11-layer slab. It is therefore obvious that experiments probing the magnetization with a different depth resolution and differently close to the Curie temperature must come to completely different answers with regard to critical exponents.

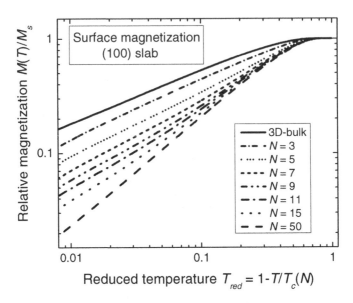

**Fig. 9.8.** Magnetization in the surface layer of slabs comprising of $N$ (100) layers of an fcc crystal as a function of the reduced temperature $T_{red}$. The reduced temperature $T_{red}$ refers to the specific Curie-temperature $T_c(N)$ of each $N$-layer slab.

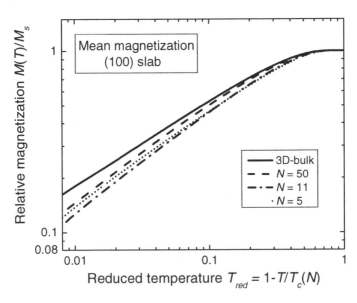

**Fig. 9.9.** Mean magnetization in slabs comprising of $N$ (100) layers of an fcc crystal as a function of the reduced temperature $T_{red}$. $T_{red}$ is calculated with the specific Curie temperature of each slab $T_c(N)$.

## 9.3 Domain Walls

### 9.3.1 Bloch and Néel Walls

In the absence of an external field, a ferromagnet has domains in which the magnetization is polarized along one of the easy axes. To go from one orientation to the next the spins need to rotate. The energy associated with the magnetic anisotropy would be minimal if that transition would occur abruptly, hence if the domain wall thickness were one lattice constant. The exchange energy, on the other hand is minimal if the spins rotate as little as possible from one site to the next. The total exchange energy is therefore minimal, in fact zero if the wall were infinitely thick. The minimum of the sum of exchange and anisotropy energy determines the actual thickness of the wall. This is the classical concept to explain the finite thickness of domain walls and to describe quantitatively the spin orientation along the path from one domain into the other. Let us assume a perfectly straight domain wall in the $yz$-plane. There are two limiting cases as to how the reorientation of the spin may proceed along the $x$-axis. The spin may rotate around the $x$-axis or may rotate within a plane containing the $x$-axis. The first type of wall is called a *Bloch wall*, the second a *Néel wall*. In a Néel wall, one has a non-vanishing divergence of the longitudinal component of the magnetization, $\nabla M_x$, while $\nabla M_x \equiv 0$ for the Bloch wall. A non-vanishing gradient $\nabla M_x$ causes a magnetic field and the interaction of that field with the magnetization adds magnetic self-energy to the energy of the wall. For bulk crystals, with dimensions large compared to the domain size, the energy associated with the $\nabla M_x$-term makes the Néel wall the less favorable choice. This is why the domain walls in the bulk

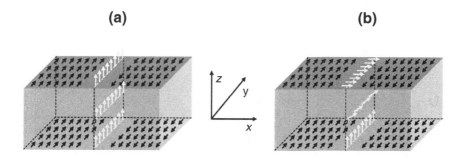

**Fig. 9.10.** To minimize depolarization energy magnetic domains are polarized parallel to the surface. This entails that the magnetization inside a Bloch wall has a magnetization component perpendicular to the surface, which adds depolarization energy to the energy of the wall (**a**). The surface energy is minimized by turning the Bloch wall into a Néel wall at the surface (**b**).

of a crystal are Bloch walls. The situation changes when the Bloch wall meets the surface of the crystal. At the surface, the domains are in general polarized parallel to the surface in order to minimize the depolarization energy. Thus, the magnetization would have a component normal the surface inside a Bloch wall. The Bloch wall would therefore have a higher surface energy than a Néel wall. The configuration of lowest energy is Bloch wall in the bulk turning into a Néel wall at the surface (Fig. 9.10).

## 9.3.2 Domain Walls in Thin Films

For thin film systems, the thickness of the film is typically small compared to the width of the domain walls. A wall between two domains with the magnetization parallel to the surface is then completely of the Néel type. If the film has tetragonal symmetry, 90° and 180° Néel walls exist, depending on the orientation of the magnetization in the adjacent domains (Fig. 9.11). If the magnetization

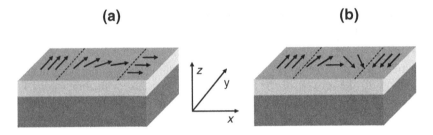

**Fig. 9.11.** Illustration of **(a)** a 90° Néel wall and **(b)** a 180° Néel wall between two domains with in-plane magnetization.

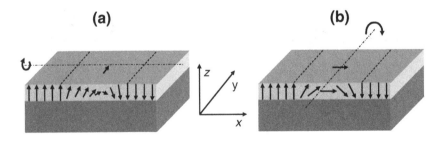

**Fig. 9.12.** Bloch **(a)** and Néel **(b)** walls between two perpendicularly magnetized domains. In both types of walls, the magnetization has an in-plane component. The depolarization energy is smaller inside the wall than in the domains.

inside the domains is perpendicular to the surface both, Bloch and Néel walls have a parallel component of the magnetization (Fig. 9.12). Correspondingly, one gains the depolarization energy in both cases and the wall energies are a little less than in the bulk. The $\nabla M_x$-contribution to the magnetic energy is larger for the Néel wall.

**(a)**                                              **(b)**

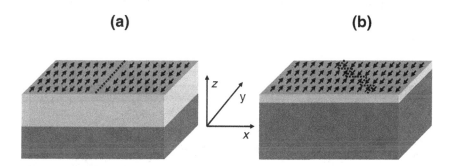

**Fig. 9.13.** (a) Domain walls in bulk crystals and in thick films run parallel to the magnetization to avoid the depolarization energy associated with *charged walls*. The depolarization energy per wall area decreases linear with the thickness when the film thickness is smaller than the wall thickness. The free energy is reduced by introducing kinks in the wall (b).

Domain walls that intersect the magnetization direction at some angle lead a magnetic $H$-field because of a non-vanishing $\nabla M$ -term, causing a contribution to the magnetic self-energy. The analog case in electrostatics is an interface at which the perpendicular component of the electrostatic polarization $P$ changes. This change leads to polarization charges at the interface. In analogy, magnetic domain walls that intersect the magnetization direction at an angle are called *charged walls*. In bulk crystals, charged walls are avoided as much as possible and the domain walls run parallel to the magnetization (Fig. 9.13a). In thin film systems, the situation is different. The wall is perfectly planar with respect to the $z$-axis as long as the film is thinner than the width of a domain wall. However, the wall may meander in the $xy$-plane. The domain wall thereby becomes a one-dimensional object. We have learned in Sect. 4.3.3 that such objects are thermodynamically always rough, as the position correlation function diverges. Furthermore, the depolarization energy associated with charged walls is very small. It does not cost much energy to make a kink in the wall even though the kink may involve more than just a monolayer (Fig. 9.13b). The free energy of a wall is therefore considerably reduced by the introduction of kinks. Thus, domain walls in thin films meander in space just as steps do, and domains in thin films assume rather irregular shapes (Fig. 9.13b).

As an example, we consider the domains in thin Co films on copper surfaces. Figure 9.14 shows SEMPA-images of (a) 5ML and (b) 9ML Co films on Cu(1 1 13) surfaces [9.29]. The magnetization was analyzed with the detector sensitive to the polarization in the horizontal direction. Black and white patches correspond to magnetization parallel and antiparallel to a horizontal axis, respectively (black and white arrows in Fig. 9.14). The entire image is made up of black and white patches. This indicates an uniaxial anisotropy with a horizontal easy axis. The steps (not visible in Fig. 9.14) also run along the horizontal direction. Steps therefore convert the tetragonal anisotropy into an uniaxial anisotropy, and the easy axis is parallel to the step direction. While the magnetization is oriented strictly along the easy axis, the domains are completely irregularly shaped and *charged walls* are abundant.

**(a)**    **(b)**

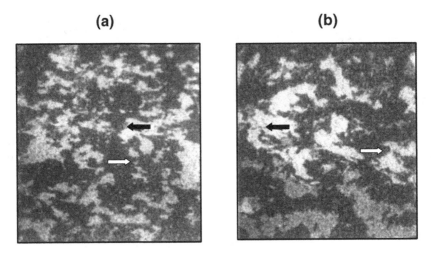

**Fig. 9.14.** SEMPA-images of **(a)** 5ML and **(b)** 9ML Co films on Cu(1 1 13) (after Berger et al. [9.29]). Image size is 500μm×500μm. The black and white arrows indicate the magnetization. The magnetization is parallel to the step orientation. The domains are irregularly shaped. *Charged walls* are everywhere.

### 9.3.3 The Internal Structure of Domain Walls in Thin Films

We now consider the rotation of the magnetization inside the domain walls. We begin with the Bloch wall between two perpendicularly magnetized domains (Fig. 9.10a) and take an fcc (100) film as an example. For this purpose, we interpret the Heisenberg operator (9.28) as a classical equation. The spin operators $S^{(i)}$ then become conventional vectors, and the product between the spin operators is a scalar product. All spins in the $yz$-plane are equal, they merely change their orientation as one progresses along the $x$-axis. We place the origin of the $x$-axis in the center of the wall and denote the orientation angle with respect to the positive $z$-axis as $\phi$. The angle $\phi$ varies from zero to $\pi$ as one moves across the wall from negative to positive $x$. In this notation the classical discrete Heisenberg equation becomes

$$H = -2JS^2 \sum_n \cos(\phi_{n+1} - \phi_n) - 2KS^2 \sum_n \cos^2 \phi_n \, . \tag{9.36}$$

The index $n$ denotes the individual lattice plane perpendicular to the $x$-axis. Each term in (9.36) is the energy per half a cubic cell. The anisotropy constant $K$ is therefore related to the uniaxial anisotropy constant $K_2$ of the hexagonal system (9.8) and the tetragonal system (9.10) by

$$2KS^2 = K_2 a_0^3 / 2 \, . \tag{9.37}$$

We expand the cosine function

$$H = JS^2 \sum_n (\phi_{n+1} - \phi_n)^2 - KS^2 \sum_n \cos 2\phi_n + \mathrm{const} \, . \tag{9.38}$$

The wall is in equilibrium if the derivatives $\partial H / \partial \phi_n$ vanish identically

$$\partial H / \partial \phi_n = JS^2 (2\phi_n - \phi_{n+1} - \phi_{n-1}) + 2KS^2 \sin 2\phi_n = 0 \, . \tag{9.39}$$

As the wall thickness is large compared to a unit cell one may consider $\phi_n$ as a continuous variable $\phi(n)$, the term in the bracket thereby become the negative of the second derivative of $\phi$ with respect to $n$,

$$\frac{\partial^2 \phi}{\partial n^2} = \frac{2K}{J} \sin 2\phi \, . \tag{9.40}$$

This is the sine-Gordon equation previously discussed in the context of structural domain walls (1.23). The solution is (cf. 1.28)

$$\phi = 2 \tan^{-1}(\exp(n/r)) \tag{9.41}$$

with

$$r = (J/4K)^{1/2} = (JS^2 / K_2 a_0^3)^{1/2} \, . \tag{9.42}$$

The width of the wall is about $w = 4r$.

Experiments typically measure the components of the magnetization parallel and perpendicular to the domain wall. These components are

$$M_\parallel \propto \cos \phi = \cos(2 \tan^{-1}(\exp(n/r))) = \tanh(n/r)$$
$$M_\perp \propto \sin \phi = \sin(2 \tan^{-1}(\exp(n/r))) = 1/\cosh(n/r) . \tag{9.43}$$

We now consider the case of a Néel wall between two domains of in plane magnetization (Fig. 9.11). We assume that the exchange and anisotropy terms in the Hamiltonian prevail so that the dipole term can be neglected. A film of a cubic material with the (100) surface parallel to the film plane has tetragonal symmetry, i.e. fourfold rotation symmetry around the $z$-axis. The classical Hamiltonian is therefore

$$H = -2JS^2 \sum_n \cos(\phi_{n+1} - \phi_n) - KS^2 \sum_n \cos 4\phi_n \ . \tag{9.44}$$

The conversion of the anisotropy constant $K$ into the anisotropy constant $K_{4\perp}$ introduced in (9.10) for the tetragonal system is

$$KS^2 = \frac{1}{8} K_{4\perp} \frac{a_0^3}{2} \ . \tag{9.45}$$

For a strictly cubic film $K_{4\perp}$ can be replaced by $K_1$. The sine Gordon equation is now

$$\frac{\partial^2 \phi}{\partial n^2} = \frac{4K}{J} \sin 4\phi \tag{9.46}$$

The solution

$$\phi = \tan^{-1}(\exp(n/r)) \ \text{with} \ r = (J/16K)^{1/2} = (JS^2 / K_{4\perp} a_0^3)^{1/2} \tag{9.47}$$

represents a 90° domain wall (Fig. 9.11). Figure 9.15 shows the polarization of secondary electrons in SEMPA images of a Néel type 180° wall in a 5.5 monolayer thick Co film on Cu(100) [9.30]. Cobalt grows as an fcc crystal in that case. Shown is the polarization component parallel to the domain boundary. The polarization has therefore a maximum in one domain, passes through to zero at the center of the wall and levels off at the negative of the initial value after passing through the wall. The solid line in Fig. 9.14 is the polarization according to (9.43). The dashed line is a numerical simulation that includes the dipolar term [9.30]. The match to the experimental data is a little better though not perfect.

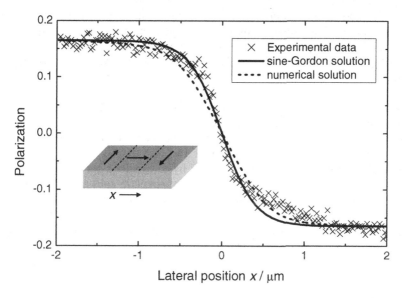

**Fig. 9.15.** Polarization of secondary electrons in a *Scanning Electron Microscope with Polarization Analysis* (SEMPA) across a 180° Néel wall (after Berger and Oepen [9.30]). The solid line is the solution of the sine-Gordon equation. The dashed line represents a numerical solution that includes the dipole term.

## 9.4 Magnetic Coupling in Thin Film Systems

### 9.4.1 Exchange Bias

In 1956 Meiklejohn and Bean discovered that the magnetic hysteresis curve of a sample of cobalt nanoparticles each covered with a cobaltous oxide was shifted on the $H$-axis   after cooling to 77 K in a strong magnetic field [9.31]. The phenomenon was attributed to the interaction between the antiferromagnetic cobaltous oxide and the ferromagnetic cobalt and was termed as *exchange anisotropy* by Meiklejohn and Bean. Today, the effect is usually called *exchange bias*. While the theoretical explanation of this effect is still controversial in detail, it is undoubtedly caused by a local exchange coupling between the ferromagnet and the antiferromagnet across the interface. The existence of the effect as such should not surprise. After all, there is chemical bonding across the interface. The macroscopic manifestation of exchange coupling across the interface, as apparent from the asymmetric hysteresis loop, is visible only after certain preparation steps. We consider the example of an antiferromagnetic substrate, e.g. NiO, which is covered with a ferromagnetic film. In general, the substrate has disordered domains and so has the ferromagnetic layer. No asymmetry of the ferromagnetic hysteresis loop results from that since an equal number of opposite domains compensates any bias imposed by one type of domains in the antiferromagnet, even if exchange coupling across the interface exists. Exchange bias is obtained if the antiferromagnet itself is polarized such that the surface layer has only one spin orientation. This can be achieved if the system is annealed above the Néel temperature of the antiferromagnetic substrate, which must lie below the Curie temperature of the ferromagnetic layer. The domains of the ferromagnetic layer order in an applied strong magnetic field. Upon cooling the system below the Néel temperature, the exchange bias takes care that the antiferromagnetic substrate orders according to the magnetization of the ferromagnetic layer. Figure 9.16 illustrates the effect. The dashed line is the hysteresis loop for the ferromagnetic film when deposited on a disordered antiferromagnetic substrate. After cooling below the Néel temperature in an $H$-field in the positive direction, the antiferromagnetic substrate develops ordered domains because of the exchange coupling to the ferromagnet (assumed to prefer the spins parallel). The exchange bias then shifts the hysteresis to the left.

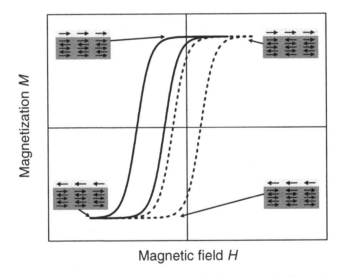

**Fig. 9.16.** Hysteresis loops for the magnetization of a ferromagnetic film exchange coupled to a disordered and ordered antiferromagnet substrate (dashed and solid lines, respectively). Ordering of the antiferromagnetic substrate is achieved by annealing the system above the Néel temperature of the substrate and cooling in a strong magnetic field (parallel to the film plane).

The technological importance of the exchange bias effect can hardly be overestimated, as it is one of the cornerstones in the construction of sensors for magnetic fields. To understand this better we consider the four-layer sandwich displayed in the inset of Fig. 9.17. The system consists of an exchange-coupled ferromagnet (ECFM) on an antiferromagnetic substrate (AF), a nonmagnetic layer (NM) and a second ferromagnetic layer (FM) on top. The FM-layer is assumed to have an in-plane easy axis. The external magnetic field is parallel to the layers. If the magnetic field is cycled between large negative and positive values, the magnetic fields are large enough to rotate the magnetization in the ECFM-layer and in the ferromagnetic top layer. The corresponding hysteresis loop is the solid line in Fig. 9.17. If the magnetic field cycle stays closer to zero, the magnetization follows the small loop indicated by the dotted line. By applying small magnetic fields, one can therefore switch the sandwich between a state where the two ferromagnetic layers have the same orientation of the magnetization and where the magnetizations have the opposite direction. Both states are stable in zero magnetic field. In this configuration, the system represents a storage device.

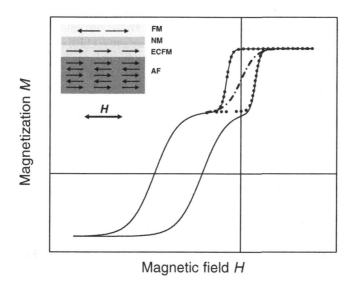

**Fig. 9.17.** Hysteresis curve of a sandwich structure consisting of an exchange-coupled ferromagnet (ECFM) on an antiferromagnetic substrate (AF), a nonmagnetic layer (NM) and a second ferromagnetic layer (FM) on top. The hysteresis loop marked by the solid line is for magnetic fields large enough to rotate the magnetization in the ECFM-layer. For smaller fields one stays inside the small loop around zero fields (dotted line). The dash-dotted line is for an FM-top layer made of a soft magnetic material (e.g. NiFe permalloy).

If the FM-top layer made of a soft magnetic material (e.g. NiFe permalloy) the orientation of the magnetization in the top layer is shifted continuously by an external field (dash-dotted line in Fig. 9.17). In this configuration, the system represents a sensor that converts an external field into a particular magnetization state. The state can be read out by the electrical resistance of the sandwich as explained in the following section.

## 9.4.2 The GMR-Effect

In 1988, two groups independently discovered that the resistance of a sandwich structure of two ferromagnetic Fe-layers separated by an antiferromagnetic Cr-interlayer depends on the relative orientation of the magnetization in the Fe-layers [9.32, 33]. The magnetoresistance defined as

$$\Delta R / R = (R_{\uparrow\downarrow} - R_{\uparrow\uparrow}) / R \tag{9.48}$$

amounted to 1.5% for a trilayer system [9.32, 33] and up to 50% for a multilayer system at low temperatures [9.32, 33]. The effect is observable because the Cr-interlayer couples the magnetization of the iron layer so that the magnetizations in the ferromagnetic layers are antiparallel in zero fields and parallel in a high external field (Fig. 9.18). We discuss the original work of Binasch et al. concerning the trilayer system in more detail. The entire thin film system was grown on a GaAs(110) substrate. The Fe-films also grow (110) oriented so that the easy [100] axis and the hard [110] axis are in the film plane. The iron films were 12 nm thick, the chromium films 1nm thick. The resistance is measured parallel to the films. In Fig. 9.18 the [100] direction is horizontal and in the plane of drawing. When an external magnetic field is applied in the [100] direction and the field is strong enough to overcome the antiferromagnetic coupling and the coercive field of the iron layer, then the magnetizations are switched to parallel orientation. The

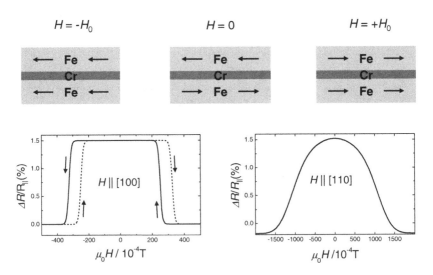

**Fig. 9.18.** Magnetoresistance in a three layer sandwich after Binasch et al. [9.32, 33]. In zero fields, the magnetizations in the two iron layers are held antiparallel via interlayer coupling through the antiferromagnetic chromium. The magnetizations become parallel by

applying external fields. The electrical resistance is lower for that state. This effect is called *Giant MagnetoResistance* (GMR).

parallel state has as 1.5% lower resistance (left graph). When the magnetic field is moved back to smaller values, the magnetizations stay parallel for a little while because of the coercive field and then switch back to the antiparallel state. If the external field is along the hard axis (perpendicular to the plane of drawing), the magnetizations of the two Fe-films are gradually rotated into the direction of the applied field until they become eventually parallel at rather high fields (right panel in Fig. 9.18). The total change in the resistance is a little larger because of the normal, anisotropic magneto resistance effect. Evidently, the magnetoresistance observed in the magnetically coupled sandwich is significantly larger than the normal magnetoresistance. Exaggeratingly it is called the *Giant Magneto-Resistance* (GMR). We note that the magnetoresistance curves are symmetric around $\mu_0 H = 0$ and that the slope of $R(H)$ is zero there. The arrangement shown in Fig. 9.18 is not yet a technical useful device. Before we can proceed to make such a device by combining GMR and exchange bias, we devote ourselves to the basic physics of the GMR effect.

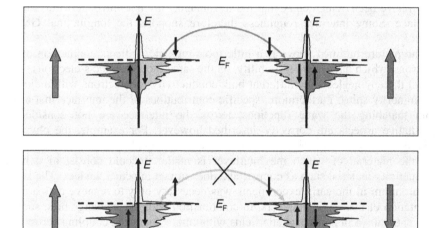

**Fig. 9.19.** Density of states of a ferromagnet (nickel) for spin parallel and antiparallel to the magnetization. The diagrams on the *left* and *right* represent the two ferromagnetic layers. In the upper pair, the magnetization in the layers is parallel and antiparallel in the lower pair. The antiparallel configuration has a higher resistance (see text).

Figure 9.19 shows the density of states of a ferromagnet (here nickel) for electrons with spins parallel and antiparallel to the magnetization. The diagrams on the left and right of each panel represent the two ferromagnetic layers. The layer magnetizations are parallel in the upper and antiparallel in the lower panels. The mean current parallel to the film plane is a weighted average over the contributions from electrons that embark on trajectories forming an angle $|\theta \leq 90°|$ with the film plane. These trajectories carry electrons from one ferromagnetic layer into the other as long as the mean free path of the electrons is larger than the distance between the ferromagnetic layers along the trajectory. The current carried by those electrons is part of the total current. We assume that the electrons do not loose their spin orientation while traversing the chromium layer. Electrical current is carried only by electrons at the Fermi level and we therefore need to consider only those. For parallel oriented magnetization, the Fermi electrons find a high density of states of the same spin orientation in the other ferromagnetic layer. They can therefore continue their path with little scattering at the interface and thereby contribute to the current. If the magnetizations are antiparallel then also the spin orientation of electrons with a high density of states at the Fermi level is antiparallel. Electron transfer from one film to the other is only from a high density of states into a low density of states of the same orientation and vice versa. This causes a larger interface resistance, which leads to an increase in the resistance of the sample. Saying that implies that electrons which have been denied entry into the ferromagnet are not mirror reflected at the interface. Some interface roughness therefore supports the longitudinal GMR-effect.

The picture outlined above is a little too simplistic. It treats conduction of d-electrons, which have a lower mobility, on the same footing as the s-electrons. The model thereby neglects the different bulk conductivity for electrons with majority and minority spins. Furthermore, specific contributions of the interface that arise from matching the wave functions across the interface are not considered. Qualitative aspects are correctly described however. For example, the physical picture outlined above does not refer to the magnetic coupling across the interlayer and the material of which the interlayer is made; it could consist of a high conductivity material such as copper, and does in fact in actual devices. The layer of chromium in the early experiments was necessary only to achieve antiparallel orientation of the magnetic layers. In combination with the exchange bias, such a state is obtained in zero magnetic fields without any magnetic coupling across the interlayer (Fig. 9.17). At the same time, the exchange bias system shown in Fig. 9.17 makes for a much more useful device as one can vary the magnetization in the outer film in small external fields, and the variation of the magnetization and therefore of the magnetoresistance is linear in the field.

A sensitive detector for small magnetic fields requires the magnetization of the outer ferromagnetic layer to change rapidly with the applied field. The dash-dotted line in Fig. 9.17 should therefore be as steep as possible. The nickel/iron alloy *permalloy* is a suitable material for the outer ferromagnetic layer. A technically advanced layer system for a sensor is shown in the inset of Fig. 9.20 [9.34]. Two very thin Co layers on either side of the Cu-interlayer act as a spin filter to enhance the magnetoresistance effect. Composite structures like that are called *spin valves*. The antiferromagnetic layer that provides the exchange bias is made of FeMn. The entire composite is grown by magnetron sputtering on an oxidized silicon wafer so that the sensor is electrically isolated and integrated into silicon device technology. Buffer and texture layers between the wafer and the active part of the sensor serve for growing smooth and well-ordered films. Figure 9.20 shows schematically the magnetization (in arbitrary units) as a function the applied field. For a sensor, only the magnetization near zero fields is of interest. There, the magnetization in the free ferromagnetic layer rises steeply in an external field (Fig. 9.20). The magnetoresistance follows with a likewise steep curve so that the entire system is a sensitive device for the detection of magnetic fields. Its sensitivity and the fact that it can be made quite small have made the GMR-sensor the present day sensor in hard disk data storage devices.

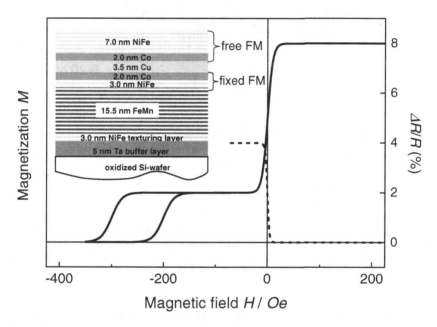

**Fig. 9.20**. Magnetization (solid line) and magnetoresistance (dashed line) of a complex layer system (schematic after Paul et al. [9.34]). The magnetic field is in Oerstedt (1Oe for $H$ is $1000/4\pi$ Am$^{-1}$).

### 9.4.3 Magnetic Coupling across Nonmagnetic Interlayers

In 1986, Grünberg et al. found that two Fe-layers separated by 10 Å thick Cr-film are antiferromagnetically coupled [9.35]. The work was instrumental for the discovery of the GMR effect, but was also a very interesting effect in its own right. An important stimulus to the field of interlayer exchange-coupling was the discovery that the exchange-coupling oscillates between ferromagnetic and antiferromagnetic coupling as a function of the interlayer thickness [9.36]. The effect can be observed with virtually any interface material [9.37]. The effect is elegantly demonstrated by shaping the interlayer as a wedge so that different positions on the surface of the ferromagnetic cover layer correspond to different thicknesses of the interlayer. The oscillations between ferromagnetic and antiferromagnetic coupling can be made visible by imaging techniques that are sensitive to the orientation of the magnetization. Figure 9.21 displays the scheme of such an experiment following Unguris et al. [9.38]. An iron whisker crystal (a crystal that is free of dislocations) serves as the substrate. The shading of the top Fe layer indicates the black and white contrast in a SEMPA image (Sect. 9.2).

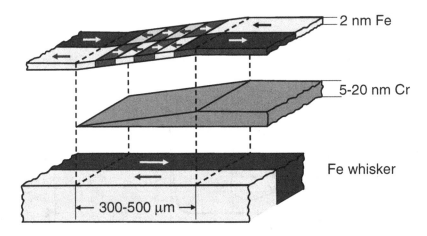

**Fig. 9.21.** Schematic picture of an experiment for the observation of the oscillations between ferromagnetic and antiferromagnetic coupling of two ferromagnets as a function the interlayer thickness (After Unguris et al. [9.38]). The shading of the top Fe layer indicates the observed black and white contrast in a SEMPA-image.

The oscillation periods depend also on the thickness of the ferromagnetic cover layer. For the rather thin Fe-layer of 2 nm a second, rapid oscillation period was observed [9.38]. Double wedge structures in which the non-magnetic as well as the magnetic layer thickness varies display a two-dimensional polarization pattern if the wedge orientations are mutually orthogonal to each other [9.39].

The interlayer exchange coupling through non-magnetic layers is caused by spin dependent quantum size oscillations [9.40]. In Sect. 8.3.1 we considered the energy associated with the oscillations in the density of states for the nonmagnetic case

$$\Delta E = \frac{1}{2\pi^3} \, \mathrm{Im} \int \mathrm{d}^2 k_\| \int\limits_{-\infty}^{E_F} \ln\left(1 - r_a r_b e^{i2k_\perp D}\right) \mathrm{d}E \; . \tag{9.49}$$

Here $r_a$ and $r_b$ are the reflection coefficients for electrons at the interfaces $a$ and $b$ of the spacer layer to the ferromagnetic layers. These reflection coefficients now depend on the spin state of the reflected electron in relation to the spin orientation within the ferromagnet, since the band structure in the ferromagnet differs for spin-up and spin-down electron. Thereby the energy becomes dependent on the relative orientation of the spins in the ferromagnets. For parallel orientation of the spins in the two magnetic layers the energy is

$$\Delta E_{\uparrow\uparrow} = \frac{1}{4\pi^3} \operatorname{Im} \int d^2k_\parallel \int\limits_{-\infty}^{E_F} \left\{ \ln\left(1 - r_a^\uparrow r_b^\uparrow e^{i2k_\perp D}\right) + \ln\left(1 - r_a^\downarrow r_b^\downarrow e^{i2k_\perp D}\right) \right\} dE . \quad (9.50)$$

For antiparallel orientation one has

$$\Delta E_{\uparrow\downarrow} = \frac{1}{4\pi^3} \operatorname{Im} \int d^2k_\parallel \int\limits_{-\infty}^{E_F} \left\{ \ln\left(1 - r_a^\uparrow r_b^\downarrow e^{i2k_\perp D}\right) + \ln\left(1 - r_a^\uparrow r_b^\downarrow e^{i2k_\perp D}\right) \right\} dE . \quad (9.51)$$

For the sake of a simpler discussion, we assume that differences between the spin-up and spin-down reflection coefficients are small.

$$\Delta r = (r^\uparrow - r^\downarrow)/2 \ll (r^\uparrow + r^\downarrow)/2 \quad (9.52)$$

The first order expansion of the energy difference between the ferromagnetic and antiferromagnetic orientation in $\Delta r$ is

$$\Delta E = \Delta E_{\uparrow\uparrow} - \Delta E_{\uparrow\downarrow} = -\frac{1}{\pi^3} \operatorname{Im} \int d^2k_\parallel \int\limits_{-\infty}^{E_F} \Delta r_a \Delta r_b e^{i2k_\perp D} dE \quad (9.53)$$

The energy $\Delta E$ oscillates between negative and positive values with an oscillation period determined by the Fermi vector $k_F$. The reason is that for $D \gg k_F^{-1}$, the integrand oscillates so rapidly that no contribution to $\Delta E$ survives except those near the Fermi cut-off. The oscillation period is therefore $2k_F D$ .

## 9.5 Magnetic Excitations

### 9.5.1 Stoner Excitations and Spin Waves

Theory and experiment discern two types of elementary excitations in solids that involve a reversal of an electron spin. In the case of *Stoner excitations*, the spin of a single electron is reversed ("flipped"). This excitation corresponds to a transition from a spin-up to a spin-down band. The second case is a collective excitation, a *spin wave* or a *magnon*. The excitation of one quantum of a magnon corresponds to a change of the magnetization equivalent to the spin flip of one electron. However, the spin flip is distributed over the entire ensemble of electrons. Figure 9.22 illustrates schematically the two excitations. In Fig. 9.22a two energy bands are shown that are separated by the $k$-dependent exchange splitting $\Delta E_{ex}(k)$. Transitions between the two bands can take place between occupied and unoccupied states. In the case shown the spin up band is completely occupied and separated from the Fermi level by an energy gap, the *Stoner gap* $\Delta$, which marks the minimum energy required to reverse the spin of an electron. Ferromagnets that possess a nonzero Stoner gap are called *strong ferromagnets*, those who do not posses a Stoner gap are called *weak ferromagnets*. This distinction is not related in any way to the saturation magnetization or the magnetic anisotropy, which would differentiate strong and weak magnets in the conventional sense of the words. Among the 3d-metals, Ni and Co are *strong ferromagnets* if the low density of s-electrons at the Fermi level is disregarded, and Fe is a *weak ferromagnet*.

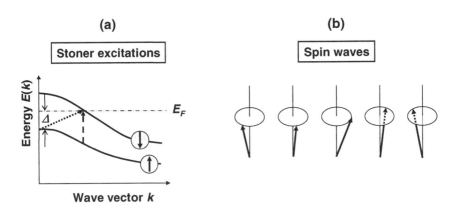

**Fig. 9.22.** Schematic illustration of **(a)** Stoner excitations and **(b)** a classical spin wave (magnon). Both excitations correspond to the flip of one spin. In the case of spin waves, the spin flip is distributed over the entire ensemble of electrons. The dashed and dotted arrows in **(a)** correspond to the minimal energies for momentum conserving and non-conserving spin flip transitions, respectively.

Figure 9.22b illustrates a spin wave in the classical limit. The spin rotates around the magnetization direction with a phase in the form of a wave. Quantum mechanically the spin wave is a particular excited state solution of the Heisenberg Hamiltonian with nearest neighbor interactions

$$H = -J\sum_{i,\delta} \vec{S}^{(i)} \cdot \vec{S}^{(i+\delta)} = -J\sum_{i,\delta} S_z^{(i)} S_z^{(i+\delta)} + \frac{1}{2}(S_+^{(i)}S_-^{(i+\delta)} + S_-^{(i)}S_+^{(i+\delta)}) \,. \quad (9.54)$$

The spin operators $S_+$ and $S_-$ flip the spin state from "down" to "up" and from "up" to "down", respectively. For the spin of one electron the operators $S_z$, $S_+$ and $S_-$ can be represented by Pauli matrices

$$S_z = \frac{1}{2}\begin{pmatrix} 1 & 0 \\ 0 & -1 \end{pmatrix}, \; S_+ = S_x + iS_y = \begin{pmatrix} 0 & 1 \\ 0 & 0 \end{pmatrix}, \; S_- = S_x - iS_y = \begin{pmatrix} 0 & 0 \\ 1 & 0 \end{pmatrix}. \quad (9.55)$$

As easily proved, these operators have the commutators

$$\begin{aligned} \left[S_+^{(i)}, S_-^{(j)}\right] &= 2S_z^{(i)}\delta_{ij} \,, \\ \left[S_+^{(i)}, S_z^{(j)}\right] &= -S_+^{(i)}\delta_{ij} \,. \end{aligned} \quad (9.56)$$

The time dependence of the operator $S_+$ is given by

$$i\dot{S}_+^{(j)} = \left[S_+^j, H\right]. \quad (9.57)$$

With the help of the commutator rule [A, BC]=[A, B]C+B[A, C] one obtains

$$i\dot{S}_+^{(j)} = J\sum_\delta \left\{ S_+^{(j)}S_z^{(j+\delta)} + S_z^{(j-\delta)}S_+^{(j)} - S_+^{(j-\delta)}S_z^{(j)} - S_z^{(j)}S_+^{(j+\delta)} \right\}. \quad (9.58)$$

One now assumes that the system is in the ferromagnetic ground state and replaces the operator $S_z$ by the scalar $S$. The Heisenberg operator becomes linear thereby,

$$i\dot{S}_+^{(j)} = JS\sum_\delta \left\{ 2S_+^{(j)} - S_+^{(j-\delta)} - S_+^{(j+\delta)} \right\}. \quad (9.59)$$

Equation (9.59) is solved with the ansatz for the spin wave

$$S_+^{(j)} = \Delta S_+ e^{-i\omega t} e^{i\mathbf{k}\cdot\mathbf{r}_j}. \quad (9.60)$$

Inserting (9.60) into (9.59) yields the spin wave frequency

$$\omega = JS\sum_{\delta}(2 - \mathrm{e}^{\mathrm{i}k\cdot r_{\delta}} - \mathrm{e}^{-\mathrm{i}k\cdot r_{\delta}}) = 2JS\left(v - \sum_{\delta}\cos k\cdot r_{\delta}\right). \tag{9.61}$$

While spin waves and their dispersion relation are most easily deduced from the Heisenberg Hamiltonian they do exist also in itinerant ferromagnets where the magnetization arises as a collective property of delocalized band electrons. Experiments using neutron scattering show reasonable sharp lines in the energy spectrum that are ascribed to spin waves [9.24, 41]. The dispersion differs from the solution of the Heisenberg model for large wave vectors because of the increasing interaction with the Stoner continuum. This is illustrated in Fig. 9.23a for the case of a strong ferromagnet and a k-independent exchange splitting between the spin-up and spin-down bands. Transitions with $\Delta k = 0$ then involve a fixed energy the exchange splitting $\Delta E_{\mathrm{ex}}$. For larger $\Delta k$, the excitation spectrum is a continuum. The spin-wave dispersion curve (fat solid line in Fig. 9.23a) dives into the continuum of Stoner excitation at a particular point of the $k$-space. The coupling to Stoner excitations affects the dispersion and imposes a strong damping on the spin waves. The interaction with the Stoner-continuum is even stronger at surfaces. The surface reduces the symmetry and momentum conservation concerns merely the parallel component $k_{\parallel}$ of the wave-vector while the perpendicular component may assume an arbitrary value. At surfaces and in thin films spin waves are therefore even more severely damped than in the bulk.

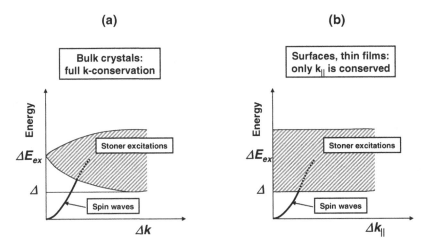

**Fig. 9.23.** Spectrum of Stoner excitations and the spin wave dispersion **(a)** in bulk crystals and **(b)** at the surface and in thin film systems, both for the case of a strong ferromagnet possessing a Stoner gap $\Delta$. For surfaces, momentum conservation holds only for the parallel component of the wave vector, which provides a larger phase space for magnon/electron interaction.

## 9.5.2 Magnetostatic Spin Waves at Surfaces and in Thin Films

In the limit of very small $k$-vectors, the frequency of exchange-coupled spin waves goes to zero proportional to $k^2$. In an external magnetic field $H_0$, the frequency merges into the frequency of the *magnetostatic spin waves*. Their $k$-independent frequency is calculated by assuming that the entire magnetization precesses around the magnetic field. The result is in SI-units [9.23]

$$\omega = \mu_0 \gamma \sqrt{H_0^2 + M_s H_0} \ . \tag{9.62}$$

Here, $\gamma$ is the gyromagnetic ratio $g\mu_B/h$.

If the solid has the form of a slab, surface waves exist in addition to the bulk spin waves. These surface spin waves are called *Damon-Eshbach waves*. Their amplitude decays into the bulk as $\exp(-k|z|)$, if the slab thickness $D$ is large enough so that $\exp(-kD) \ll 1$. Otherwise, the modes on the two surfaces couple to each other. As the $k$-vector is small compared to a vector of the reciprocal lattice Damon-Eshbach waves do not reflect genuine surface and thin film properties, analogous to long wave-length surface plasmons (Sect. 8.1) and Fuchs-Kliewer surface phonons (Sect. 7.1.7). The frequency of Damon-Eshbach waves is

$$\omega_{DE} = \mu_0 \gamma (M_s / 2 + H_0) \ . \tag{9.63}$$

Damon-Eshbach waves have the interesting property that they travel only in one direction: Viewed from the tip of the magnetic vector they circulate around the slab in clockwise direction (Fig. 9.24, for a derivation of (9.63) and further details see [9.23]).

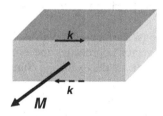

**Fig. 9.24.** Direction of Damon-Eshbach magnetostatic surface waves relative to the orientation of the magnetization.

Damon-Eshbach waves can be observed by the optical technique of *Brillouin scattering*. Because of their handedness, Damon-Eshbach waves appear either as Stokes-lines or as Anti-Stokes lines, depending on the scattering geometry (see e.g. [9.23], p.221ff.).

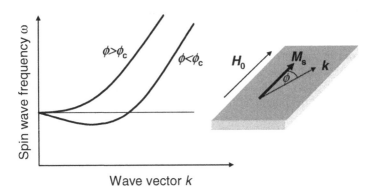

**Fig. 9.25.** Dispersion of spin waves in a thin slab in the transition region between the magnetostatic and the exchange-coupling regime.

The normally $k$-independent frequency of magnetostatic spin waves can acquire dispersion because of $k$-dependent magnetic dipole interactions in thin films. This leads to an interesting transition regime between magnetostatic and exchange coupled spin waves [9.42, 43]. The initial dispersion is linear in $k$ and negative (positive) if the angle $\phi$ between the magnetization and the $k$-vector is smaller (larger) than a critical wave vector $\phi_c$ (Fig. 9.25). The negative dispersion for some directions opens a channel for the decay of ferromagnetic resonance oscillations into magnons of the same frequency. This yields a damping of ferromagnetic resonance modes in thin film systems, which comes in addition to the conventional *Gilbert damping* of the ferromagnetic resonance in bulk systems.

### 9.5.3 Exchange Coupled Surface Spin Waves

Just as for surface phonons, surface spin waves localize at the surface on an atomic scale when the $k$-vector approaches the boundary of the Brillouin zone. Their dispersion therefore reflects the exchange coupling and magnetic moment of the surface atoms. Probing for localized surface spin waves requires particles that have a high cross section with matter, much higher than neutrons. Electrons are suitable for that purpose. However, the inelastic cross-section for spin waves lies about 2-3 orders of magnitude below the cross-section for phonon excitations [9.44]. Despite several attempts, short wavelength spin waves had therefore escaped detection for a long time. In order to be able to discriminate spin waves against vibrational modes one must demonstrate the dependence of the intensity of energy losses on the spin orientation of the scattered electron. A feasible way is to scatter spin-polarized electrons from a magnetically polarized sample and measure the asymmetry of the cross-section with respect to the spin orientation of the electron beam as a function of the energy loss. This technique is called *Spin Polarized Electron Energy Loss Spectroscopy* (SPEELS). As discussed in Sect. 9.2 strained

GaAsP negative electron affinity cathodes are employed in the production of spin polarized electrons. A special layout of the electron spectrometer serves for the suitable orientation of the electron spin perpendicular to the scattering plane and a high intensity of the electron beam while maintaining sufficient energy resolution of 20-40 meV (Fig. 9.26) [9.45].

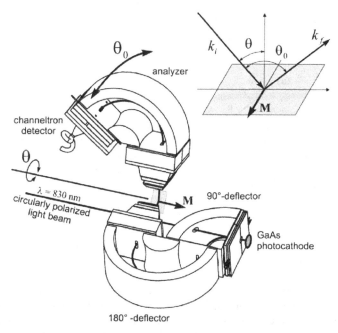

**Fig. 9.26.** Electron spectrometer for inelastic scattering of spin polarized low energy electrons from surfaces. Longitudinal polarized electrons are photoemitted from a GaAsP cathode and energy-selected in a specially designed electrostatic double-pass deflector featuring a total deflection angle of 90°. The spin polarization is thus perpendicular to the scattering plane [9.45].

Figure 9.27 shows selected results of the first successful experiment for inelastic scattering of electrons from spin waves [9.46]. Electrons were scattered from an eight monolayer thick fcc-cobalt film, which was pseudomorphic with the Cu(100) substrate. The scattering plane was oriented along the [110]-direction. Spectra in Fig. 9.27a are shown for the spin-up and spin-down channels. As usual, spin-up refers to the orientation of the majority spins in the ferromagnet. If the spin of the incoming electron has spin-down orientation, it can transfer its spin, energy and wave vector, and its spin orientation flips upwards in the scattering process. If the incoming electron has spin-up orientation no inelastic scattering from spin waves occurs since a spin wave in a ferromagnet involves always involves a transition from spin-up to spin-down orientation and the total spin must be conserved in the scattering. The spectra $I_\uparrow$ and $I_\downarrow$ as shown in Fig. 9.27a are calculated from the observed spectra $I'_\uparrow$ and $I'_\downarrow$ in order to correct for the finite polarization P < 1 of the cathode.

$$I_{\uparrow(\downarrow)} = \left[ I'_{\uparrow(\downarrow)}(P+1) + I'_{\downarrow(\uparrow)}(P-1) \right] / 2P . \qquad (9.64)$$

The increased noise in the spectra around the spin wave energy arises from that procedure.

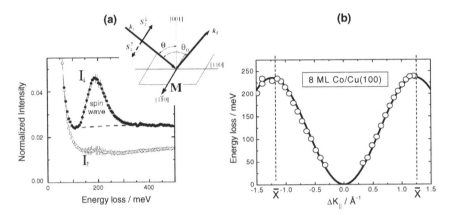

**Fig. 9.27.** (a) Spin-up and spin-down electron energy loss spectra for an 8ML pseudomorphic Co film on Cu(100). The momentum transfer to the spin wave is $k = 0.87$ Å$^{-1}$ along the [110] direction. The inset displays the scattering geometry. (b) Spin wave dispersion of the film. The spin wave is a surface spin wave because of the horizontal tangent at the boundary $\overline{X}$ of the surface Brillouin zone [9.46].

Figure 9.27b displays the dispersion of the spin wave. The data can be nicely matched to an analytical expression for the surface spin wave dispersion of an fcc-structure [9.47]

$$\hbar\omega(k) = 8JS(1 - \cos ka_{nn}) \qquad (9.65)$$

in which $a_{nn} = a_0/\sqrt{2}$ is the nearest neighbor distance. From the fit one obtains JS = 15 meV. The surface wave has its maximum frequency at $ka_{nn} = \pi$, the $\overline{X}$ - point of the surface Brillouin zone. The dispersion is quite distinct from the dispersion of a bulk spin wave traveling along the [110] direction as calculated from (9.61) to

$$\hbar\omega_{bulk}(k) = 4JS(5 - \cos ka_{nn} - 4\cos ka_{nn}/2) . \qquad (9.66)$$

There is a small correction to the frequency for a thin film due to the missing nearest neighbors in adjacent layers for the two outer layers, which, however, is of no concern here. The bulk waves have their maximum frequency at $ka_{nn} = 2\pi$, a point which is equivalent to the X-point. (Note, the $k$-vector of a bulk wave in the [110] direction goes from $\Gamma$ over K to the X-point of an adjacent bulk Brillouin zone. The same point is reached by moving from $\Gamma$ to X along the [001] direction). The fact that the experimental data display a zero slope at the $\overline{X}$-point of the surface Brillouin zone therefore proves that indeed surface spin waves are excited (cf. Fig. 7.2).

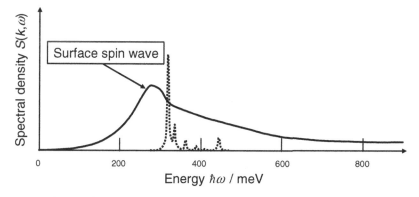

**Fig. 9.28.** Spectral density of spin flip excitations at $k_{\parallel} = 0.6\pi/a_{nn}$ for an 8 ML Co film on Cu(100). The solid line represents the combined excitation spectrum of spin waves and Stoner excitations calculated with a realistic representation of the electronic structure of cobalt [9.49]. The dashed line is a *frozen magnon* calculation with a nearest neighbor Heisenberg Hamiltonian.

In addition to the spin wave, the spectrum in Fig. 9.27a shows a continuum of Stoner excitations. Furthermore, the energy loss due to spin wave excitations is quite broad. The width increases with increasing momentum transfer so that near the zone boundary the spin wave signal is hardly distinct from an onset of a Stoner continuum. This demonstrates that for itinerant ferromagnets spin waves and Stoner excitations are closely coupled. Fits of experimental data to spin wave solutions of the Heisenberg Hamiltonian, which yields discrete δ-function shaped magnon modes, are therefore not very meaningful, even though they might work technically to describe the dispersion. An appropriate theory has to take the itinerant nature of the magnetism in 3d-metals and thus the electronic band structure properly into account. A first attempt in this direction is the work of Costa, Muniz and Mills [9.48, 49]. As an example of their results Fig. 9.28 shows the calculated spectral density of the excitation spectrum of the surface layer of an 8ML thick Co film on Cu(100) at $k_\parallel = 0.6\pi/a_{nn}$. Only the surface mode stands out as a discernible, albeit broad feature. Bulk spin waves of the slab and Stoner excitation are submerged in a broad continuum. A *frozen magnon* calculation using the Heisenberg Hamiltonian produces the spectral density shown by the dashed line, which bears little resemblance to the experimental spectrum (Fig. 9.27).

# 10. Diffusion at Surfaces

## 10.1 Stochastic Motion

Diffusion phenomena at surfaces range from the random walk of single atoms to mass transport on macroscopic length scales which involves hundreds, even thousands of different individual processes with merely a few of them being rate-determining. Their nature remains frequently unknown. A theoretical description of diffusive mass transport has to take the various length scales and the different levels of knowledge into account. Transport over large distances and across many atom layers is best described in a coarse-grained view with the thermodynamic and statistical models developed in chapters 4 and 5, supplemented by however rudimentary knowledge of the processes on the atomic level. We begin the presentation with the random walk of single atoms on flat surfaces, a process that is well understood from an atomic point of view.

### 10.1.1 Observation of Single Atom Diffusion Events

Experimental studies of the random walk of single atoms became possible rather early in the history of surface science. In 1951 Erwin Müller invented the *Field Ion Microscope* (FIM) [10.1], an instrument that permits the real space observation of surfaces with atomic resolution. The experimental set-up is extremely simple, though quite some expertise is involved in the preparation of field emission tips. Figure 10.1 shows the basics of the experimental equipment. A sharp tip made from a single crystal wire is mounted on a liquid nitrogen cooled cryostat inside a UHV-vessel. The UHV-vessel is backfilled with a dilute helium gas. A positive potential is applied to the tip. He-atoms are ionized in the high-electric field at the tip, preferably at atomic protrusions such as steps, corners and single adatoms. The positive $He^+$-ions are accelerated towards the phosphorous screen along a radial trajectory whose direction reflects the position on the surface at which the ionization had occurred. The image on the phosphorous screen is therefore a magnified image of the surface. The magnification factor is the ratio of the radii of the screen and the tip, and hence of the order of $10^7$. Although the use of the field ion microscope for diffusion studies was already proposed by E. W. Müller in 1957 [10.2] it took almost a decade until Ehrlich and Hudda published the first systematic study of single atom diffusion [10.3]. Figure 10.2 shows an example from the review of G. Kellogg [10.4]. The FIM images display a rhodium tip with a (100) surface at the apex. A single platinum atom has been deposited onto the tip.

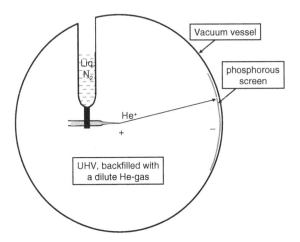

**Fig. 10.1.** Scheme of a *Field Ion Microscope* (FIM). A sharp single crystal tip with a tip radius of 50-200 Å is fixed to a liquid-nitrogen cooled cold finger. Instruments that are more elaborate use liquid helium cooled tips. The image is generated by He-atoms, which are ionized at atomic protrusions on the tip and accelerated towards a screen to form a magnified image of the protrusions there.

Quantitative studies of the random walk of single atoms are performed by the so-called *cook-and-look* technique: At liquid nitrogen temperature, the single atom stays in fixed position forever. Diffusion is initiated by heating the tip quickly to a particular temperature while the imaging electric field is switched off, so that the diffusion process is not affected by the high electric field. After some time at the higher temperature, the tip is cooled down to liquid nitrogen temperature and the FIM-image is observed. If the annealing temperature was high enough, the atom may be found at the nearest neighbor site of its original position or at sites farther away (Fig. 10.2 b, c, and d). The great advantage of the cook-and-look technique is that the diffusion process is not affected by the imaging technique. This is more difficult to ensure for the other available technique, scanning tunneling microscopy. There, cook-and-look is not possible, as one cannot find the same site and the same atom on the surface after a thermal circle. Furthermore, the drift induced by thermal cycles would render observation impossible for some time. The random walk of atoms must therefore be observed with the sample and the STM stabilized at a particular temperature. Tip effects on the diffusion process cannot be excluded systematically. The absence of a tip-induced effect (if so) can be established by running a series of images representing a certain time span, but with a different number of images taken within the time span.

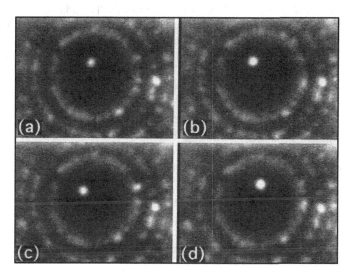

**Fig. 10.2.** A series of FIM images showing a single Pt adatom on a Rh(100) tip. The images are recorded at 77 K. Adatom motion is induced by heating the tip to 345 K for 30 s between each image (Courtesy of Gary Kellogg).

The observation of random walk processes on surfaces yield important information on the atomic nature of the diffusion process. We illustrate this with the example of the random walk of a single atom on a square lattice such as represented by (100) surfaces of cubic materials. The extension to trigonal (111) surfaces is straightforward. On both types of surfaces, the diffusion is isotropic. The movement of one atom from one site to the next may proceed via two different processes. One possibility is the hopping of the adatom from one site to the next across the bridging site as the intermediate state (Fig. 10.3a).

**(a)**          **(b)**

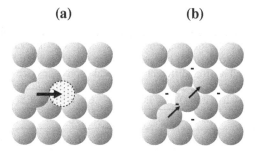

**Fig. 10.3.** Illustration of two different transport mechanisms that take an adatom from one site to the next equivalent site: **(a)** the hopping process and **(b)** the exchange process where the adatom dives into the surface and pushes the atom in the surface layer upwards into the adjacent site. Note that the transport is in different directions in the two cases.

The other process involves an exchange of atom positions. The adatom dives into the surface and, in a concerted motion, the atom that was originally in the surface plane moves up into the next-nearest neighbor site along the [100] direction (Fig. 10.3b). The two processes are easily distinguished by the fact that in the latter case only every second site is visited by the adatom. The sites denoted by "-" in Fig. 10.3b are avoided. The visited sites form a c(2×2) pattern. The clearest experimental evidence for the mechanism of adatom diffusion is therefore from FIM investigations. Figure 10.4 shows the *site visiting maps* for the diffusion of Pt and Pd on Pt(100) after Kellogg et al. [10.5]. For Pt the visited sites form a c(2×2) pattern. The diffusion mechanism is therefore of the exchange type, contrary to the case of Pd diffusion on the same surface.

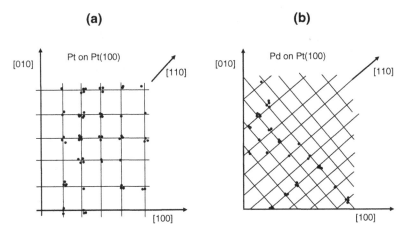

**Fig. 10.4**. Site visiting map of (**a**) Pt atoms and (**b**) Pd atoms in FIM images of a Pt(100) tip. The orientation and the lattice constant identify the diffusion mechanisms as exchange for Pt atoms and hopping for Pd atoms. The distortions of the lattices reflect the distortions of the FIM images (after Kellog et al. [10.5]).

**Table 10.1**. Activation energies for self-diffusion on some fcc(100) surfaces; "hop" stands for adatom hopping, "ex" for exchange of adatoms, and "vac" for vacancy diffusion. The activation energies are approximate insofar as different values have been calculated from experimental data as well as from different theoretical approaches (see e.g. [10.12]).

| Surface | Rh(100) a | Ir(100) b | Pt(100) c | Cu(100) d | Cu(100) f | Au(100) g |
|---------|-----------|-----------|-----------|-----------|-----------|-----------|
| Mechanism | hop | ex. | ex. | vac. | hop | hop |
| $E_a$ / eV | 0.88 | 0.93 | 0.47 | 0.42 | 0.44 | 0.50 |

(a: [10.6], b: [10.7], c: [10.8], d: [10.9], f: [10.10], g: [10.11])

Table 10.1 lists three cases where exchange diffusion was established using FIM. The transport mechanism in surface self-diffusion is not always via adatoms. On Cu(100) e.g., the easiest pathway is mass transport via vacancies. On the other hand, theory and (indirect) experimental evidence point towards hopping of adatoms as the dominant mass transport mechanism on the unreconstructed Au(100) surface [10.11]. At present, there seems to be no systematic understanding as to which transport mechanism should occur when and where.

### 10.1.2 Statistics of Random Walk

We consider the random walk of atoms on surfaces with square symmetry. Because of the stochastic nature of the process the random walk in the two mutual orthogonal directions are independent of each other. It suffices therefore to consider random walk in one dimension. We denote the sites and the jumps by the indices $i$ and $j$, respectively, both starting from zero. After a first jump the atom is either at $i = -1$ or at $i = +1$ with a probability 1/2. The position of the atom is $x_{j=1} = \pm l$, with $l$ the jump length. After a second jump ($j = 2$) the atom resides with a probability 1/4 at $i = 2$, hence at $x_{j=2} = \pm 2l$ and with the probability $1/4 + 1/4 = 1/2$ at $i = 0$. The general rule for the probability to find the atom at site $x_j$ after $j$ jumps is

$$w_{ij} = \frac{1}{2}(w_{i+1, j-1} + w_{i-1, j-1}) . \tag{10.1}$$

For a particular number of jumps, e.g. $j = 2$, one finds that the mean square distance from the origin is

$$\left\langle (x_j - x_0)^2 \right\rangle = l^2 \sum_i w_{ij}\, i^2 = l^2 j . \tag{10.2}$$

By complete induction, one can prove this law to hold for any number of jumps. With the progression law (10.1) one obtains

$$\sum_i w_{i, j+1}\, i^2 = \frac{1}{2} \sum_i (w_{i+1, j} + w_{i-1, j}) i^2 = \frac{1}{2} \sum_i w_{ij} \left[ (i-1)^2 + (i+1)^2 \right]$$
$$= \sum_i w_{ij}\, i^2 + \sum_i w_{ij} = j + 1 . \tag{10.3}$$

As one can show (10.2) to be correct for a particular number of jumps $j$, eq. (10.3) proves that (10.2) holds also for $j+1$ jumps, and so forth. The total number of jumps $j$ can be expressed as the product of a jump rate and time. We will later introduce a theory that permits the calculation of the jump rate $\nu$ in each identical direction of which we have two here. Thus we write $j = 2\nu t$. The mean distance from the origin after a time $t$ is therefore

$$\left\langle (\Delta x)^2 \right\rangle = \left\langle (\Delta y)^2 \right\rangle = 2\bar{l}^2 \nu t .$$ (10.4)

We have introduced a mean jump distance $\bar{l}$ to account for the possibility that the jump distance need not always be the distance from one site to the next possible site but could include so called *long jumps*. Since the random walk along the $x$ and $y$ directions are independent of each other the mean distance $\Delta r$ from the origin after a time $t$ is

$$\left\langle (\Delta r)^2 \right\rangle = \left\langle (\Delta x)^2 \right\rangle + \left\langle (\Delta y)^2 \right\rangle = 4\bar{l}^2 \nu t .$$ (10.5)

The product of the squared mean jump length and the jump rate in one direction

$$D^* = \bar{l}^2 \nu$$ (10.6)

is the *jump diffusion coefficient $D^*$ [17]*, also named *tracer diffusion coefficient* because it is measured in diffusion experiments with individually marked radioactive isotopes (*tracers*). For a triangular lattice (a (111) surface) one needs to consider that the jumps into the four orthogonal cartesian directions progress differently far. By averaging the squares of the jump length one obtains

$$D^* = \frac{3}{4} \bar{l}^2 \nu .$$ (10.7)

Evidently, one cannot determine whether the atom transport involves long jumps from the observation of the mean square displacement $\Delta r$ in the long time limit. However, the existence of long jumps can be proven with the help of site visiting maps at short times. FIM is the ideal instrument for that purpose. The first experimental evidence of long jumps from FIM images came from the one dimensional diffusion along the [111] direction on the W(211) surface (Fig. 10.5a) [10.13]. Figure 10.5b and 10.5c show the number of observations for an adatom displacement in units of the nearest neighbor distance along the [111] direction ($a_{nn} = \sqrt{3}/2a_0$, with $a_0$ the lattice constant). The experimental data (shaded column) is compared to the probability distribution derived under the assumption that the probability for a jump length of twice the nearest neighbor distance $w_2$ is zero (white columns) and under the assumption that the jumps over twice and three times the nearest neighbor distance ($w_3$) have a nonzero probability. The probability distributions for the various cases can be calculated analytically from an ex-

---

[17] The jump diffusion coefficient is sometimes defined as $D^* = \bar{l}^2 \nu_0 / 4$, with $\nu_0$ the total jump rate.

pression, which involves Bessel-functions of the second kind [10.13]. For the limited number of jumps it is just as easy to calculate the probability distribution from the recursion

$$w_{ij} = \frac{w_1}{2}(w_{i+1,j-1} + w_{i-1,j-1})$$

$$+ \frac{w_2}{2}(w_{i+2,j-1} + w_{i-2,j-1}) + \frac{w_3}{2}(w_{i+3,j-1} + w_{i-3,j-1})... \qquad (10.8)$$

For the strongly bound W atoms on W(211) (for which the activation barrier for diffusion is high, $E_{act} = 0.83$ eV) the distribution is compatible with nearest neighbor jumps (Fig. 10.5b). For the weakly bound Pd the activation barrier is $E_{act} = 0.31$ eV. The probability distribution can only be fit successfully if one assumes a high probability of long jumps ($w_2/w_1 = 0.21$ and $w_3/w_1 = 0.14$).

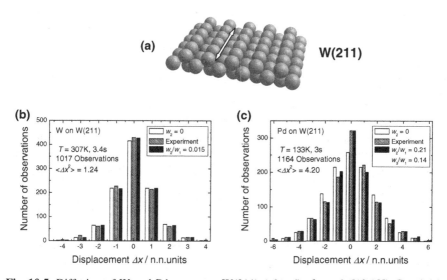

**Fig. 10.5.** Diffusion of W and Pd atoms on W(211) (after Senft et al. [10.13]). Panel **(a)** shows a ball picture of the W(211) surface. Diffusion is solely along the [111] direction in the temperature range of interest here (white arrow). For W atoms, the probability $w_2$ of long jumps with twice the nearest neighbor distance is small **(b)**. The distribution calculated with $w_2/w_1 = 0.015$ is compatible with the data as well as the distribution with $w_2 = w_3 = 0$. For Pd on the other hand the experimental result can only be fitted to a distribution calculated with significant contributions from long jumps **(c)**.

## 10.1.3 Absolute Rate Theory

The *absolute rate theory*, also known as *transition state theory*, is an attempt to describe kinetic processes by equilibrium properties of the system in its ground state (or a metastable steady state) and in one particular activated state, called the transition state. The theory was originally designed to describe rate constants for complex systems with many degrees of freedom and provides a simple rational for the general observation that the speed of kinetic processes can be described by a rate constant $k$ which consists of exponential term and a prefactor [10.14]. We have encountered a special form of the transition state theory already in the context of adsorption-desorption in Sect. 6.3.2. Applied to the diffusion of single atoms on a flat surface we can write the jump rate $\nu$ introduced in (10.4) as

$$\nu(T) = \nu_0(T)\, e^{-\frac{E_{\text{act}}}{k_B T}} . \tag{10.9}$$

In general, the prefactor $\nu_0$ depends also on the temperature $T$ in a power law form. The transition state theory links in a rather general way the temperature dependence of $\nu_0$ and the activation energy to calculable properties of the surface.

The transition state theory in its conventional form considers the Gibbs free energy $G$ of the system along a *minimal path* that leads from the initial state to a final state. Traveling along that path involves a concerted variation of the entire ensemble of general coordinates $\{q\}$ and their conjugated momenta $\{p\}$ in such a way that $G(\{q\},\{p\})$ stays minimal with reference to all adjacent paths. This minimal path is described by only one coordinate, the *reaction coordinate* $q_r$ and its canonical conjugate $p_r$ (Fig. 10.6). For the special application to diffusion on surfaces, the conventional Gibbs free energy has to be replaced by the surface excess thermodynamic potential that is appropriate to the boundary conditions (cf. Chapt. 4). For diffusion on surfaces in vacuum, the external pressure is zero and the macroscopic strain on the solid surface remains constant. The appropriate thermodynamic potential is therefore the Helmholtz surface free energy $F^{(s)}$. For surfaces held at constant potential in an electrolyte of fixed concentration, the Helmholtz free energy is to be replaced by the product of the *surface tension* $\gamma$ and the area $A$ as defined in eq. (4.19) in Sect. 4.2.3. In order to have the following equations in a general form we denote the relevant thermodynamic potential as the *free energy* $G(\{q\},\{p\})$ but keep in mind that $G(\{q\},\{p\})$ is something very different, depending on the specific problem. The point on the path where $G$ has a maximum is the *transition state*. The free energy $G$ at this point without the kinetic energy associated with $p_r$ is $G(q^+)$. Because of the construction of the minimal path the transition state represents a maximum of $G$ with respect to the reaction coordinate and a minimum with respect to all other adjacent paths, hence a saddle point. For the example of hopping diffusion of an adatom from one fourfold site to the next on a (100) surface the transition state is the two-fold bridge site (Fig. 10.3). In that case, the reaction coordinate would involve only the

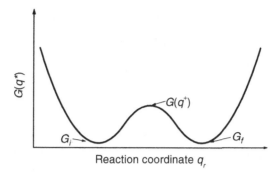

**Fig. 10.6.** Schematic representation of $G$ along the reaction path for the case of diffusion. $G$ is equal in the initial and final state in that case.

x- and z-coordinate of the adatom. In reality, the positions of neighboring surface atoms around the adatom yield to the changing bonds with the adatom, and the vibration spectrum also changes as the adatom moves along the minimum path. Hence, $G$ is a quite complex entity even in a comparatively simple case of diffusion. If one assumes that the system moves adiabatically along the minimum path the jump rate $v$ can be expressed in terms of the difference of $G$ in the transition state and the ground state. The kinetic energy along the reaction path is $p_r^2/2\mu$ with $\mu$ a reduced mass and the conjugate momentum $p_r = \mu\dot{q}_r$. The number of quantum states in the phase volume $dp_r dq_r$ is $dp_r dq_r/h$ with $h$ the Planck constant. The transition state theory assumes that the transition state is in equilibrium with the initial state. The probability to find the system in the transition state is then

$$w(q_r, p_r)\,dq_r\,dp_r = \frac{e^{-\frac{G(q^+)}{k_BT}}\,e^{-\frac{p_r^2}{2\mu k_BT}}}{e^{-\frac{G_i}{k_BT}}}\frac{dq_r\,dp_r}{h} \qquad (10.10)$$

The second term in the nominator $\exp(-p_r^2/2\mu k_BT)$ is the contribution to $G$ by the momentum along the reaction coordinate. The transition rate $v$ is given by the integral over all probabilities $w(q_r, p_r)$ multiplied with the velocity along the path, which is $\dot{q}_r$

$$v = \int_0^\infty dp_r\, w(q_r, p_r)\, \dot{q}_r = \frac{k_BT}{h}e^{-\frac{\Delta G}{k_BT}} \qquad (10.11)$$

with

$$\Delta G = G(q^+) - G_i . \tag{10.12}$$

If $\Delta G$ contains only energetic terms, the prefactor is

$$v_0 = k_B T / h . \tag{10.13}$$

We have encountered this prefactor already in the context of desorption (Sect. 6.3.2). Its value at 300 K is $6.25 \times 10^{12}$ s$^{-1}$. The fact that the prefactor is of the order of a vibration frequency has frequently lead to the misunderstanding that it represents an attempt frequency. Accordingly, (10.9) was misinterpreted as the product of an attempt frequency with the probability of successful attempts given by the Boltzmann factor. This is a gross misunderstanding insofar as the ground state, being a quantum mechanical eigenstate, does not couple to the transition state being another quantum mechanical eigenstate. The transition occurs via the thermal population of the entire ensemble of eigenstates including the ones having energies above the transition state. In these delocalized states an adatom, described by a localized wave packet moves into the neighboring sites by virtue of its group velocity. We see that long jumps are a natural consequence of the mechanism: The longer the lifetime of the atom in an excited state, the larger the probability of long jumps. Quite generally the lifetime is the shorter the higher the energy above the ground state. Hence, long jumps are rare when the activation energies are high (cf. Fig. 10.5). The fact that the interpretation of $v_0$ as an attempt frequency is erroneous becomes obvious also if one calculates $v_0$ in specific models. A simple one is being considered now.

### 10.1.4 Calculation of the Prefactor

The simplest conceivable model for diffusion is a particle in a one-dimensional parabolic potential (Fig. 10.7). For the calculation of the vibrational entropy, we take the vibrational eigenstates in the harmonic approximation

$$\omega = 2\pi \sqrt{(E^+ - E_g)/2a_{nn}^2 M} \tag{10.12}$$

where $a_{nn}$ is the distance between the minima in the potential, $M$ is the mass of the atom and $E^+$ and $E_g$ are the energies in the transition state and the ground state, respectively. With this frequency the difference $\Delta G$ is

$$\Delta G = E^+ - E_g + k_BT \ln \sum_n \exp(-\hbar\omega(n+1/2)/k_BT)$$

$$= E^+ - (E_g + \frac{1}{2}\hbar\omega + k_BT \ln(1 - e^{-\hbar\omega/k_BT}) \ . \tag{10.14}$$

The prefactor is therefore

$$\nu_0 = \frac{k_BT}{h}\left(1 - e^{-\frac{\hbar\omega}{k_BT}}\right) \ . \tag{10.15}$$

In the limit $k_BT \ll \hbar\omega$, the prefactor remains $k_BT/h$ and the "attempt frequency" $\omega$ does not enter the prefactor at all. In the reverse case $k_BT \gg \hbar\omega$, the prefactor becomes temperature independent and equal to the "attempt frequency"

$$\nu_0 = \omega/2\pi = \sqrt{(E^+ - E_g)/2a_{nn}^2 M} \ . \tag{10.16}$$

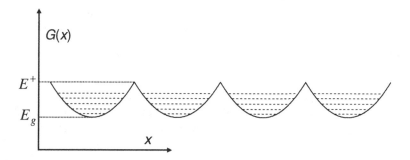

**Fig. 10.7.** A particle in a one-dimensional parabolic potential is the simplest model for diffusion. The dashed lines indicate schematically the vibrational levels of the particle in its binding site.

We apply this model to the hopping diffusion on a Cu(100) surface. The activation barrier is $(E^+ - E_g) = 0.44$ eV (Table 10.1) from which one calculates $\hbar\omega = 9.3$ meV. The solid line in Fig. 10.8 shows the prefactor as calculated from the 1D-model. The dashed and dash-dotted lines are the high and low temperature limits according to the model. Also shown is the theoretical result of Ulrike Kür-pick (dotted line) [10.10]. Her analysis took the vibration entropies in the activated state and the ground state fully into account. While the results of the full theory were only given down to a temperature of 100 K one can argue that ultimately the prefactor obtained from the full theory must also vanish at $T = 0$ K. As seen from

Fig. 10.8, the simple model reproduces the prefactor of the full theory within a factor of two. The same holds for the (110) and (111) surface of Cu and for the same ensemble of low index surfaces of Ni (Table 10.2).

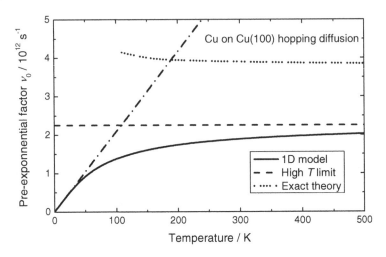

**Fig. 10.8.** Prefactor as calculated from the 1D-model (solid line). The high and low temperature limits according to the model are drawn as dashed and dash-dotted lines, respectively. Also shown is the result of a theory that takes the vibration states in the ground state and the transition state correctly into account (dotted line) [10.10].

**Table 10.2.** Activation energies for diffusion $E_{act}$ in eV and prefactors $\nu_{0,th}$ in $10^{12}$ s$^{-1}$ for the low index surfaces of Cu and Ni according to Ulrike Kürpick [10.10] together with the values for $\nu_{0,mod}$ calculated from (10.15). The work of Kürpick quotes prefactors for the diffusion coefficient which are converted into the prefactors $\nu_0$ according (10.6) and (10.7).

| Surface | $E_{act}$(Cu) | $\nu_{0,th}$(Cu) | $\nu_{0,mod}$(Cu) | $E_{act}$(Ni) | $\nu_{0,th}$(Ni) | $\nu_{0,mod}$(Ni) |
|---|---|---|---|---|---|---|
| (100) | 0.44 | 3.8 | 2.3 | 0.68 | 5.9 | 3.0 |
| (110) | 0.25 | 1.7 | 1.7 | 0.39 | 2.3 | 2.3 |
| (111) | 0.042 | 0.74 | 1.2 | 0.063 | 1.2 | 1.6 |

Several lessons may be learned these considerations. Firstly, in analyzing experimental data on diffusion the prefactor may be taken as constant. The situation is here distinctly different from the prefactor in desorption (Sect. 6.3.2). Secondly, the prefactor should be in the range of $10^{12}$-$10^{13}$ s$^{-1}$. Converted to the prefactors of the diffusion coefficients $D_0$, this corresponds to values of the order of $D_0 \approx 10^{-3}$ cm$^2$ s$^{-1}$. The values listed for a large number of single atom diffusion events are in that ballpark [10.4]. If one finds prefactors that differ by orders of magnitude, one might question the theoretical foundation of the analysis. This comment applies in particular to the analysis of data where diffusion on terraces enters indirectly through certain model assumptions. Examples are data on Ostwald ripening, step fluctuations and nucleation (see Sects. 10.4, 10.5, 11.1). Finally, one may use the high temperature limit of the prefactor of the 1D-model (10.16) as a rule of the thumb to estimate the prefactor from the activation energy, the atom mass and the hopping distance.

### 10.1.5 Cluster and Island Diffusion

Single atoms or single atom vacancies are the species which carry the atom transport in the equilibration of rough surfaces, because they require lower formation energies than units of two or more atoms. Frequently, but not always, single atoms and single atom vacancies are the fastest moving species. However, cluster of two or more atoms, even large islands diffuse as well. Such processes are particular important during epitaxial growth on flat surfaces. Two or more atoms form stable nuclei on surfaces during growth and the motion of these nuclei contributes to the coarsening process. In some cases, clusters may diffuse even faster than single atoms. Figure 10.9 shows two diffusion mechanisms of dimers on (100) surfaces. The dimer in Fig. 10.9a moves by shearing, the dimer in 10.9b moves by exchange. The exchange involves a concerted motion of three atoms with the surrounding atoms relaxing their positions adiabatically during the process. Depending on the material, both mechanisms can have a lower activation energy than single atom motion [10.15, 16]. Diffusion by shearing is also pathway with a low activation energy for larger clusters. Figure 10.10 shows the shearing motion of a tetramer on a (100) surface after Zhu-Pei Shi et al. [10.15].

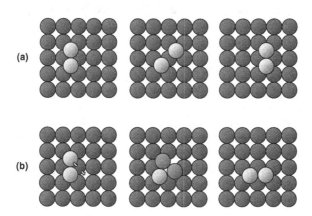

**Fig. 10.9.** Diffusion mechanisms of dimers on (100) surfaces. Dimer shearing **(a)** has a lower activation energy than single atom hopping on Cu(100) [10.15]. The same holds for exchange diffusion **(b)** on Pt(100) [10.16].

**Fig. 10.10.** Diffusion of a tetramer by shearing. The two atoms in the ellipse move together in a concerted motion. The activation energy for this cluster diffusion process is significantly lower than for diffusion by evaporation of an atom (after Zhu-Pei Shi et al. [10.15]).

On (111) surfaces the motion of large islands may also proceed via formation of dislocations. This process is particular effective for strained heteroepitaxial islands on (111) surfaces. Figure 10.11 illustrates the process on a small island. In the left panel the atoms of the island reside in fcc-sites. In the center panel, the island has a partial dislocation in the center. The atoms in right half of the island have assumed hcp-sites. The energy for the formation of a dislocation is particularly low if the island is elastically compressed, i.e. if its natural lattice constant is larger than the lattice constant of the substrate material. The activation energy of diffusion via the motion of dislocations depends also critically on the number of atoms in the island [10.17, 18].

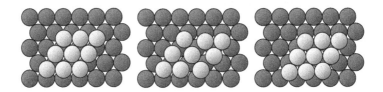

**Fig. 10.11.** Island diffusion by generation and motion of a partial dislocation. The mechanism is effective for large islands, in particular when the island is elastically compressed because of a misfit between the lattice constant of the island and the substrate.

Considerably experimental and theoretical effort has been devoted to the investigation of the *Brownian motion* of large islands. First observations of the Brownian motion of islands were reported by Karina Morgenstern et al. [10.19]. The term Brownian motion refers to the random walk of the center of gravity of large islands or vacancy islands across the surface such that the center of gravity $r(t)$ obeys

$$\left\langle (r(t-t_0)-r(t_0))^2 \right\rangle = 4D|t-t_0|. \tag{10.17}$$

The origin of the Brownian motion is the stochastic fluctuations of the center of gravity due to diffusion of atoms from one side of the island to the other. A theory based on the continuum description of equilibrium shape fluctuations (cf. Sect. 4.3.8) predicts a scaling law for the island diffusion coefficient $D$ [10.20] according to which $D$ should scale as

$$D \propto R^{-\beta} \tag{10.18}$$

where $R$ is the mean radius of the island with the exponent $\beta$ depending on the rate limiting process in the fluctuations of the shape. If the shape fluctuates because of diffusion along the periphery then $\beta = 3$. If terrace diffusion is rate limiting, then $\beta = 2$, and $\beta = 1$ should hold if the attachment-detachment process of atoms from the periphery is the rate limiting process. Because of the scaling law, small islands should diffuse faster than large islands. This scaling is qualitatively consistent with experimental observations. Since small islands diffuse quite rapidly, the diffusion can lead to coalescence events in which two islands merge to become one. Island coalescence can be the prevailing coarsening mechanism on surfaces. The continuum theory based on shape fluctuations as well as other theories fails completely on the quantitative side. Careful and detailed experimental studies show [10.21] that the scaling exponent for Cu(111) and Ag(111) vacancy islands is fractional ($\beta \approx 1.5$) over a wide range of island sizes and temperatures. The reason for the failure of existing theories is presently not understood. The recently discovered considerable elastic strain field even around homoepitaxial islands (Sect. 3.4.1) may play a role.


## 10.2 Continuum Theory of Diffusion

### 10.2.1 Transition from Stochastic Motion to Continuum Theory

The transition from stochastic motion to continuum theory of diffusion is made by defining the coverage $\Theta(x,y)$ as the density $\rho(x,y)$ of diffusing species multiplied by the area $\Omega_s$ of an adsorption site

$$\Theta(x, y) = \Omega_s \rho(x, y). \tag{10.19}$$

In the spirit of a continuum approach, we assume that the concentration from site to site varies merely by an infinitesimally small amount. On a square lattice the variation of the coverage in the time interval $dt$ is then

$$\dot{\Theta}(x, y)\,dt = v\,dt\big[\big(\Theta(x+l, y) - \Theta(x, y)\big) - \big(\Theta(x, y) - \Theta(x-l, y)\big)\big]$$
$$+ v\,dt\big[\big(\Theta(x, y+l) - \Theta(x, y)\big) - \big(\Theta(x, y) - \Theta(x, y-l)\big)\big]$$

$$\cong v\,ldt\left\{\left.\frac{\partial\Theta}{\partial x}\right|_{x+l/2}-\left.\frac{\partial\Theta}{\partial x}\right|_{x-l/2}+\left.\frac{\partial\Theta}{\partial y}\right|_{y+l/2}-\left.\frac{\partial\Theta}{\partial y}\right|_{y-l/2}\right\}$$

$$\cong l^2 v\,dt\left(\left.\frac{\partial^2\Theta}{\partial x^2}\right|_{x,y}+\left.\frac{\partial^2\Theta}{\partial y^2}\right|_{x,y}\right).$$

$$(10.20)$$

Here, $l$ is again the jump length which can be replaced by the mean jump length $\bar{l}$. In writing (10.20), we have assumed that coverage is small. After introducing the concentration one obtains

$$\dot{\rho}(x,y)=v\bar{l}^2\Delta\rho(x,y)=D^*\Delta\rho(x,y) \qquad (10.21)$$

where $\Delta$ denotes the Laplace operator. Stationary diffusion profiles therefore obey the Laplace equation

$$\Delta\rho(x,y)=0. \qquad (10.22)$$

The time derivative of the concentration is related to the divergence of the particle current density $j(x,y)$, which is defined as the number of particles per cross section length and time that flow from one length element into the next. The change of the concentration of particles in a length element $dx$ per time equals the difference in the current density of influx and outflow

$$j_x(x+dx)-j_x(x)=-\dot{\rho}(x)\,dx \qquad (10.23)$$

so that

$$\nabla j=-\dot{\rho}. \qquad (10.24)$$

From (10.21) follows

$$j=-D\nabla\rho. \qquad (10.25)$$

This equation is known as the first *Fick's law of diffusion*. We have omitted now the asterisk in the notation of the diffusion coefficient since (10.25) is a general linear equation for the diffusion current for arbitrary coverages and the diffusion coefficient can depend on the coverage in a very complex way that is not covered by the single atom jump model. The diffusion coefficient $D$ is called *chemical diffusion coefficient* or sometimes *Fick's law diffusion coefficient*.

Fick's law is a special form of a general transport equation that relates the particle current to the gradient in the chemical potential

$$j = -L_T^{(t)} \nabla \mu . \tag{10.26}$$

Here $L_T^{(t)}$ is a transport coefficient for diffusion on terraces, which can be expressed in terms of microscopic parameters of certain models (Sect. 10.2.4). Eq. (10.26) can be related to (10.25) if one considers the chemical potential to be a function of the coverage

$$j = -L_T^{(t)} \frac{\partial \mu}{\partial \Theta} \nabla \Theta = -L_T^{(t)} \Omega_s \frac{\partial \mu}{\partial \Theta} \nabla \rho \tag{10.27}$$

The chemical diffusion coefficient $D$ is therefore

$$D = L_T^{(t)} \Omega_s \frac{\partial \mu}{\partial \Theta} . \tag{10.28}$$

The relation between the chemical diffusion coefficient $D$ and the jump coefficient $D^*$ is obtained by considering the chemical potential of a non-interacting dilute ($\Theta \ll 1$) lattice gas (Sect. 5.4.1, eq. 5.95)

$$\mu = \mu_0 + k_B T \ln \Theta . \tag{10.29}$$

The chemical diffusion coefficient $D$ becomes

$$D = L_T^{(t)} \Omega_s \frac{k_B T}{\Theta} . \tag{10.30}$$

For a non-interacting lattice gas the diffusion coefficient must be equal to the jump diffusion coefficient $D^* = v \bar{l}^2$. By comparison, one obtains the transport coefficient of a dilute lattice gas

$$L_T^{(t)} = v \bar{l}^2 \frac{\Theta}{\Omega_s k_B T} \tag{10.31}$$

After inserting (10.31) into (10.27) one obtains

$$j = -v \bar{l}^2 \frac{\Theta}{k_B T} \frac{\partial \mu}{\partial \Theta} \nabla \rho . \tag{10.32}$$

Hence the relation between the jump diffusion coefficient $D^*$ and the chemical diffusion coefficient $D$ is

$$D = D^* \frac{\partial(\mu / k_B T)}{\partial \ln \Theta} .$$

(10.33)

This relation is known as the *Darken equation*. While the derivation here was only for the trivial case of a dilute non-interacting lattice gas, it can be shown that the relation holds in general. For the derivation of the equation as well as of many other useful relations in the context of diffusion, see e.g. [10.22].

## 10.2.2 Smoothening of Rough Surfaces

Standard methods of surface preparation frequently involve sputtering by rare gas atoms (Sect. 2.2.3). The locations of arrival on the surface and therefore the associated removal of mass are stochastically distributed. Prolonged sputtering leads to a roughening of the surface. To heal the damage and to remove the roughness each sputtering cycle is followed by an annealing step. We study the smoothening process via surface diffusion in a continuum approximation with the help of the general transport equation (10.26). To do so we need to assume that the chemical potential is a continuously differentiable function. In other words, the surfaces must be rough in the thermodynamic sense, i.e. must not have facets due to cusps in the chemical potential (Sect. 4.3.1). The driving force for smoothening by diffusion is the gradient of the chemical potential according to the Herring-Mullins equation (Sect. 4.3.2, eq. 4.63).

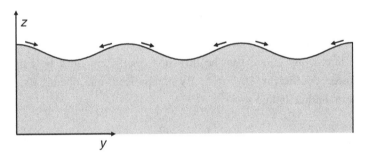

**Fig. 10.12.** A cosine surface profile. The arrows mark the directions of diffusion from regions of high to low chemical potential.

Since an arbitrary roughness profile of a surface can be decomposed into its Fourier-components, it suffices to study the decay of a one-dimensional cosine profile (Fig. 10.12)

$$z(x) = z_0 \cos qy$$

(10.33)

With the one-dimensional Herring-Mullins equation for small curvatures

$$\mu = -\Omega \, \tilde{\gamma} \, \frac{\partial^2 z}{\partial y^2} \tag{10.34}$$

one obtains the particle current density $j$

$$j = L_T^{(t)} \Omega \, \tilde{\gamma} \, \frac{\partial^3 z}{\partial y^3} \,, \tag{10.35}$$

in which $\Omega$ is the atomic volume and $\tilde{\gamma}$ the surface stiffness. As the orientation dependence of the surface tension is generally small, the stiffness $\tilde{\gamma}$ is frequently equated with the surface tension $\gamma$. The sign in (10.34) refers to a surface of a solid in the negative half space so that a bump in the surface has a positive chemical potential. A second equation follows from the continuity condition (10.24). A gradient in the flux of atoms must lead to a reduction of the local height,

$$\frac{\partial j}{\partial y} = -\frac{\partial \rho(y,t)}{\partial t} = -\frac{1}{\Omega} \frac{\partial z(y,t)}{\partial t} \,. \tag{10.36}$$

The profile therefore decays according to

$$\frac{\partial z(y,t)}{\partial t} = -L_T^{(t)} \, \Omega^2 \, \tilde{\gamma} \, \frac{\partial^4 z(y,t)}{\partial y^4} \,. \tag{10.37}$$

Each Fourier component decays as

$$z(t) = z_0(t_0) e^{-\frac{t-t_0}{\tau(q)}} \tag{10.38}$$

with

$$\tau^{-1}(q) = L_T^{(t)} \, \Omega^2 \, \tilde{\gamma} \, q^4 \,. \tag{10.39}$$

This remarkable result tells us that roughness on a short length scale anneals out quickly. However, it takes impractically long times or very high temperatures to smoothen roughness on a long length scale. Applied to the problem of sputtering and annealing, this means that one can expect to heal sputter damage effectively when the total dose amounts to a removal of merely a few monolayers, as the roughness introduced by that is only on a lateral length scale of a few atoms. Prolonged and repeated sputter-annealing cycles that involve a removal of many layers on the other hand, may lead to irreparable damage. Roughness on a long length scale can often be noticed visually by the naked eye. The mirror finish of the sample has given way to an orange skin appearance. Because of the danger of sputter damage beyond repair, alternative methods of preparation were introduced if possible. An example is the Si(111) surfaces. In the early days of surface science, it was customary to prepare Si(111) by sputter-annealing cycles. Much better results are however obtained by wet chemical cleaning, thermal oxidation and a short flash in UHV to remove the oxide (Sect. 2.2.3).

For very high temperatures, profile decay can also proceed via exchange with the surrounding vapor phase through evaporation/condensation or by exchange with the bulk by evaporation/condensation of defects, vacancies in practice. If the exchange with the vapor phase or the bulk is determined by the attachment/detachment process then the current density leading to decay is simply proportional to the local chemical potential

$$j = -L_T^{(b)} \mu \qquad (10.40)$$

in which $L_T^{(b)}$ is a rate constant characterizing the detachment process. The profile decay then obeys

$$z(t) = z_0(t_0) e^{-\frac{t-t_0}{\tau(q)}} \qquad (10.41)$$

with

$$\tau^{-1}(q) = L_T^{(b)} \Omega^2 \tilde{\gamma} q^2 . \qquad (10.42)$$

Roughness on a long length scale anneals out much more quickly in this case. Excessive sputter damage can therefore be removed by annealing to temperatures just below the melting point where most solids have a high equilibrium concentration of bulk vacancies. However, the method works only, if the surface of interest does not undergo a roughening transition up to the melting temperature.

There is also an intermediate case where the rate constant is proportional to $q^3$ when the surface is in local equilibrium with the gas-phase or the vacancy concentration. Mullins has discussed all these scenarios theoretically as early as 1959, long before atomic scale experiments became available [10.23]. The corresponding theoretical models have later been rediscovered and outlined in the context of step fluctuations and 2D Ostwald ripening processes. We shall consider these different scenarios in the theory of step fluctuations (Sect. 10.5). A special case is the frequently observed decay of a non-equilibrium protrusion in a step to which we turn now.

### 10.2.3 Decay of Protrusions in Steps
### and Equilibration of Islands after Coalescence

Protrusions in steps form temporarily during coarsening when an island merges with a step. The decay of such protrusion with time can be studied with the help of STM images. Figure 10.13 shows an example of four consecutive STM images of a Cu(111) surface taken at 303 K [10.24]. Each image represents a time span of only 30 s. Equilibration proceeds therefore very fast on Cu(111) even at room temperature. The islands and vacancy islands seen in the images were formed during deposition of several monolayers of copper. In the course of a ripening process, the islands migrate across the surface in a random walk. In Fig. 10.13a one island has come close to a [1$\bar{1}$0]-oriented step and has merged with the step in Fig. 10.13b, causing temporarily a protrusion in the step.

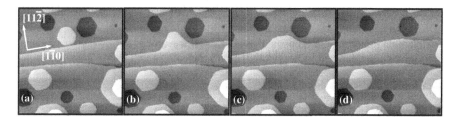

**Fig. 10.13**. A series of 90 nm × 90 nm STM images of a Cu(111) surface taken 303 K (after M. Giesen and G. Schulze Icking-Konert [10.24]). The time difference between images is about 30 s. The island seen in (**a**) merges with the step due to the random walk of islands and the resulting protrusion decays because of diffusion along the step edge.

The theory of profile decay outlined in the preceding section applies also to this one-dimensional case. Instead of the surface stiffness, now the line stiffness $\tilde{\beta}$ is the driving force. We note that the line stiffness cannot be equated, not even approximately with the line tension $\beta$ (Sect. 5.3.2). In keeping with the nomenclature introduced in Sect. 4.3.7 we denote the profile function as $x(y)$ where $x$ and $y$ are the directions perpendicular and parallel to the step direction. Contrary to the preceding case of an assumed very flat profile in the surface, we have now large angles $\theta$ with the [1$\bar{1}$0]-orientation of dense atom packing. The exact expression for the chemical potential in that case is

$$\mu = -a_\parallel a_\perp \tilde{\beta}(\theta) \frac{\partial^2 x / \partial y^2}{(1 + (\partial x / \partial y)^2)^{3/2}} \qquad (10.43)$$

with $a_\perp$ and $a_\parallel$ the distance between atom rows on (111) surfaces and the atomic length unit parallel to the step orientation (atom diameter), respectively. Integration of the differential equation for the profile decay requires the knowledge of the dependence of the line stiffness on the angle and a numerical solution of the resulting differential equation. The task is greatly facilitated, however, by making use of the fact that the protrusion in the step originated from the consumption of an island, which had an equilibrium shape. The "corners" and angles of the protrusion shown in Fig. 10.13b are those of the island equilibrium shape shown in Fig. 10.13a. Equilibrium shape means that the chemical potential is constant along the perimeter. The shape of the initial protrusion is therefore such that $\mu$ is at least approximately constant. It may be therefore permitted to assume that the simple expression for the chemical potential near the $[1\bar{1}0]$-orientation holds everywhere so that

$$\mu = -a_\parallel a_\perp \tilde{\beta}(\theta = 0)\, \partial^2 x / \partial y^2 . \tag{10.44}$$

For the line tension of a step on the (111) surface along the direction of dense packing we have derived in Sect. 5.3.2, eq. (5.84)

$$\tilde{\beta}_{\text{hex}}(\theta = 0) = \frac{k_B T a_\parallel}{a_\perp^2} e^{\varepsilon_k / k_B T} \tag{10.45}$$

with $\varepsilon_k$ the kink energy. For a rectangular lattice, we had (5.83)

$$\tilde{\beta}_{\text{rect}}(\theta = 0) = \frac{k_B T a_\parallel}{2 a_\perp^2} e^{\varepsilon_k / k_B T} \tag{10.46}$$

with $a_\perp = a_\parallel$ for a square lattice like the fcc(100) surface. In analogy to the derivation above (10.34 - 10.39) one finds that each Fourier-component of the profile function decays with the time constant

$$\tau^{-1}(q) = L_T^{(\text{st})} a_\perp^2 a_\parallel^2 \tilde{\beta} q^4 \tag{10.47}$$

where $L_T^{(\text{st})}$ is the one-dimensional transport coefficient describing diffusion along step edges and $\tilde{\beta}$ is the line stiffness of the corresponding lattice. The relation $\tau \propto q^{-4}$ is well fulfilled for the decay of the Fourier-components proving that indeed the rate determining transport mechanism is along the step edges (Fig. 10.14).

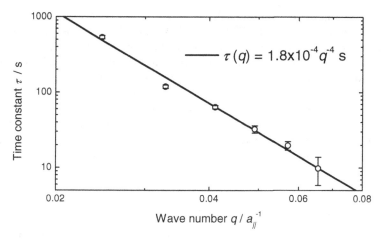

**Fig. 10.14.** Time constant $\tau(q)$ of the Fourier-component of the step profile function displayed in Fig. 10.13. The $q^{-4}$-dependence shows that the rate determining transport mechanism is diffusion along the step (after M. Giesen and G. Schulze Icking-Konert [10.24]).

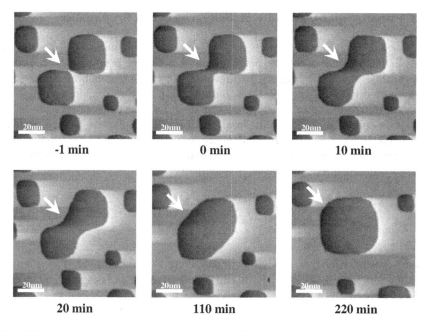

**Fig. 10.15.** Coalescence of two vacancy islands on Cu(100) at 323 K (After Ikonomov et al. [10.25]).

Shape equilibration is also observed when two islands coalesce on surfaces. Figure 10.15 shows a series of excerpts of STM images of a Cu(100) surface on which vacancy islands were produced by mild sputtering [10.25]. Occasionally two islands meet and coalesce due to the random walk of these islands on surfaces and a dumb-bell shaped island forms temporarily. The neck widens quickly as atoms move from regions of high to low chemical potential. Eventually the equilibrium shape is established again. Experiments show that both the width of the neck and the length of the dumb-bell approach the final equilibrium in an exponential manner with a characteristic time constant $\tau$. The time constant depends on the length scale $L$ (the diameter of the resulting equilibrium shape, e.g.) according to

$$\tau \propto L^{\alpha}. \tag{10.48}$$

The exponent follows from the rate determining transport mechanism in the same way as the exponents of the relaxation time of the Fourier-components of step profiles. For Cu(100) steps mass transport along the step edges dominates and the exponent is therefore $\alpha = 4$.

### 10.2.4 Asaro-Tiller-Grinfeld Instability and Crack Propagation

An enormous analytical and predictive power of theory is occasionally gained if one relinquishes comprehension on an atomic scale. There is no better demonstration for this than with concepts that develop from the combination of elasticity theory with diffusion. We consider a solid or a thin solid film under compressive stress parallel to the surface. The stress can arise from an external pressure or, in case of an epitaxially grown film, from misfit with the substrate lattice constant. The chemical potential of a surface then acquires an additional term from the elastic energy. For a (thermodynamic rough) surface curved in the $z,y$ plane and stressed along the $y$ axis the chemical potential is

$$\mu = \Omega \left( \frac{1-\nu^2}{2Y} \tau_{yy}^2 - \gamma \kappa \right) \tag{10.49}$$

where $\kappa$ is the curvature, $\tau_{yy}$ is the stress, $\nu$ the Poisson number, $Y$ is Young's modulus (cf. Sect. 3.64) and $\gamma$ is the surface tension. The elastic term is always positive whereas the curvature term depends on the sign of the curvature. We consider the consequences of the stress term in (10.49) for a sinusoidal profile

$$z = z_0 \sin qy \tag{10.50}$$

For this profile, the total free surface energy per area is higher than for a flat surface by the amount

$$\Delta\gamma_{\text{surf}} = \gamma\left(\frac{1}{\lambda}\int_0^{\lambda} z(y)\sqrt{1+z'^2}\,dx-1\right)=\gamma\left(\frac{z_0 q}{4}\right)^2 \tag{10.51}$$

In the absence of the stress term, this energy drives the profile decay as discussed above. Because of the stress term, the elastic energy may be larger in the valleys than on the hills. Then the amplitude of the profile grows in time so that a flat surface becomes unstable. A full study of the spatiotemporal evolution of a surface profile under stress requires the solution of a nonlinear integral-differential equation. However a qualitative criterion for the stability or instability of a flat surface can be derived by considering the contributions to the energy per area [10.26].

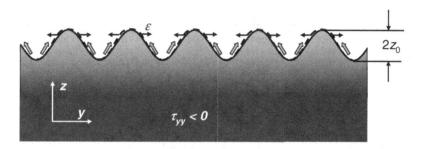

**Fig. 10.16.** A thin film or bulk system under compressive stress with a sinusoidal surface profile. The energy gain associated with the outward relaxation in the upper hill region (black arrows) may outweigh the energy expenditure for the increased surface energy, which gives rise to the *Asaro-Tiller-Grinfeld* (ATG) instability. The height of the profile grows by a surface diffusion flux (gray block arrows) from the regions of high strain to the regions of low strain.

For a solid under stress, we have two contributions to the change in the energy from elasticity. One arises from the elastic relaxation near the top of the hills in the profile (Fig. 10.16). We estimate this energy per area by

$$\Delta\gamma_{\text{relax}} = -z_0 \tau_{yy}\varepsilon \tag{10.61}$$

where $\varepsilon$ is the relative relaxation of the compressed solid. This relaxation occurs only in the upper section of the profile not at the bottom. The relaxation therefore imprints a periodic strain field into the solid. The extension of that field is of the order of the inverse wave vector of the profile. The elastic energy per area associated with the periodic strain field is therefore estimated as

$$\Delta\gamma_{\text{strain}} = \frac{Y}{2(1-v^2)}\frac{\varepsilon^2}{q}.$$ 

(10.53)

The sum of the two elastic contributions has a minimum at

$$\varepsilon = z_0\tau_{yy}q(1-v^2)/Y$$ 

(10.54)

which fixes the strain relaxation for a given wave vector. The sum of all contribution to the energy per area is

$$\Delta\gamma_{\text{tot}}(q) = z_0^2\left(\frac{\gamma}{4}q^2 - \frac{\tau_{yy}^2(1-v^2)}{2Y}q\right).$$ 

(10.55)

This energy is negative for small enough wave vectors $q$, which means that energy is gained by an accidental perturbation if $q$ is below a critical value $q_{\text{crit}}$ that is given by

$$q_{\text{crit}} = \frac{2\tau_{yy}^2(1-v^2)}{\gamma Y}.$$ 

(10.56)

The energy gain has its maximum at half the critical wave vector $q_{\text{max}} = q_{\text{crit}}/2$. Accidental perturbations with $q < q_{\text{crit}}$ grow by surface diffusion from the regions of high strain at the bottom of the profile (gray block arrows in Fig. 10.16) to the region of low elastic strain at the top. In other words, the mass flow is reversed compared to the normal situation. This is the *Asaro-Tiller-Grinfeld* (ATG) instability [10.27, 28]. We note that the instability occurs independently of the sign of the stress.

The ATG instability gains practical importance if the wavelength of a growing perturbation is in the range of nanometers. Otherwise, the time scales for diffusion are too long and the energy associated with (10.55) becomes too small. The stress must therefore exceed a certain limit before the ATG instability becomes noticeably. For a wavelength of $\lambda_{\text{crit}} = 100$ nm, e.g., a typical surface tension of $\gamma = 2$ N/m, and an elastic constant of $Y/(1-v^2) = 10^{11}$ N/m$^2$ the critical stress amounts to 2.5 GPa. Stress of this order of magnitude builds up in epitaxial layers with misfits of a few percent. Such films are therefore prone to show ATG instability if held at higher temperatures to allow for sufficient surface diffusion [10.29].

The growth of the ATG instability shows a highly nonlinear behavior due to the nonlinear nature of the diffusion equation and the nonlinearity of the chemical potential in terms of the height profile (10.43). Some of that nonlinearity is revealed even in the simple scaling considerations above: the energy gain increases quadratic with the depth of the protrusion (10.55). Because of the nonlinearity of the problem, an initially sinusoidal profile develops a cusp at the bottom that quickly turns into a crack like feature, which (for a semi-infinite solid) grows with ever increasing speed. This is demonstrated in Fig. 10.17, which partly reproduces a selection from the profiles calculated by Yang and Srolovitz [10.30]. The times are given in units of

$$t_0 = \lambda^4 / L_T^{(t)} \Omega^2 \gamma \tag{10.57}$$

The cusp develops at about $t = t_0$ and quickly turns into a grove, then into a crack. The dramatic increase in speed is highlighted by the very small time difference between the two last profiles which differ merely by $0.002t_0$. Concomitant with the increase in speed the radius of the crack at the tip decreases to eventually collapse into a singularity. This behavior is at variance with the experimental observation that cracks propagate with a defined speed that depends on the driving force and saturates at a fraction of the speed of the Rayleigh wave (Sect. 7.1.4). It was therefore not clear for some time whether crack development and propagation can be understood as an ATG instability. Brener and Spatschek [10.31] showed that the singular behavior of conventional ATG-theory is removed by introducing a kinetic energy term into the expression for the chemical potential (10.49). In their theoretical approach, the crack propagates with a finite speed that is a function of the driving force. The ATG-instability as such reappears in their theory as a bifurcation of cracks.

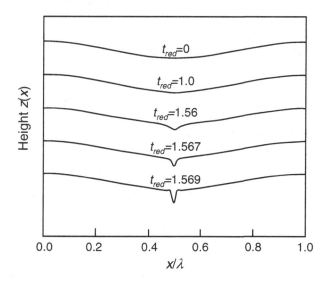

**Fig. 10.17.** The growth of a crack-like feature in the sinusoidal height profile of a solid under stress that amounts to $\approx 1.7 \, \tau_{crit}$ (after Young and Srolovitz [10.30]).

It may be surprising that the phenomenon of crack propagation that most people would intuitively associate with bond breaking rather than with diffusion should boil down to a stress-induced diffusion problem. After all, cracks develop and progress at low temperatures, where there is no diffusion. However, as Brener and Spatschek pointed out, an enormous energy is released at the progressing tip of the crack that should bring the local temperature at the tip close to the melting temperature, independently of the temperature of the environment. Surface diffusion right at the tip should therefore by quite high. Furthermore, the mathematics of surface diffusion that enters the theory may stand for a wider range of transport mechanisms that include bulk diffusion and plastic flow of material.

## 10.3 The Ehrlich-Schwoebel Barrier

### 10.3.1 The Concept of the Ehrlich-Schwoebel Barrier

The elegance and simplicity of the theory of profile decay is owed to the fact that all atomic processes are hidden in the transport coefficient $L$ which is assumed to be independent of the azimuthal and polar orientation. In particular, the latter assumption cannot be justified on general grounds. A surface that is vicinal to a low index orientation consists of terraces and monatomic steps. We exclude the possibility of faceting and the formation of step bunches for the moment. A profile on the surface is then equivalent to a variation in the local density of steps (Fig. 10.18). The decay of a profile on a surface therefore requires atom transport not only across the terraces but also across steps, in other words *intralayer* and *interlayer* transport.

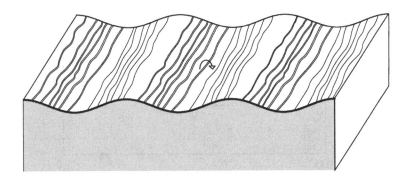

**Fig. 10.18.** Microscopic view on an undulated surface. Profile decay requires *intralayer* as well as *interlayer* mass transport across step edges (arrow). Interlayer mass transport requires to overcome an activation barrier which is often larger than the activation barrier for intralayer transport. The additional activation energy is called ES-barrier.

The activation barriers for intralayer and interlayer transport are in general different. The effective diffusion coefficient must therefore depend on the local concentration of steps, hence on the local orientation of the surface. The transport coefficient $L_T$ is independent of orientation only if traversing the steps either involves the same activation barrier than diffusion on the terraces or if the activation barrier for interlayer transport is small and the concentration of step small. The difference (!) in the activation barrier for transport across a step edge and on terraces is called the *step edge barrier* or *Ehrlich-Schwoebel barrier* ("ES-barrier" in the following) after the first authors of two papers that appeared in 1966. The paper of Ehrlich and Hudda describes experimental FIM-observations on the reduced diffusion across steps [10.3]. The second paper by Schwoebel and Shipsey postulates the existence of a step edge barrier and discusses its consequences for epitaxial growth [10.32]. This truly remarkable paper anticipates many experimental and theoretical developments of the decades that followed. Figure 10.19 is adapted from Fig. 1 of that paper and shows schematically the potential for an atom near a step edge. The figure is actually quite ingeniously designed, as it suggests the correct physics for the wrong reason: The ES- barrier appears to arise from the low coordination of the adatom when it "rolls" over the step site.

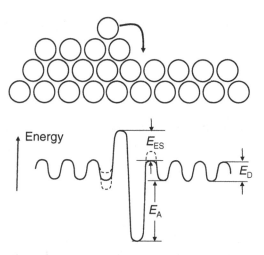

**Fig. 10.19.** Schematic drawing of the potential for an atom near a step edge with an ES-barrier $E_{ES}$. $E_A$ is the binding energy to a step site and $E_D$ the activation energy for terrace diffusion. The dashed lines represent a potential with an attachment barrier and with a higher or lower binding energies at the last binding site at the upper step edge. All three alternatives may exist independently. An ES-barrier may exist also for vacancy diffusion.

A three-dimensional picture reveals however that the nearest neighbor coordination in the transition state at the upper step edge is just the same as in the transition state on the terrace. The binding energy in the transition state at the step edge is lower than on a terrace only because of the fewer number of next-nearest neighbors compared to the transition state in terrace diffusion. An alternative interpretation of the lower binding energy is that embedding energy of the atom in the transition state at the step edge is lower because of the lower electron density there. The higher binding energy $E_A$ at the lower edge of a step follows from the larger number nearest neighbors. The dashed lines in Fig. 10.19 show the possibility of an attachment barrier for adatoms approaching the step from the lower terrace and a site of higher binding energy next to the upper step edge. All three alternatives may exist independently of each other.

The set of three energies $E_{ES}$, $E_D$, and $E_A$ is the minimal set of energies and activation energies required for the description of profile decay and epitaxial growth. The one-dimensional potential suffices for the rationalization of some phenomena. Other phenomena such as certain growth instabilities require the consideration of the full three-dimensional potential landscape.

## 10.3.2 Mass Transport on Stepped Surfaces

We consider the effect of an ES-barrier on the transport coefficient $L_T$ and the effective diffusion coefficient $D_{eff}$ on a stepped surface. The problem is easily treated by analogy to an electrical network consisting resistive elements $i$ in series for which the conductance is known [10.33]. The total conductance $R_{tot}^{-1}$ is

$$R_{tot}^{-1} = \frac{1}{\sum_i 1/R_i^{-1}} \, . \tag{10.58}$$

The result for $R_{tot}^{-1}$ is independent of the order in which the individual conductors appear in the sequence. By analogy we can write for the mean jump rate

$$\langle \Gamma \rangle = \frac{1}{\frac{1}{N} \sum_{i=1}^{N} \frac{1}{\Theta_i \Gamma_{i,i+1}}} \, . \tag{10.59}$$

Here, $\Theta_i$ is the occupation probability of site $i$ and $\Gamma_{i,i+1}$ the jump rate from site $i$ to site $i+1$. To calculate $\langle \Gamma \rangle$ one may either consider transport from left to right or right to left in Fig. 10.19. Choosing the latter we notice that the coverage at the lower edge of the step site is $\Theta_i^{(s)} \equiv 1$. For the $n_s$ step sites we have therefore

$$\Theta_i^{(s)} \Gamma_{i,i+1}^{(s)} = \nu_0^{(s)} e^{-(E_A + E_D + E_{ES})/k_B T} \tag{10.60}$$

with $\nu_0^{(s)}$ the prefactor for jumping over the step edge. For the $n_t = N - n_s$ terrace sites the coverage $\Theta_i^{(t)}$ is the equilibrium coverage of the diffusing species

$$\Theta_{eq}^{(t)} = e^{-E_A/k_B T} .\tag{10.61}$$

This equation holds under the assumption that $\Theta_i^{(t)} \ll 1$ which is extremely well fulfilled in all realistic situations. The jump rate on the terrace is

$$\Gamma_{i,i+1}^{(t)} = v_0^{(t)} e^{-E_D/k_B T}\tag{10.62}$$

with $v_0^{(t)}$ the prefactor for terrace diffusion. Summing up one obtains for $\langle \Gamma \rangle$

$$\langle \Gamma \rangle = e^{-E_A/k_B T} v_0^{(t)} e^{-E_D/k_B T} \left( 1 - c_s + c_s \frac{v_0^{(s)}}{v_0^{(t)}} e^{E_{ES}/k_B T} \right)^{-1}\tag{10.63}$$

where $c_s = n_s/N$ is the step concentration. The mean jump rate can also be expressed in terms of the mean jump diffusion coefficient $D_{eff}^{(t)}$ of the diffusing species

$$D_{eff}^{(t)} = l^2 \langle \Gamma \rangle e^{E_A/k_B T} = l^2 \langle \Gamma \rangle / \Theta_{eq}^{(t)}\tag{10.64}$$

in which $\Theta_{eq}^{(t)}$ is the equilibrium concentration on the terraces. We note in passing that in three dimensions the energy, which determines the equilibrium concentration on terraces, is the work required to bring the diffusing species from a kink site to a terrace (Sect. 4.3.4, eq. 4.72). This issue is not well represented by the one-dimensional potential in Fig. 10.19. We see from (10.62) that for $E_{ES} = 0$ or small step concentrations $c_s$, $D_{eff}^{(t)}$ equals the tracer diffusion coefficient $D^* = \bar{l}^2 v$ on the terraces introduced in (10.6) with the jump length $l$ equal to the distance between one site and the next. If the ES-barrier is large, the effective diffusion coefficient is determined by $E_{ES}$ but also by the step concentration or by the slope of the profile. With the help of (10.31), we can express the general transport coefficient $L_T^{(t)}$ in terms of $D_{eff}^{(t)}$

$$L_T^{(t)} = D_{eff}^{(t)} \frac{\Theta_{eq}^{(t)}}{\Omega_s k_B T} = \frac{l^2 \langle \Gamma \rangle}{\Omega_s k_B T} .\tag{10.65}$$

The transport coefficient depends on the step concentration and thus on the slope of the profile if $E_{ES}$ is not small. This complicates the solution of the macroscopic equation for profile decay (10.37). Experimental results on profile decay are therefore difficult to interpret in terms of microscopic parameters, even in simple models.

## 10.3.3 The Kink Ehrlich-Schwoebel Barrier

The equivalent of an ES-barrier exists also in mass transport along step edges. There, transport of atoms (or vacancies) around a kink may be hindered by an additional activation barrier relative to the activation barrier for transport along straight steps. In analogy to the Ehrlich-Schwoebel barrier at steps, the barrier at kinks is called the *kink Ehrlich-Schwoebel barrier* or *kink-rounding barrier*. Theoretical papers refer to the phenomenon mostly as *K̲ink E̲hrlich-S̲chwoebel E̲ffect* (KESE). The potential for atom transport along a step is schematically the same as for step crossing (Fig. 10.19). Mass transport along steps is generally believed to occur via adatoms at step edges and we discuss only this case in the following. The energy $E_A$ is then the difference in binding energy of an atom at a kink site and the straight step and determines the equilibrium concentration of atoms at steps $\Theta_{eq}^{(st)}$.

$$\Theta_{eq}^{(st)} = e^{-E_A/k_B T}, \quad \text{if } \Theta_{eq}^{(st)} \ll 1.   \tag{10.66}$$

In the nearest neighbor bond-breaking model, $E_A$ equals twice the kink energy $\varepsilon_k$. The mean diffusion coefficient for transport along steps is calculated the same way as for interlayer transport on vicinal surfaces. We note, however, that $\Theta_{eq}^{(st)}$ is not quite as small as the equilibrium concentration on terraces. There is furthermore the reduced dimension. Two adatoms at a step site have good chance to meet before they recombine with existing kinks and form a nucleus of a short step with two new kinks. Contrary to the transport across steps, transport along step edges involves therefore perpetual kink generation and annihilation, rather than just transport across an existing kink structure. With this caveat one can write

$$\langle \Gamma \rangle = \Theta_{eq}^{(st)} D_{eff}^{(st)} / l^2   \tag{10.67}$$

where $D_{eff}^{(st)}$ is the effective diffusion coefficient along steps,

$$D_{eff}^{(st)} = l^2 v_0^{(st)} e^{-E_d^{(st)}/k_B T} \left(1 - P_k + P_k \frac{v_0^{(k)}}{v_0^{(st)}} e^{E_{ES}^{(k)}/k_B T}\right)^{-1}.   \tag{10.68}$$

Here, $v_0^{(st)}$ and $E_d^{(st)}$ are the prefactor and the activation energy for diffusion along the straight step, $v_0^{(k)}$ and $E_{ES}^{(k)}$ the prefactor and the activation energy for rounding a kink site, and $P_k$ is the concentration of kinks. As for diffusion on stepped surfaces, we have the complication that transport along kinked steps depends on the kink concentration. We have argued in Sect. 10.3.2 that the chemical potential of the protrusion arising from the incorporation of an island can be treated as that of a step running essentially along the direction of dense packing. For the analysis of protrusion decay it might be consequent to take the kink concentration $P_k$ as that of a step in equilibrium. The transport coefficient for transport along step edges then becomes

$$L_T^{(st)} = D_{eff}^{(st)} \frac{\Theta_{eq}^{(st)}}{a_\parallel k_B T}$$

$$= \frac{l^2}{a_\parallel k_B T} v_0^{(st)} e^{-(E_A + E_d^{(st)})/k_B T} \left( 1 + 2 \frac{v_0^{(k)}}{v_0^{(st)}} e^{(E_{ES}^{(k)} - \varepsilon_k)/k_B T} \right)^{-1}. \tag{10.69}$$

This relation gives us the possibility to express the characteristic time for the decay of a non-equilibrium bump in a step (10.47)

$$\tau^{-1}(q) = l^2 a_\parallel^2 q^4 v_0^{(st)} e^{-(E_A + E_d^{(st)} - \varepsilon_k)/k_B T} \left( 1 + 2 \frac{v_0^{(k)}}{v_0^{(st)}} e^{(E_{ES}^{(k)} - \varepsilon_k)/k_B T} \right)^{-1}. \tag{10.70}$$

### 10.3.4 The Atomistic Picture of the Ehrlich-Schwoebel Barrier

Transport of atoms across a step edge may proceed in many different ways and it is not at all clear what atomic process should have the lowest possible ES- barrier and should therefore be rate determining in experiments. Figure 10.20 displays some commonly considered possibilities for steps on fcc(100) and fcc(111) surfaces. The three cases, hopping over the step edge, exchange at a straight step, and exchange at a kink site, represent a minimum set of possibilities which doubles already in the case of (111) surfaces because of the crystallographic different A- and B-steps. By inspection of Fig. 10.20, one can easily envision further possibilities. Moreover, there are steps of different orientation and the mass transport may be via vacancies rather than by adatoms. To complicate the issue even further, it is not only the activation barrier, which decides what the easiest pathway is. The binding energy in the initial state before the jump also counts because it determines the population in that site.

For example, in an ab-initio calculation Feibelman has found an extremely low ES-barrier of 0.02 eV for exchange crossing of the A-step on Pt(111) (XA, last column in Table 10.3). However, by mapping out the entire landscape of binding energies Feibelman also found that the last position before an XA-jump has a 0.2eV lower binding energy so that the product of equilibrium coverage and jump rate still calls for an appreciable ES-barrier.

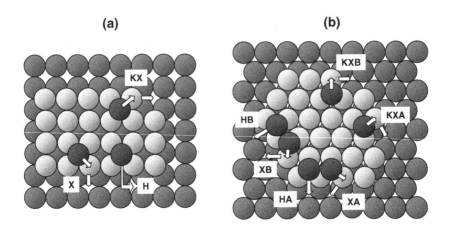

**Fig. 10.20.** A basic set of possibilities for adatoms to traverse densely packed steps on **(a)** fcc(100) and **(b)** fcc(111) surfaces. "H" stands for hopping, "X" for exchange, and "KX" for exchange at a kink site. On the (111) surface the type of step (A or B) is added to the notation.

In order to establish the activation energy for a certain pathway in a total energy calculation one has to establish the minimal path in the sense of the transition state theory (Sect. 10.1.3), in other words one has to find the minimal energy with respect to the coordinates of several atoms. This amounts to a substantial effort. In most cases semi-empirical model potentials such as provided by the *embedded atom method* (EAM) or the *effective medium theory* (EMT) were employed. Given the smallness of the energy differences involved, the predictive value of such model potentials remains questionable, however. A comparison of the penultimate and the last row in Table 10.3 illustrates the point: EAM model potentials predict that the lowest energy path on Pt(111) should be exchange crossing of the B-step [10.34] in gross disagreement with ab-initio theory [10.35]. Table 10.3 also contains the ES-barriers for hopping "H", exchange "X" and exchange at kinks "KX" for steps on Cu(100), Ag(100) and Ni(100) surfaces calculated in the EMT model [10.36]. The study shows that while hopping and exchange crossing of straight steps require an ES-barrier, no such barrier is involved at kink sites. If correct, this would mean that the step-crossing rate should depend on the step orientation.

The energies listed in Table 10.3 are just energies, not free energies. Concerning the effect of an ES-barrier, e.g. on the development of the surface morphology during epitaxial growth, the prefactor that is determined by the entropic contribution to the free energy in the transition state is of equal importance.

**Table 10.3.** ES-barriers in eV for hopping "H", for exchange "X", and exchange at kink sites "KX" for steps on some fcc(100) surfaces and of the A- and B-steps on Ag(111) and Pt(111). TH and TV denote the activation energies for hopping diffusion of adatoms and vacancies, respectively, and SV and KV the activation barriers for the filling of a vacancy next to a step by atoms from a straight step and a kink site. The set is a somewhat arbitrary selection from a large volume of calculations. Concerning the most common metal systems, a more complete listing that includes experimental data up to 2001 was provided by M. Giesen [10.12].

|     | Cu(100)a | Ag(100)a | Ni(100)a |      | Ag(111)b | Pt(111)c | Pt(111)d |
| --- | -------- | -------- | -------- | ---- | -------- | -------- | -------- |
| H   | 0.179    | 0.114    | 0.289    | HA   | 0.44     |          | 0.24     |
| X   | 0.232    | 0.149    | 0.113    | AB   | 0.43     |          | 0.49     |
| KX  | -0.021   | -0.044   | -0.160   | XA   | 0.31     | 0.22     | 0.02     |
| TH  | 0.399    | 0.367    | 0.631    | XB   | 0.06     | 0.10     | 0.35     |
| TV  | 0.482    | 0.412    | 0.655    | KXA  | 0.19     |          |          |
| SV  | 0.166    | 0.170    | 0.178    | KXB  | 0.05     |          |          |
| KV  | 0.143    | 0.198    | 0.230    |      |          |          |          |

(a: [10.36], b: [10.37], c: [10.34], d: [10.35])

U. Kürpick has addressed this problem for straight steps on Ir(111) surfaces [10.38]. She found that on A-steps the activation energy is lower for hopping than for exchange (0.90 eV vs. 1.58 eV). However, the prefactor for exchange is higher by a factor of 35 so that exchange prevails at higher temperatures and hopping at lower. On the B-steps, the situation is reversed. Activation energies for kink sites are not known presently. Even less is known about kink Ehrlich-Schwoebel barriers. In this somewhat unsatisfactory situation, it is pleasing that there are experiments from which activation energies for terrace diffusion and for the lowest energy path across and along steps can be determined. These experiments are Ostwald ripening in defined geometries, the decay of stacks of islands and step equilibrium fluctuations. The experiments carry the additional advantage that they are not restricted to surfaces in vacuum.

# 10.4 Ripening Processes in Well-Defined Geometries

## 10.4.1 Ostwald Ripening in Two-Dimensions

The term *Ostwald ripening* stands for coarsening processes in an ensemble of particles of different sizes [10.39]. Coarsening occurs because particles of different size have a different Gibbs-Thomson chemical potential (Sect. 4.3.2). The equilibration process may be through evaporation/condensation or via diffusion. A typical Ostwald ripening situation occurs after nucleation. The broad distribution of initial sizes equilibrates towards a more homogeneous distribution. Simultaneously, the number of particles per area or volume shrinks because small particles disappear at early times due to their high chemical potential, and the mean particle size increases. The theory describing the ripening of the particle size-distribution is complex because each particle sees another environment of particles around it and each individual decay or growth process has an effect on all other particles [10.40]. With respect to surfaces, the material has been reviewed in 1992 by Zinke-Allmang et al. [10.41].

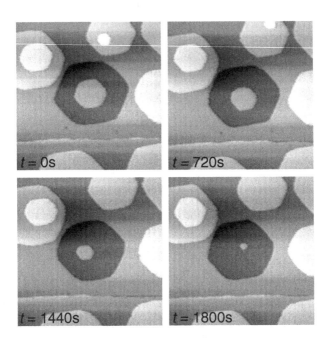

**Fig. 10.21.** A series of STM images of about 60 nm × 60 nm of a Cu(111) surface showing an adatom island inside a vacancy island at different times *t*. The temperature was 303 K. Vacancy islands and adatom islands where produced by sputtering and subsequent evaporation of Cu. The images are excerpts from a movie of a larger area from which these particular frames were selected for quantitative analysis (after Schulze Icking-Konert [10.12, 43].

The complexity of the problem however results entirely from the many particle aspect. The ripening of a single particle in a well-defined environment is a much simpler problem. With the help of certain preparation steps and the STM as means for observation it is possible to study the decay of an island or a cluster on a surface under well-defined conditions. Figure 10.21 shows a Cu(111) surface with an adatom island that stays approximately in the center of a vacancy island until is disappears by evaporation of adatoms to the terrace which surrounds the island. The atoms attach to the perimeter of the vacancy island, which shrinks in size accordingly. The system was prepared by sputtering off less than a monolayer to produce a distribution of vacancy islands. On that surface a sub-monolayer amount of copper was evaporated which produced adatom islands on the surface. Some of these islands reside inside a vacancy, most of those (by virtue of the properties of the nucleation process, Sect. 11.1.1) near the center of the vacancy island. During decay the adatom islands undergoes a Brownian motion with respect to its position (Sect. 10.1.5). By chance, some of the adatom islands stay close to the center of the vacancy islands during their entire life. Those islands are the ideal objects for quantitative studies. The frames shown in Fig. 10.21 are selected excerpts from a movie consisting of many images from a larger area. All islands have the equilibrium shape, save for fluctuation and therefore a defined chemical potential. The advantage of a quantitative study of island decay in a vacancy island is that the boundary of the vacancy island provides a defined chemical potential. Because of the nearly round shape of the islands and because of the nearly centrosymmetric geometry, the diffusion problem can be analytically treated in cylindrical coordinates with circular islands. Studies of this type, at first without the vacancy island, were introduced by K. Morgenstern, G. Rosenfeld and G. Comsa on Ag(111) surfaces [10.42].

As the islands in Fig. 10.21 have their equilibrium shape one can attribute a single Gibbs-Thomson chemical potential to an island that depends solely on the size. By applying the Gibbs-Wulff theorem, we have derived in Sect. 4.3.3 the chemical potential of a monolayer adatom island as

$$\mu = \Omega_s \, \beta_0 \, / \, y_0 \qquad \qquad (10.71)$$

where $y_0$ is the distance of the point of least curvature to the center and $\beta_0$ is the line tension of the step at the point of least curvature. By definition, the chemical potential of a straight step is set to zero. Since the chemical potential is uniform the same relation holds for A- and B-steps on (111) surfaces. In the case of Cu(111) surfaces however, the line tension is nearly the same for A- and B-steps and the islands are (truncated) hexagons. The chemical potential of a vacancy island has the same form but with a negative curvature term.

$$\mu = -\Omega_s \, \beta_0 \, / \, y_0 \qquad \qquad (10.72)$$

These chemical potentials determine the equilibrium concentration of the diffusing species on the surface right next to the edge of the perimeter. To simplify the discussion we assume that the diffusing particles are adatoms on the surface that detach from the island perimeter. The argument can be pursued the same way if the diffusion current is carried by vacancies. In all realistic situations, the equilibrium coverage of adatoms on the terraces $\Theta_{eq}$ is extremely small. A realistic order of magnitude for the equilibrium coverage with adatoms on Cu(111) is $\Theta_{eq} = 10^{-12}$. The chemical potential of these adatoms is therefore that of an ideal lattice gas

$$\mu = \mu_0 + k_B T \, \ln \Theta_{eq} \, . \qquad \qquad (10.73)$$

By definition, $\mu_0$ is the chemical potential of the adatoms in equilibrium with a straight step, which is given by is the work that is required to bring an atom from a kink site to a terrace site. In terms of the static potential defined before $\mu_0 = E_A$. Hence, the equilibrium coverage at the perimeter of an island is

$$\Theta_{eq} = e^{-\frac{\mu_0}{k_B T} \pm \frac{\Omega_s \beta_0}{y_0 k_B T}} .\qquad(10.74)$$

The plus sign stands for an adatom island and the minus sign for a vacancy islands. Diffusion from the center island to the perimeter is governed by the Laplace equation (10.21) which reads in terms of the coverage

$$\dot{\Theta}(x, y) = D^* \Delta\Theta(x, y) .\qquad(10.75)$$

Because of the low concentration, the relevant diffusion constant is the jump diffusion constant $D^*$. The flux of atoms from the center island changes only very slowly in time. The diffusion problem is therefore solved by the stationary diffusion equation, which is the Laplace equation. Because of the centrosymmetric of the problem, we introduce cylinder coordinates.

$$\Delta\Theta(r) = \frac{\partial^2\Theta}{\partial r^2} + \frac{1}{r}\frac{\partial\Theta}{\partial r} = 0 \qquad(10.76)$$

The solution has the form

$$\Theta = c_1 \ln r + c_2 .\qquad(10.77)$$

The constants $c_1$ and $c_2$ are given by the coverages at the inner island with radius $r_i$ and the outer vacancy island with the radius $r_o$.

$$c_1 = -\frac{\Theta(r_i) - \Theta(r_o)}{\ln(r_o / r_i)} \qquad c_2 = \Theta(r_o) - c_1 \ln r_o \qquad(10.78)$$

The inner island decays because of a net current flow, which is the difference between the attachment current and detachment current. The attachment current density $j_{att}$ is given by the rate of successful jumps towards the island per length unit. With the jump rate $\nu(T)$ and the fraction of successful jumps denoted by the *sticking coefficient* $s_i$ one obtains

$$j_{att} = s_i \nu(T)\Theta(r_i) / a_\| .\qquad(10.79)$$

The sign in (10.79) is chosen such that a current toward the island are counted as positive. In equilibrium, the attachment current and the detachment current are equal. The detachment current is therefore

$$j_{\text{det}} = -s_i v \Theta_{\text{eq}}(r_i)/a_\parallel \tag{10.80}$$

The net current density is therefore

$$j_{\text{net}} = j_{\text{att}} + j_{\text{det}} = s_i \, v(T)(\Theta(r_i) - \Theta_{\text{eq}}(r_i))/a_\parallel = v(T) \, \nabla \Theta(r_i) \tag{10.81}$$

The second part of the equation is the condition that the net flux from the island must be carried away by diffusion. From (10.81), (10.77) and (10.78) one obtains

$$\Theta(r_i) = \Theta(r_o) + \frac{\Theta_{\text{eq}}(r_i) - \Theta(r_o)}{1 + a_\parallel / s_i r_i \ln(r_o / r_i)} \tag{10.82}$$

A relation equivalent to (10.81) holds for the net current density at the perimeter of the vacancy island, from which follows

$$\Theta(r_o) = \Theta(r_i) + \frac{\Theta_{\text{eq}}(r_o) - \Theta(r_i)}{1 + a_\parallel / s_o r_o \ln(r_o / r_i)} \tag{10.83}$$

with $s_o$ the sticking coefficient at the outer boundary. If the boundary is an ascending step as for the case of an island in a vacancy island then $s_i = s_o$. By inserting (10.83) into (10.82) one obtains $\Theta(r_i)$ and by inserting that result into the first part of (10.81) one arrives at the following expression for the net current density

$$j_{\text{net}} = -v(T)\frac{1}{r_i}\frac{\Theta_{\text{eq}}(r_i) - \Theta_{\text{eq}}(r_o)}{\ln(r_o / r_i) + a_\parallel / s_i r_i + a_\parallel / s_o r_o} \, . \tag{10.84}$$

The number of atoms in the center island therefore decays with a rate

$$\begin{aligned}
\frac{dN}{dt} &= -2\pi v(T)\frac{\Theta_{\text{eq}}(r_i) - \Theta_{\text{eq}}(r_o)}{\ln(r_o / r_i) + a_\parallel / s_i r_i + a_\parallel / s_o r_o} \\
&= -2\pi v_0 e^{-E_d^{(t)}/k_B T} e^{-\mu_0/k_B T}\frac{e^{\zeta \Omega_s \beta_0 / r_i k_B T} - e^{-\zeta \Omega_s \beta_0 / r_o k_B T}}{\ln(r_o / r_i) + a_\parallel / s_i r_i + a_\parallel / s_o r_o} \, .
\end{aligned} \tag{10.85}$$

We have introduced the equilibrium coverages and scaling factor $\zeta$ that relates the radii $r$ to $y_0$. If the radii $r$ are chosen to describe hexagonal islands of the same area as a circle then

$$\zeta_{\text{hex}} = \left(\frac{6}{\pi\sqrt{3}}\right)^{1/2} \cong 1.05 \tag{10.86}$$

$$\zeta_{\text{sq}} = \left(\frac{4}{\pi}\right)^{1/2} \cong 1.128 \tag{10.87}$$

for hexagonal and square shaped islands, respectively. In (10.85) we have also expressed the rate in terms of an activation barrier and a prefactor. For the relation between the number of particles in the island and the equivalent radius appearing in the centrosymmetric diffusion problem it suffices to calculate $N$ for circular islands.

$$\frac{dN}{dt} = \frac{2\pi r_i}{\Omega_s} \frac{dr_i}{dt} \tag{10.88}$$

Depending on the magnitude of the sticking coefficient, one distinguishes two cases that are considered in the next sections.

### 10.4.2 Attachment/Detachment Limited Decay

If $s \ll 1$ one can neglect the logarithm in the denominator of (10.85) and obtains

$$\frac{dN}{dt} = -2\pi\nu_0 e^{-E_d^{(t)}/k_BT} e^{-\mu_0/k_BT} \frac{e^{\zeta\Omega_s\beta/r_i k_BT} - e^{-\zeta\Omega_s\beta/r_0 k_BT}}{a_\parallel/s_i r_i + a_\parallel/s_0 r_0} . \tag{10.89}$$

The attachment/detachment limited decay is usually discussed by making two further assumptions. The assumptions are that $r_0 \gg r_i$ and that the Gibbs-Thomson exponents are small. As $a_\parallel\beta \approx 10k_BT$ at room temperature, the radius $r$ must be large compared to ten atom diameters to have the latter assumption fulfilled. This means the island should more than about 1000 atoms.

$$\frac{dN}{dt} = -\frac{2\pi s_i \zeta\Omega_s \beta}{a_\parallel k_BT} \nu_0 e^{-E_d^{(t)}/k_BT} e^{-\mu_0/k_BT} \tag{10.90}$$

In the attachment/detachment limited case, the decay rate of the number of atoms in the island or of the island area is therefore independent of the island size. The rate increases as the island becomes very small because of the exponential form of the Gibbs-Thomson factor. The independence of the rate on the size is considered as being indicative of detachment/attachment limited decay.

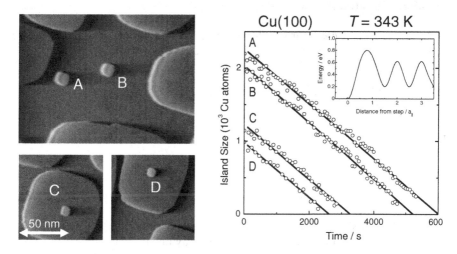

**Fig. 10.22.** Excerpts from STM images of a Cu(100) surface taken at 343 K and the decay rate as function of time. The inset shows the potential for vacancy migration and incorporation into an ascending step. The constancy of the rate is indicative of detachment/attachment limited decay. The fact that islands C and B decay with the same rate as the islands A and B shows that no ES-barrier exists (after Klünker et al. [10.9]).

An example for attachment/detachment limited decay is Cu(100). Figure 10.22 shows STM images of an ensemble of islands. All islands decay with same rate, which remains constant through their entire life. Since an attachment barrier for adatoms would be at variance with theory and furthermore difficult to understand, it was concluded that mass transport on Cu(100) proceeds via vacancies [10.9]. The attachment of a vacancy to an ascending step, i.e. the filling of the vacancy by a step atom involves the motion of atoms over sites of lower coordination, which gives rise to an activation barrier (see inset, compare also Table 10.3). The results in Table 10.3 assign a higher activation energy to vacancy diffusion compared to adatom diffusion, but the formation energy is much lower for vacancies, presumably because a considerable relaxation of the atoms surrounding the vacancy. The islands C and B can decay only via interlayer transport. Their decay rate is the same as for the islands A and B, which can send their atoms to an ascending step. Hence, there is no ES-barrier for this decay, which is again consistent with mass transport through vacancies. The small islands on top of another island decay because vacancies are generated at the boundary of the island below. The vacancies migrate to the boundary of the decaying island where they are filled by an atom from the edge. No explicit interlayer transport and thus no ES-barrier is involved in the process.

### 10.4.3 Diffusion Limited Decay

If the sticking coefficients are of the order of one, the terms containing $s$ can be neglected in the denominator of (10.85) and one obtains

$$\frac{dN}{dt} = -2\pi\nu_0 e^{-E_d^{(1)}/k_BT} e^{-\mu_0/k_BT} \frac{e^{\zeta\Omega_s\beta/r_i k_BT} - e^{-\zeta\Omega_s\beta/r_0 k_BT}}{\ln(r_0/r_i)}. \qquad (10.91)$$

In the limit of small exponents this becomes

$$\frac{dN}{dt} \cong -\frac{2\pi\,\zeta\Omega_s\,\beta}{a_\parallel k_BT}\nu_0 e^{-E_d^{(1)}/k_BT} e^{-\mu_0/k_BT} \frac{1}{r_i(t)\ln(r_0/r_i(t))}. \qquad (10.92)$$

In most of the past literature, the equation was approximated further by neglecting the logarithmic term in the denominator. The decay rate is then

$$\frac{dN}{dt} \propto (t_f - t)^{2/3} \qquad (10.93)$$

with $t_f$ the time where the island vanishes. This is however an oversimplification.

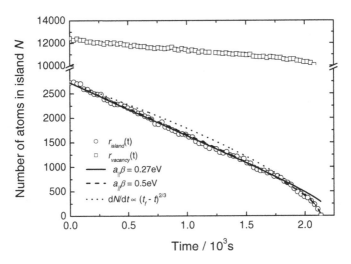

**Fig. 10.23.** Decay rate of the center island depicted in Fig. 10.21 (circles) and of the surrounding vacancy island (squares). The dotted line is the $(t_f - t)^{2/3}$ time dependence which is commonly attributed to diffusion limited decay. The curvature is much too large, however. The solid and dashed lines result from a numerical integration of (10.90) with two different step line tensions.

In reality, the decay rate is a much less curved function. Figure 10.23 shows the experimental decay rate of the mean radius of the center island in Fig. 10.21 as circles [10.12, 43] together with (10.93) as a dotted line. The latter is fitted to match the first and last experimental point. The experimental decay curve is hardly distinct from a linear decay save for the last data points before the island vanishes. The solid line is the result of a numerical integration of (10.91) with the radius of the vacancy island and the line tension taken as ingredients from experimental data [10.44]. The numerical integration describes the initial shape of the curve quite well, but the final decay is not well represented. The entire curve is well portrayed by a numerical integration if $a_{\parallel}\beta$ is set heuristically to 0.5 eV. This peculiar behavior has been found consistently for all islands on Cu(111). The reason for the failure of the theory is not clear. One possibility is that the curvature is due to the elastic energy associated with the strain fields even around homoepitaxial islands (Sect. 3.4.1, Fig. 3.18). Another possibility is that the effective chemical potential of small islands is larger as they may not retain their equilibrium shape during their final decay. Since the decay curve does not deviate much from a linear slope for the most part of the lifetime of an island, the shape of a decay curve is not a safe indicator for the decay mechanism. A much better indicator is the presence of absence of an influence of the environment on the decay rate of an island.

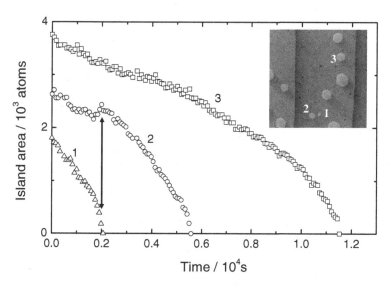

**Fig. 10.24.** Interaction between islands during diffusion limited decay. During the final moments of its life, the chemical potential of island "1" becomes very large. If the decay is diffusion limited then this large chemical potential transfers immediately into a larger adatom concentration around this island. The normal decay of neighboring islands is interrupted by a *hick-up* (after Schulze-Icking-Konert et al. [10.45]).

Figure 10.24 illustrates the effect for diffusion-limited decay. The figure shows the decay curves of three islands in an ensemble of several islands on a stepped surface (see insert). The smallest island denoted as #1 disappears first. During the final stages of its life, the island possesses a high chemical potential. In the case of diffusion-limited decay, the concentration of adatoms on the terrace next to the island perimeter is in equilibrium with the chemical potential of the island and therefore becomes suddenly very large at the end of the life. Because of the high adatom concentration, neighboring islands interrupt their normal decay and start growing for a short while until the dying island has disappeared completely. This effect does not occur in attachment/detachment limited decay since there the adatom concentration stays practically unaffected by the chemical potential of a single island. The presence or absence of *hick-ups* is therefore a safe indicator for the decay mechanism.

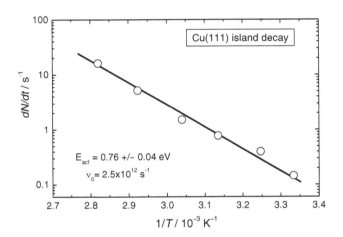

**Fig. 10.25.** Arrhenius plot for the decay rate of islands of a standard size on Cu(111) surfaces (after Schulze-Icking-Konert et al. [10.45]). The activation energy is the sum of the activation energies for diffusion on the terraces and the formation energy of adatoms (10.93).

By measuring the decay rate of islands at different temperatures, one can determine the activation energy for the decay process. For diffusion-limited decay with adatoms as the transporting particles the activation energy is

$$E_{act} = E_d^{(t)} + E_{ad} \tag{10.94}$$

with $E_d^{(t)}$ the activation energy for terrace diffusion and $E_{ad}$ the formation energy of an adatom on the terrace from a kink site. In case of attachment/detachment limited decay, the activation energy of the attachment process adds to the sum

(10.94). In other words, the total activation energy is the energy barrier for a complete detachment process (see inset in Fig. 10.22). Figure 10.25 shows an Arrhenius plot for the decay of islands on the Cu(111) surface [10.45] from which one obtains $E_{act} = 0.76$ eV and a prefactor of about $2.5 \times 10^{12}$ s$^{-1}$.

### 10.4.4 Extension to Noncircular Geometries

The quantitative analysis of island decay is one of the few methods to establish formation and diffusion energies from experiment. Since it is not always possible to prepare surfaces with islands centered in vacancy islands, it should be useful to consider the feasibility of quantitative analysis for other geometries. One possibility is to study the final decay of an island that, by chance, is surrounded by a wreath of larger adatom islands. An example is shown in Fig. 10.26. During the last stage of the decay, the decay rate of the encircled island is as if the islands were placed in a vacancy island whose radius is the mean distance of the surrounding island from the center. The chemical potential of the "vacancy island" is approximately than of a straight step. One can define a normalized decay rate as

$$\frac{dN}{dt}\bigg|_{norm} = \frac{dN}{dt} r_i(t) \ln\left(\langle r_0 \rangle / r_i(t)\right) \propto \exp\{-E_{act}/k_BT\}$$

$$\text{with } \langle r_0 \rangle = \sum_j r_{0,j}$$

(10.95)

The sum extends of the radial distances to the neighboring islands $r_{0,j}$ shown in Fig. 10.26.

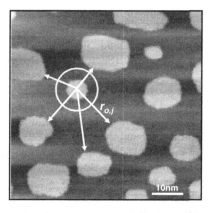

**Fig. 10.26.** STM image of a Au(100) electrode in 50 mM $H_2SO_4$ after lifting the reconstruction by a positive potential sweep (Sect. 4.3.2). The encircled island is smaller than the surrounding islands and decays therefore. During the last stage of its decay, the rate is as if the islands were placed in a vacancy island of the mean radius defined above (courtesy of Margret Giesen).

The normalized decay rate defined in (10.95) is time independent and proportional to the activation term. Hence, the activation energy follows from an Arrhenius plot of the normalized decay rate. If the database consists of a large number of STM images of the same area a more elaborate procedure is fruitful. Shapes and positions of the initial island configuration are transferred into a computer program. The sizes of the islands determine their chemical potential and thus define the adatom concentrations at the perimeters, modulo the formation energy. These concentrations set the boundary conditions for the Laplace equation, which is solved numerically. The gradient of the concentration integrated around the perimeter of each island defines the current away or towards each island, modulo the jump frequency. With these currents the size of the islands after the next time step are calculated. Again, the chemical potentials are determined from the sizes, and so forth. Line tension and the activation term are fitted until the optimum agreement with the experimental decay curves is achieved. The method works very well if one takes the size of outer islands in the series of STM frames completely from experiment. The method also works well if the islands reside on relatively narrow terraces such as shown in the inset of Fig. 10.24, since the steps set defined boundary conditions. One can even determine the ES-barrier that way by incorporating a boundary condition of flux conservation in the spirit of (10.81), rather than by fixing the concentration at the step edges [10.45]. A better experimental access to the ES-barrier is provided by studying the interlayer mass transport in stacks of islands.

## 10.4.5 Interlayer Transport in Stacks of Islands

Once the parameters governing the decay of islands in vacancy islands are known the (effective) Ehrlich-Schwoebel barrier can be determined from the decay of islands that reside on top of another island. Stacks of islands develop in homoepitaxial growth if an ES-barrier exists (Sect. 11.1.3). In brief, the reason is that because of the existence of an ES-barrier the adatom concentration on an existing island becomes large during growth so that a second island nucleates on top of an existing island, and so forth. Figure 10.27 shows the morphology of a Cu(111) surface after deposition of about 20 monolayers of Cu at room temperature. The surface exposes up to eight layers, and the top four to five layers are organized in stacks. The lowest island in the stack can donate its atoms to an ascending step and therefore decays without interlayer transport involved. All higher layers in the stack can decay only by interlayer transport.

The decay of the top layer island is described by (10.85) if the sticking coefficient to the outer perimeter $s_o$ is interpreted as the probability to surmount the ES-barrier relative to the probability for diffusion on a terrace. We denote this probability as

$$s = s_0 \, e^{-E_{ES}/k_B T} \tag{10.96}$$

where $s_0$ is the ratio of the prefactors for jumping over ES-barrier $v_0^{(ES)}$ and the prefactor for terrace diffusion $v_0$

$$s_0 = v_0^{(ES)} / v_0. \qquad (10.97)$$

The decay rate for the top layer island in a stack is therefore described by

$$\frac{dN_{top}}{dt} = -2\pi v_0 e^{-E_d^{(t)}/k_BT} e^{-\mu_0/k_BT} \frac{e^{\zeta\Omega_s\beta/r_{top}k_BT} - e^{-\zeta\Omega_s\beta/r_{2nd}k_BT}}{\ln(r_{2nd}/r_{top}) + a_\parallel/s\,r_{2nd}} \qquad (10.98)$$

where $r_{top}$ and $r_{2nd}$ are the time dependent radii of the top layer island and the next layer island below, respectively.

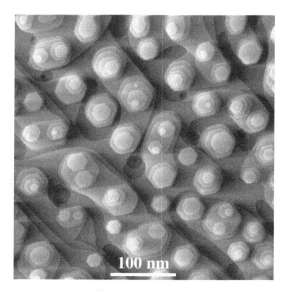

**Fig. 10.27.** STM image of a Cu(111) surface after deposition of about 20 monolayers of copper (courtesy of Margret Giesen). The decay of the stacks of islands offers rich opportunities to study interlayer mass transport quantitatively.

With the time dependence of $r_{2nd}(t)$ taken from experiment and the rest of the parameters from the experiments on island decay in a vacancy island, the coefficient $s$ can be determined by fitting a numerical integration of (10.98) to the experiment. Figure 10.28 shows an Arrhenius plot of results obtained that way [10.46]. The activation barrier is determined to be about 0.22 eV. This number is a weighted average over the many possibilities for interlayer transport (Fig. 10.20) and therefore an effective barrier. The prefactor $s_0$ is about $15$[18].

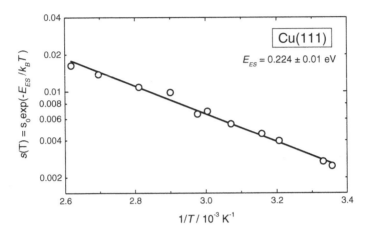

**Fig. 10.28.** Arrhenius plot of the *sticking factor s,* which describes the reduction of inter-layer mass transport compared to terrace diffusion. The activation energy is the Ehrlich-Schwoebel barrier $E_{ES} = 0.224$ eV [10.46].

### 10.4.6 Atomic Landslides

The existence of an ES-barrier is of great importance for the morphology of epi-taxially grown films as well as for some growth instabilities to be discussed in Chapter 11. It was therefore quite a surprise to see that the ES-barrier is bypassed under certain circumstances [10.47]. Figure 10.29 shows the decay of a stack of two islands on Cu(111). Because of the ES-barrier, the top layer island decays with a much slower rate than the islands below. At a particular point in time, the boundary of the upper islands appears to touch the boundary of the lower island (frame II in Fig. 10.29). Then the decay rate increases by about two orders of magnitude. A short time later (frame III) the upper island detaches from the boundary of the island below and the decay rate turns back to normal. Finally, the upper island touches the boundary of the lower again and the rapid decay contin-ues until the upper island has disappeared. Because of the rapid decay of the upper island atoms accumulate in the island below and its size increases until the upper island has vanished.

Two things are remarkable about this rapid decay process. One is the large increase in the decay rate by a factor that is approximately $1/s$, hence as if the ES-

---

[18] In the original work the factor was quoted, apparently erroneously, as $s_0 = 3.5$.

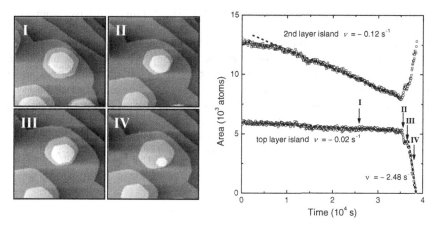

**Fig. 10.29.** Rapid interlayer transport by *atomic landslides* on Cu(111) at $T = 314$ K [10.47]. The STM images on the *left* each cover an area of 80 nm×80 nm. Frame **(I)**: The upper island in the stack of two resides near the center of the island below and the decay rate is small due to the ES-barrier. **(II)** The rate increases by more than an order of magnitude when the upper island touches the boundary of the lower island. **(III)** The upper island has detached itself from the boundary of the lower and the decay is slow again. **(IV)** The upper island stays at the boundary of the lower and decays rapidly until it has disappeared.

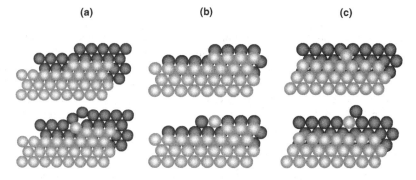

**Fig. 10.30.** Processes that might cause the observed rapid decay: Process **(a)** and **(b)** involve a transfer of atoms from a kink in the upper island to a kink site in the lower island [10.48, 49] , process **(c)** involves merely straight steps in close proximity [10.50].

barrier had ceased to exist. Secondly, the fact that the upper island remains perpetually at the boundary of the lower islands requires explanation. The rapid decay occurs only when the boundary of the upper island is close to the boundary of the lower. The transfer of atoms from the upper island to the lower should immediately increase the distance between the boundaries, hence the process should quench itself. The continuation of the process over a long time connected with the ubiquity of the rapid decay appears to call for a short-range attractive interaction between steps that comes on top of the long-range elastic repulsion (Sect. 3.4.2). The nature of this attraction is not understood at present.

Several propositions have been made for the atomic processes that might lead to rapid decay [10.48-50]. All propositions have in common that the transport of atoms is directly from the step of the upper island to the step edge of the lower island without the formation of an adatom on the terrace as an intermediate step. Which of the various processes should be rate determining is difficult to decide as not only the magnitude of the activation barrier counts but also the frequency by which one or the other configuration is realized and reestablished, as well as the completely unknown prefactors that enter the rate.

## 10.4.7 Ripening at the Solid/Electrolyte Interface

The rates of many ripening phenomena on solid electrodes have been found to depend on the electrode potential. The thereby induced smoothening of the surface has been termed "electrochemical annealing" [10.51] in order to stress the apparent similarity between the effect of the potential and a raise in temperature. Early quantitative investigations on electrochemical annealing concern the smoothening of rough gold and platinum electrodes [10.52-54]. More recently, the filling of STM-induced indentations and the decay of deposited islands were studied on Ag(100) and Au(100) electrodes [10.55-57]. Quantitative investigations on structurally well-defined systems include equilibrium fluctuations of monatomic steps on Cu and Ag surfaces [10.58-61]. All studies report an approximately exponential increase of the rate of transport processes with the electrode potential. Considering the very different nature of the investigated processes, which involve different combinations of atomic processes with different activation energies one is lead to the conclusion that there should be a common, rather general cause for the exponential dependence, irrespective of the specific process, surface, and electrolyte. In Sect. 4.3.6, we have shown that the common cause rests in the thermodynamic conditions of charged electrolyte surfaces held at constant potential. Because of this thermodynamic constraint, all defect formation energies and all activation energies for diffusion are renormalized by an electrostatic energy term that involves the difference in charge density on surfaces with and without defects (4.81, 4.82).

Ostwald ripening in two-dimensions offers an excellent opportunity to study the effect of the electrode potential in a quantitative manner. Au(111) and Au(100) surfaces are ideal test cases in that regard as the lifting of the reconstruction produces islands on the surface (Sect. 4.2.3) with a non-uniform size distribution which undergoes Ostwald ripening as time proceeds. The inset in Fig. 10.31 shows an STM image of Au(100) in 50 mM $H_2SO_4$ after lifting the reconstruction. The image is the same as in Fig. 10.26. The normalized decay rate (10.95) during the final decay of islands such as the encircled one is plotted as a function of the electrode potential. Beyond a certain threshold, the decay rate increases exponentially with the potential. The activation energy $E_{act}$ for the decay process is the sum of the formation energy of the diffusing species (here adatoms again) and the activation energy for diffusion. Following the argument in Sect 4.3.6 the formation energy $\mu_0$ is

$$\mu_0(\phi) = \mu_0(\phi_{pzc}) - \frac{p_z}{\varepsilon_0}\sigma_0(\phi) \qquad (10.99)$$

Here, $p_z$ is the dipole moment of the adatom in its ground state, $\sigma_0$ is the charge density on a flat (100) surface, $\varepsilon_0$ is the absolute permeability, $\phi$ the electrode potential and $\phi_{pzc}$ the potential of zero charge. The activation energy for diffusion is correspondingly

$$E_{diff}^{(t)}(\phi) = E_{diff}^{(t)}(\phi_{pzc}) - \frac{\Delta p_z}{\varepsilon_0}\sigma_0(\phi) \qquad (10.100)$$

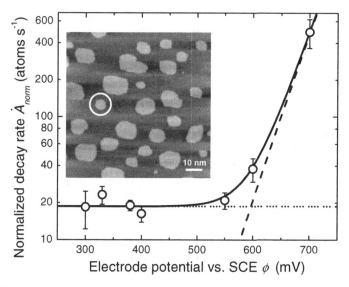

**Fig. 10.31.** Normalized decay rate (see text) at room temperature as a function of the electrode potential (with reference to the saturated calomel electrode, SCE). The dotted line is the mean of the data points between 300 and 400 mV. The dashed line is calculated in the model when fitted to the data point at 700 mV. The solid curve is the sum of the dotted and dashed line (see text). The inset is the same as Fig.10.24. It shows a STM image of one atom layer high islands on Au(100) in 50 mM $H_2SO_4$ that are produced by lifting of the reconstruction [10.11].

Here, $\Delta p_z$ is the difference between the dipole moment of the adatom in the transition state and the ground state. The potential dependence of the sum of $\mu_0$ and $E_{\text{diff}}^{(t)}$ is therefore solely determined by the product of the charge density and the dipole moment of the adatom in the transition state.

$$\mu_0(\phi) + E_{\text{diff}}^{(t)}(\phi) = \mu_0(\phi_{\text{pzc}}) + E_{\text{diff}}^{(t)}(\phi_{\text{pzc}}) - \frac{p_z^+}{\varepsilon_0}\sigma_0(\phi) \qquad (10.101)$$

Here $p_z^+$ is the sum of the dipole moment of the adatom $p_z$ and the difference between the dipole moments of the adatom in the transition state and the ground state $\Delta p_z$, hence is the dipole moment in the transition state. The dashed line in Fig. 10.31 is calculated with that dipole moment taken from ab-initio theory ($p_z^+ = 0.144$ eÅ) and the experimental charge density [10.11]. According to the theory, the islands should not decay at all at lower potentials. The observed finite rate does require a different diffusion mechanism.

The theoretical considerations in Sect. 4.3.2 that lead to (10.101) and (10.100) excluded specific adsorption of ions. Eq. (10.101) remains nevertheless a good approximation even in the presence of specific adsorption if the dipole moment in the transition state is as large as it is [10.11].

## 10.5 The Time Dependence of Step Fluctuations

### 10.5.1 The Basic Phenomenon

Scanning tunneling microscopy has opened our eyes not only to many static structural and morphological features on surfaces but also to previously unknown manifestations of transport processes. One of these manifestations is the appearance of monatomic steps in STM images. Figure 10.32 illustrates the point. At 290 K steps on the Cu(1 1 19) surface display some kinks as indicated by the white arrows as well as several sudden displacements of the step from one scan line to the next. The latter features result from the fact that the position of a kink in the step has moved through the position of the scan line during the time span from one scan line to the next ($\tau_{\text{sl}} \approx 10^2$ ms). The frequency of these jumps increases with temperature. In Fig. 10.32b taken at 320 K geometric kinks are no longer visible. Yet, the jumps in the position of step from one scan line to the next still amount to one atom unit. At even higher temperatures, one or more kinks move through the scan line in a time span that is short compared to the time required to scan one atom length ($\tau_{\text{at}} \approx 0.2$ ms). The apparent step jumps in the STM image are larger than one atom distance and are no longer quantized in integer values of atomic units. All information on the spatial structure of kinks is lost in such images (Fig. 10.32c). The image would appear the same if a single scan line were scanned

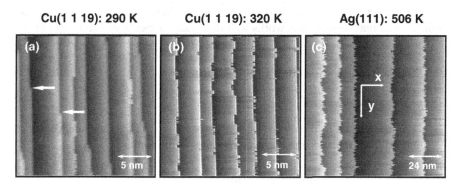

**Fig. 10.32.** STM images of vicinal copper surfaces (**a, b**) and (**c**) of a nominally flat Ag(111) surface with some steps. Scan time for the 512×512 pixel images was of the order of a minute (courtesy of Margret Giesen [10.12]).

repeatedly and the results were assembled in an $x,t$-image rather than an $x,y$-image. The distribution of the apparent position of the steps is a Gaussian in that case. Steps that appear as in Fig. 10.32 (b) or (c) have been termed *frizzy steps*.

There are several methods to investigate frizzy steps in a quantitative manner. The simplest is by the width of the Gaussian distribution. Dietterle et al. analyzed the step dynamics of Ag(111) electrodes in sulfuric acid electrolytes as a function of the potential this way [10.62]. An alternative is to study the time intervals of the telegraph noise in images like Fig. 10.32b [10.63]. The frequency of the occurrence of particular time intervals between two jumps has two characteristic time ranges. For short times, the jumps are due to return events. A kink moves by emitting an adatom to a step site. This atom returns after a brief random walk along the step, and the kink is back in its previous position. The long range is due to annihilation and creation of kinks or due to the capture of the adatom by another kink. The mathematical analysis of the telegraph noise is complicated [10.63] and experiments are limited to a narrow temperature window where steps display this type of noise.

By far the most studies of step fluctuations focus on the time structure of the mean square displacement of the step position (Sect. 4.3.7)

$$G(t,t';y,y') = \left\langle \left(x(t,y) - x(t',y')\right)^2 \right\rangle. \tag{10.102}$$

Because time and spatial structure of the mean square displacement interfere only in a very small temperature range (10-20 K), it suffices to study the time dependence. Experimentally it is possible to focus entirely on the time dependence by scanning repetitively the same scan line and determine the function $G(t)$ thereby.

$$G(t) = \left\langle \left(x(t) - x(0)\right)^2 \right\rangle \tag{10.103}$$

The analysis of step fluctuations by means of the $G(t)$-function is typically per-formed on vicinal surfaces and works well in a wide temperature range [10.64].

## 10.5.2 Scaling Laws for Step Fluctuations

### General considerations

The time dependence of the mean square displacement and the dependence on the step-step distance $L$ obeys universal scaling laws of the form

$$G(t) = \left\langle (x(t) - x(0))^2 \right\rangle = c(T)L^\delta t^\alpha . \tag{10.104}$$

The exponents $\alpha$ and $\delta$ depend on the atom transport processes and $c(T)$ is a tem-perature dependent factor that contains, among other quantities, the diffusion coefficient for the relevant process. The $G(t)$-function can be derived from the continuum description of the step position as a function of time as introduced in Sect. 4.3.7 (see also 10.3.2). The fluctuations result from the stochastic transport of atoms along or to and from step sites which, within the continuum approach, is described by a noise term [10.65]. The mathematics of the derivation is however rather cumbersome; in particular as different scenarios for mass transport require separate treatment. It is much easier and more pleasing (for most of us) to derive the $t$- and $L$-dependence of $G(t)$ from scaling considerations [10.66]. The results not only agree with the exact treatment within factors of the order of one but also the meaning of the physical quantities that enter becomes clear from the deriva-tion[19].

The scaling laws are derived by considering a grand canonical ensemble of $\langle N \rangle$ particles that is in equilibrium with a *reservoir* and exchanges particles with the reservoir through a *bottleneck* (*pipe* in the seminal paper of Pimpinelli et al. [10.66]) (Fig. 10.33). The mean square of the fluctuations of the particles in the system after a time $t$ is equal to the number of particles that pass through the bot-tleneck during that time,

$$\left\langle (\Delta N(t))^2 \right\rangle = \frac{N_b}{\tau_b} t . \tag{10.105}$$

Here, $N_b$ is the number of particles in the bottleneck and $\tau_b$ is their mean residence time. The system is the fluctuating step. As a matter of convenience, we chose the metric such that distances are in atomic units, the origin of the fluctuations at $x = 0$, and the reference value of $y$ is $y = 0$. The spatial fluctuations are then written as

---

[19] I would like to acknowledge enlightening discussions with D. E. Wolf who introduced me to these scaling laws.

$$\left\langle x^2 \right\rangle = \frac{k_B T}{\tilde{\beta}} y \, , \tag{10.106}$$

in which $\tilde{\beta}$ is the step line tension (cf. 4.95). We consider a fluctuation of the length $y$. The mean square of the number of particles that cause the fluctuation of the step with an amplitude $x$ over a length $y$ is

$$\left\langle (\Delta N(t))^2 \right\rangle = \left\langle x^2 \right\rangle y^2 = \frac{k_B T}{\tilde{\beta}} y^3 = \frac{N_b}{\tau_b} t \, . \tag{10.107}$$

This is the starting equation for the consideration of a variety of scenarios with different physical realizations of the reservoir and the bottleneck.

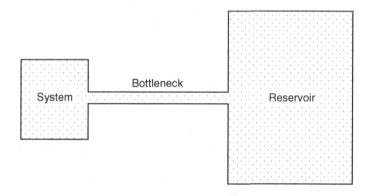

**Fig. 10.33**. Scaling laws for step fluctuation can be derived by considering the step as a system that exchanges particles with an infinite reservoir via a bottleneck.

### Case I - Diffusion along steps

The fluctuations of steps on metal surfaces are mostly due to the stochastic diffusion processes along the steps (case I in Fig. 10.34). The step acts therefore simultaneously as reservoir and bottleneck. The particles are the diffusing species, which we take as adatoms at step sites (as opposed to vacancies in steps). The number of particles in the bottleneck $N_b$ is the concentration of the adatoms in equilibrium $\Theta_{eq}^{(st)}$ multiplied by the length $y$

$$N_b = \Theta_{eq}^{(st)} y \, . \tag{10.108}$$

Their residence time is

$$\tau_b = y^2 / D_{\text{eff}}^{(\text{st})} \tag{10.109}$$

with $D_{\text{eff}}^{(\text{st})}$ the effective diffusion coefficient introduced in Sect. 10.3.3. By inserting (10.108) and (19.100) in (10.107) one obtains

$$y^4 = \frac{\tilde{\beta}}{k_B T} \Theta_{\text{eq}}^{(\text{st})} D_{\text{eff}}^{(\text{st})} t \ . \tag{10.110}$$

The result for the mean square fluctuation in atomic units $\hat{G}(t)$ is therefore

$$\hat{G}(t) = \langle x^2 \rangle = \left( \frac{k_B T}{\tilde{\beta}} \right)^{3/4} (\Theta_{\text{eq}}^{(\text{st})} D_{\text{eff}}^{(\text{st})} t)^{1/4} \ . \tag{10.111}$$

This is the famous $t^{1/4}$-power law for the step fluctuations that is found mostly for steps on metal surfaces.

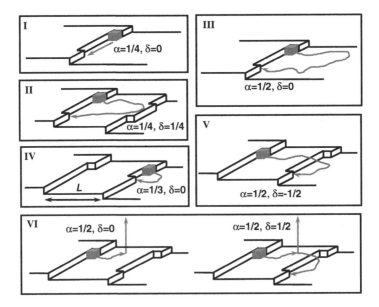

**Fig. 10.34.** The various scenarios for step fluctuations discussed in the text.

By introducing the transport coefficient in atomic units

$$L_T^{(st)} = \Theta_{eq}^{(st)} D_{eff}^{(st)} / k_B T \tag{10.112}$$

one may write (10.111) in a form that is independent of the nature of the species that carry the diffusion current,

$$\hat{G}(t) = \frac{k_B T}{\tilde{\beta}} (\tilde{\beta} L_T^{(st)} t)^{1/4} . \tag{10.113}$$

In 10.3.3, we have derived the transport coefficient for adatoms in the presence of kinks in the step and a kink ES-barrier. After inserting (10.69) one obtains for $G(t)$

$$\hat{G}(t) = \left(\frac{k_B T}{\tilde{\beta}}\right)^{3/4} \left( v_0^{(st)} e^{-(E_A + E_d^{(st)})/k_B T} (1 - P_k + P_k \frac{v_0^{(k)}}{v_0^{(st)}} e^{E_{ES}^{(k)}/k_B T})^{-1} \right)^{1/4} . \tag{10.114}$$

As before, $P_k$ is the kink concentration. For steps running in the direction of dense packing $P_k \cong 2 \exp(-\varepsilon_k / k_B T)$. Otherwise, $P_k$ is given by the step orientation. We note further that the quantity $\tau_{fl} = 1/ \tilde{\beta} L_T^{(st)}$ is a characteristic time constant for the fluctuations. The time constant is identical to the relaxation time $\tau(q)$ of a Fourier-component $q$ of a non-equilibrium protrusion in a step (Sect. 10.2.3) when $q$ is expressed in atomic units $\hat{q}$.

$$\tau_{fl} = 1/ \tilde{\beta} L_T^{(st)} = \tau(q)\hat{q}^4 \tag{10.115}$$

Step equilibrium fluctuations and step relaxations are two sides of the same medal [10.24].

### Case II - Terrace diffusion on vicinal surfaces with ES-barrier

The $t^{1/4}$-power law is not a unique indicator for step diffusion. One more scenario leads to the same time dependence. This is fluctuations caused by exchange of atoms with terraces when the diffusion on terraces is the rate-determining step and if there are other steps in close proximity with a large ES-barrier for interlayer diffusion (Fig. 10.34, case II). Because of the ES-barrier, the atoms eventually have to return to the same step from which they originated. The number of atoms in the bottleneck is then

$$N_b = \Theta_{eq}^{(t)} L y \tag{10.116}$$

with $L$ the distance between steps and $\Theta_{\text{eq}}^{(t)}$ the equilibrium concentration of atoms on terraces. The residence time on the terrace as the bottleneck is

$$\tau_{\text{b}} = y^2 / D^{(t)} \qquad (10.117)$$

with $D^{(t)}$ the diffusion coefficient for terrace diffusion. By the procedures described above, we obtain for the correlation function

$$\hat{G}(t) = \left(\frac{k_{\text{B}}T}{\tilde{\beta}}\right)^{3/4} (\Theta_{\text{eq}}^{(t)} D^{(t)})^{1/4} L^{1/4} t^{1/4}, \qquad (10.118)$$

or in terms of the transport coefficient for terrace diffusion in atomic units (10.30)

$$\hat{G}(t) = \frac{k_{\text{B}}T}{\tilde{\beta}} (\tilde{\beta} L_{\text{T}}^{(t)})^{1/4} L^{1/4} t^{1/4}. \qquad (10.119)$$

Case II is distinct from case I by its $L$-dependence. The exponents $\alpha$ and $\delta$ are both 1/4. The derivation of this correlation function assumes that the diffusing atoms on the terrace are repelled from the ES-barrier during their migration path before they re-enter the step. This can be the case only if $L$ is small. Eq. (10.119) holds therefore only for small $L$. Yet our simple derivation provides no hint as to how small $L$ has to be.

### Case III - Terrace diffusion in the detachment limit

This case holds for fluctuations that are caused by the exchange with terraces if the exchange is limited by the detachment process. The detachment of atoms from steps occurs at the kink sites. These kink sites are therefore the bottleneck in this problem. The number of particles in the bottleneck is

$$N_{\text{b}} = P_{\text{k}} y \qquad (10.120)$$

and the residence time is

$$\tau_{\text{b}} = P_{\text{k}} \tau_{\text{A}} \qquad (10.121)$$

with $\tau_{\text{A}}$ the mean time between the emission (or recapture) of terrace atoms by kink sites. By noting that $P_{\text{k}} = k_{\text{B}}T / \tilde{\beta}$ (Sect. 4.3.7, 5.2.1) one obtains

$$\hat{G}(t) = \left(\frac{P_{\text{k}}}{\tau_{\text{A}}}\right)^{1/2} t^{1/2}. \qquad (10.122)$$

This special case can also be treated by considering that the kinks in the step perform a random walk on the step edge [10.67]. The exact equation for $G$ obtained that way has an additional factor $(2/\pi)^{1/2}$ in it.

### Case IV - Slow terrace diffusion

Now we have

$$N_b = \Theta_{eq}^{(t)} y^2 \text{ and } \tau_b = y^2 / D^{(t)} \tag{10.123}$$

and thus

$$\hat{G}(t) = \left(\frac{k_B T}{\tilde{\beta}}\right)^{2/3} (\Theta_{eq}^{(t)} D^{(t)})^{1/3} t^{1/3} . \tag{10.124}$$

This $t^{1/3}$ power law was first observed on Cu(39 39 37) surfaces (B-steps) in contact with a HCl electrolyte [10.68].

### Case V - Exchange of atoms between steps

If the distance between steps is small, atoms that detach from a step are more likely to attach to the adjacent step than to return to the same step. The number of atoms in the bottleneck and their residence time are then

$$N_b = \Theta_{eq}^{(t)} L y \text{ and } \tau_b = L^2 / D^{(t)} , \tag{10.125}$$

so that

$$\hat{G}(t) = \left(\frac{k_B T}{\tilde{\beta}}\right)^{1/2} (\Theta_{eq}^{(t)} D^{(t)})^{1/2} L^{-1/2} t^{1/2} . \tag{10.126}$$

This case has the same time dependence as case III, is however distinct by its $L$-dependence.

### Case VI - Exchange with terraces and the surrounding bulk phase

This case is important for surfaces in equilibrium with a surrounding bulk phase. The latter can be the bulk material underneath the surface. At high temperatures, the bulk possesses a high concentration of vacancies and the exchange of step atoms with the substrate by bulk vacancy annihilation can give rise to step fluctuations. Another possibility is surfaces in equilibrium with a gas phase of its own material and, finally, a surface in contact with an electrolyte that contains ions of the substrate. We assume that the exchange with the bulk phase is via the atoms on the terraces. Otherwise, the situation would be as in case III. An atom that has detached from a step can travel a path $\Lambda$ before it is immersed into the bulk phase. The path length $\Lambda$ is given by the diffusion constant and the residence time on the surface $\tau_r$.

$$\Lambda = (D^{(t)}\tau_r)^{1/2} \tag{10.127}$$

Depending whether $\Lambda$ is smaller or larger than the distance between steps one has for the number of atoms in the bottleneck (Fig. 10.34, case VI)

$$N_b = \Theta_{eq}^{(t)} y\Lambda \quad \Lambda < L \tag{10.128}$$

$$N_b = \Theta_{eq}^{(t)} yL \quad \Lambda > L . \tag{10.129}$$

Correspondingly, the step correlation functions for the two situations are

$$\hat{G}(t) = \left(\frac{k_B T}{\tilde{\beta}}\right)^{1/2} \Theta_{eq}^{(t)1/2} \left(\frac{D^{(t)}}{\tau_r}\right)^{1/4} t^{1/2} \quad \Lambda < L \tag{10.130}$$

$$\hat{G}(t) = \left(\frac{k_B T}{\tilde{\beta}}\right)^{1/2} \left(\frac{\Theta_{eq}^{(t)}}{\tau_r}\right)^{1/2} L^{1/2} t^{1/2} \quad \Lambda > L . \tag{10.131}$$

### 10.5.3 Experiments on Step Fluctuations

By measuring the time and step-distance dependences one can uniquely establish to which case the fluctuation belongs and therefore the prevailing transport mechanism causing the fluctuations. Experiments on step fluctuations yield therefore important information on transport processes at surfaces and the activation energies involved. Frequently the study of step fluctuations is one of the few or even the sole method to learn something about a particular transport process. With very few exceptions, experiments on step fluctuation were performed using *Scanning Tunneling Microscopy* (STM) [10.12]. Alternatives are *Reflection Electron Microscopy* (REM) [10.69] and *Low Energy Electron Microscopy* (LEEM). Figure 10.35 shows STM data from the first reported study of its kind with steps on Cu(1 1 19) surfaces [10.64]. Plotted is the mean square deviation of the distance between two adjacent steps on the surface. For uncorrelated step motion, this is equivalent to twice the mean square displacement for a single step. Step fluctuations can safely be regarded as uncorrelated as long as the amplitude of the fluctuation is small compared to the distance between steps and if the relevant diffusion mechanism does not involve exchange between steps (as in case V). At the temperature of 362 K conventional $x,y$ images are completely dominated by the time structure of the fluctuations. The important aspect of the $y$-coordinate is only that is corresponds simultaneously to an increase in time. Conventional and $x,t$ images (*time images*) yield the same result. The $G(t)$ function shows the $t^{1/4}$-time dependence expected for fluctuations dominated by mass transport along the step edge. Figure 10.36 shows the Arrhenius-plot of $G(t)$ obtained from several different Cu vicinal surfaces [10.70]. The plot extends over three orders of magnitude. The very large fluctuations at high temperatures require large step separations and are therefore obtained only from images of the Cu(1 1 79) surface for which the mean step separation is about 100 Å. The fact that all data points fit to the same Arrhenius-plot demonstrates that the fluctuations are independent of the mean step separation. The fluctuations are therefore really of the type I. The activation energy is $E_{act} = 0.324 \pm 0.008$ eV.

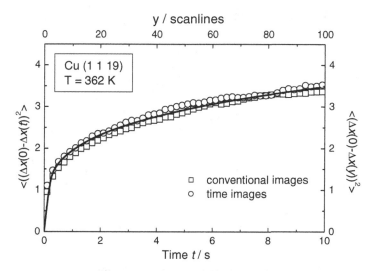

**Fig. 10.35.** The mean square separation in atom units between two steps on Cu(1 1 9) in $x,t$-images (*bottom* and *left* scale) and in $x,y$-images (*top* and *right* scale) [10.64].

The activation energy can be further evaluated with the help of (10.114). For steps on (100) surfaces oriented along the direction of dense packing the activation energy for the kink concentration $P_k$ and for $k_B T / \tilde{\beta}$ is the kink energy (Sect. 5.2.1) which is $\varepsilon_k = 0.128$ eV for Cu(100) steps (cf Sect. 4.3.7, Fig. 4.23). The formation energy of an adatom from a kink site $E_A$ is approximately twice the kink energy. Depending on the existence of a kink ES-barrier the activation energy is either (10.114)

$$4E_{act} = 5\varepsilon_k + E_{diff}^{(st)}, \quad E_{ES}^{(k)} < \varepsilon_k$$

$$E_{diff}^{(st)} \cong 0.65\,eV$$

(10.132)

or

$$4E_{act} = 4\varepsilon_k + E_{diff}^{(st)} + E_{ES}^{(k)}, \quad E_{ES}^{(k)} > \varepsilon_k$$

$$E_{diff}^{(st)} + E_{ES}^{(k)} = 0.78\,eV$$

(10.133)

The existence of a kink ES-barrier on Cu(100) surfaces is still under debate [10.71]. Hence, we have to leave the result as it stands.

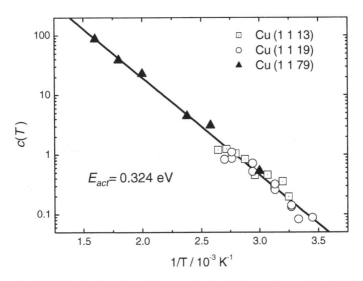

**Fig. 10.36.** Arrhenius-plot of the magnitude of the step fluctuations [10.70].

Interesting transitions from one time dependence to another have been observed for steps on Ag(111) surfaces. Figure 10.37 shows log/log plots of $G(t)$ for Ag(111) surfaces in vacuum and in contact with an electrolyte. In vacuum, one finds $\alpha = 1/4$ for temperatures below 450 K. As for most metal surfaces, the prevailing mass transport leading to step fluctuations is diffusion alongside steps. Between 450 K and 500 K the slope changes to $\alpha = 1/2$. Since the fluctuations do not depend on the step-step distance $L$, the slope is indicative of case III: exchange of atoms with terraces with the detachment/attachment being the rate-determining step [10.12]. A similar change in slope has also been found on Cu(21 21 23) [10.24].

When surfaces are in contact with an electrolyte (here an aqueous electrolyte containing $H_2SO_4$ and $CuSO_4$) the step fluctuations depend on the potential. At negative potentials with respect to the saturated calomel electrode (SCE) the fluctuation function $G(t)$ is proportional to $t^{1/4}$ as in vacuum and for the same reason (Fig. 10.37b). At +80 meV, the step fluctuations have increased in magnitude and the time dependence obeys a $t^{1/2}$-power law. Now the magnitude of fluctuations depends on the step-step distance with an exponent that is, within the error $\delta = 1/2$ (Fig. 10.38a). Evidently, we have now a transition to scenario described as case VI where the steps exchange atoms with terraces and the terraces with the surrounding bulk phase, and where the diffusion length $\Lambda$ is larger than the step-step distance $L$ (10.131). The surrounding bulk phase is the electrolyte, which entails that the surface must exchange atoms with the $Ag^+$-ions in solution. A view on the potential dependence of the magnitude of the fluctuation in Fig.10.38b shows that this exchange with the electrolyte occurs shortly before dissolution of the silver electrode. The initiation of the dissolution process itself can be seen in STM as a sudden recession of the steps when the potential is raised over a certain limit. The experiments demonstrate that the transport processes leading to fluctuations and the eventual dissolution follow the same kinetic path: atoms detach from kink sites to become adatoms on terraces from where they leave as solvated ions.

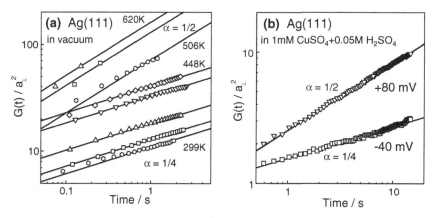

**Fig. 10.37.** The time dependence of the mean square displacement $G(t)$ of B-steps on Ag(111) **(a)** in vacuum and **(b)** in a 50 mM $H_2SO_4$ electrolyte to which 1 mM $CuSO_4$ was added (courtesy of Margret Giesen, [10.12]). The electrode potentials are with reference to a saturated calomel electrode (SCE). In both cases the exponent $\alpha$ changes from $\alpha = 1/4$ to $\alpha = 1/2$, albeit for different reasons.

The exponential increase of the fluctuations with the potential can be understood in a similar way as the increase in the speed of Ostwald ripening. An important contribution to the magnitude of the fluctuations comes from the equilibrium concentrations of adatoms on terraces $\Theta_{eq}^{(t)}$ (10.131). With (10.99) this concentration depends on the surface charge density as

$$\Theta_{eq}^{(t)} = \Theta_{eq}^{(t)}(\phi_{pzc})e^{\frac{p_z}{\varepsilon_0}\sigma_0(\phi)} \tag{10.125}$$

where $p_z$ is the dipole moment of the adatom and $\sigma_0(\phi)$ is the charge density on the surface (see also Sect. 4.3.6). In a small potential range, the charge density increase can be written as the product of the capacitance and the variation of the potential $\phi$ so that an exponential increase results for $\Theta_{eq}^{(t)}$ and for the magnitude of the fluctuations. As for Ostwald ripening of Au(100) electrodes the effect is so dramatic because of the large dipole moments of adatoms.

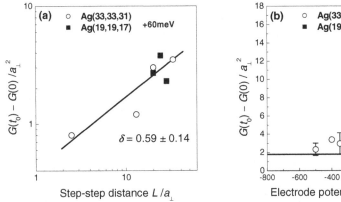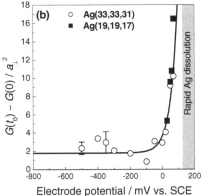

**Fig. 10.38.** Step fluctuations of silver Ag(33,33,31) and Ag(19,19,17) electrodes in a 50 mM $H_2SO_4$ electrolyte with 1 mM $CuSO_4$; **(a)** the dependence on the step-step distance together with the time exponent $\alpha = 1/2$ (Fig. 10.37b) shows that the fluctuations are of the type VI. **(b)** In the range where the exponent is $\alpha = 1/2$ the magnitude of the fluctuations increases exponentially with the potential. Hence, the activation energy for the process must decrease approximately linearly with the potential (courtesy of Margret Giesen, [10.12]).

# 11. Nucleation and Growth

In the last two decades, the interest has shifted away from plain surfaces to the properties of special systems that involve surfaces, interfaces, thin films and nanostructures. Those systems are prepared by using classical, mostly single crystal surfaces as templates on which other materials are deposited from UHV, a gas phase or a liquid. Atomic level control is required in the manufacturing process, which inevitably involves growth processes. The importance of growth phenomena on the entire field covered in this volume is emphasized by the fact that the phrases *epitaxial growth, epitaxially grown films* or *epitaxy* have been used in all but two of the preceding chapters. It is therefore about time to explain these phrases and devote a chapter to this topic.

The word *epitaxy* stems from the Greek words $\varepsilon\pi\iota$ and $\tau\alpha\xi\iota\varsigma$ ("epi" and "taxis") meaning "on", "on top" and "positioning", "line-up" (of soldiers), respectively. Growth on a substrate of the same material is called *homoepitaxy*; growth on a substrate of a different material is named *heteroepitaxy*. If material is deposited onto a flat, step-free surface then the deposited atoms first have to aggregate into stable nuclei that can grow by capturing further atoms. The formation of nuclei requires a surface concentration of atoms that exceeds the equilibrium concentration, a *supersaturation*. If the growth is layer-by-layer, then nucleation has to repeat after completion of each layer. Growth on vicinal surfaces on the other hand is nucleationless: the deposited atoms attach to the step edges and the growth is by step advance. Thus, the field is naturally divided into growth that requires nucleation and growth that does not.

Like always in thermodynamics it is important to realize which quantity is kept constant by external parameters. This holds also for the thermodynamics of nucleation. In epitaxial growth in UHV by *Molecular Beam Epitaxy* (MBE) the flux of particles towards the surface is externally controlled and typically kept constant. The deposition temperature is usually chosen such that no re-evaporation occurs, which means the net flux towards the surface and thereby the mean growth rate is controlled externally. The same condition of constant flux may apply to *Chemical Vapor Deposition* (CVD) from the gas phase. While the flux is kept constant, the supersaturation may vary. The supersaturation is high before nucleation has occurred and decreases dramatically after stable nuclei are formed. The mean supersaturation is typically high in those experiments.

In deposition from a liquid phase, in *Liquid Phase Epitaxy* (LPE) including *electrodeposition* or *electroplating*, and possibly also in CVD the supersaturation is controlled by external parameters, e.g. the by the electrode potential while the flux of atoms to the surface (in electrochemical deposition the ion current) varies.

## 11.1 Nucleation and Growth Under Controlled Flux

### 11.1.1 Nucleation

By way of introduction we consider a series of three STM images of an Ag(111) surface on which several monolayers of Ag atoms have been deposited (Fig. 11.1). The three images differ in temperature while the flux of atoms during deposition was kept constant. Between 300 K and 360 K the density of (top layer) island decreases from 30 to 2 per frame, and the island sizes increase accordingly.

$T = 300K$          $T = 340K$          $T = 360K$

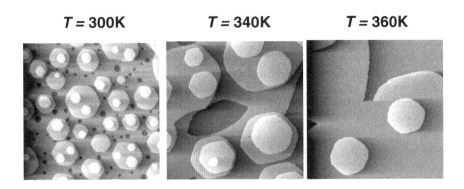

**Fig. 11.1**. 190 nm × 190 nm STM images of Ag(111) after deposition of a few monolayers of Ag at different temperatures with the same flux. The lower density of islands at higher temperatures is due to the faster diffusion (courtesy of Margret Giesen, unpublished).

To understand this temperature dependence we consider a surface exposed to a constant flux of atoms. Re-evaporation is neglected[1]. The deposited atoms diffuse about on the surface the faster, the higher the temperature is. When two atoms meet, they temporarily form a cluster. Depending on the temperature, this two-atom cluster may already be a stable nucleus for further growth, that is, its decay time would be longer than the mean time until a next adatom arrives and attaches to the cluster. At higher temperatures, a cluster of two atoms may not be stable and would decay before the next atom arrives. Then possibly a cluster of three or four atoms could be stable in the sense described above. The largest cluster, which is not quite stable, is called the *critical cluster*. Its size depends primarily on the temperature, but also on the flux since the flux determines the mean time between atoms arriving at a particular cluster: the higher the flux, the smaller the critical size. As clusters up to the critical size come and go their mean concentration is in equilibrium with the lattice gas of adatoms. The formation of those clusters from adatoms is formally a homonuclear reaction of the type

---

[1] Silver at 300 K evaporates one atom every $10^{25}$ years!

$$A^{(1)} + A^{(1)} + A^{(1)} + ... \Leftrightarrow A^{(i)} \qquad (11.1)$$

where $A^{(1)}$ denotes the adatoms and $A^{(i)}$ a cluster of $i$ adatoms. Equilibrium requires that the chemical potentials on both sides are identical, whence we have the relation (cf. Sects. 4.3.4 and 5.4)

$$i(E^{(1)} + k_B T \ln \Theta_1) = E^{(i)} + k_B T \ln \Theta_i . \qquad (11.2)$$

Here $E^{(1)}$ and $E^{(i)}$ are the ground state energies of the adatoms and the cluster of $i$ atoms, and $\Theta_1$ and $\Theta_i$ are the fractional coverages per surface site which we address as *densities* in the following. In writing the equilibrium condition above, we have assumed that the coverages are small and we have neglected vibration contributions to the chemical potentials. From (11.2) we obtain the coverage with clusters containing $i$ atoms

$$\Theta_i = (\Theta_1)^i e^{\frac{iE^{(1)} - E^{(i)}}{k_B T}} = (\Theta_1)^i e^{\frac{E_i}{k_B T}} . \qquad (11.3)$$

This equation is called *Walton relation* after D. Walton [11.1]. We note that $E_i$ is positive because of the formation of bonds between the atoms. The number of bonds depends on the size and shape of the cluster. On an fcc (100) surface e.g., clusters which are squares or nearly squares are particular stable while clusters with 2, 3, 5, 7,... atoms are least stable.

We now establish the rate equation for the adatom density after a flux of atoms $F$ is turned on at $t = 0$. ($F$ is defined as the number of arriving atoms per surface site and time). At $t = 0$ the time derivative of the density of adatoms on the surface is equal to the flux. During an initial period, the steady state sequence of clusters up to the critical size is quickly generated by consuming adatoms. After this initial period, adatoms are consumed only in as much they are captured by clusters of the critical size and stable clusters. We have therefore the rate equation

$$\frac{d\Theta_1}{dt} = F - \nu' \Theta_1 \Theta_s - \nu' \Theta_1 \Theta_i . \qquad (11.4)$$

Here, $\nu'$ describes the rate of successful attachments to a cluster if a terrace site adjacent to the cluster is occupied. This rate is of the order of the jump diffusion rate. It is lower if a barrier for attachment exists or if under growth condition atoms are evaporated back into the lattice gas from a fraction of sites at the cluster perimeter, e.g. from straight steps. The attachment rate can also be higher than the jump rate if the critical cluster is large so that the number of sites from which attachments can occur becomes large. Evidently, the solution of (11.4) calls for a steep rise in the density of adatoms and a subsequent decay to a small number when stable clusters are formed. At that stage, $\Theta_i$ is practically zero because $\Theta_1$ is small (11.4). Equation (11.4) then has the stationary solution

$$\Theta_1(t \to \infty) = \frac{F}{v'\Theta_s} . \tag{11.5}$$

The number of stable nuclei increases by adding a single atom to a critical nucleus. The growth rate of the density of stable nuclei is therefore

$$\frac{d\Theta_s}{dt} = v'\Theta_1\,\Theta_i = v'\Theta_1\,(\Theta_1)^i\,e^{\frac{E_i}{k_BT}} = v'\left(\frac{F}{v'\Theta_s}\right)^{i+1} e^{\frac{E_i}{k_BT}} . \tag{11.6}$$

We have inserted the stationary solution for $\Theta_1$. Equation (11.6) can be integrated directly. The solution for $i > 1$ is

$$\Theta_s = \left[\Theta(i+2)\right]^{\frac{1}{i+2}}\left(\frac{F}{v'}\right)^{\frac{i}{i+2}} e^{\frac{E_i}{(i+2)k_BT}} . \tag{11.7}$$

We have replaced the product of the flux and time by the total coverage $\Theta$ with atoms (in any form)

$$\Theta = Ft . \tag{11.8}$$

For $i = 1$ one obtains

$$\Theta_s = 3\Theta\left(\frac{F}{v'}\right)^{\frac{1}{3}} . \tag{11.9}$$

As expected, the density of stable nuclei, and thus the density of islands on the surface, depends on the ratio of the flux and the diffusion constant. The exponent is the smallest for $i = 1$ and raises up to one for $i \to \infty$. The diffusion constant varies exponentially with the temperature. This explains why the island densities in Fig. 11.1 decrease so rapidly with temperature.

According to (11.7) and (11.9), the density of nuclei and therefore the density of two-dimensional islands should increase linearly with the total coverage. This is however, an artifact of the simple ansatz. To show that, let us consider the case where the critical nucleus is the adatom itself ($i = 1$). The model assumes that two adatoms have a finite chance of meeting each other independently where they are deposited. In reality, two atoms that are deposited on opposite sides of an already existing island or with even several islands between them have practically no chance to meet. They will attach to one of the already existing islands, which means, that nucleation stops completely after some time; merely the already existing islands grow.

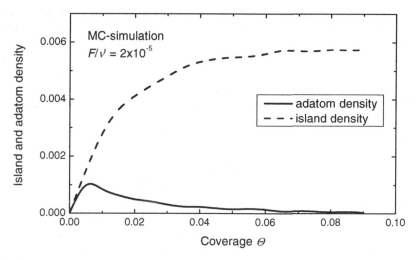

**Fig. 11.2.** Fractional coverage of the surface with islands and adatoms obtained from a Monte-Carlo simulation in which the critical nucleus is assumed to be $i = 1$ (dashed and solid line, respectively). The assumed ratio of flux to jump rate is $F/v' = 2\times10^{-5}$.

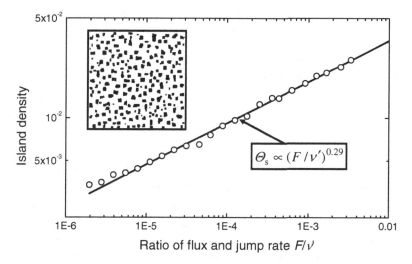

**Fig. 11.3.** Island density as function of the ratio of flux and jump rate according to a Monte-Carlo simulation for $i = 1$ (circles). Compared to the analytical expression obtained from the rate equations (11.9) the Monte-Carlo simulation produces a somewhat smaller exponent (0.29 instead of 1/3). The inset shows islands on a square lattice obtained by the MC-simulation with $F/v' = 2\times10^{-5}$.

We illustrate the point with a simple Monte-Carlo simulation in which the critical nucleus is assumed to be $i = 1$. The Monte-Carlo simulation is performed on a square surface with a maximum of four bonds to neighboring atoms in the spirit of the TSK-model (Sect. 5.2). Condition for a successful jump is that the number of bonds becomes larger or stays equal after the jump. This condition is equivalent to a low temperature as bonds, once established do not break again. Allowed jumps are both along the <10> and the <11> directions. The latter provision ensures the formation of compact islands despite the low temperature.

Figure 11.2 shows the result of the simulation for a ratio $F / v' = 5 \times 10^4$. The adatom density rises quickly initially, passes through a maximum at less than 1/100 of a monolayer and drops down to a very small value afterwards. The density of stable nuclei increases up to a saturation level of $\Theta_s \cong 0.057$ and is not proportional to the total coverage $\Theta$. The functional dependence of the island density on the ratio $F / v'$ (Fig. 11.3) is approximately described by (11.9). The slightly smaller exponent ($\approx 0.29$) is partly because of stochastic coalescence of islands. As remarked before the prefactor in equation (11.9) is not well reproduced by the Monte-Carlo simulation.

Equation (11.9) has sometimes been employed to determine the diffusion coefficient for adatoms on surfaces (after establishing the size of the critical nucleus from the dependence of the density of nuclei on the flux). The method has several shortcomings. One stems from the fact the (11.9) is only crudely recovered in simulations. The details depend on the model employed. Another shortcoming is that the simple nucleation theory does not consider repulsive interactions between the adatoms that arise from surface state confinement discussed in Sect. 8.3.4 or from elastic repulsive interactions (Sect. 3.4.2).

Equation (11.9) defines a mean distance between the stable nuclei as

$$\overline{d}_{nucl} = (\Omega_s / \Theta_s)^{1/2} = \left(\frac{\Omega_s}{3\Theta}\right)^{1/2} \left(\frac{v'}{F}\right)^{1/6} \tag{11.10}$$

This mean distance or nucleation length $\overline{d}_{nucl}$ is a useful quantity for qualitative considerations. For example, nucleation on the terraces of a stepped surface occurs if the terrace width is larger than the nucleation length $\overline{d}_{nucl}$. On perfectly regular vicinal surfaces, nucleation on all terraces would occur simultaneously below a certain temperature, while above that temperature the deposited atoms would migrate to the step edges, leading to *step flow growth*. On real vicinal surfaces, the width of a terrace may accidentally exceed the nucleation length and nucleation occurs on that particular terrace. Figure 11.4 shows an example. The image was obtained after evaporation of 50 monolayers at 313 K with a flux of 0.08 s$^{-1}$. The distance between the islands on the terrace is nearly equal and corresponds to a nucleation length of about 60 nm. If one identifies $v'$ with the jump rate on terrace one obtains from Table 10.2 $v' = 1.6 \times 10^{11} s^{-1}$. From Fig. 11.2 we take that the nucleation saturates at a coverage of about 0.1. The calculated nucleation length is then $\bar{d}_{nucl} \approx 49\,nm$ , which is in good agreement with the experiment.

**Fig. 11.4**. STM Image (195 nm×195 nm) of a vicinal Cu(111) surface on which the width of one terrace has accidentally exceeded the nucleation length (courtesy of Margret Giesen, unpublished). Nucleation occurs on that terrace, followed by nucleation of further islands on top of each other as the growth proceeds because of the Ehrlich-Schwoebel barrier on Cu(111) surfaces (Sect. 10.3).

### 11.1.2 Growth Without Diffusion

Because of the importance of cold-deposited films in practical applications, but also for its tutorial value it is useful to study the case of *hit and stick growth*, i.e. the case where the deposited atoms stay put at the sites where they have arrived. There has been some debate in the past as to whether accommodation of the kinetic energy of the arriving atoms may possibly take some time during which the arriving atoms, powered by their kinetic energy, may jump from the site of arrival to other sites in the vicinity. The effect has been observed for dissociative adsorption of oxygen molecules as the atoms from one molecule are found at some, though small distance apart [11.2]. Molecular dynamic calculations on the other hand, show that metal atoms evaporated onto a metal surface have no transient mobility [11.3].

Figure 11.5 displays a schematic side view of a cold-deposited film with a mean total coverage of about two monolayers. The morphology of such films can be characterized by various measures: the fractional coverages with single atoms in the layers, the total coverage $\Theta_n$ in each layer $n$, and the fraction $\Theta_n^{(\text{open})}$ in each layer that is open to further exposure. These coverages are easily calculated in a Monte-Carlo simulation. The result is plotted in Fig. 11.6 for a total mean coverage up to six monolayers. The fractional coverage with single atoms in the first layers starts at zero, rises up to about 7% and drops quickly thereafter as the first layer gets filled (Fig. 11.6a). With increasing filling of the first layer, single atoms are deposited in the second layer. The fraction of single atoms in the second layer rises up to a maximum of about 5.5% and eventually drops to zero as the second layer is filled. Each consecutive curve peaks at a lower value and the curve becomes broader since the consecutive layers fill more gradually. The reason is that the second layer cannot fill up completely unless the first layer is filled, the third layer not unless the second is filled, and so forth. Each layer fills more gradually than the layer immediately below. This is demonstrated in Fig. 11.6b. Figure 11.6c shows the fractional area that is open. The layer denoted as "0" is the substrate itself.

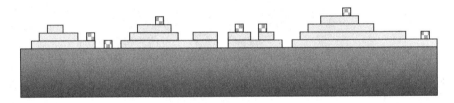

**Fig. 11.5**. Schematic side view of a cold-deposited film. The atoms stay in the sites where they have arrived. The checkerboard-patterned squares symbolize single atoms.

One can easily show that the fractions of open areas in each layer $\Theta_n^{(\text{open})}$ obey a Poisson distribution. The increase in coverage in the $n$th layer is the flux per site multiplied with the fractional open area in the layer below.

$$\frac{d\Theta_n}{dt} = F \, \Theta_{n-1}^{(\text{open})} \equiv F\left(\Theta_{n-1} - \Theta_n\right) \tag{11.11}$$

The sum of the growth rates in all layers is the flux itself

$$\sum_{n=1}^{\infty} \frac{d\Theta_n}{dt} = F \; . \tag{11.12}$$

This equation is fulfilled by the ansatz

$$\frac{\mathrm{d}\Theta_n}{\mathrm{d}t} = F \frac{(Ft)^{n-1}}{(n-1)!} \mathrm{e}^{-Ft} \tag{11.13}$$

since

$$\sum_{n=1}^{\infty} \frac{(Ft)^{n-1}}{(n-1)!} \equiv \sum_{n=0}^{\infty} \frac{(Ft)^n}{n!} = \mathrm{e}^{Ft} . \tag{11.14}$$

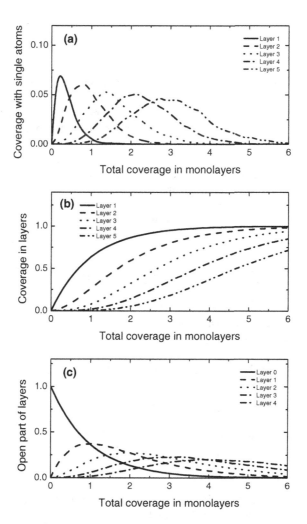

**Fig. 11.6.** Coverages with single atoms in layers (**a**), total coverages in layers (**b**) and the open part of layers versus the total coverage (**c**) when diffusion of atoms is completely suppressed. For simplicity, the simulation assumes a simple cubic structure with nearest-neighbor bonds (Kossel crystal), but the general result does not depend on the model.

Combining (11.13) and (11.11) yields

$$\Theta_n^{(\text{open})} = \frac{(Ft)^n}{n!} e^{-Ft} . \tag{11.15}$$

From (11.13) and (11.11) the total coverage in each layer is obtained in closed form as

$$\Theta_n = 1 - \sum_{m=1}^{n} \frac{(Ft)^{m-1}}{(m-1)!} e^{-Ft} . \tag{11.16}$$

With the exception of the concentration of single atoms all curves in Fig. 11.6 are represented by the analytical expressions (11.15) and (11.16). We note that these two expressions are derived under the sole assumption that an atom, which has arrived in layer n, stays in that layer. No assumption is made about diffusion in the layer. The expressions for the fractional open areas and the total coverage are therefore salient also for large diffusivity within a layer and a large Ehrlich-Schwoebel barrier (Sect. 10.3) that prevents interlayer transport. The morphology of the surface is quite different in that case, however. We return to this issue shortly.

From (11.15) one calculates easily the mean roughness of the surface. The mean layer $\bar{n}$ open to further deposition is as it must be

$$\bar{n} = \sum_{n=0}^{\infty} n\Theta_n^{(\text{open})} = \sum_{n=0}^{\infty} n\frac{(Ft)^n}{n!} e^{-Ft} = Ft . \tag{11.17}$$

The last step follows from differentiating

$$\frac{\partial}{\partial F} \sum_{n=0}^{\infty} \frac{(Ft)^n}{n!} = \frac{\partial}{\partial F} e^{Ft} \Rightarrow \sum_{n=0}^{\infty} n\frac{(Ft)^n}{n!} = Ft\, e^{Ft} . \tag{11.18}$$

Proceeding further one obtains

$$\sum_{n=0}^{\infty} n^2 \frac{(Ft)^n}{n!} = \left\{ Ft + (Ft)^2 \right\} e^{Ft} . \tag{11.19}$$

Braced with (11.18) and (11.19) one obtains for the root mean square roughness (rms-roughness) $w_{\text{rms}}$ in units of the step height

$$w_{\text{rms}} / h = \sqrt{\left\langle (n-\bar{n})^2 \right\rangle} = \sqrt{\left\langle n^2 - \bar{n}^2 \right\rangle} = \sqrt{\bar{n}} = \sqrt{Ft} . \tag{11.20}$$

The rms-roughness therefore increases with the square root of the total coverage. Equation (11.20) is obeyed even if the mean coverage is below a monolayer. We are not dealing with an asymptotic behavior in this case. This becomes apparent also from the simulation. Figure 11.7 displays the rms-roughness as obtained from a Monte-Carlo simulation in a double logarithmic plot that extends over two orders of magnitude in total deposition ($0 < Ft < 10$). Figure 11.7 displays also the step density a function of the exposure. Here, the condition of no diffusion enters significantly. As the coverage increases, the asymptotic value of four atomic step lengths per surface site is approached gradually.

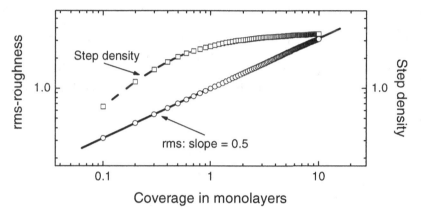

**Fig. 11.7.** Log/log-plot of rms-roughness and step density on surfaces versus coverage. The rms-roughness is in units of the atomic step height $h$. The step density is in atom length per surface site. The limit of four length units per site is asymptotically approached as the coverage increases. The condition of no diffusion is important only for the step density.

## 11.1.3 Growth with Hindered Interlayer Transport

As seen from the step density in Fig. 11.7 the condition of no diffusion produces a lateral roughness on the scale of a single atom and the rms-roughness of the height is given by the Poisson distribution of open layers (11.15). The Poisson distribution is maintained even in case of rapid *intralayer* diffusion as long as the Ehrlich-Schwoebel barrier (Sect. 10.3) blocks *interlayer* diffusion. The intralayer diffusion has a large effect on the lateral scale of morphological features on the surface. If the growth process begins with a flat surface template then the lateral length scale in the first deposited layer is given by the nucleation length $\bar{d}_{nucl}$ which was defined by (11.10) for the case of single-atom critical nuclei. As growth progresses, atoms are deposited on the open parts of the template from where they diffuse to the perimeter of the nuclei, effecting their lateral growth so that the nuclei turn into larger two-dimensional islands. As these islands grow, an increasing fraction of the atoms arrives on top of the islands, which leads to nucleation in the second layer. Simultaneously, the uncovered template area shrinks. Fewer atoms are deposited directly on the template. Consequently, the lateral growth of the islands slows down. The procedure repeats in the third, fourth, fifth, etc. layer. In the end, pyramids cover the surface whose lateral distance is established by the nucleation in the first layer and whose height is determined by the deposited amount. In case of rapid diffusion along the perimeter of the 2D-islands, their shape approximates the equilibrium shape of islands modulo some distortions caused by the varying distances to the adjacent pyramids. Figure 11.8 shows an STM image of a Pt(111) surface after deposition of 40 monolayers of Pt at 440 K in the presence of a CO ambient pressure [11.4]. CO adsorbed at the step edges enlarges the Ehrlich-Schwoebel barrier. All pyramids are alike, save for the distortions caused by the random distribution of stable nuclei in the first layer. A salient feature of the shape of the pyramids is a steep slope near the bottom, a smaller slope in the middle and again a steep slope before the pyramid ends with a flat top. Except for the flat top, the shape of the pyramids is entirely determined by the size and shape of bottom layer and the number of deposited layers in combination with the condition of blocked interlayer diffusion. From (11.15) with (11.11) we obtain the recursion

$$\Theta_n = \Theta_{n-1} - \frac{(Ft)^n}{n!} e^{-Ft} .$$

(11.21)

The linear dimension of the layers in the stack therefore obey

$$\frac{R_n}{R_0} = \sqrt{\frac{R_{n-1}^2}{R_0^2} - \frac{(Ft)^n}{n!} e^{-Ft}}$$

(11.22)

where $R_0$ is the linear dimension of the island at the bottom. By using this recursion, the linear dimensions of the layers are easily calculated for any given total exposure $Ft$. In the limit of very large exposures, the Poisson distribution in (11.21) is approximated by a Gaussian and the functional dependence of the linear dimension as well as the height function can be expressed in closed form in terms of the error function [11.5].

Figure 11.8b shows a top view on the stack of 2D-islands for a total exposure of 40 monolayers calculated with (11.22). To simulate the experimental island shapes the islands are drawn as hexagons with rounded corners. Figure 11.8c represents the calculated cross section of the pyramid. With the exception of the sharp tip, the calculated shape and the number visible layers are akin to the observed pyramids. The sharp tip follows from the assumed complete blocking of interlayer transport. It is easily seen that even a mild relaxation of the blocking condition must lead to flat tops. The very small islands at the top have an extremely high Gibbs-Thomson chemical potential (Sect. 10.4.1) causing a high concentration of adatoms on the terraces. The high 2D-pressure of adatoms leads to a downward current even in presence of an Ehrlich-Schwoebel barrier. The current is the larger, the smaller the islands are. The very small islands forming the tip therefore decay rapidly leaving the top of the pyramids flat. The experimentally observed flat tops in Fig. 11.8 (and in Fig. 11.4) therefore result from the finiteness of the Ehrlich-Schwoebel barrier. In combination with Monte-Carlo simulations or alternatively with the help of the analytical solutions of the Ostwald-ripening problem for a stack of islands (Sect. 10.4.5) the magnitude of the Ehrlich-Schwoebel barrier can be determined from experimental data (see cf. [11.4]).

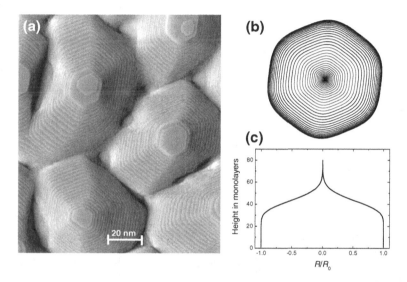

**Fig. 11.8.** (a) STM image of a Pt(111) surface after deposition of 40 ML of platinum at 440 K in the presence of $1.9 \times 10^{-9}$ mbar CO which increases the Ehrlich-Schwoebel barrier (after M. Kalff et al. [11.4]). (b) Top view on hexagonal symmetric pyramids as calculated from (11.22) for an infinitely high Ehrlich-Schwoebel barrier. The solid lines mark the steps. (c) Cross section through the calculated shape.

### 11.1.4 Growth with Facile Interlayer Transport

Thermodynamics predicts that homoepitaxial growth should proceed in a layer-by-layer fashion (Frank-van-der-Merwe growth, cf. Sect. 4.2.5). One might therefore expect that homoepitaxial growth under a constant flux of atoms should render that growth mode if intralayer and interlayer diffusion is fast. This is not so, however large one chooses the ratio of diffusion and flux $\nu'/F$! The reason rests in the increasing nucleation length $d_{nucl}$ for large ratios $\nu'/F$ (11.10). The more rapid

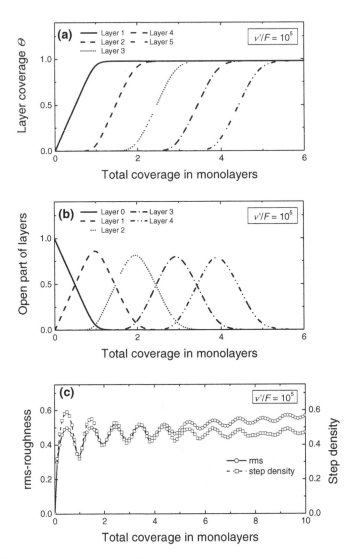

**Fig. 11.9**. Growth with unhindered interlayer transport and moderately rapid diffusion $(v'/F = 10^5)$. **(a)** Coverage in layers 1-5 vs. the total coverage, **(b)** fraction of open area in layers 0-4, **(c)** rms-roughness and step density.

the diffusion, the larger the islands become before they coalesce. Even though the mean diffusion length in the time required for completing a monolayer is necessarily larger than the nucleation length and therefore larger than the size of the islands (if $\nu'$ is of the order of the diffusion jump rate), nucleation in the next layer is not suppressed. Because of the nature of the random walk, atoms arriving near the center of an island have a much higher chance to meet each other and to form a stable nucleus than to diffuse to the perimeter of the island where they would contribute to the lateral growth of the island. Consequently, one has always three layers open in homoepitaxial growth under constant flux. This is illustrated with results obtained from Monte Carlo simulations in Fig. 11.9. The critical nucleus is again assumed to be a single atom and the ratio of attachment rate and flux is chosen $\nu'/F = 10^5$. Fig. 11.9a displays the coverage in each layer $\Theta_n$ as a function of the total coverage ($=Ft$) up to the equivalent of five monolayers. The overlap in layer coverages is clearly visible and becomes more pronounced as the total coverage increases. Even more instructive is the development of the coverage of open layers ($\Theta_n^{(\text{open})}$), which shows that two to three layers are open at any total coverage (Fig. 11.9b). Figure 11.9c shows the rms-roughness and the step density. The step density has an asymptotic limit that depends on the ratio $\nu'/F$. The rms-roughness rises continuously. Here, the slope depends on the ratio $\nu'/F$. Step density and rms-roughness exhibit damped oscillations with minima and maxima at full and half monolayer coverages, respectively. The damping is smaller the faster the diffusion is.

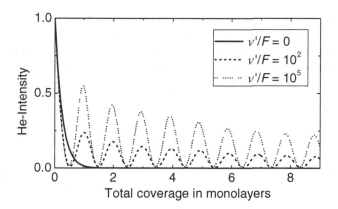

**Fig. 11.10.** Calculated intensities for a beam of He-atoms reflected in anti-phase condition from a growing surface.

The oscillations in the step density, in the rms-roughness and other related oscillations can be observed experimentally ex-situ by *Surface X-Ray Diffraction* (SXRD) and in-situ by *Reflection High Energy Electron Diffraction* (RHEED) (Sect. 2.2.2), *Medium Energy Electron Diffraction* (MEED) and *Helium Atom*

*Scattering* (HAS) (cf. Sect. 7.3). The oscillations observed in Helium scattering are particularly easy to interpret as a flat surface reflects the He-atoms like a mirror [11.6]. If the reflection angle is chosen such that the interference is destructive for He-atoms reflected from two terraces separated by a monolayer height (*antiphase condition*) then the intensity in the direction of mirror reflection is

$$I_{He} = \left| \sum_{n=0}^{\infty} \Theta_n^{(open)} e^{i\pi n} \right|^2 . \tag{11.23}$$

Figure 11.10 shows the calculated intensities for several $v'/F$ using the coverages of open layers obtained by Monte-Carlo simulation. For the case $v'/F = 0$ (alternatively complete blocking of interlayer transport, but rapid intralayer transport) an analytical expression for the intensity is obtained by inserting the Poisson distribution (11.15) for $\Theta_n^{(open)}$,

$$I_{He} = \left| \sum_{n=0}^{\infty} \frac{(Ft)^n}{n!} e^{-Ft} e^{i\pi n} \right|^2 = \left| \sum_{n=0}^{\infty} \frac{(-Ft)^n}{n!} e^{-Ft} \right|^2 = e^{-4Ft} . \tag{11.24}$$

The solid line in Fig. 11.10, although resulting from Monte-Carlo simulations with $v'/F = 0$ equals the exponential decay obtained analytically (11.24).

An example of an experimental result is shown in Fig. 11.11 with the data of Kunkel et al. referring to homoepitaxy of Pt on Pt(111) at three different temperatures [11.7]. At 621 K (Fig. 11.11a), the temperature is high enough to overcome the Ehrlich-Schwoebel barrier and the growth is layer-by-layer. At variance with the calculation, the intensity minima are not at zero. The effect results from the finite angular resolution of the scattering experiments, in other words, from the finite *coherence length* or *transfer width*. Equation (11.23) describes the intensity at exactly the mirror reflection angle. As we are dealing with interference, intensity cannot disappear. Rather, the angle-integrated reflected intensity stays constant. If the intensity is zero at the mirror reflection angle because of the antiphase condition, it appears on both sides of the mirror angle. The angular separation is the smaller the larger the terraces are. From a certain terrace width onwards, the angular spread of the He-beam and the finite acceptance angle of the detector render the observation of oscillations impossible and the intensity is given by the mean reflectivity of the terraces. The zero offset of the intensity in Fig. 11.11a therefore stems from terraces that are larger than the transfer width of the instrument.

At temperatures around 400 K Kunkel et al. observed a continuous decay in the reflected intensity as expected for blocked interlayer transport (11.24). The decay is not is not quite as rapid as predicted for the reasons discussed above. At even lower temperature (275 K) the oscillations return, albeit strongly damped indicating an approximate layer-by-layer growth. The phenomenon has been called *reentrant layer-by-layer growth*. Interlayer transport is less effectively suppressed at those low temperatures since the islands assume a ramified shape because of slow diffusion along the step edges. The long perimeter and the many kink sites with presumably lower Ehrlich-Schwoebel barriers (cf. Sect. 10.3.4) may permit some interlayer transport even at those low temperatures. An alternative interpretation is that the atoms have a high chance to arrive at positions directly at the step edge from where they are *downward funneled* into the next layer below [11.8, 9].

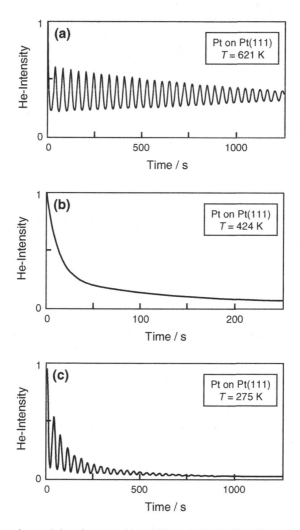

**Fig. 11.11.** Experimental data for deposition of Pt on Pt(111) after Kunkel et al. [11.7]. The deposition rate was between 1/40 and 1/36 monolayers per second. (**a**) Layer-by-layer growth, (**b**) growth with blocked interlayer transport, (**c**) reentrant layer-by-layer growth at low temperatures.

## 11.2 Nucleation and Growth
## under Chemical Potential Control

In nucleation and growth of solid phases under chemical potential control, the concentration of adatoms on the surface rather than the flux towards the surface is kept constant during the nucleation process. The chemical potential of adatoms is either controlled by the concentration of reactive particles in the surrounding phase as e.g. in *Liquid Phase Epitaxy* (LPE) and frequently also in *Chemical Vapor Deposition* (CVD), or by the potential of the solid electrode as in *electrodeposition* (*electroplating*). A necessary condition for a constant surface concentration during a nucleation process is the rapid exchange of atoms or molecules with the surrounding phase while the net flux towards the surface remains flexible. Two issues are of prime interest in this context. One is the size and shape of the nuclei, in particular whether the nuclei are two- or three-dimensional. This question is most straightforwardly answered with the help of a thermodynamic approach. The second issue is the nucleation rate to which we attend later.

### 11.2.1 Two-Dimensional Nucleation

For simplicity, we consider first two-dimensional homoepitaxial growth. As the concentration of adatoms is typically very small, the surplus chemical potential beyond equilibrium associated with the adatoms is well described by the non-interacting lattice gas model (cf. Sects. 5.4.1, 4.3.4).

$$\Delta\mu = k_{\mathrm{B}}T \ln(\rho / \rho_{\mathrm{eq}}) = k_{\mathrm{B}}T \ln(\Theta / \Theta_{\mathrm{eq}}) \tag{11.25}$$

The concentration can be expressed either as the number of atoms per area $\rho$, or as the fractional occupation of sites $\Theta$. The ratio $\rho / \rho_{\mathrm{eq}}$ or $\Theta / \Theta_{\mathrm{eq}}$ is called *supersaturation*. If the surrounding phase that controls the supersaturation on the surface is an ideal gas then the supersaturation equals the ratio of the pressure to the equilibrium vapor pressure of the material ($p/p_{\mathrm{eq}}$). The chemical potential difference $\Delta\mu$ described by (11.25) is addressed as *supersaturation potential*. For surfaces in contact with an electrolyte the supersaturation potential is directly related to the overpotential $\eta$ (with respect to the Nernst potential)

$$\Delta\mu = ze|\eta| \tag{11.26}$$

Here, $z$ is the charge number of the ions and $e$ the elementary charge. Positive metal ions are deposited for electrode voltages potential $\phi$ negative of the Nernst potential, whence the modulus in (11.26).

To simplify the discussion of two-dimensional nucleation we assume nuclei in the form of round 2D-islands so that the additional free energy of an island consisting of $g$ atoms compared to the flat surface is

$$\Delta F_g = 2\pi\, r_g\, \beta = 2\beta\left(\pi\Omega_s g\right)^{1/2}, \tag{11.27}$$

with $\beta$ the step line tension and $\Omega_s$ the area of an atom. The Gibbs free energy associated with that island is

$$\Delta G_g = 2\beta(\pi\Omega_s g)^{1/2} - \Delta\mu\, g. \tag{11.28}$$

The Gibbs free energy of islands depends therefore on the size as well as on the supersaturation potential (Fig. 11.12).

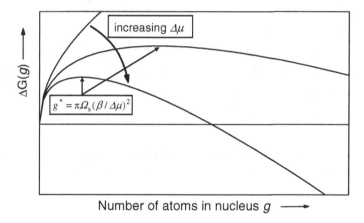

Fig. 11.12. Gibbs free energy of 2D-nuclei (islands) as function of size (schematic). The shape of the curve depends on the supersaturation $\Delta\mu$. Nuclei beyond the critical size given by the condition $d\Delta G_g/dg = 0$ grow.

Islands grow if the addition of atoms reduces $\Delta G_g$, they decay when $\Delta G_g$ becomes smaller by removing atoms. The condition

$$\frac{\mathrm{d}\Delta G_g}{\mathrm{d}g} = \beta(\pi\Omega_s)^{1/2}\, g^{-1/2} - \Delta\mu = 0 \tag{11.29}$$

defines the *critical nucleus*. This critical nucleus is a different entity than the critical nucleus of the kinetic approach described in Sect. 11.1.1, as the latter did not depend explicitly on the supersaturation. The number of atoms in the critical nucleus here is

$$g^* = \frac{\pi \Omega_s \beta^2}{\Delta \mu^2} .$$
(11.30)

The Gibbs free energy of the critical nucleus is

$$\Delta G_{\text{crit}} = \pi \Omega_s \beta^2 / \Delta \mu .$$
(11.31)

This critical energy plays the role of an activation barrier for nucleation and is therefore the decisive quantity in nucleation processes.

## 11.2.2 Two-Dimensional Nucleation in Heteroepitaxy

In heteroepitaxial growth, additional terms contribute to the energy of nuclei. The mismatch of the lattice constants between substrate and deposit gives rise to strain energy if nuclei grow pseudomorphic. Further contributions arise from the surface and interface tensions (Fig. 11.13). To first order, the elastic energy is proportional to the area $A$ of the nucleus and the square of the misfit strain $\varepsilon_{\text{mf}}$ (Sect. 3.3.2)

$$U_{\text{elast}} = A C \varepsilon_{\text{mf}}^2 .$$
(11.32)

The constant $C$ characterizes the elastic properties of the monolayer. If one assumes the monolayer to behave as an elastic isotropic film then

$$C = \frac{t}{s_{11} + s_{22}} = \frac{Y t}{1 - \nu} .$$
(11.33)

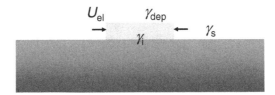

**Fig. 11.13.** The figure illustrates the additional contributions to the energy of a nucleus in heteroepitaxial growth: $U_{\text{el}}$ is the elastic energy due to the misfit strain, $\gamma_s$, $\gamma_i$ and $\gamma_{\text{dep}}$ are the substrate surface tension, the interface tension, and the surface tension of the deposit, respectively.

Here $t$ is the thickness of a monolayer, $Y$ is the Young modulus and $\nu$ is the Poisson number (3.63). In introducing these quantities, we must remember that the elastic constants as well as the natural lattice constants of monolayers may deviate considerably from bulk systems so that (11.33) yields merely a crude estimate when bulk quantities are inserted. In addition to the elastic energy, which is proportional to the area, there are contributions that scale with the perimeter and the aspect ratios of the shape (cf. Sect. 3.4). However, for nucleation these contributions play only a minor role. They become important for the shape, in particular of 3D-nuclei, as we shall see shortly.

The energy associated with the differences in the surface tension is

$$\Delta F_{\text{tension}} = A(\gamma_{\text{dep}} + \gamma_i - \gamma_s). \tag{11.34}$$

We have considered this term earlier in the context of the various growth modes in Sect. 4.2.5. Layer-by-layer growth (= Franck-van-der-Merwe growth) is preferred energetically if $\Delta F_{\text{tension}} < 0$, growth in three-dimensional islands (= Vollmer-Weber-growth) if $\Delta F_{\text{tension}} > 0$. Two-dimensional nucleation is therefore of interest if $\Delta F_{\text{tension}} < 0$. With (11.32) and (11.34) the Gibbs free energy becomes

$$\Delta G(g) = 2\beta(\pi\Omega_s g)^{1/2} - g\left(\Delta\mu - \Omega_s(\gamma_{\text{dep}} + \gamma_i - \gamma_s + C\,\varepsilon_{\text{mf}}^2)\right). \tag{11.35}$$

The additional energetic terms contribute to the effective supersaturation potential. Without further ado, we obtain the energy of the critical nucleus as

$$\Delta G_{\text{crit}} = \frac{\pi\Omega_s\beta^2}{\Delta\mu - \Omega_s(\gamma_{\text{dep}} + \gamma_i - \gamma_s + C\,\varepsilon_{\text{mf}}^2)}. \tag{11.36}$$

The elastic energy is positive definite. The surface tension term can be either positive or negative. The sum of the two terms can also be either negative or positive. If it is negative, $\Delta\mu$ must exceed a minimum value in order to have nucleation at all,

$$\Delta\mu > \Omega_s(\gamma_{\text{dep}} + \gamma_i - \gamma_s + C\,\varepsilon_{\text{mf}}^2). \tag{11.37}$$

If the lattice mismatch is small or the adsorption is not very site-specific, the elastic energy becomes small. If furthermore the sum of the tensions is negative then $\Delta G_{\text{crit}}$ is finite and possibly small even when the supersaturation $\Delta\mu$ is zero.

$$\Delta G_{\text{crit}}(\Delta\mu = 0) = \frac{\pi\beta^2}{\gamma_s - \gamma_i - \gamma_{\text{dep}}}, \quad \gamma_s - \gamma_i - \gamma_{\text{dep}} > 0 \tag{11.38}$$

Consider e.g. the growth of Ag(111) ($\gamma_{dep} \cong 1.2$ N/m) on W(110) ($\gamma_s \cong 4.0$ N/m)! If one neglects the interface tension and inserts the line tension from Table 4.2, one obtains the energy of the critical nucleus as $\Delta G_{crit} = 0.11$ eV at $\Delta\mu = 0$. Thus, a silver layer should grow on tungsten with arbitrarily small supersaturation even at room temperature. Experiments corroborate this conclusion.

### 11.2.3 Three-Dimensional Nucleation

The three-dimensional nucleation that occurs in Volmer-Weber and in Stranski-Krastanov growth (Sect. 4.2.5) is considerably more complicated. The crystalline anisotropy of the surface tension becomes an important factor as it determines the shape and the overall energy of the nuclei. The energy of a nucleus is therefore not simply proportional to its volume and its surface but depends on the ratios of the heights to the lateral extensions and on the type and relative size of the facets. Likewise, does the strain energy in a pseudomorphic nucleus depend on the aspect ratio since the misfit strain relaxes when the height of the nucleus becomes large. Last, not least the crystalline anisotropy of the elastic constant should not be neglected if one is interested in quantitative statements. Size, shape and energy of the nuclei are therefore system specific, depend on the supersaturation itself and may change as more material is deposited. With these caveats in mind, it is nevertheless illuminating to study nucleation in the simplest of all physical "realizations" of nucleation, which is nucleation in the Kossel-model. A Kossel-crystal represents atoms as cubes that are bonded to each other by their six faces (Fig. 11.14). The surfaces of substrate and deposit have surface energies $\varepsilon_s$ and $\varepsilon_{dep}$, respectively, and the interface energy per atom is $\varepsilon_i$. The energy required to

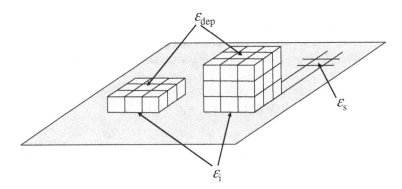

**Fig. 11.14.** 2D- and 3D-nuclei of a Kossel-crystal where the atoms are represented by cubes with bonds between the faces. Surface energies per atom of substrate and deposit are $\varepsilon_s$ and $\varepsilon_{dep}$, $\varepsilon_i$ is the interface energy per atom.

break a bond of a substrate material is therefore $2\varepsilon_s$. Its cohesive energy is $E_{coh} = 6\varepsilon_s$. A realistic number for $\varepsilon_s$ is therefore 0.5-1.0 eV if the Kossel-crystal should represent a real solid. Figure 11.14 shows 2D-and 3D-nuclei of 9 and 27 atoms, respectively. As we are interested in the energies of the critical nuclei, we compare only energies of square planar and cubic 3D-nuclei. We need not consider double-layer 2D-nuclei, e.g., as (for a Kossel-crystal) their energy is necessarily larger than the energy of single-layer nuclei and scales in the same way with the number of atoms.

The Gibbs free energies of a square 2D-nucleus containing $g$ atoms is

$$\Delta G^{(2D)} = 4\varepsilon_{dep}g^{1/2} + (\varepsilon_{dep} + \varepsilon_i - \varepsilon_s)g - \Delta\mu g \tag{11.39}$$

The first term stand for the boundary energy, the second for the surface/interface energies. We can also introduce the elastic energy of a homogeneously distorted nucleus into the model by adding a term that is proportional to the area, hence to the number of atoms. For 2D-nuclei, the elastic energy has the quality of an interfacial energy and it can be viewed as being part of it. For 3D-nuclei, we need to add an extra term $\varepsilon_{el}$ to account for the elastic energy. We keep in mind, however, that the elastic energy $\varepsilon_{el}$ is much smaller than the other energies. Consider for example the epitaxial growth of Ge(100) on Si(100). The misfit strain is $\varepsilon_{mf}$ is 4.2%. With the elastic constant $Y = 1.03\times10^{11}$ N/m of Ge and the Poisson number $\nu = 0.27$ one obtain $\varepsilon_{el} = 0.035$ eV.

The Gibbs free energies of a 3D-nucleus containing $g$ atoms is

$$\Delta G^{(3D)} = (5\varepsilon_{dep} + \varepsilon_i - \varepsilon_s)g^{2/3} - (\Delta\mu - \varepsilon_{el})g . \tag{11.40}$$

We see that the energies $\varepsilon_{dep}$, $\varepsilon_s$, $\varepsilon_i$, and $\varepsilon_{el}$ now enter in a qualitative different manner! This has noteworthy consequences for the dependence of the preferred nucleation mode on the energies. For an easier discussion, we normalize critical energy and supersaturation as

$$\tilde{G}_{crit} = G_{crit} / \varepsilon_{dep} \qquad \Delta\tilde{\mu} = \Delta\mu / \varepsilon_{dep} . \tag{11.41}$$

The critical energies and critical number of atoms in the nuclei are

$$\Delta\tilde{G}_{crit}^{(2D)} = \frac{4}{\Delta\tilde{\mu} - 1 + (\varepsilon_s - \varepsilon_i)/\varepsilon_{dep}} , \quad g^{*(2D)} = \frac{4}{(\Delta\tilde{\mu} - 1 + (\varepsilon_s - \varepsilon_i)/\varepsilon_{dep})^2} \tag{11.42}$$

$$\Delta\tilde{G}_{crit}^{(3D)} = \frac{4}{27}\frac{(5 + (\varepsilon_i - \varepsilon_s)/\varepsilon_{dep})^3}{(\Delta\tilde{\mu} - \varepsilon_{el}/\varepsilon_{dep})^2} , \quad g^{*(3D)} = \frac{8}{27}\frac{(5 + (\varepsilon_i - \varepsilon_s)/\varepsilon_{dep})^3}{(\Delta\tilde{\mu} - \varepsilon_{el}/\varepsilon_{dep})^3} . \tag{11.43}$$

We discuss several scenarios, depending on the magnitudes of the energies.

**Case I:** $\varepsilon_{dep} = \varepsilon_s$, $\varepsilon_i = 0$, $\varepsilon_{el} = 0$

This case corresponds to homoepitaxy. The energies and size of critical nuclei are

$$\varepsilon_{dep} = \varepsilon_s, \varepsilon_i = 0 : \Delta\tilde{G}_{crit}^{(2D)} = \frac{4}{\Delta\tilde{\mu}}, \qquad g*^{(2D)} = \frac{4}{\Delta\tilde{\mu}^2}$$

$$\Delta\tilde{G}_{crit}^{(3D)} = \frac{256}{27}\frac{1}{\Delta\tilde{\mu}^2}, \qquad g*^{(3D)} = \frac{8}{27}\frac{5^3}{\Delta\tilde{\mu}^2} .$$

(11.44)

In Fig. 11.15a the energies are plotted as solid and dashed lines and the critical sizes as dash-dotted and dotted lines, for 2D- and 3D-case respectively. The thin horizontal line marks a boundary $g* = 10$ below which the model is not meaningful. According to Fig. 11.15a two-dimensional nucleation is preferred.

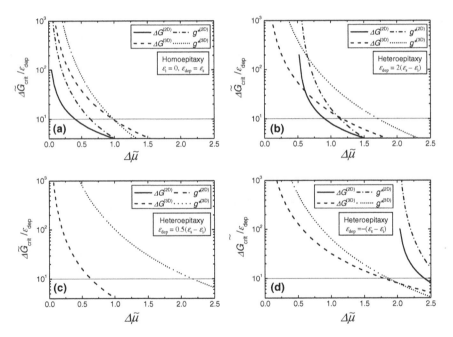

**Fig. 11.15.** Gibbs free energy and size of the critical nucleus versus supersaturation in the Kossel-model. Energies are normalized to $\varepsilon_{dep}$ (see text for discussion).

**Case II:** $\varepsilon_{dep} > (\varepsilon_s - \varepsilon_i)$, $\varepsilon_{el} = 0$

This case corresponds to heteroepitaxy for which according to the Bauer-criterion Vollmer-Weber growth should occur under equilibrium conditions (Sect. 4.2.5). As an example, we choose $\varepsilon_{dep} > 2(\varepsilon_s - \varepsilon_i)$. The energies and sizes of the critical nuclei are

$$\varepsilon_{dep} = 2(\varepsilon_s - \varepsilon_i): \ \Delta\tilde{G}_{crit}^{(2D)} = \frac{4}{\Delta\tilde{\mu} - 0.5}, \ g*^{(2D)} = \frac{4}{(\Delta\tilde{\mu} - 0.5)^2}$$

$$\Delta\tilde{G}_{crit}^{(3D)} = \frac{27}{2}\frac{1}{\Delta\tilde{\mu}^2}, \ g*^{(3D)} = \frac{8}{27}\frac{(5.5)^3}{\Delta\tilde{\mu}^3}.$$

(11.45)

These functions are plotted in Fig. 11.15b. For small supersaturations 3D-nucleation is the preferred growth mode; 2D-nucleation takes over at larger supersaturations. If elastic energy is added, the pole for $\Delta\tilde{G}_{crit}^{(3D)}$ shifts away from the origin of $\Delta\tilde{\mu}$. The crossover to 2D-nucleation shifts to lower supersaturations.

**Case III:** $\varepsilon_{dep} < (\varepsilon_s - \varepsilon_i)$, $\varepsilon_{el} = 0$

This case covers the growth of low surface tension material on substrates with a high surface tension. The equilibrium growth mode is the Frank-van-der-Merwe growth. As an example we choose $\varepsilon_{dep} = 0.5(\varepsilon_s - \varepsilon_i)$. The energies of the critical nuclei are

$$\varepsilon_{dep} = 0.5(\varepsilon_s - \varepsilon_i): \ \Delta\tilde{G}_{crit}^{(2D)} = \frac{4}{\Delta\tilde{\mu} + 1}, \ g*^{(2D)} = \frac{4}{(\Delta\tilde{\mu} + 1)^2}$$

$$\Delta\tilde{G}_{crit}^{(3D)} = \frac{4}{\Delta\tilde{\mu}^2}, \ g*^{(3D)} = \frac{8}{27}\frac{7^3}{\Delta\tilde{\mu}^3}.$$

(11.46)

The functional dependences are shown in Fig. 11.15c. 2D-nucleation is preferred for all relevant supersaturations.

**Case IV:** $\varepsilon_i - \varepsilon_s > 0$

This scenario corresponds to a situation where the deposit rather creates a surface of its own than to form bonds with the substrate, a situation of complete non-wetting. We choose $\varepsilon_{dep} = \varepsilon_i - \varepsilon_s$ as an example and obtain

$$\varepsilon_{\text{dep}} = \varepsilon_{\text{i}} - \varepsilon_{\text{s}} : \quad \Delta \tilde{G}_{\text{crit}}^{(2D)} = \frac{4}{\Delta \tilde{\mu} - 2}, \quad g^{*(2D)} = \frac{4}{(\Delta \tilde{\mu} - 2)^2}$$

$$\Delta \tilde{G}_{\text{crit}}^{(3D)} = \frac{32}{\Delta \tilde{\mu}^2}, \quad g^{*(3D)} = \frac{8}{27} \frac{6^3}{\Delta \tilde{\mu}^3} . \tag{11.47}$$

The result is plotted in Fig. 11.15d. Three-dimensional nucleation is preferred as expected.

## 11.2.4 Theory of Nucleation Rates

This section considers the rate at which atoms or, speaking in general terms, monomers are incorporated into stable nuclei when the external source keeps the surface concentration constant. The thermodynamic picture developed above yields an expression for the equilibrium number of clusters containing $g$-atoms (or monomers)

$$n_g^{(\text{eq})} = n_0 e^{-\frac{\Delta G(g)}{k_{\text{B}} T}} \tag{11.48}$$

where $n_0$ is the number of possible nucleation sites on a given surface[20]. We keep in mind in the following that $n_g^{(\text{eq})}$ is the equilibrium number for a fixed super-saturation potential. With the notion that clusters grow continuously once they have surpassed the critical size (Fig. 11.12) one might infer that the growth rate is the number of critical nuclei multiplied by the flux of monomers into those nuclei. Accordingly, the growth rate $I$ should be proportional to

$$I \propto n_{g^*}^{(\text{eq})} \tag{11.49}$$

In reality, the problem is more complex. All cluster of arbitrary size are assembled from monomers. The critical cluster containing $g^*$ monomers is generated by adding one monomer to the $(g^*-1)$-cluster, the $(g^*-1)$-cluster is generated by adding a monomer to the $(g^*-2)$-cluster, and so forth. Adding one monomer to a cluster of $g$ monomers not only generates a $(g^*+1)$-cluster; it also removes one cluster from the ensemble of clusters of size $g$. These statements are cast into mathematics by considering the growth rate of the number of clusters with $g$ monomers,

$$\frac{\partial n_g}{\partial t} = C_{g-1} n_{g-1} + E_{g+1} n_{g+1} - C_g n_g - E_g n_g . \tag{11.50}$$

---

[20] We note that in homogeneous nucleation, e.g. water condensation, $n_0$ is replaced by the number of existing monomers as these monomers are the condensation sites.

$C_g$ and $E_g$ are the condensation and evaporation rates into and from clusters of the size $g$, respectively. In words, the number of clusters of size $g$ increases by condensing monomers into clusters of size $g-1$ and by evaporating monomers from clusters of size $g+1$; the number of clusters of size $g$ is reduced by evaporation or by condensation of monomers. The overall condensation process requires a stream that carries monomers from clusters of size $g$ to $g+1$. The reverse process involves evaporation from clusters of size $g+1$. We can therefore define a condensation current at the cluster of size $g$ as

$$I_g = C_g N_g - E_{g+1} n_{g+1} . \tag{11.51}$$

Equation (11.50) can be written as the difference between the currents $I_{g-1}$ and $I_g$,

$$\frac{\partial n_g}{\partial t} = I_{g-1} - I_g . \tag{11.52}$$

In the very beginning of a nucleation process the number of clusters of all sizes changes rapidly. Under the condition of constant supersaturation, a steady state is reached quickly however with all $I_g$ identical and positive. One is primarily interested in this steady-state nucleation. We discuss two solutions to this problem. One involves a mapping of the problem onto the statistical problem of a random walker. The other one is based on an ingenious steady-state solution of coupled rate equations. Both solutions turn out to be numerical (almost) identical although it is not obvious why that should be so.

### Solution based on the statistics of a random walker

The solution focuses on the nucleation current into the critical cluster. The current can be written as the product of the attachment rate $C_{g*}$ to the critical cluster, the number of critical clusters in steady state $n_{g*}^{(st)}$, and the probability $P_{no}(g^*)$ that a cluster will never return to the critical stage, once a monomer has been added,

$$I = C_{g*} n_{g*}^{(st)} P_{no}(g^*) . \tag{11.53}$$

It can be shown that the steady state number of clusters of any size $g$ are given by the equilibrium number multiplied by the probability $P_{ret}(g)$ that the cluster of size $g$ returns to the state of a monomer,

$$n_{g*}^{(st)} = n_{g*}^{(eq)} P_{ret}(g^*) . \tag{11.54}$$

For the (non-elementary) proof, the reader is referred to the study of White [11.10]. The return probability $P_{ret}(g)$ is

$$P_{\mathrm{ret}}(g) = \sum_{j=g}^{\infty} \prod_{k=1}^{j} \frac{E_k}{C_k} \bigg/ \left\{ 1 + \sum_{j=1}^{\infty} \prod_{k=1}^{j} \frac{E_k}{C_k} \right\}. \tag{11.55}$$

The ratios of evaporation and condensation rates are > 1 for clusters smaller than the critical size, = 1 for clusters of the critical size and < 1 for clusters larger than the critical size.

The condition that the incorporation of a monomer into the critical cluster must be a permanent one is identical to the requirement that the cluster never returns to the critical size $g^*$. This no-return probability is

$$P_{\mathrm{no}}(g^*) = \left\{ 1 + \sum_{j=g^*+1}^{\infty} \prod_{k=g^*+1}^{j} \frac{E_k}{C_k} \right\}^{-1}. \tag{11.56}$$

The total rate $I$ by which monomers nucleate is therefore

$$I = C_{g^*} P_{\mathrm{no}}(g^*) P_{\mathrm{ret}}(g^*) n_0 e^{-\frac{\Delta G_{\mathrm{crit}}}{k_B T}}. \tag{11.57}$$

For a numerical solution of the expression for the probabilities, one needs the ratios of evaporation and condensation for cluster of size $g^*$. With reference to the discussion on Ostwald ripening (Sect. 10.4.1), we can write the evaporation current density at the perimeter of the cluster as

$$j_{\mathrm{evaporation}} = vs\,\Theta^{(\mathrm{eq})}(r_g) = vs\,\Theta^{(\mathrm{eq})}(\infty)e^{\frac{\mu_{\mathrm{GT}}}{k_B T}} \tag{11.58}$$

where $\Theta^{(\mathrm{eq})}(r_g)$ is the equilibrium concentration of monomers at the periphery of a cluster of radius $r_g$ (possessing the Gibbs-Thomson chemical potential), $s$ is a sticking coefficient and $v$ a jump frequency. The equilibrium concentration at the periphery of the cluster can be expressed in terms of the equilibrium concentration on a flat surface $\Theta^{(\mathrm{eq})}(\infty)$ and the Gibbs-Thomson chemical potential of the cluster (11.58). The current density for condensation is

$$j_{\mathrm{condensation}} = vs\,\Theta = vs\,\Theta^{(\mathrm{eq})}(\infty)S \tag{11.59}$$

with $\Theta$ the actual concentration on the surface and $S$ the supersaturation. The ratio of evaporation and condensation is therefore

$$\frac{E_g}{C_g} = S^{-1} e^{\frac{\mu_{\mathrm{GT}}}{k_B T}}. \tag{11.60}$$

For spherical (3D) and circular (2D) clusters (11.60) can be reduced to a very simple expression. With the identity

$$e = S^{\frac{1}{\ln S}} \qquad (11.61)$$

and

$$\mu_{GT}(g) = \beta \Omega_s / r_g \;,\; r_{g*} = \beta \Omega_s / k_B T \ln S \qquad (11.62)$$

for the 2D-case and

$$\mu_{GT}(g) = 2\gamma\Omega / r_g \;,\; r_{g*} = 2\gamma\Omega / k_B T \ln S \qquad (11.63)$$

for spherical clusters, one obtains

$$\frac{E_g}{C_g} = S^{-1} S^{(r_{g*}/r_g)} = S^{-1} S^{(g*/g)^\alpha} \qquad (11.64)$$

where $\alpha$ is 1/2 for circular 2D-clusters and 1/3 for spherical clusters. With (11.64) the products and sums in (11.55) and (11.56) are easily evaluated in terms of $g^*$ and $S$. Good numerical accuracy is achieved when the upper limit $\infty$ is replaced by a number of the order of 10–100 above $g^*$.

### Solution based on coupled rate equations

The derivation of the steady state condensation current by playing with the set of rate equations (11.51) is much more elementary, thanks to some ingenious tricks. Our presentation follows the very tutorial papers of McDonald [11.11, 12].

The steady state condition requires that all $I_g$ in (11.51) are identical, hence

$$\begin{aligned}
I &= C_1 n_1^{(st)} - C_2 n_2^{(st)} = C_2 n_2^{(st)} - C_3 n_3^{(st)} = \dots \\
&= C_g n_g^{(st)} - E_{g+1} n_{g+1}^{(st)} \;.
\end{aligned} \qquad (11.65)$$

The problem in calculating the nucleation current is that the steady state numbers $n_g^{(st)}$ are unknown. They differ from the equilibrium numbers $n_g^{(eq)}$ except in the limit of zero current. For zero current, we have

$$C_g n_g^{(eq)} = E_{g+1} n_{g+1}^{(eq)} \;. \qquad (11.66)$$

Using (11.66), equation (11.65) can be cast into the form

$$I = C_g n_g^{(eq)} \left\{ n_g^{(st)} / n_g^{(eq)} - (E_{g+1} / C_g)(n_{g+1}^{(st)} / n_{g+1}^{(eq)}) \right\}$$
$$= C_g n_g^{(eq)} (n_g^{(st)} / n_g^{(eq)} - n_{g+1}^{(st)} / n_{g+1}^{(eq)}) .$$

(11.67)

We rearrange this set of equations to

$$I / C_1 n_1^{(eq)} = n_1^{(st)} / n_1^{(eq)} - n_2^{(st)} / n_2^{(eq)}$$
$$I / C_2 n_2^{(eq)} = n_2^{(st)} / n_2^{(eq)} - n_3^{(st)} / n_3^{(eq)}$$
$$\vdots \quad \vdots \quad \vdots \quad \vdots \quad \vdots$$
$$I / C_{G-2} n_{G-2}^{(eq)} = n_{G-2}^{(st)} / n_{G-2}^{(eq)} - n_{G-1}^{(st)} / n_{G-1}^{(eq)}$$
$$I / C_{G-1} n_{G-1}^{(eq)} = n_{G-1}^{(st)} / n_{G-1}^{(eq)} - n_G^{(st)} / n_G^{(eq)} .$$

(11.68)

By summing up the left and the right hand sides one obtains

$$I = \frac{n_1^{(st)} / n_1^{(eq)} - n_G^{(st)} / n_G^{(eq)}}{\sum_{g=1}^{G-1} 1 / C_g n_g^{(eq)}} .$$

(11.69)

We see that the rearrangement has spirited away all of the unknown steady state numbers save for those of the clusters of size one, the monomers, and size $G$. At first sight, not much seems to be gained from that; however, we remember that we have imposed the condition of constant supersaturation. The number of monomers is therefore independent of the magnitude of the nucleation current, whence

$$n_1^{(st)} / n_1^{(eq)} = 1 .$$

(11.70)

To take care of the second term in the nominator of (11.69) we invoke a kind of Maxwell demon and request that this demon keeps the supersaturation constant by breaking up clusters of size $G$ into monomers to feed them back into the system. This artifice, invented by Szilard, maintains the steady state in the system while keeping the number of monomers in the system finite and constant. It also imposes the boundary condition

$$n_G^{(st)} = 0 .$$

(11.71)

When $G$ is significantly larger than the critical size $g^*$ the resulting current $I$ becomes independent of the choice of $G$ which is the a posteriori justification for the artifice. With (11.70) and (11.71), the current is expressed solely in terms of equilibrium properties and condensation rates,

$$I = \left[ \sum_{g=1}^{G-1} \frac{1}{C_g n_g^{(eq)}} \right]^{-1} . \tag{11.72}$$

The further evaluation of the sum (11.72) is straightforward. For not too small critical clusters, we can convert the sum into an integral.

$$I \cong \left[ \int_1^G \frac{dg}{C_g n_g^{(eq)}} \right]^{-1} = \left[ n_0^{-1} \int_1^G \frac{e^{+\frac{\Delta G(g)}{k_B T}}}{C_g} dg \right]^{-1} \tag{11.73}$$

$\Delta G$ is maximal at the critical size $g^*$ (Fig. 11.12). The exponential function has therefore a sharp maximum at $g^*$. The integral is evaluated by expanding $\Delta G$ around the maximum at $g^*$ and by retaining the first two non-vanishing terms.

$$\Delta G(g) = \Delta G_{crit} + \frac{1}{2} \frac{\partial^2 \Delta G}{\partial g^2} (g - g^*)^2 ... \tag{11.74}$$

The linear term vanishes by definition of $\Delta G(g^*)$. The second derivative of $\Delta G$ is necessarily negative. With the abbreviation

$$Q = -\frac{1}{2} \frac{\partial^2 \Delta G}{\partial g^2} \tag{11.75}$$

the inverse nucleation current becomes

$$I^{-1} = \frac{e^{\frac{\Delta G_{crit}}{k_B T}}}{C_{g^*} n_1} \int_{-\infty}^{+\infty} e^{-Qx^2/2k_B T} dx \tag{11.76}$$

Because of the sharp peak of the integrand at $x = (g - g^*) = 0$ the boundaries can be extended to infinity. The final solution for the current is

$$I = \left( \frac{Q}{2\pi k_B T} \right)^{1/2} C_{g^*} n_0 e^{-\frac{\Delta G_{crit}}{k_B T}} . \tag{11.77}$$

The first prefactor in (11.77) is known as the *Zeldovich-factor Z* [11.13]. By comparing (11.77) with (11.57) we note that this factor (if everything is correct!) should be identical to the product of the two probabilities $P_{no}(g^*)$ and $P_{ret}(g^*)$,

$$Z = \left( \frac{Q}{2\pi k_B T} \right)^{1/2} = P_{no}(g^*)P_{ret}(g^*) . \tag{11.78}$$

This is a remarkable result, far from being obvious. In order to test the equality (11.78) we calculate $Z$ for spherical 3D-clusters and circular 2D-clusters. The second derivative of $\Delta G$ for 2D-clusters is obtained from (11.28). The second derivative of $\Delta G$ for 3D-clusters is obtained from the analogous equation for spheres. It is convenient to replace the line tension and the surface tension by the supersaturation $S$. The results are then

$$Z^{(2D)} = \left( \frac{\ln S}{4\pi g^*} \right)^{1/2} \equiv \left( \frac{\Delta G_{crit}}{2\pi k_B T g^*} \right)^{1/2} \equiv \left( \frac{\Delta\mu^3}{4\pi^2 \Omega_s \beta^2 k_B T} \right)^{1/2} \tag{11.79}$$

$$Z^{(3D)} = \left( \frac{\ln S}{6\pi g^*} \right)^{1/2} \equiv \left( \frac{\Delta G_{crit}}{3\pi k_B T (g^*)^2} \right)^{1/2} \equiv \left( \frac{\Delta\mu^4}{64\pi^2 k_B T \Omega^2 \gamma^3} \right)^{1/2} . \tag{11.80}$$

As before, $\beta$ and $\gamma$ are line and surface tension of the nuclei and $\Omega_s$ and $\Omega$ are the surface area of an atom and the atom volume, respectively. The second forms of the Zeldovich-factors are conventionally quoted; the first forms are convenient for the intended numerical comparison; the third forms are useful for calculations of the dependence of nucleation prefactors on experimental parameters.

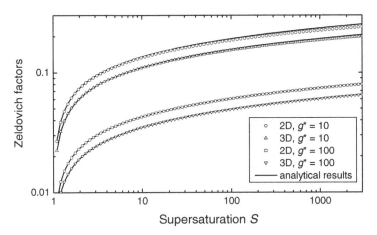

**Fig. 11.16.** Zeldovich-factors for critical clusters of size 10 and 100 calculated from the random walk theory of White [11.10] with (11.55), (11.56) and (11.64) and from the analytical expressions (11.79) and (11.80).

Figure 11.16 shows the Zeldovich-factors calculated numerically from (11.55, 11.56) with the help of (11.64) as symbols. The solid lines are obtained from (11.79) and (11.80). For $g* = 100$ numerical and analytical results are identical within the numerical accuracy. For the smaller critical cluster, there is a small deviation presumably because the replacement of the sum (11.72) by the integral (11.73) and the approximation made in solving the integral work better for larger critical clusters. The numerical agreement was already noticed by White [11.10], but remains mysterious in view of the completely different mathematics. From the standpoint of a physicist, the agreement of the two approaches is pleasing though.

### 11.2.5 Rates for 2D- and 3D-Nucleation

This section condenses what we have learned about nucleation rate theory into expressions for 2D- and 3D-nucleation and discusses the result in the perspective of experimental data. We focus on homoepitaxial growth of element crystals. The monomers are then adatoms of the same type as the substrate atoms. We first re-write (11.77) into a current density $J$ by replacing the number of nucleation sites by their number per area $n_s$. For a perfect surface, one can identify the concentration of nucleation sites with the concentration of sites for adatoms.

$$J = C_{g*} Z n_s e^{-\frac{\Delta G_{crit}}{k_B T}} \tag{11.81}$$

This equation is now further evaluated for several cases.

### *Case I: 2D-nucleation and surface diffusion*

The condensation rate $C_{g*}$ is calculated as for the case of Ostwald ripening (10.80, 11.59)

$$
\begin{aligned}
C_{g*} &= 2\pi (r*/a_\parallel)(\Theta - \Theta_{eq}(r*)) s v_0 e^{-\frac{E_d}{k_B T}} \\
&\cong 2\pi \frac{\Omega_s \beta}{a_\parallel \Delta \mu}(\Theta - \Theta_{eq}(\infty)) s v_0 e^{-\frac{E_d}{k_B T}} .
\end{aligned} \tag{11.82}
$$

Here, $a_\parallel$ is the atom diameter, $s$ is a sticking coefficient, $v_0$ is the prefactor for diffusion, $E_d$ is the activation energy for diffusion and $\Theta$ and $\Theta_{eq}$ are the actual and the equilibrium coverage with adatoms per site next to the critical nucleus. The dependence of the equilibrium coverage on $r*$ can be neglected so that $\Theta_{eq}(r*)$ can be replaced by $\Theta_{eq}(\infty)$. By definition, the coverage $\Theta$ is the product of the equilibrium coverage $\Theta_{eq}(\infty)$ and the supersaturation $S$. The equilibrium coverage $\Theta_{eq}(\infty)$ is given by the energy $E_A$ to create an adatom from a kink site (Sect. 1.3.2, eq. 1.33), so that

$$\Theta - \Theta_{\text{eq}}(\infty) = (S-1)e^{-\frac{E_A}{k_B T}} = (e^{\frac{\Delta\mu}{k_B T}} - 1)e^{-\frac{E_A}{k_B T}}. \tag{11.83}$$

After inserting (11.82), (11.83), (11.79) and (11.31) we obtain from (11.81)

$$J^{(2D,\text{diff})} = J_0^{(2D,\text{diff})} e^{-\frac{\pi\Omega_s\beta^2}{\Delta\mu\, k_B T}} \qquad \text{with}$$

$$J_0^{(2D,\text{diff})} = 2\pi \frac{\Omega_s\beta}{a_\parallel \Delta\mu} n_s s v_0 e^{-\frac{E_d + E_A}{k_B T}} (e^{\frac{\Delta\mu}{k_B T}} - 1)\left(\frac{\Delta\mu^3}{4\pi^2\Omega_s\beta^2 k_B T}\right)^{1/2}. \tag{11.84}$$

For clarity, we have abstained from reducing fractions. The behavior of the nucleation current as function of the supersaturation potential is dominated by the exponential term since the supersaturation potential $\Delta\mu$ enters as a factor (!) to the temperature there. This calls for an extremely steep increase in the rate in a small chemical potential range. Experimentally this increase has the appearance of a sudden onset of nucleation. Compared to that, the variation of the other factors on the supersaturation potential is small. This holds even for the exponential term $\exp(\Delta\mu/k_B T)$ that arises directly from the supersaturation. In many relevant cases, $\Delta\mu$ is smaller or comparably with $k_B T$ so that the term depends only weakly on $\Delta\mu$. During the onset of nucleation, the prefactor $J_0$ is practically independent of $\Delta\mu$ and is therefore rightly treated as a constant prefactor in the analysis of experimental data. Equation (11.84) permits the calculation of that prefactor from parameters obtained from other experiments; here from studies of Ostwald ripening (Sect. 10.4) and island equilibrium fluctuations (Sect. 4.3.8).

### Case II: 2D-nucleation and direct growth from the ambient phase

We consider the case when atoms are directly incorporated into the edge of a growing 2D-nucleus from the ambient phase without intermediate surface diffusion. The condensation rate should be given by the number of sites next to the step edge multiplied with the flux onto these sites from the ambient phase. For vapor-phase deposition, one can make the ansatz that the rate is the flux per area multiplied by the area of the atoms adjacent to the step edge. The flux density can be taken from kinetic gas theory (2.3). With that, one obtains

$$J^{(2D,\text{direct})} = J_0^{(2D,\text{direct})} e^{-\frac{\pi\Omega_s\beta^2}{\Delta\mu\, k_B T}} \qquad \text{with}$$

$$J_0^{(2D,\text{direct})} = \frac{\Omega_s\beta}{a_\parallel \Delta\mu} \frac{2\pi a n_s p_{\text{eq}} (e^{\frac{\Delta\mu}{k_B T}} - 1)}{(2\pi m k_B T)^{1/2}} \left(\frac{\Delta\mu^3}{4\pi^2\Omega_s\beta^2 k_B T}\right)^{1/2}. \tag{11.85}$$

Here, $a$ is an atomic dimension and $m$ is the mass of the deposited atoms. The pressure in the ambient phase is replaced by the product of the equilibrium pressure $p_{eq}$ and the supersaturation.

The equilibrium pressure contains the cohesive energy of the solid as an activation term (Sect. 5.1.3, eq. 5.17). Since $E_{coh}$ is of the order of several eV the activation term is many orders of magnitude smaller than the corresponding factor in (11.84) that stemmed from the equilibrium concentration of monomers and their diffusion towards the nucleus. Nucleation by direct deposition from the gas phase onto the edges of a growing 2D-cluster cluster is therefore negligible for homoepitaxial growth from the vapor phase. The argument may not hold for deposition from a high-pressure gas or from an electrolyte solution. It is therefore useful to consider the relative importance of nucleation by surface diffusion and direct deposition on more general grounds. The derivation of the rate equation (11.84) requires that the incorporation process does not significantly reduce the concentration next to a growing nucleus. Each critical nucleus captures atoms by the rate

$$\tau_{capt}^{-1} = J_0^{(2D,diff)} / n_s \tag{11.86}$$

which sets the time scale for diffusion. During the time between two capture events, the atom can diffuse the distance (measured in atom units)

$$\hat{L}_D = (\tau_{cap} \hat{D})^{1/2} = \left( \tau_{cap} \nu_0 \exp(-E_D / k_B T) \right)^{1/2} \tag{11.87}$$

with $\nu_0$ and $E_D$ the prefactor and the activation energy for diffusion, respectively. The atoms within a distance $\hat{L}_D$ from the perimeter of the nucleus have a chance to arrive at the perimeter before the next capturing event takes place. The area of a circle with the radius $\hat{L}_D$ therefore feeds the nucleation process by diffusion. If the nucleus grows from the ambient phase directly it captures atoms at most by its own area $\pi(\hat{r}*)^2$. Hence, nucleation via diffusion requires

$$\pi \hat{L}_D^2 \gg \pi(\hat{r}*)^2 = g* . \tag{11.88}$$

This places an upper bound on $E_D$ for diffusion-controlled nucleation.

$$E_D < k_B T \ln(k_B T n_s / \pi h g * J_0) . \tag{11.89}$$

We have replaced $\nu_0$ by $k_B T/h$. With this condition, we can calculate the upper limit for the diffusion activation energy from the experimental value of $J_0$. For electrodeposition of silver typical experimental values of $J_0$ are in the range between $10^{10}\,cm^{-2}s^{-1}$ and $10^{15}\,cm^{-2}s^{-1}$ [11.14]. Assuming $g* = 100$ and $T = 300$ K, one obtains upper limits of $E_D = 1.1$ eV and $0.6$ eV, for $J_0 = 10^{10}\,cm^{-2}s^{-1}$ and $J_0 = 10^{15}\,cm^{-2}s^{-1}$ respectively. Since activation energies for diffusion on terraces are typically below this range, nucleation should occur via surface diffusion.

It is difficult to decide experimentally between diffusion controlled and direct nucleation from the ambient phase since all basic relations, as e.g. the dependence of the growth rate on the overpotential, are either identical or experimentally indistinguishable. Guided by the interpretation of electrochemical deposition experiments on the growth of step bunches researchers in the 1970-ties were inclined to give preference to nucleation by direct attachment (see [11.14] for details). In the light of our present understanding, the arguments are not compelling, however. More importantly, studies on step fluctuations on Ag(111) electrodes in 1mM CuSO$_4$+0.05 M H$_2$SO$_4$ close to the point of dissolution have shown that steps exchange atoms with the terraces and the terraces with the electrolyte (Fig. 10.37) [11.15, 16]. Direct exchange with the electrolyte could be excluded at least for this system, as it would lead to a different time dependence of the fluctuations (Fig. 10.34).

### Case III: 3D-nucleation in the Kossel-model

In Sect. 11.2.2, we have calculated $\Delta G_{crit}$ in the Kossel model for various cases and have argued for either 2D- or 3D-nucleation depending on the value of $\Delta G_{crit}$. Since $\Delta G_{crit}$ enters the activation energy of the nucleation rate (11.81) the argument should be sound. It is nevertheless interesting to compare 2D- and 3D-nucleation rates quantitatively. We consider the example of homoepitaxial growth. Because of the argument above, we focus on attachment by surface diffusion. The nucleation rate for 2D-growth is

$$J^{(2D,K)} = \frac{8\varepsilon}{\Delta\mu} s v_0 e^{-\frac{E_d+E_A}{k_BT}} (e^{\frac{\Delta\mu}{k_BT}} - 1) \left(\frac{\Delta\mu^3}{32\pi\varepsilon^2 k_BT}\right)^{1/2} n_s e^{-\frac{4\varepsilon^2}{\Delta\mu k_BT}} . \quad (11.90)$$

Here, $\varepsilon$ is the parameter representing the step energy per atom. The corresponding equation for the growth of 3D-nuclei via surface diffusion is

$$J^{(3D,K)} = \frac{2^{10}}{9} \frac{\varepsilon^2 n_s}{\Delta\mu^2} s v_0 e^{-\frac{E_d+E_A}{k_BT}} (e^{\frac{\Delta\mu}{k_BT}} - 1) \left(\frac{9\Delta\mu^4}{2^{11}\pi\varepsilon^3 k_BT}\right)^{1/2} e^{-\frac{256\varepsilon^3}{27\Delta\mu^2 k_BT}} . \quad (11.91)$$

To be reasonably consistent with experiment the parameter $\varepsilon$ should now represent the surface energy per atom.

### Numerical examples

It may be refreshing at this point to introduce some numbers into the game and calculate the growth rate for the various mechanisms with a particular example in mind. We choose the Ag(111) surface, as for this surface the sum of the energies $E_d + E_A$, the prefactor and the line tension are known from experiment ($E_d + E_A = 0.71$ eV, $v_0 = 1\times10^{12}$ s$^{-1}$ [11.17], $a_\parallel\beta = 0.23$ eV [11.18]).

We consider first deposition from the vapor phase. The condition of constant su-persaturation requires that the exchange flux of atoms between the vapor phase and the surface is much larger than the flux into nuclei. This calls for substrate temperatures where the vapor pressure of the substrate is appreciable. At $T = 950$ K e.g. the equilibrium pressure of silver is $1.3 \times 10^{-4}$ Pa (Fig. 5.1) corre-sponding to a flux of $10^{14}$ atoms cm$^{-2}$s$^{-1}$, which is large enough to divert atoms into nuclei at a considerable rate without disturbing the steady state concentration of adatoms on the surface.

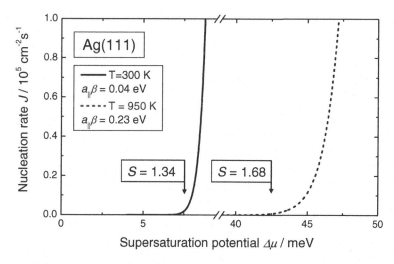

**Fig. 11.17.** 2D-nucleation rates on Ag(111) for two different step line tensions $a_{\parallel}\beta$ and temperatures to represent typical situations for MBE-growth ($T = 950$ K) and electrodeposi-tion ($T = 300$ K), dashed and solid line respectively. The curves are calculated from (11.84) with $E_d + E_A = 0.71$ eV, $\nu_0 = 1 \times 10^{12}$ s$^{-1}$ [11.17], $a_{\parallel}\beta = 0.23$ eV [11.18] and $a_{\parallel}\beta = 0.04$ eV [11.14]. The latter represents the step line tension of Ag(111)-electrodes near the Nernst potential in 6 M AgNO$_3$. The flags mark the supersaturation at the points of onset.

The dashed line in Fig. 11.17 shows the result as obtained from equation (11.84). Table 11.1 lists characteristic parameters of the nuclei for an arbitrarily chosen fixed nucleation rate of $J = 10^6$ cm$^{-2}$s$^{-1}$. The critical number of atoms $g^* = 60$ is in a range where the continuum model should be applicable. The same calculation for $T = 300$ K yields $g^* = 3$ (actually a fractional number) which is outside the realm of the continuum model. Moreover, the condition of constant supersatura-tion cannot be maintained in that case as the vapor pressure of silver is essentially zero. Hence, from a very different standpoint, we recover that vapor-phase growth at room temperature is under constant flux, not constant supersaturation. This is different in electrodeposition. Nucleation experiments on silver electrodes show that the step line tension is much smaller than in vacuum. Values of $a_{\parallel}\beta = 0.04$ eV and $a_{\parallel}\beta = 0.035$ eV have been quoted for Ag(111) and Ag(100) in AgNO$_3$ solution, respectively [11.14]. The smaller values for the line tension are qualitatively consistent with the fact that the Nernst potential is far to the positive of the pzc (Sect. 4.3.5, eq. 4.76). The solid line in Fig. 11.17 is obtained for Ag(111) when $E_d + E_A$ and $\nu_0$ are taken as for surfaces in vacuum. The prefactor of $J_0 \cong 10^{15}$ cm$^{-2}$s$^{-1}$ thereby obtained is about four orders of magnitude larger than measured on Ag(111) electrodes (Table 5.3 in [11.14]) leading to the not unreasonable suggestion that diffusion may be hindered by the presence of the concentrated electrolyte. The reduction in the diffusion coefficient (or the higher activation energy) is still well below the estimated limit for attachment by diffusion.

**Table 11.1.** Characteristic nucleation parameters for nucleation via surface diffusion calculated for a fixed nucleation rate of $J = 10^6$ cm$^{-2}$s$^{-1}$; $a_\parallel \beta$ is the step line tension, $J_0$ the prefactor, $g^*$ the size of the critical nucleus in atoms, $Z$ the Zeldovich-factor, $\Delta\mu$ the supersaturation potential, and $S$ the supersaturation.

| System | $T$/K | $a_\parallel \beta$, $\varepsilon$/eV | $J_0$/cm$^{-2}$s$^{-1}$ | $g^*$ | $Z$ | $\Delta\mu$/eV | $S$ |
|---|---|---|---|---|---|---|---|
| Ag(111) | 950 | 0.23 | $4.5\times10^{23}$ | 58 | 0.029 | 0.0498 | 1.84 |
| Ag(111) | 300 | 0.23 | $1.6\times10^{19}$ | 3 | 0.45 | 0.211 | 3450 |
| Ag(111) | 300 | 0.04 | $4.3\times10^{14}$ | 46 | 0.024 | 0.00972 | 1.46 |
| Kossel2D | 300 | 0.23 | $8.1\times10^{19}$ | 3 | 0.35 | 0.256 | 19400 |
| Kossel3D | 300 | 0.6 | $1.6\times10^{34}$ | 3 | 0.61 | 1.105 | $4\times10^{18}$ |
| Kossel2D | 950 | 0.23 | $1.1\times10^{24}$ | 55 | 0.024 | 0.0622 | 2.14 |
| Kossel3D | 950 | 0.6 | $4.1\times10^{27}$ | 11.5 | 0.14 | 0.709 | 5750 |

Table 11.1 also compares 2D- and 3D-nucleation with the help of the Kossel model. The free parameter is the energy $\varepsilon$ per face of the elemental cubes. To be in keeping with the experimental data on silver $\varepsilon$ is taken as 0.23 eV for 2D-nucleation where it represents a step line tension and as 0.6 eV in 3D-nucleation where it represents a surface energy per atom. The table shows that 3D-nucleation is very unfavorable compared to 2D-nucleation. This result was already obtained in 11.2.3 by comparing critical Gibbs energies, but is accentuated here because of the different values for $\varepsilon$ in the two cases.

## 11.2.6 Nucleation Experiments at Solid Electrodes

The solid/electrolyte interface is an ideal playground for controlled studies on nucleation since the chemical potential of the deposit is given the overpotential $\eta$ multiplied by the elementary charge and the charge number of the ions (11.26). If the electrolyte is of high enough concentration, its resistance is low and the chemical potential can be varied within nanoseconds. It is therefore possible to create a single nucleus by a applying a short pulse of a voltage above threshold, or a small number of nuclei, at will. With a continuing smaller applied overpotential, these nuclei grow. Experiments are frequently performed with the *double pulse technique*, which is illustrated in (Fig. 11.18).

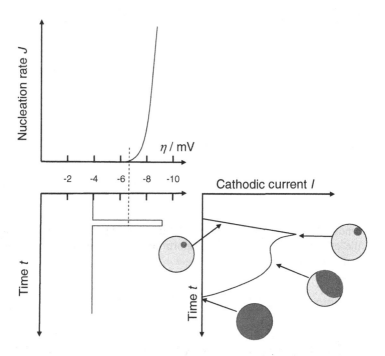

**Fig. 11.18.** Illustration of the double pulse method (see text for discussion, after Budevski et al. [11.14]).

A constant overpotential $\eta$ that is large enough to continue growth of an existing nucleus is superseded by a short overpotential pulse, large enough to initiate nucleation on a defect free surface. A single nucleus can be generated by matching time duration and amplitude of the pulse. Once generated, the nucleus grows under condition of constant supersaturation. The number of added atoms per time is proportional to the length of the perimeter. Assuming circular islands with a radius

$\hat{r}$ (in atom diameters) the rate by which adatoms on the surface are attached to the island perimeter is (cf. 11.84)

$$\frac{dN}{dt} = 2\pi\,\hat{r}(t)\,s\,v(T)\,\Theta_{eq}(\infty)(e^{ze|\eta|/k_B T} - 1)$$

$$\cong 2\pi\,\hat{r}(t)\,s\,v(T)\,\Theta_{eq}(\infty)\,ze|\eta|/k_B T \quad . \tag{11.92}$$

As before, $s$ is the sticking coefficient, $v(T)$ is a jump frequency, $ze$ the charge of the ion as it is discharged on the surface, and $\eta$ the overpotential. The Gibbs-Thomson pressure can be neglected, as we are interested in the growth rate when the island is of macroscopic dimensions (several micrometers). After substituting the radius by the number of atoms in the island $N$

$$\hat{r} = (N/\pi)^{1/2} \tag{11.93}$$

one obtains

$$\frac{dN}{dt} = cN^{1/2} \quad \text{with} \quad c = 2\sqrt{\pi}\,s\,v(T)\,\Theta_{eq}\,ze|\eta|/k_B T \tag{11.94}$$

which after integration yields

$$\frac{dN}{dt} = \frac{1}{2}c(t-t_0) \quad . \tag{11.95}$$

We have neglected a brief induction period in the integration. After the induction period, the current $I = ze\,dN/dt$ rises proportional to the time as long as the island growth is unhindered,

$$I = \frac{(ze)^2\sqrt{\pi}\,s\,v(T)\,\Theta_{eq}|\eta|}{k_B T}(t-t_0) \quad . \tag{11.96}$$

In the electrochemical literature, the prefactor in (11.96) is mostly expressed in terms of the *exchange current density* $j_0$ and a *diffusion length* $\lambda_s$ [11.19]. This latter form would be meaningful if the exchange of adatoms on the surface with ions in solution would be rate determining. To be consistent with the basic assumption of our discussion on the nucleation theory we have considered the limit where the exchange with the solution is infinitely fast (constant chemical potential on the surface) and the growth rate is limited by the attachment rate.

The linear rise of the current is shown in the right panel of Fig. 11.18. The rise continues until the boundary of the growing island (colored in dark gray in Fig.11.18) touches the boundary of the crystal face (light gray) on which it is growing. Upon contact with the boundary of the crystal, the current drops sharply first, then more smoothly when a larger portion of the growing island has merged with the boundary of the crystal. Depending on the initial position of the nucleus, the current may increase a little before it eventually turns down to zero when the monolayer island fills the surface completely. Experiments of this kind permit the determination of the growth speed as function of the overpotential. Furthermore, the number of nuclei generated by the pulse can be determined from the number of peaks in the current. An alternative to counting the peaks in the current is to let nucleated cluster grow until they become visible in a microscope. The dependence of the nucleation rate on the overpotential is thereby measured. By plotting the logarithm of the nucleation rate versus the inverse of the overpotential or the square of the overpotential 2D- and 3D nucleation are distinguished. Figure 11.19 shows the nucleation rate on an Ag(100) electrode [11.14, 20]. The linear slope in the $\ln J(|\eta|^{-1})$ -plot proves 2D-nucleation (11.84, 11.90). Assuming square shaped crystals, the step energy per atom $a_\parallel\beta$ is determined to 31 meV (11.44). The critical sizes vary between 28 atoms at $|\eta|^{-1} = 85$ $V^{-1}$ and 70 atoms at $|\eta|^{-1} = 135$ $V^{-1}$.

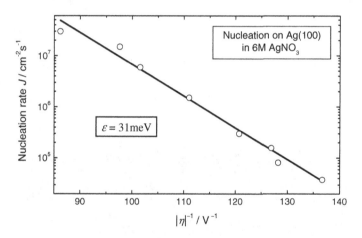

**Fig. 11.19.** Nucleation rate on Ag(100) electrodes in 6 M AgNO$_3$ as function of the inverse of the overpotential $|\eta|$. The linear dependence proves 2D-nucleation (11.84, 11.90)

Nucleation on surfaces is a stochastic process. At small overpotentials when nucleation is a rare event the number of nucleation events $m$ in a given time span obeys Poisson statistics,

$$P(m) = \frac{(JAt)^m}{m!} e^{-JAt} .$$   (11.97)

The stochastic generation of nuclei can be observed directly in the current when a perfect single crystal electrode is held at constant overpotential slightly below the onset of rapid nucleation. Figure 11.20 shows a recorder trace of nucleation events on Ag(100) electrodes at the overpotential of $\eta = -6\,\text{meV}$ where nucleation is a rare event.

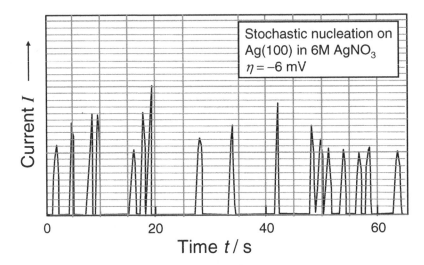

**Fig. 11.20.** Recorder trace of current pulses due to the stochastic nucleation events on Ag(100) in 6 M AgNO$_3$ at $\eta = -6$ mV (after Obretenov et al. [11.21], see also [11.14]).

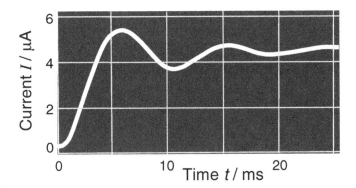

**Fig. 11.21.** Oscillogram of the cathodic current for multilayer growth on a perfectly flat Ag(100) surface in 6 M AgNO$_3$ (after Bostanov et al. [11.19, 22]. At any time, the current is proportional to the total step length on the surface. Comparison to Fig. 11.9 instructs us that the minima mark the completion of monolayers.

Multilayer growth can be investigated nicely by analyzing the current at constant overpotential. Figure 11.21 shows a redrawn oscillogram of the current when an Ag(100) electrode in 6 M AgNO$_3$ is held at constant overpotential $\eta = -14$ mV. At this relatively high overpotential, nuclei are generated progressively and they continue to grow on the surface. At any time, the current is proportional to the total length of steps on the surface. The initial increase is quadratic in time. While a direct Monte-Carlo simulation of the nucleation process is impossible because of the extremely small adatom concentration, one may compare to the step length in Fig. 11.9 obtained by Monte-Carlo simulations that referred to constant flux and to a critical nucleus of size $g^* = 1$. The initial increase in the step length in Figs. 11.9c and the damped oscillations are rather similar to the current shown in Fig. 11.21. According to Fig. 11.9 the total step length should pass through a maximum near 50% coverage of the first monolayer, show a first minimum at monolayer coverage, and undergo a series of oscillations thereafter, a result that presumably transfers to the situation here. The comparison has to be taken with a grain of salt, however, because of the different thermodynamic conditions in the two cases.

## 11.3 Nucleation and Growth in Strained Systems

Experimental studies on nucleation and growth in strained systems have revealed a large variety of phenomena [11.23]. Many of them came rather unexpected. Some have found applications in technology. Again, tunneling microscopy, and lately force microscopy played a key role in the experiments as scanning probe techniques are able to image individual clusters with atomic resolution. Nearly all observations have been made under the thermodynamic condition of constant flux. This boundary condition of constant flux makes it intrinsically difficult to decide whether a particular structure results from kinetics or represents equilibrium. Often enough both aspects are intertwined in a complex way, which has occasionally triggered heated debates in the literature. Out of the many examples, we discuss a few tutorial ones where the role of kinetics and equilibration is believed to be understood.

### 11.3.1 2D-Nucleation on Strained Layers

Under constant flux, the density of nuclei is determined by the ratio of flux and diffusion constant (11.7). Since the activation barrier for diffusion is very sensitive to changes in the interatomic distances, the density of nuclei changes with the strain in substrate surface layer. A tutorial demonstration of the effect of strain was published by Meyer et al. for Ni-layers deposited on Ru(0001) [11.24]. The Ni-layers on Ru(0001) grow in (111) orientation. The nearest neighbor distances of ruthenium and nickel are 2.7 Å and 2.49 Å, respectively. Because of the large lattice mismatch, nickel grows pseudomorphic only in the submonolayer regime.

After completion of the first monolayer, additional Ni-atoms are incorporated into the first layer. This leads to a reconstruction with domains in which most of the Ni-atoms reside alternatively in fcc- or hcp-sites. Upon deposition of further layers, the Ni-film contracts to approximately the bulk lattice constant, which causes a height corrugation pattern. Figure 11.22 shows an STM-image of a nominally 2.5 ML films that displays patches of 1, 2, 3, and 4 monolayer thickness after annealing. The thicknesses are marked by white numbers. Upon further deposition of 0.05 ML of nickel, 2D-nuclei are generated. Their density is the highest on the monolayer film and decreases as the Ni-film grows thicker. This suggests an increase of the diffusion constant with the film thickness. The lines where steps existed on the Ru(0001) substrate (white dashed lines) are decorated with an extra dense row of nuclei. While the experiments of Meyer et al. nicely demonstrate the effect of strain on nucleation, they are not amenable to quantitative analysis because the strain in the film has lateral and vertical components, is not uniform and its magnitude is not known quantitatively.

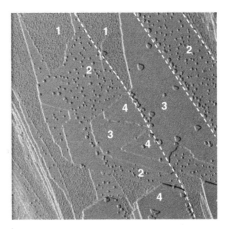

**Fig. 11.22.** STM-image (courtesy of R. J. Behm) of an annealed Ni-film on a Ru(0001) substrate onto which an additional amount of 0.05 monolayers of Ni was deposited at 300 K. Image size is 400 nm × 400 nm. The numbers indicate the thickness of the Ni-film in monolayers. The dashed lines mark some of the steps on the underlying ruthenium substrate, ascending from left to right. The higher nucleation density on thinner films is indicative of a smaller diffusion coefficient (after Meyer et al. [11.24]).

Inspired by nucleation experiments of Ag on Ag(111), on Pt(111) and on a monolayer of Ag on Pt(111) of Brune et al. [11.25] Ratsch et al. calculated the binding energies of Ag-atoms in fcc and bridge sites on uniformly strained Ag-films [11.26]. Their results are displayed in Fig. 11.23. Both binding energies increase linearly with positive strain albeit with a different slope. The increase in the binding energy with larger lateral interatomic distances is a consequence of the non-directional metallic bond that can be understood in terms of the *embedded atom model*. According to this model, a major part of the binding energy of an adatom on a metal surface is the *embedding energy*, which depends on the collective charge density at the position of the atom. For larger lateral distances between the surface atoms, the adatom is drawn closer to the surface so that it is located in a larger electron density and is therefore more strongly bound. Because of the nonlinear functional dependence of the embedding energy with the charge density, the effect strain on the binding energy is larger the more the atom is immersed into the surface; hence, the larger shift in the binding energy for the fcc-site compared to the bridge site. The activation energy for diffusion is roughly equal to the difference of the energy in the bridge site and the fcc-site and increases therefore with positive strain (lower panel of Fig. 11.23).

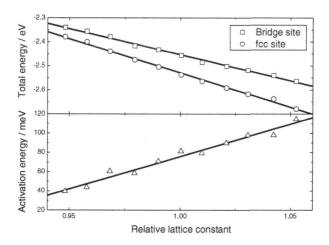

**Fig. 11.23.** *Top panel*: Total energy of silver atoms in the fcc-sites and bridge-sites of an Ag(111) film as function of the surface lattice constant. *Bottom panel*: Differences in the energies of bridge and fcc-sites, the activation energy for diffusion (after C. Ratsch et al. [11.26]).

The dependence of the diffusion barrier on the strain has some interesting consequences on the nucleation process when a lattice mismatched thin film exhibits a regular pattern of dislocations at the interface to the substrate. The STM image in Fig. 11.24a shows the dislocation network of a two-monolayer Ag-film on a Pt(111) surface [11.27]. The dislocations are displayed as black lines. They separate small and medium sized triangular areas in which the silver atoms reside in hcp-sites from larger hexagonal areas where the atoms sit in fcc-sites. Nucleation upon further deposition of silver follows the registry of the pattern (Fig. 11.24b). Nucleation occurs only within the distorted hexagons because atoms have a higher binding energy in the fcc-sites. The dislocation lines act as repulsive barriers for diffusion, which promotes nucleation in the center of the fcc-areas.

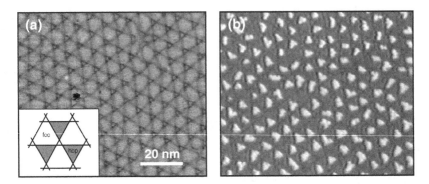

**Fig. 11.24.** (a) Dislocation network of a two-monolayer Ag-film on Pt(111). Within the distorted hexagons, the Ag atoms reside in fcc-sites. (b) Nucleation of the next silver layer follows the dislocation pattern; nucleation occurs in the center of the fcc-areas (courtesy of Harald Brune [11.27]).

## 11.3.2 3D-Nucleation on Strained Layers

A system that has been studied extensively, both for technological relevance and fundamental interest, is the growth of Ge and $Ge_xSi_{1-x}$ alloys on Si(100). The growth is of the Stranski-Krastanov type: the first layers grow pseudomorphic, after that 3D-nucleation takes place. The nuclei display a rather interesting and complex shape, which changes with coverage from square-shaped *huts* over rectangular huts to *domes*. Figure 11.25 shows STM images of these hut and dome clusters. The names "hut" and "dome" are somewhat misleading as they suggest a larger height to width ratio than the clusters actually have. The huts are formed from four {105}-facets. Hence, the angle with the surface is merely 11.3°. The domes have somewhat steeper sides. The {113}-facets form angles of 25.2° with the surface plane.

It is probably as much a mission impossible as a fruitless endeavor to attempt a complete understanding of this nucleation phenomenon. Nevertheless, several qualitative and semi-quantitative models have been proposed. The huts are initially square-shaped, consistent with the fourfold symmetry of the surface and the principle of minimization of the surface energy. The comparatively low surface energy of the {105}-facets results from a specific reconstruction that involves a combination of vacancies and dimer bonds to minimize the number of dangling bonds [11.29]. The rate determining step in hut growth is the nucleation of steps on {105}-facets which grow from bottom to the top. The activation energy for the step-nucleation process increases with the length of the step. This leads to self-limited growth and also to a bifurcation of the aspect ratio since steps nucleate easier on already shorter sides of the huts [11.30, 31].

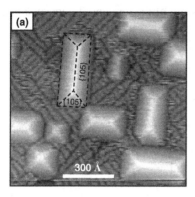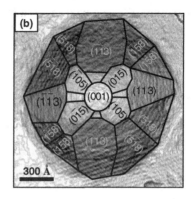

**Fig. 11.25.** (a) *Hut* cluster of Ge formed on a pseudomorphic Ge-layer on Si(100). The facets display {105}-orientation. (b) A multifaceted *dome* cluster (courtesy of Bert Voigtländer, see also [11.28]).

The transition to domes is induced by the elastic deformation energy. The steeper sides of the dome allow for a better relaxation of the strain in the 4% lattice-mismatched Ge-cluster which reduces strain energy. The balance of elastic energy relaxation and the surface energy of three different facets has been modeled by Daruka et al. in a two-dimensional model for the cluster shapes [11.32]. Figure 11.26a shows the cross section through the model cluster displaying a (100)-surface designated as "0" and two more facets, a flatter one and a steeper one designated as "1"and "2", respectively. Depending on the ratio of the projected lengths $L_1/L_0$ and $L_2/L_0$, the cluster assumes one of the 6 forms denoted by the numbers 1, 2, 3, 1', 2', and 3'. Which form is the most stable one depends on the total surface energy and the elastic relaxation energy. The surplus surface energy of the cluster (per length perpendicular to the plane of drawing) is

$$E_\mathrm{s} = 2\gamma_\mathrm{p1}L_1 + 2\gamma_\mathrm{p2}L_2 \tag{11.98}$$

where $\gamma_\mathrm{p1}$ and $\gamma_\mathrm{p2}$ are the projected surface energies as defined in (4.45). For the calculation of the elastic relaxation energy the reader is referred to the original literature [11.32, 33]. The result of the calculation of Daruka et al. can be condensed into a phase diagram spanned by the coordinates $r = \gamma_\mathrm{p1}/\gamma_\mathrm{p2}$ and a reduced volume $V_\mathrm{red}$. The reduced volume is defined as

$$V_\mathrm{red} = V\left[\tau_{xx}^2(1-\nu)\tan^2\theta_1 / 2\pi\mu\right]^2 / \left|\gamma_\mathrm{p}\right|^2 \tag{11.99}$$

where $V$ is the volume per length perpendicular to the plane of drawing, $\tau_{xx}$ is the misfit stress parallel to the surface, $\nu$ the Poisson number, $\mu$ the shear modulus of the substrate and $\theta_1$ is the angle of facet 1 with the (100) plane (Fig. 11.26a). The reduced volume $V_{red}$ can be taken as a measure of the amount of deposited germanium.

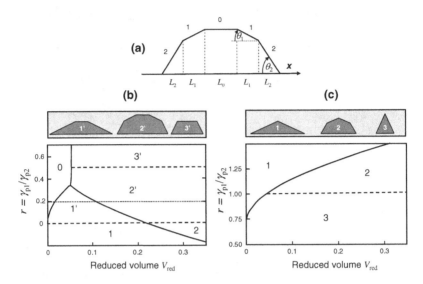

**Fig. 11.26.** (a) Basic shape of a three-facet cluster to model the growth of Ge on Si(100). (b) Phase diagram for $\gamma_{p2} > 0$. (c) Phase diagram for $\gamma_{p2} < 0$. The form of the most stable cluster is indicated by the numbers 1, 2, 3, 1', 2' and 3'. The number 0 stands for a flat layer.

The phase diagrams in Fig. 11.26 show the stability ranges for six different cluster forms and the flat layer designated by the number 0. The panel (b) and (c) are for the cases $\gamma_{p2} > 0$ and $\gamma_{p2} < 0$, respectively. By following the dotted line in panel (b) one comes closest to the experimentally observed shapes. At small total coverages, one is in the range where the pseudomorphic flat layer is stable. For higher coverages, the transition to the form 1' allows for some relaxation of the elastic energy. It is intuitively clear that a better elastic relaxation is achieved when the cluster assumes the form with the steeper facet but at $r = \gamma_{p1}/\gamma_{p2} = 0.2$ the surface energy of facet "2" is much higher so that the balance is in favor of the "hut"-cluster. Only with further increasing volume and correspondingly further increasing total strain energy, a transition to forms with the steeper facets takes place. For larger surface energies $\gamma_{p1}$ of the less steep facet the pseudomorphic layer transforms directly into clusters with the steeper facet.

The qualitative agreement of the prediction of the model along the line $r = \gamma_{p1}/\gamma_{p2} = 0.2$ with the experimental observation lends some credibility to the notion that the observed cluster shapes are equilibrium shapes. However, we should not shut our eyes to the oversimplifications of the model: it is two-dimensional, considers only two facets and neglects elastic anisotropy. On the experimental side, we have no clues as to the surface energies of the various facets. Furthermore, alloy formation between germanium and silicon cannot be excluded.

# 11.4 Nucleation-Free Growth

Crystals may grow continuously without nucleation if monatomic steps persist on the growing surface. There are two possibilities for persisting steps: Surfaces with screw dislocations and vicinal surfaces (see Sect. 1.3.1, Fig. 1.41 and Fig. 1.30). The importance of screw dislocations for the growth of crystals was first discussed by Frank [11.34]. The quantitative theory of spiral growth was developed by Burton, Cabrera and Frank [11.35, 36]. Presently, the interest focuses on vicinal surface because they offer unique possibilities for the engineering of thin film systems that are structurally characterized on an atomic level. Ideally, the growth on vicinal surfaces proceeds as *step flow growth*: the atoms land on the terraces, diffuse to the steps and are incorporated there whereby the steps advance. In the following section, we look into the details of that process.

## 11.4.1 The Steady State Concentration Profile

We consider an ideal vicinal surface with equally spaced, perfectly straight steps separated by terraces of width $L$. If one assumes that the steps stay straight at all times, the problem reduces to a one-dimensional one (Fig. 11.27a). The surface be exposed to a flux $F$ of atoms, which after deposition diffuse to the step edges. The step edges themselves are assumed perfect sinks for atoms approaching from both sides (no Ehrlich-Schwoebel barrier). The concentration of adatoms at the step positions is therefore equal to the equilibrium concentration $\Theta_{eq}$. We also allow for a re-evaporation of the deposited atoms. The rate by which the coverage $\Theta$ changes is

$$\frac{\partial \Theta}{\partial t} = F + a^2 v_d \frac{\partial^2 \Theta}{\partial x^2} - v_e \Theta \qquad (11.100)$$

Here, $F$ is defined as the flux per adatom site, $v_d$ and $v_e$ are the hopping and the evaporation rate, $a$ is the distance between adatom sites, and $x$ is the coordinate along the surface. The origin of $x$ is chosen to be at the center between two steps. We are interested in the steady state solution

$$\frac{\partial \Theta}{\partial t} = 0 = F + a^2 v_d \frac{\partial^2 \Theta}{\partial x^2} - v_e \Theta . \qquad (11.101)$$

The general solution of (11.101) is a linear combination of the hyperbolic cosine and sine functions. For the specific boundary condition of steps being perfect sinks the solution is

$$\Theta(x) = \frac{F}{\alpha^2 v_\mathrm{d}} \frac{\cosh(\alpha L/2a) - \cosh(\alpha x/a)}{\cosh(\alpha L/2a)} + \Theta_\mathrm{eq} \frac{\cosh(\alpha x/a)}{\cosh(\alpha L/2a)} \quad (11.102)$$

with

$$\alpha = \sqrt{v_\mathrm{e}/v_\mathrm{d}} . \quad (11.103)$$

In the limit of zero evaporation ($\alpha \to 0$) the solution is simply a parabola, regardless of the boundary conditions at $x = \pm L/2$. For the boundary condition chosen here, the solution is

$$\Theta = \Theta_\mathrm{eq} + \frac{F}{8a^2 v_\mathrm{d}} (L^2 - 4x^2) . \quad (11.104)$$

The solutions for various ratios of evaporation rate and diffusion rate are shown in Fig. 11.27. For larger evaporation rates, the profile acquires a flat top since the evaporation is the faster the larger the coverage is.

As discussed in Sect. 10.3, transport across a step is frequently hindered by the Ehrlich-Schwoebel barrier. The rate of incorporation of atoms into the growing step from the upper terrace is smaller in that case. The boundary condition that determines the steady state profile becomes asymmetric. Instead of a boundary condition on the concentration, we have the condition that the flux must be continuous. The flux across the step edge given by the hopping rate over the ES-barrier must match the diffusion flux towards the step.

$$\Theta(L/2)v_\mathrm{ES} = a v_\mathrm{d} |\nabla \Theta|_{L/2} \quad (11.105)$$

Here $v_\mathrm{ES}$ is the hopping rate over the ES-barrier that is assumed to exist at $x = +L/2$ and $a$ is the distance between atom sites. The general solution of (11.101) with such boundary conditions becomes rather clumsy. The analytical expression for the coverage is simple if one assumes a high ES-barrier ($v_\mathrm{ES} = 0$) at $x = +L/2$ and zero evaporation rate ($v_\mathrm{d} = 0$),

$$\Theta(x) = \Theta_\mathrm{eq} + \frac{F}{2a^2 v_\mathrm{d}} \left[ L^2 - (x - L/2)^2 \right]. \quad (11.106)$$

This parabola has its apex at $x = +L/2$, so that the concentration gradient and thus the flux towards the descending step is zero.

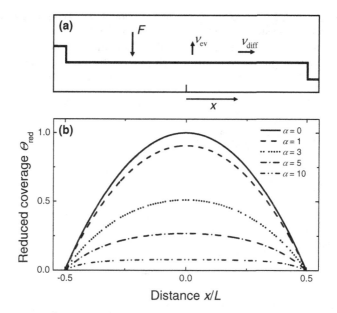

**Fig. 11.27.** (a) The one-dimensional model for step flow growth. (b) Coverage profiles for several $\alpha = (v_e/v_d)^{1/2}$. Coverages are shown in reduced units defined as $\Theta_{red} = 8 v_d \Theta/F$. The equilibrium coverage is assumed to be negligibly small. For small $v_e$ the profile is parabola, for larger evaporation rates the profile acquires a flat top.

## 11.4.2 Step Flow Growth

We calculate the advancement of individual steps due to the atom flow from the upper and lower terrace towards the steps. We denote steps and the upper adjacent terraces by the index $n$ (Fig. 11.28b). The step $n$ advances with the speed

$$\frac{\partial x_n}{\partial t} = a^2 v_d \left( |\nabla \Theta|_{+L_n/2} + |\nabla \Theta|_{-L_{n+1}/2} \right). \tag{11.107}$$

In the absence of an Ehrlich-Schwoebel barrier one obtains from the symmetric concentration profile (11.104)

$$\frac{\partial x_n}{\partial t} = F \left( \frac{L_n}{2} + \frac{L_{n+1}}{2} \right). \tag{11.108}$$

On an ideal vicinal surface, all steps advance with the same speed. Equation (11.108) appears trivial, has however an important and nontrivial implication. If the step $n$ should be lagging behind for any reason, the terrace width $L_{n+1}$ increases and the terrace width $L_n$ decreases by the same amount $\Delta L$. According to (11.108) the speed by which the step $n$ progresses stays the same. Hence, there is nothing to stabilize the initially regular step array. Incidental fluctuations increase in time and the surface becomes macroscopically rough as growth proceeds. The roughness resulting from a fluctuating step density is illustrated in Fig. 11.28a.

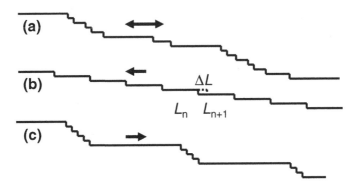

**Fig. 11.28.** Stable and unstable step flow growth. **(a)** Without Ehrlich-Schwoebel barrier, **(b)** with Ehrlich-Schwoebel barrier and **(c)** if the attachment rate from the upper terrace is larger (negative ES-barrier). The arrows mark the direction of the mean adatom current.

In case of a high Ehrlich-Schwoebel steps advance only by attachment of atoms from the lower terrace next to the step edge. The steps advance with the rate that is proportional to the width of the lower adjacent terrace (11.106.

$$\frac{\partial x_n}{\partial t} = a^2 v_d \left| \nabla \Theta \right|_{-L_{n+1}/2} = F L_{n+1} \tag{11.109}$$

If now the step lags behind for some reason the lower terrace increases in width and the speed of the step increases, and vice versa. The Ehrlich-Schwoebel barrier therefore stabilizes equal terrace widths (Fig.1.28b). An initially ideal vicinal surface stays ideal. An initial irregular step array should turn into one with equal step-step distance, according to the one-dimensional model at least.

In case of a negative Ehrlich-Schwoebel barrier or if an attachment barrier exists for atoms from the lower terrace, the attachment is mostly from the upper terrace. Assuming for the purpose of illustration that only atoms from the upper terrace are incorporated into the steps the concentration profile is

$$\Theta(x) = \Theta_{eq} + \frac{F}{2a^2 v_d} \left[ L^2 - (x + L/2)^2 \right]. \tag{11.110}$$

The step $n$ advances then with the speed

$$\frac{\partial x_n}{\partial t} = a^2 v_d \left| \nabla \Theta \right|_{+L_n/2} = FL_n \qquad (11.111)$$

If the step $n$ lags behind, the terrace width $L_n$ now becomes smaller and the speed reduces even further. A regular vicinal surface is therefore unstable with respect to step bunching (Fig.11.28c).

The result of this section can be summarized by considering the direction of the mean current flow of adatoms on the surface. In the absence of an Ehrlich-Schwoebel barrier and an attachment barrier (symmetric concentration profile), an equal number of atoms diffuses to the step edges from either side. The mean adatom current is therefore zero (arrow in Fig. 11.28a). In case of an Ehrlich Schwoebel barrier the mean adatom current is oriented upwards (arrow in Fig. 11.28b). Ideal step flow growth occurs only then. If the Ehrlich-Schwoebel barrier is negative or if an attachment barrier exists, the mean current is oriented downwards leading to step bunching.

### 11.4.3 Meander-Instability of Steps

The simple one-dimensional model of step flow growth takes it for granted that steps move as rigid entities with the same speed everywhere along the step. This is too a simplistic viewpoint. In 1990, Bales and Zangwill predicted that an initially ideal vicinal surface should be subject to a step *meandering instability* [11.37]. Because of that discovery, meandering instabilities in the course of step flow growth are quite generally addressed as *Bales-Zangwill instabilities*. The first experimental evidence for a meandering instability came from Helium-diffraction experiments of Schwenger et al. on copper vicinal surfaces in 1997 [11.38]. Later the same group published STM images of copper surfaces with uniformly meandering steps and performed a detailed analysis of the characteristic wavelength of meandering as function of the temperature [11.39, 40]. Since then several theoretical papers have dealt with the topic using analytical approaches as well as computer simulations.

Fig. 11.29 shows an STM image of a Cu(21 21 23) surface after deposition of 600 monolayers of copper at 313 K. The Cu(21 21 23) surface is vicinal to Cu(111) with a terrace width of 21+2/3 atom rows (Sect. 1.3.1). The initially straight steps (on the scale shown in the figure) show an in-phase meandering with a more or less uniform wavelength that depends on the temperature and the flux of atoms [11.39, 40]. Because of the large amplitude of the meander, the steps display nearly straight segments in directions of dense atom packing at 60° off the original horizontal step orientation. Some segments of the steps resemble therefore the island equilibrium shapes on Cu(111) surfaces (Sect. 4.3.3).

**Fig: 11.29**. STM image (400 nm×400 nm) of a Cu(21 21 23) surface after deposition of 600 monolayers of copper at 313 K (courtesy of M. Giesen, from the thesis of G. Schulze Icking-Konert [11.45]).

The development of a quantitative theory of the meandering instability is still under debate [11.41-44]. With the concepts developed earlier in this volume, it is nevertheless straightforward to elucidate the origin of the instability. The considerations below are equivalent to a *linear stability analysis*. Let us assume that steps on a vicinal surface have a sinusoidal undulation (Fig. 11.30) and ask the question under what circumstances the amplitude of the undulation grows. For simplicity, we assume a perfectly blocking Ehrlich-Schwoebel barrier so that the diffusion current is only towards the ascending step (block-arrows in Fig. 11.30a). The dashed lines connecting the two steps divide the terrace between the two steps into regions of equal terrace area. The vertical solid lines divide the step lengths into equal segments. Because of the difference in length $\Delta$ between the convex and the concave parts of the step, the convex part (denoted as #1 in Fig. 11.30a) is fed by a larger adatom current than the concave part (denoted as #2). The amplitude of the undulation must therefore grow in time. This is the origin of the instability. Against that growth works the diffusion process that transports atoms from regions of high chemical potential (convex curvature) to regions of low chemical potential (concave curvature).

In order to cast these qualitative statements into a quantitative model we describe the undulation as

$$x(y,t) = x_0(t)\sin qy \ . \tag{11.112}$$

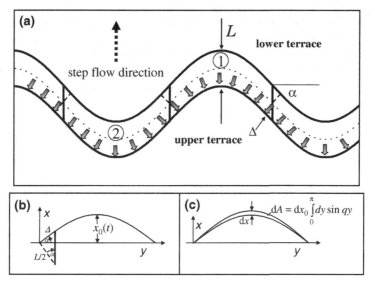

**Fig. 11.30.** (a) Illustration of the diffusion flux on a vicinal surface with sinusoidal steps in presence of an Ehrlich-Schwoebel barrier. (b), (c) Visual aids for the derivation of (11.112).

The number of atoms per time that arrive in section #1 exceeds the number of atoms arriving in section #2 by the amount

$$dN = dt \, FL2\Delta / \Omega_s \tag{11.113}$$

where $F$ is the flux of deposited atoms per site, $L$ the distance between the steps, and $\Omega_s$ the area per adatom. From Figs. 11.30b and 11.30c we take

$$2\Delta = L x_0(t) q \quad \text{and} \quad dx_0 = \frac{q}{2} \Omega_s dN \,, \tag{11.114}$$

so that

$$\frac{dx_0}{dt} = \frac{1}{2} x_0 FL^2 q^2 - \frac{1}{\tau(q)} x_0 \tag{11.115}$$

The second term in (11.115) describes the decay of the profile by diffusion (Sect. 10.2.3). The $q$-dependence of $\tau(q)$ obeys a power law of the type

$$\tau^{-1}(q) = c_\alpha(T) q^\alpha \,. \tag{11.116}$$

The exponent $\alpha$ depends on the type of diffusion process that dominates the profile decay. For terrace diffusion with the detachment from the step being rate determining finally, the exponent is $\alpha = 2$ (cf. Sect. 10.2.2, eq. 10.42). For profile decay via terrace diffusion with terrace diffusion being rate determining, the exponent $\alpha$ is $\alpha = 3$. This case was (implicitly) discussed as case IV in Sect. 10.5.2 in the context of step fluctuation. For diffusion along the step edge one has $\alpha = 4$ (10.47). For the case $\alpha = 2$ one has the stability criterion for the flux

$$ F < \frac{1}{2c_\alpha(T)L^2} \, . \tag{11.117} $$

If the flux is larger, the amplitude of the undulation grows exponentially with a rate that is proportional to $q^2$.

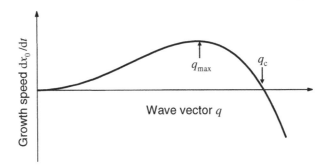

**Fig. 11.31.** Growth speed of a uniform sinusoidal deformation of the steps on a vicinal surface as function of the wave vector $q$ when diffusion along steps dominates the equilibration process ($\alpha = 4$).

In the other two cases, the growth rate is always positive for small $q$ and passes through a maximum at $q_{max}$. The steps are stable only for wave vectors beyond a critical wave vector $q_c$. Figure 11.31 shows the growth rate vs. $q$ for the case $\alpha = 4$. For large exposures, the fastest growing mode wins over all other modes and dominates the morphology. The observed undulations such as shown in Fig. 11.29 reveal a defined wave pattern.

The wave vector of maximum growth speed and the critical wave vector are

$$ q_{max}^{\alpha-2} = \frac{FL^2}{\alpha c_\alpha(T)}, \quad \alpha = 3,4 \tag{11.118} $$

$$ q_c^{\alpha-2} = \frac{\alpha}{2} q_{max}^{\alpha-2}, \quad \alpha = 3,4 \, . \tag{11.119} $$

For diffusion along steps, we can express the time constant and the coefficient $c_\alpha(T)$ in terms of the one-dimensional transport coefficient $L_T^{(st)}$ and the step stiffness $\tilde{\beta}$ (10.47).

$$c_{\alpha=4}(T) = L_T^{(st)} \, \Omega_s^2 \, \tilde{\beta} \tag{11.120}$$

After converting the transport coefficient into the product of the mean diffusion coefficient $D_{eff}^{(st)}$ and the concentration of diffusing adatoms $\Theta_{eq}^{(st)}$, one obtains the wavelength of the mode with the fastest growth as

$$\lambda_{max} = 4\pi \left( \frac{D_{eff}^{(st)} \Theta_{eq}^{(st)} \tilde{\beta} \Omega_s^2}{FL^2 a_\| k_B T} \right)^{1/2} \tag{11.121}$$

The considerations above predict a meandering instability for arbitrary small fluxes at any temperature. To become observable in real systems the surface must be free of defects on a length scale that exceeds $\lambda_{max}$. For the instability shown in Fig. 11.29, this would imply less than one defect per $10^5$ surface atoms, which is rarely achieved. Quantitative observations on $\lambda_{max}$ versus temperature and flux where therefore performed at lower temperatures in the range $4 < \lambda_{max} < 40$ nm [11.40]. These quantitative studies performed on Cu(1 1 n) vicinal surfaces have not confirmed the predicted scaling of $\lambda_{max}$ with the flux as $F^{-1/2}$ (11.121). The experimental data would rather be consistent with $\lambda_{max} \propto F^{-1/4}$. This indicates that the specific mechanism outlined above is probably not operative on copper surface. Recent theories have attributed the meandering instability on Cu-vicinals to a process of one-dimensional nucleation at the step edges [11.39, 40, 44, 46].

# Appendix: Surface Brillouin Zones

## (0001) Surface Brillouin Zone for hcp

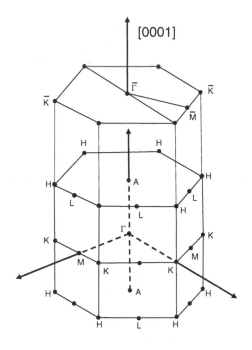

## Surface Brillouin Zones for fcc

**(001)**          **(110)**          **(111)**

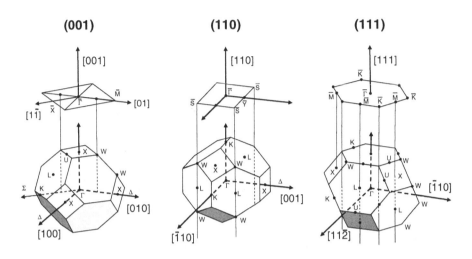

## Surface Brillouin Zones for bcc

**(001)**          **(110)**          **(111)**

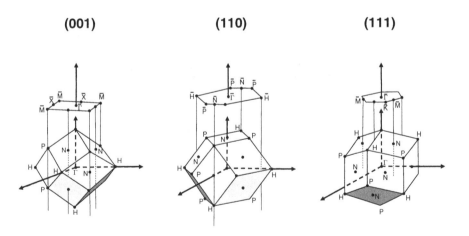

# References

## Chapter 1

1.1 P. R. Watson, M. A. v. Hove, K. Hermann, National Institute of Standards and Technology Database 42, 2001.

1.2 H. E. Farnsworth, R. E. Schlier, T. H. George, R. M. Burger: J. Appl .Phys. **29** (1958) 1150.

1.3 J. J. Lander, J. Morrison, F. Unterwald: Rev. Sci. Instrum. **33** (1962) 782.

1.4 K. Heinz: Reports Progress in Physics **58** (1995) 637.

1.5 J. B. Pendry, *Low Energy Electron Diffraction*, Academic Press, New York 1974.

1.6 M. A. v. Hove, W. H. Weinberg, C. M. Chan, *Low Energy Electron Diffraction: Experiment, Theory and Surface Structure Determination, Vol. 6*, Springer, Berlin, Heildelberg, New York 1986.

1.7 H. Ibach, H. Lüth, *Solid-State Physics - An Introduction to Principles of Materials Science*, Springer-Verlag, Heidelberg 2003.

1.8 R. Feidenhans'l: Surf. Sci. Rep. **10** (1989) 105.

1.9 G. Prévot, B. Croset, Y. Girard, A. Coati, Y. Garreau, M. Hohage, L. D. Sun, P. Zeppenfeld: Surf. Sci. **549** (2004) 52.

1.10 B. Croset, Y. Girard, G. Prévot, M. Sotto, Y. Garreau, R. Pinchaux, M. Sauvage-Simkin: Phys. Rev. Lett. **88** (2002) 056103.

1.11 J. Wang, B. M. Ocko, A. J. Davenport, H. S. Isaacs: Phys. Rev. B **46** (1992) 10321.

1.12 E. A. Wood: J. Appl. Phys. **35** (1964) 1306.

1.13 V. Fiorentini, M. Methfessel, M. Scheffler: Phys. Rev. Lett. **71** (1993) 1051.

1.14 C. E. Bach, M. Giesen, H. Ibach, T. L. Einstein: Phys. Rev. Lett. **78** (1997) 4225.

1.15 A. Filippetti, V. Fiorentini: Surf. Sci. **377-379** (1997) 112.

1.16 E. Santos, W. Schmickler: Chem. Phys. Lett. **400** (2004) 26.

1.17 H. Ibach: Surf. Sci. Rep. **35** (1999) 71.

1.18 E. Lang, K. Müller, K. Heinz, M. A. vanHove, R. J. Koestner, G. A. Somorjai: Surf. Sci. **127** (1993) 347.

1.19 G. Ritz, M. Schmid, P. Varga, A. Borg, M. Ronning: Phys. Rev. B **56** (1997) 10518.

1.20 O. Schaff, A. K. Schmid, N. C. Bartelt, J. d. l. Figuera, R. Q. Hwang: Mat. Sci. Engin. A **319-321** (2001) 914.

1.21 S. Narasimhan, D. Vanderbilt: Phys. Rev. Lett. **69** (1992) 1564.

1.22 D. L. Abernathy, S. G. J. Mochrie, D. M. Zehner, G. Grübel, D. Gibb: Phys. Rev. B **45** (1992) 9272.

1.23 S. G. J. Mochrie, D. M. Zehner, B. M. Ocko, D. Gibbs: Phys. Rev. Lett. **64** (1990) 2925.

1.24 B. M. Ocko, D. Gibbs, K. G. Huang, D. M. Zehner, S. G. J. Mochrie: Phys. Rev. B **44** (1991) 6429.

1.25    C.-M. Chan, M. A. vanHove: Surf. Sci. **171** (1986) 226.

1.26    V. Blum, L. Hammer, K. Heinz, C. Franchini, J. Redinger, K. Swamy, C. Deisl, E. Bertel: Phys. Rev. B **65** (2002) 165408.

1.27    E. Vlieg, I. K. Robinson, K. Kern: Surf. Sci. **233** (1990) 248.

1.28    H. Landskron, N. Bickel, K. Heinz, G. Schmidtlein, K. Müller: J. Phys. C. Cond. Mat. **1** (1989) 1.

1.29    G. Schmidt, H. Zagel, H. Landskron, K. Heinz, K.Müller, J. B. Pendry: Surf. Sci. **271** (1992) 416.

1.30    J. J. Lander, G. W. Gobeli, J. Morrison: J. Appl .Phys. **34** (1963) 2298.

1.31    F. G. Allen, G. W. Gobeli: Phys. Rev. **127** (1962) 150.

1.32    H. Ibach: Phys. Rev. Lett. **27** (1971) 253.

1.33    G. Chiarotti, S. Nannarone, R. Pastore, P. Chiaradia: Phys. Rev. B **4** (1971) 3398.

1.34    H. Froitzheim, H. Ibach, D. L. Mills: Phys. Rev. B **11** (1975) 4980.

1.35    H. Sakama, A. Kawazu, K. Ueda: Phys. Rev. B **34** (1986) 1367.

1.36    K. Takayanagi, Y. Tanishiro, S. Takahashi, M. Takahashi: Surf. Sci. **164** (1985) 367.

1.37    S. Y. Tong, H. Huang, C. M. Wei, W. E. Packard, F. K. Men, G. Glander, M. B. Webb: J. Vac. Sci. Technol. A **6** (1988) 615.

1.38    S. Y. Tong, H. Huang, C. M. Wei: Chem. Phys. of Sol. Surf. VIII **22** (1990) 395.

1.39    D. J. Chadi: Phys. Rev. Lett. **43** (1979) 43.

1.40    H. Over, J. Wasserfall, W. Ranke, C. Ambiatello, R. Sawitzki, D. Wolf, W. Moritz: Phys. Rev. B **55** (1997) 4731.

1.41    K. Heinz, G. Schmidt, L. Hammer, K. Müller: Phys. Rev. B **32** (1985) 6214.

1.42    H. P. Bonzel, C. R. Helms, S. Kelemen: Phys. Rev. Lett. **35** (1975) 1237.

1.43    S. Jakubith, H. H. Rotermund, W. Engel, A. von Oertzen, G. Ertl: Phys. Rev. Lett. **65** (1990) 3013.

1.44    H. H. Rotermund, S. Jakubith, A. von Oertzen, G. Ertl: Phys. Rev. Lett. **66** (1991) 3083.

1.45    H. S. Taylor: Proc. Roy. Soc. (London) Ser. A **108** (1925) 105.

1.46    W. K. Burton, N. Cabrera, F. C. Frank: Phil. Trans. Roy. Soc. London Ser. A **243** (1951) 299.

1.47    S. V. Dijken, H. J. W. Zandvliet, B. Poelsema: Surface Review and Letters **5** (1998) 15.

1.48    Z. Zhang, W. Fang, M. G. Lagally: Surf. Rev. Lett. **3** (1996) 1449.

1.49    B. S. Swartzentruber, Y.-W. Mo, R. Kariotis, M. G. Lagally, M. B. Webb: Phys. Rev. Lett. **65** (1990) 1913.

1.50    B. N. J. Persson: Surf. Sci. Rep. **15** (1992) 1.

1.51    P. Bak: Rep. Prog. Phys. **45** (1982) 587.

1.52    H. Brune, H. Röder, C. Boragno, K. Kern: Phys. Rev. B **49** (1994) 2997.

1.53    J. d. l. Figuera, K. Pohl, O. R. d. l. Fuente, A. K. Schmid, N. C. Bartelt, C. B. Carter, R. Q. Hwang: Phys. Rev. Lett. **86** (2000) 3819.

1.54    J. F. Wolf, H. Ibach: Appl. Phys. A **52** (1991) 218.

1.55    B. Müller, B. Fischer, L. Nedelmann, A. Fricke, K. Kern: Phys. Rev. Lett. **76** (1996) 2358.

1.56    C. Steimer, M. Giesen, L. Verheij, H. Ibach: Phys. Rev. B **64** (2001) 085416.

1.57    P. Stoltze: J. Phys.: Cond. Mat. **6** (1994) 9495.

1.58    S. M. Foiles, M. I. Baskes, M. S. Daw: Phys. Rev. B **33** (1986) 7983.

1.59    S. Dieluweit, H. Ibach, M. Giesen, T. L. Einstein: Phys. Rev. B **67** (2003) R 121410.

1.60    M. Giesen, S. Dieluweit: Journal of Molecular Catalysis A **216** (2004) 263.

1.61    G. A. Attard, C. Harris, E. Herrero, J. Feliu: Farady Discuss. **121** (2002) 253.

1.62    T. Flores, S. Junghans, M. Wuttig: Surf. Sci. **371** (1997) 14.

1.63    R. v. Gastel, E. Somfai, S. B. v. Albada, W. v. Saarloos, J. W. M. Frenken: Surf. Sci. **521** (2002) 10.

1.64    M. Henzler: Surf. Rev. Lett. **4** (1997) 489.

1.65    E. W. Müller: Z. Phys. **131** (1951) 136.

1.66    G. L. Kellogg: Surf. Sci. Rep. **21** (1994) 1.

1.67    G. Binnig, H. Rohrer, C. Gerber: Appl. Phys. Lett. **40** (1982) 178.

1.68    J. Frohn, J. F. Wolf, K. Besocke, M. Teske: Rev. Sci. Instrum. **60** (1989) 1200.

1.69    C. J. Chen, *Introduction to Scanning Tunneling Microscopy, Vol. Oxford University Press*, New York, Oxford 1993.

1.70    J. Li, R. Berndt, W.-D. Schneider: Phys. Rev. Lett. **76** (1996) 1888.

1.71    B. F. U. Kürpick: Surf. Sci. **460** (1999).

1.72    L. Bartels, G. Mayer, K.-H. Rieder: J. Vac. Sci. Technol. A **16** (1998) 1047.

1.73    G. Meyer, L. Bartels, S. Zöphel, K. H. Rieder: Appl. Phys. A **68** (1999) 125.

1.74    L. Gross, F. Moresco, L. Savio, A. Gourdon, C. Joachim, K.-H. Rieder: Phys. Rev. Lett. **93** (2004) 056103.

1.75    F. Moresco, G. Meyer, K.-H. Rieder, H. Tang, A. Gourdon, C. Joachim: Phys. Rev. Lett. **86** (2000) 672.

1.76    M. F. Crommie, C. P. Lutz, D. M. Eigler: Science **262** (1993) 218.

1.77    E. J. Heller, M. F. Crommie, C. P. Lutz, D. M. Eigler: Nature **369** (1994) 464.

1.78    O. M. Magnussen, B. M. Ocko, R. R. Adzic, J. X. Wang: Phys. Rev. B **51** (1995) 5510.

1.79    H. Ibach, S. Lehwald: Surf. Sci. **91** (1980) 187.

1.80    S. Meng, L. F. Xu, E. G. Wang, S. Gao: Phys. Rev. Lett. **89** (2002) 176104.

1.81    H. Ogasawara, B. Brena, D. Nordlund, M. Nyberg, A. Pelmenschikov, L. G. M. Pettersson, A. Nilsson: Phys. Rev. Lett. **89** (2002) 276102.

1.82    A. Glebov, A. P. Graham, A. Menzel, J. P. Toennies: J. Chem. Phys. **106** (1997) 9382.

1.83    S. Haq, J. Harnett, A. Hodgson: Surf. Sci. **505** (2002) 171.

1.84    G. Held, D. Menzel: Surf. Sci. **316** (1994) 92.

1.85    J. Weissenrieder, A. Mikkelsen, J. N. Andersen, P. J. Feibelman, G. Held: Phys. Rev. Lett. **93** (2004) 196102.

1.86    R. Guidelli, W. Schmickler: Electrochim. Acta **45** (2000) 2317.

# Chapter 2

2.1    M. Grundner, H. Jacob: Appl. Phys. A **39** (1986) 73.

2.2    H. Ubara, T. Imura, A. Hiraki: Solid State Commun. **50** (1984) 673.

2.3    J. M. C. Thornton, R. H. Williams: Semicond. Sci. Technol **4** (.1989) 847.

2.4    D. Gräf, M. Grundner, R. Schulz: J. Vac. Sci. Technol. A **7** (1989) 808.

2.5    G. W. Trucks, K. Raghavachari, G. S. Higashi, Y. J. Chabal: Phys. Rev. Lett. **65** (1990) 504.

2.6    J. M. C. Thornton, R. H. Williams: Physica Scripta **41** (1990) 1047.
2.7    L. Kleinman: Phys. Rev. B **3** (1971) 2982.
2.8    J. C. Ashley, R. Ritchie: Phys. Stat. Sol. **62** (1974) 253.
2.9    H. Ibach, in H. Ibach (Ed.): *Electron Spectroscopy for Surface Analysis*, Springer, Berlin, Heidelberg, New York 1977, p. 1.
2.10   H. Gant, W. Mönch: Surf. Sci. **105** (1981) 217.
2.11   M. Klasson, A. Berndtsson, J. Hedman, R. Nilsson, R. Nyholm, C. Nordling: J. Electr. Spectr. Rel. Phen. **3** (1974) 427.
2.12   J. C. Tracy: J. Vac. Sci. Technol. **11** (1974) 280.
2.13   K. Brüggemann: private communication (2004).
2.14   E. Taglauer: Appl. Phys. A **51** (1990) 238.
2.15   H. F. Winters, P. Sigmund: J. Appl. Phys. **45** (1974) 4760.
2.16   K. G. Tschersich: J. Appl. Phys. **87** (2000) 2565.
2.17   K. G. Tschersich, V. v. Bonin: J. Appl. Phys. **84** (1998) 4065.

## Chapter 3

3.1    D. Marx, M. E. Tuckerman, M. Parrinello: J. Phys.:Cond. Mat. **12** (2000) A153.
3.2    W. Schmickler, *Interfacial Electrochemistry*, Oxford University Press, Oxford 1995.
3.3    P. Hohenberg, W. Kohn: Phys. Rev. **136** (1964) B864.
3.4    W. Kohn, L. J. Sham: Phys. Rev. **140** (1965) A1133.
3.5    N. D. Lang, W. Kohn: Phys. Rev. B **1** (1970) 4555.
3.6    N. D. Lang, W. Kohn: Phys. Rev. B **3** (1971) 1215.
3.7    J. E. Müller, H. Ibach: (to be published).
3.8    F. G. Allen, G. W. Gobeli: Phys. Rev. **127** (1962) 150.
3.9    H. Ibach, H. Lüth, *Solid-State Physics - An Introduction to Principles of Materials Science*, Springer-Verlag, Heidelberg 2003.
3.10   G. Gouy: J. Phys. (Paris) **9** (1910) 457.
3.11   D. L. Chapman: Philos. Mag. **25** (1913) 475.
3.12   C. W. Leitz, M. T. Currie, M. L. Lee, Z.-Y. Cheng, D. A. Antoniadis, E. A. Fitzgerald: J. Appl .Phys. **92** (2001) 3745.
3.13   P. Drescher, e. al.: J. Appl. Phys. A **63** (1996) 203.
3.14   D. Sander: Rep. Prog. Phys. **62** (1999) 809.
3.15   R. Mahesh, D. Sander, S. M. Zharkov, J. Kirschner: Phys. Rev. B **68** (2003) 045416.
3.16   C.-T. Wang, *Applied Elasticity*, McGraw-Hill, New York 1953, p. 276.
3.17   H. Ibach: Surf. Sci. Rep. **29** (1997) 195.
3.18   G. G. Stoney: Proc. R. Soc. London Ser. A **82** (1909) 172.
3.19   K. Dahmen, H. Ibach: Surf. Sci. **446** (2000) 161.
3.20   K. Dahmen, H. Ibach, D. Sander: J. Magn. Mag. Mat. **231** (2001) 74.
3.21   V. S. Stepanyuk, D. I. Bazhanov, A. N. Baranov, W. Hergert, P. H. Dederichs, J. Kirschner: Phys. Rev. B **62** (2000) 15398.
3.22   O. V. Lysenko, V. S. Stepanyuk, W. Hergert, J. Kirschner: Phys. Rev. Lett. **89** (2002) 126102.

3.23   L. D. Landau, E. M. Lifshitz, *Lehrbuch der Theoretischen Physik, Bd. VII Elasti-zitätstheorie*, Akademie, Berlin 1991.
3.24   V. I. Marchenko, A. Y. Parshin: Sov. Phys. JETP **52** (1980) 129.
3.25   L. E. Shilkrot, D. J. Srolovitz: Phys. Rev. B **53** (1996) 11120.
3.26   G. Prévot, C. Cohen, D. Schmaus, P. Hecquet, B. Salanon: Surf. Sci. **506** (2002) 272.
3.27   G. Prevot, B. Croset: Physical Review Letters **92** (2004) 256104.
3.28   D. Sander, H. Ibach, in H. P. Bonzel (Ed.): *Landolt-Börnstein New Series, Vol. III/42A2*, Springer, Berlin 2002.
3.29   O. L. Alerhand, D. Vanderbilt, R. D. Meade, J. D. Joannopoulos: Phys. Rev. Lett. **61** (1988) 1973.
3.30   K. Kern, H. Niehus, A. Schatz, P. Zeppenfeld, J. Goerge, G. Comsa: Phys. Rev. Lett. **67** (1991) 855.
3.31   P. Zeppenfeld, M. A. Krzyzowski, C. Romainczyk, R. David, G. Comsa, H. Röder, K. Bromann, H. Brune, K. Kern: Surf. Sci. **342** (1995) L1131.
3.32   G. Boishin, L. D. Sun, M. Hohage, P. Zeppenfeld: Surf. Sci. **512** (2002) 185.
3.33   F. M. Leibsle, A. W. Robinson: Phys. Rev. B **47** (1993) 15865.
3.34   B. Croset, Y. Girard, G. Prévot, M. Sotto, Y. Garreau, R. Pinchaux, M. Sauvage-Simkin: Phys. Rev. Lett. **88** (2002) 056103.
3.35   R. v. Gastel, R. Plass, N. C. Bartelt, G. L. Kellogg: Phys. Rev. Lett. **91** (2003) 055503.
3.36   S. Narasimhan, D. Vanderbilt: Phys. Rev. Lett. **69** (1992) 1564.
3.37   J. Tersoff, C. Teichert, M. G. Lagally: Phys. Rev. Lett. **76** (1996) 1675.
3.38   C. Teichert, J. C. Bean, M. G. Lagally: Appl. Phys. A **67** (1998) 675.
3.39   C. Teichert: Appl. Phys. A **76** (2003) 653.
3.40   C. Bombis, M. Moiseeva, H. Ibach: Phys. Rev. B **72** (2005) 245408.

## Chapter 4

4.1    J. W. Gibbs, *On the Equilibrium of Heterogeneous Substances, Vol. 1*, Longmans, Green, New York 1928.
4.2    H. Ibach: Surf. Sci. Rep. **29** (1997) 195.
4.3    D. Sander, H. Ibach, in H. P. Bonzel (Ed.): *Landolt-Börnstein New Series, Vol. III/42A2*, Springer, Berlin 2002.
4.4    C. Herring, in R. Gomer (Ed.): *Structure and Properties of Solid Surfaces*, The University of Chicago Press, Chicago 1953, p. 5.
4.5    R. Shuttleworth: Proc. Phys. Soc. A **63** (1950) 445.
4.6    R. Monnier, J. P. Perdew, D. C. Langreth, J. W. Wilkins: Phys. Rev. B **18** (1978) 656.
4.7    C. Herring, in W. E. Kingston (Ed.): *The Physics of Powder Metallurgy*, McGraw-Hill, New York 1951, p. 143.
4.8    H. Ibach, E. Santos, W. Schmickler: Surf. Sci. **540** (2003) 504.
4.9    M. I. Haftel, M. Rosen: Surf. Sci. **523** (2003) 118.
4.10   D. M. Kolb: Prog. Surf. Sci. **51** (1996) 109.
4.11   C. E. Bach, M. Giesen, H. Ibach, T. L. Einstein: Phys. Rev. Lett. **78** (1997) 4225.
4.12   K. P. Bohnen, D. M. Kolb: Surf. Sci. **407** (1998) L629.

4.13   Z. Shi, J. Lipkowski, S. Mirwald, B. Pettinger: J. Electroanal. Chem. **396** (1995) 115.
4.14   H. Ibach: Surf. Sci. **556** (2004) 71.
4.15   N. D. Lang, W. Kohn: Phys. Rev. B **3** (1971) 1215.
4.16   H. Ibach: Electrochim. Acta **45** (1999) 575.
4.17   N. D. Lang, W. Kohn: Phys. Rev. B **1** (1970) 4555.
4.18   E. Bauer: Z. Krist. **110** (1958) 372.
4.19   C. Herring: Phys. Rev. **82** (1951) 87.
4.20   H. P. Bonzel: Phys. Rep. **385** (2003) 1.
4.21   M. Giesen: Prog. Surf. Sci. **68** (2001) 1.
4.22   H.-C. Jeong, E. D. Williams: Surf. Sci. Rep. **34** (1999) 171.
4.23   V. L. Pokrovsky, A. L. Talapov: Phys. Rev. Lett. **42** (1979) 65.
4.24   E. E. Gruber, W. W. Mullins: J. Phys. Chem. Solids **28** (1967) 875.
4.25   Z. Wang, P. Wynblatt: Surf. Sci. **398** (1998) 259.
4.26   A. Bartolini, F. Ercolessi, E. Tosatti: Phys. Rev. Lett. **63** (1989) 872.
4.27   M. Giesen, C. Steimer, H. Ibach: Surf. Sci. **471** (2001) 80.
4.28   J. Ikonomov, K. Starbova, M. Giesen: to be published.
4.29   R. Smoluchowski: Phys. Rev. **60** (1941) 661.
4.30   K. Besocke, B. Krahl-Urban, H. Wagner: Surf. Sci. **68** (1977) 39.
4.31   J. Lecoeur, J. Andro, R. Parsons: Surf. Sci. **114** (1982) 320.
4.32   A. Hamelin, L. Stoicoveciu, L. Doubova, S. Trasatti: Surf. Sci. **201** (1988) L498.
4.33   H. Ibach, W. Schmickler: Phys. Rev. Lett. **91** (2003) 016106.
4.34   H. Ibach, M. Giesen, W. Schmickler: J. Electroanal. Chem. **544** (2003) 13.
4.35   S. Baier, H. Ibach, M. Giesen: Surf. Sci. **573** (2004) 17.
4.36   M. Giesen-Seibert, F. Schmitz, R. Jentjens, H. Ibach: Surf. Sci. **329** (1995) 47.
4.37   J. M. Bermond, J. J. Metois, J. C. Heyraud, C. Alfonso: Surf. Sci. **331** (1995) 855.
4.38   C. Steimer, M. Giesen, L. Verheij, H. Ibach: Phys. Rev. B **64** (2001) 085416.
4.39   S. V. Khare, S. Kodambaka, D. D. Johnson, I. Petrov, E. Greene: Surf. Sci. **522** (2002) 75.
4.40   C. Steimer: Thesis D82 RWTH Aachen (2001).
4.41   C. Bombis, H. Ibach: Surf. Sci. **564** (2004) 201.
4.42   S. Dieluweit, H. Ibach, M. Giesen: Faraday Discussions. **121** (2002) 27.
4.43   S. Kodambaka, V. Petrova, S. V. Khare, D. D. Johnson, I. Petrov, J. E. Greene: Phys. Rev. Lett. **88** (2002) 146101.

# Chapter 5

5.1    *CRC Handbook of Chemistry and Physics, Vol. 84*, CRC Press, Cleveland, Ohio 2003.
5.2    B. N. J. Persson: Surf. Sci. Rep. **15** (1992) 1.
5.3    B. S. Swartzentruber, Y.-W. Mo, R. Kariotis, M. G. Lagally, M. B. Webb: Phys. Rev. Lett. **65** (1990) 1913.
5.4    E. E. Gruber, W. W. Mullins: J. Phys. Chem. Solids **28** (1967) 875.
5.5    C. Jayaprakash, C. Rottman, W. F. Saam: Phys. Rev. B **30** (1984) 6549.
5.6    M. Giesen: Prog. Surf. Sci. **68** (2001) 1.
5.7    T. L. Einstein, O. Pierre-Louis: Surf. Sci. **424** (1999) L299.

5.8    B. Joós, T. L. Einstein, N. C. Bartelt: Phys. Rev. B **43** (1991) 8153.
5.9    M. Giesen: Surf. Sci. **370** (1997) 55.
5.10   M. Giesen, T. L. Einstein: Surf. Sci. **449** (2000) 191.
5.11   H. L. Richards, S. D. Cohen, T. L. Einstein, M. Giesen: Surf. Sci. **453** (2000) 59.
5.12   J. Villain, D. R. Grempel, J. Lapujoulade: J. Phys. F **15** (1985) 809.
5.13   H.-C. Jeong, E. D. Williams: Surf. Sci. Rep. **34** (1999) 171.
5.14   A. Kara, S. Durukanoglu, T. S. Rahman: Phys. Rev. B **53** (1996) 15489.
5.15   A. Kara, S. Durukanoglu, T. S. Rahman: J. Chem. Phys **106** (1997) 2031.
5.16   S. Durukanoglu, T. S. Rahman: Surf. Sci. **409** (1998) 395.
5.17   A. Kara, P. Staikov, T. S. Rahman: Phys. Rev. B **61** (2000) 5714.
5.18   S. Durukanolu, A. Kara, T. S. Rahman: Phys. Rev. B **67** (2003) 235405.
5.19   G. Schulze Icking-Konert, M. Giesen, H. Ibach: Phys. Rev. Lett. **83** (1999) 3880.
5.20   H. P. Bonzel, A. Emundts: Phys. Rev. Lett. **84** (2000) 5804.
5.21   H. J. W. Zandvliet, O. Gurlu, B. Poelsema: Phys. Rev. B **64** (2001) 073402.
5.22   R. K. P. Zia, J. E. Avron: Phys. Rev. B **25** (1982) 2042.
5.23   R. K. P. Zia: J. Stat. Phys. **45** (1986) 801.
5.24   M. Giesen, C. Steimer, H. Ibach: Surf. Sci. **471** (2001) 80.
5.25   M. Giesen-Seibert, F. Schmitz, R. Jentjens, H. Ibach: Surf. Sci. **329** (1995) 47.
5.26   W. Feller, *An Introduction to Probability Theory and Its Applications, Vol. I*, Wiley, New York 1968.
5.27   R. Berndt, J. P. Toennies, C. H. Wöll: J. Electr. Spectr. Rel. Phen. **44** (1987) 65.
5.28   A. Grossmann, W. Erley, H. Ibach: Phys. Rev. Lett. **71** (1993) 2078.
5.29   A. Grossmann, W. Erley, H. Ibach: Appl. Phys. A **57** (1993) 499.
5.30   T. L. Hill: J. Chem. Phys. **14** (1946) 441.
5.31   J. J. Doll, W. A. Steele: Surf. Sci. **44** (1974) 449.
5.32   L. C. Isett, J. M. Blakely: Surf. Sci. **47** (1975) 645.
5.33   S. Patrykiejew, S. Sokolowski, K. Binder: Surf. Sci. Rep. **37** (2000) 207.
5.34   R. J. Behm, K. Christmann, G.-. Ertl: Surf. Sci. **99** (1980) 320.
5.35   K. Binder, D. P. Landau: Surf. Sci. **108** (1981) 503.
5.36   J. Wintterlin, J. Trost, S. Renisch, R. Schuster, T. Zambelli, G. Ertl: Surf. Sci. **394** (1997) 159.

## Chapter 6

6.1    R. Miranda, S. Daiser, K. Wandelt, G. Ertl: Surf. Sci. **131** (1983) 61.
6.2    K. Kern, R. David, P. Zeppenfeld, G. Comsa: Surf. Sci. **195** (1988) 353.
6.3    N. D. Lang: Phys. Rev. Lett. **46** (1981) 842.
6.4    J. E. Müller: Phys. Rev. Lett. **65** (1990) 3021.
6.5    B. Hall, D. L. Mills, P. Zeppenfeld, K. Kern, U. Becher, G. Comsa: Phys. Rev. B **40** (1989) 6326.
6.6    J. L. F. Da Silva, C. Stampfl, M. Scheffler: Phys. Rev. B **72** (2005) 075424.
6.7    P. A. Dowben: CRC Crit. Rev. Solid State Mater. Sci. **13** (1987) 191.
6.8    S. Meyer, D. Diesing, A. Wucher: Phys. Rev. Lett. **93** (2004) 137601.
6.9    M. Bonn, A. W. Kleyn, G. J. Kroes: Surf. Sci. **500** (2002) 475.
6.10   A. Gross, *Theoretical Surface Science*, Springer, Berlin, Heidelberg, New York 2003.

6.11   A. R. Miller, *The Adsorption of Gases on Solids*, Cambridge University Press, Cambridge 1949.

6.12   H. Ibach, W. Erley, H. Wagner: Surf. Sci. **92** (1980) 29.

6.13   R. H. Fowler, E. A. Guggenheim, *Statistical Thermodynamics*, Cambridge University Press, Cambridge 1965.

6.14   A. N. Frumkin: Z. Phys. **35** (1926) 792.

6.15   J. Suzanne, J. P. Coulomb, M. Bienfait: Surf. Sci. **44** (1974) 141.

6.16   G. Ehrlich: J. Chem. Phys. **36** (1962) 1499.

6.17   D. A. King: Surf. Sci. **307** (1994) 1.

6.18   A. Stuck, C. E. Wartnaby, Y. Y. Yeo, J. T. Stuckless, N. Al-Sarraf, D. A. King: Surf. Sci. **349** (1996) 229.

6.19   P. Debye, E. Hückel: Z. Physik **24** (1923) 185.

6.20   E. Budevski, G. Staikov, W. J. Lorenz, *Electrochemical Phase Formation and Growth*, VCH, Weinheim, New York 1996.

6.21   C. Sánchez, E. Leiva: J. Electroanal. Chem. **458** (1998) 183.

6.22   C. G. Sánchez, M. G. Del Popolo, E. P. M. Leiva: Surf. Sci. **421** (1999) 59.

6.23   B. G. Bravo, S. L. Michelhaugh, M. P. Soriaga, I. Villegas, D. W. Suggs, J. L. Stickney: J. Phys. Chem. **95** (1995) 5245.

6.24   B. M. Ocko, G. M. Watson, J. Wang: J. Phys. Chem. **98** (1994) 897.

6.25   O. M. Magnussen, B. M. Ocko, R. R. Adzic, J. X. Wang: Phys. Rev. B **51** (1995) 5510.

6.26   Z. Shi, J. Lipkowski, S. Mirwald, B. Pettinger: J. Electroanal. Chem. **396** (1995) 115.

6.27   J. B. Taylor, I. Langmuir: Phys. Rev. **44** (1933) 423.

6.28   G. Ehrlich: J. Phys. Chem. **60** (1956) 1388.

6.29   H. Steininger, S. Lehwald, H. Ibach: Surf. Sci. **123** (1982) 105.

6.30   P. A. Redhead: Vacuum **12** (1962) 203.

6.31   R. J. Behm, K. Christmann, G. Ertl, M. A. Van Hove: J. Chem. Phys. **73** (1980) 2984.

6.32   Y. Y. Yeo, L. Vattuone, D. A. King: J. Chem. Phys. **106** (1997) 1990.

6.33   R. J. Behm, K. Christmann, G.-. Ertl: Surf. Sci. **99** (1980) 320.

6.34   K. H. Allers, H. Pfnür, D. Menzel: Surf. Sci. **291** (1993) 167.

6.35   J.-S. McEwen, S. H. Payne, H. J. Kreuzer, M. Kinne, R. Denecke, H.-P. Steinrück: Surf. Sci. **545** (2003) 47.

6.36   F. M. Propst, T. C. Piper: J. Vac. Sci. Technol. **4** (1967) 53.

6.37   G. Blyholder: J. Phys. Chem. **68** (1964) 2772.

6.38   H. Ibach, G. A. Somorjai: Appl. Surf. Sci. **3** (1979) 293.

6.39   D. A. Wesner, G. Pirug, F. P. Coenen, H. P. Bonzel: Surf. Sci. **178** (1986) 608.

6.40   G. Pirug, H. P. Bonzel: Surf. Sci. **199** (1988) 371.

6.41   W. Erley, H. Wagner, H. Mach: Surf. Sci. **80** (1979) 612.

6.42   L. D. Mapledoram, M. P. Bessent, A. Wander, D. A. King: Chem. Phys. Lett. **228** (1994) 527.

6.43   M. E. Davila, M. C. Asensio, D. P. Woodruff, K.-M. Schindler, P. Hofmann, K.-U. Weiss, R. Dippel, P. Gardner, V. Fritzsche, A. M. Bradshaw: Surf. Sci. **311** (1994) 337.

6.44   M. Kittel, R. Terborg, M. Polcik, A. M. Bradshaw, R. L. Toomes, D. P. Woodruff, E. Rotenberg: Surf. Sci. **511** (2002) 34.

6.45   K. F. Peters, C. J. Walker, P. Steadman, O. Robach, H. Isem, S. Ferrer: Phys. Rev. Lett. **86** (2001) 5325.
6.46   J. D. Batteas, A. Barbieri, E. K. Starkey, M. A. v. Hove, G. A. Somorjai: Surf. Sci. **313** (1994) 341.
6.47   P. Hofmann, K.-M. Schindler, S. Bao, V. Fritsche, A. M. Bradshaw, D. P. Woodruff: Surf. Sci. **337** (1995) 169.
6.48   N. Materer, A. Barbieri, D. Gardin, U. Starke, J. D. Batteas, M. A. v. Hove, G. A. Somorjai: Surf. Sci. **303** (1994) 319.
6.49   L. D. Mapledoram, A. Wander, D. A. King: Chem. Phys. Lett. **208** (1993) 409.
6.50   J. Schmidt, C. Stuhlmann, H. Ibach: Surf. Sci. **284** (1993) 121.
6.51   S. Andersson, C. Nyberg, C. G. Tengst}}l: Chem. Phys. Lett. **104** (1984) 305.
6.52   J. E. Müller, J. Harris: Phys. Rev. Lett. **53** (1984) 2493.
6.53   J. E. Müller, in H. P. Bonzel, A. M. Bradshaw, G. Ertl (Eds.): *Physics and Chemistry of Alkali Metal Adsorption, Vol. 57*, Elsevier, Amsterdam 1989.
6.54   A. Michaelides, V. A. Ranea, P. L. d. Andres, D. A. King: Phys. Rev. Lett. **90** (2003) 216102.
6.55   G. B. Fisher, J. L. Gland: Surf. Sci. **94** (1980) 446.
6.56   G. B. Fisher, B. A. Sexton: Phys. Rev. Lett. **44** (1980) 683.
6.57   J. W. Döbereiner: Schweigg. J. **39** (1823) 1.
6.58   S. Völkening, K. Bedürftig, K. Jacobi, J. Wintterlin, G. Ertl: Phys. Rev. Lett. **83** (1998) 2672.
6.59   H. Ibach, H. Wagner, D. Bruchmann: Solid State Commun. **42** (1982) 457.
6.60   N. Franco, J. Chrost, J. Avila, M. C. Asensio, C. Muller, E. Dudzik, A. J. Patchett, I. T. McGovern, T. Giebel, R. Lindsay: Appl. Surf. Sci. **123-124** (1998) 219.
6.61   J. E. Demuth, H. Ibach, S. Lehwald: Phys. Rev. Lett. **40** (1978) 1044.
6.62   D. P. Land, W. Erley, H. Ibach: Surf. Sci. **289** (1993) 237.
6.63   W. Widdra, A. Fink, S. Gokhale, P. Trischberger, D. Menzel, U. Birkenheuer, U. Gutdeutsch, N. Rösch: Phys. Rev. Lett. **80** (1998) 4269.
6.64   U. Birkenheuer, U. Gutdeutsch, N. Rosch, A. Fink, S. Gokhale, D. Menzel, P. Trischberger, W. Widdra: J. Chem. Phys. **108** (1998) 9868.
6.65   P. Baumgaertel, R. Lindsay, O. Schaff, T. Giessel, R. Terborg, J. T. Hoeft, M. Polcik, A. M. Bradshaw, M. Carbonne, M. N. Piacastelli, R. Zanoni, R. L. Toomes, D. P. Woodruff: New J. Phys. **1** (1999) 20.
6.66   H. Ibach, S. Lehwald: J. Vac. Sci. Technol. **18** (1981) 625.
6.67   W. Erley, A. M. Baro, H. Ibach: Surf. Sci. **120** (1982) 273.
6.68   S. Bao, P. Hofmann, K.-M. Schindler, V. Fritzsche, A. M. Bradshaw, D. P. Woodruff, C. Casado, M. C. Asensio: Surf. Sci. **323** (1995) 19.
6.69   P. R. Watson, M. A. v. Hove, K. Hermann, National Institute of Standards and Technology Database 42 2001.
6.70   S. Lehwald, H. Ibach, J. E. Demuth: Surf. Sci. **78** (1978) 577.
6.71   O. Schaff, V. Fernandez, P. Hofmann, K.-M. Schindler, A. Theobald, V. Fritzsche, A. M. Bradshaw, R. Davis, D. P. Woodruff: Surf. Sci. **348** (1996) 89.
6.72   A. Barbieri, M. A. Van Hove, G. A. Somorjai: Surf. Sci. **306** (1994) 261.
6.73   J. Kröger, D. Bruchman, S. Lehwald, H. Ibach: Surf. Sci. **449** (2000) 227.
6.74   J. Topping: Proc. Roy. Soc. London A **114** (1927) 67.
6.75   J. E. Müller, K. Dahmen, H. Ibach: Phys. Rev. B **66** (2002) 235407.

6.76   A. T. Loburets, I. F. Lyuksytov, A. G. Naumovets, V. V. Poplavski, Y. S. Vedula, in H. P. Bonzel, A. M. Bradshaw, G. Ertl (Eds.): *Physics and Chemistry of Alkali Metal Adsorption, Vol. 57*, Elsevier, Amsterdam 1989, p. 91.

6.77   M. Ondrejcek, V. Cháb, W. Stenzel, M. Snábl, H. Conrad, A. M. Bradshaw: Surf. Sci. **331-333** (1995) 764.

6.78   H. P. Bonzel, A. M. Bradshaw, G. Ertl, *Materials Science Monographs, Vol. 57*, Elsevier, Amsterdam 1989, p. 486.

6.79   A. Auerbach, K. F. Freed, R. Gomer: J. Chem. Phys. **86** (1987) 2356.

6.80   B. Voigtländer, S. Lehwald, H. Ibach: Surf. Sci. **208** (1989) 113.

6.81   J. Kröger, S. Lehwald, H. Ibach: Phys. Rev. B **55** (1997) 10895.

6.82   M. J. Puska, R. M. Nieminen, M. Mannien, B. Chakraborty, S. Holloway, J. K. Norskov: Phys. Rev. Lett. **51** (1983) 1081.

6.83   M. P. Bessent, P. Hu, A. Wander, D. A. King: Surf. Sci. **325** (1995) 272.

6.84   L. Wenzel, D. Arvanitis, W. Daum, H. H. Rotermund, J. Stoehr, K. Baberschke, H. Ibach: Phys. Rev. B **36** (1987) 7689.

6.85   A. L. D. Kilcoyne, D. P. Woodruff, A. W. Robinson, T. Lindner, J. S. Somers, A. M. Bradshaw: Surf. Sci. **253** (1991) 107.

6.86   J. E. Müller, M. Wuttig, H. Ibach: Phys. Rev. Lett. **56** (1986) 1583.

6.87   H. Ibach: Surf. Sci. Rep. **29** (1997) 195.

6.88   H. C. Zeng, K. A. R. Mitchell: Surf. Sci. **239** (1990) L571.

6.89   M. Wuttig, R. Franchy, H. Ibach: Surf. Sci. **213** (1989) 103.

6.90   T. Lederer, D. Arvanitis, G. Comelli, L. Tröger, K. Baberschke: Phys. Rev. B **48** (1993) 15390.

6.91   M. C. Asensio, M. J. Ashwin, A. L. D. Kilcoyne, D. P. Woodruff, A. W. Robinson, T. Lindner, J. S. Somers, D. E. Ricken, A. M. Bradshaw: Surf. Sci. **236** (1990) 1.

6.92   D. E. Gardin, J. D. Batteas, M. A. Van Hove, G. A. Somorjai: Surf. Sci. **296** (1993) 25.

6.93   C. Klink, I. Stensgaard, F. Besenbacher, E. Laegsgaard: Surf. Sci. **342** (1995) 250.

6.94   M. Foss, R. Feidenhans'l, M. Nielsen, E. Findeisen, R. L. Johnson, T. Buslaps, I. Stensgaard, F. Besenbacher: Phys. Rev. B **50** (1994) 8950.

6.95   H. Ibach, H. D. Bruchmann, H. Wagner: Appl. Phys. A **29** (1982) 113.

6.96   U. Höfer, A. Puschmann, D. Coulman, E. Umbach: Surf. Sci. **211-212** (1989) 948.

6.97   T. Engel: Surf. Sci. Rep. **18** (1993) 93.

## Chapter 7

7.1   M. Born, R. Oppenheimer: Ann. Phys. (Leipzig) **84** (1927) 457.

7.2   R. E. Allen, G. P. Alldredge, F. W. d. Wette: Phys. Rev. B **4** (1971) 1648.

7.3   R. E. Allen, G. P. Alldredge, F. W. d. Wette: Phys. Rev. B **4** (1971) 1661.

7.4   R. E. Allen, G. P. Alldredge, F. W. d. Wette: Phys. Rev. B **4** (1971) 1682.

7.5   W. Daum, C. Stuhlmann, H. Ibach: Phys. Rev. Lett. **60** (1988) 2741.

7.6   T. S. Rahman, D. L. Mills, J. E. Black, J. M. Szeftel, S. Lehwald, H. Ibach: Phys. Rev. B **30** (1984) 589.

7.7   J. A. Gaspar, A. G. Eguiluz: Phys. Rev. B **40** (1989) 11976.

7.8   R. F. Wallis: Prog. Surf. Sci. **4** (1973) 233.

7.9   L. Rayleigh: Proc. London Math. Soc. **17** (1887) 4.

7.10  L. D. Landau, E. M. Lifshitz, *Lehrbuch der Theoretischen Physik, Bd. VII Elasti-zitätstheorie*, Akademie, Berlin 1991.

7.11  D. P. Morgan: International Journal of High Speed Electronics and Systems **10** (2000) 553.

7.12  R. Heid, K.-P. Bohnen: Phys. Rep. **387** (2003) 151.

7.13  P. Giannozzi, S. d. Gironcoli, P. Pavone, S. Baroni: Phys. Rev. B **43** (1991) 7231.

7.14  S. M. Foiles, M. I. Baskes, M. S. Daw: Phys. Rev. B **33** (1986) 7983.

7.15  S. Durukanoglu, T. S. Rahman: Surf. Sci. **409** (1998) 395.

7.16  S. Durukanolu, A. Kara, T. S. Rahman: Phys. Rev. B **67** (2003) 235405.

7.17  A. Kara, S. Durukanoglu, T. S. Rahman: J. Chem. Phys **106** (1997) 2031.

7.18  A. Kara, P. Staikov, T. S. Rahman: Phys. Rev. B **61** (2000) 5714.

7.19  W. Kohn: Phys. Rev. Lett. **2** (1959) 393.

7.20  R. E. Peierls, *Quantum Theory of Solids*, Clarendon Press 1955.

7.21  E. Hulpke, J. Lüdecke: Phys. Rev. Lett. **68** (1992) 2846.

7.22  E. Hulpke, J. Lüdecke: Surf. Sci. **287/288** (1993) 837.

7.23  M. Balden, S. Lehwald, H. Ibach: Phys. Rev. B **53** (1996) 7479.

7.24  J. Kröger, S. Lehwald, H. Ibach: Phys. Rev. B **55** (1997) 10895.

7.25  E. Rotenberg, S. D. Kevan: Phys. Rev. Lett. **80** (1998) 2905.

7.26  J. Kröger, T. Greber, J. Osterwalder: Phys. Rev. B **61** (2000) 14146.

7.27  H. Ibach, in O. Madelung (Ed.): *Festkörperprobleme/Advances in Solid State Physics, Vol. 11*, Vieweg/Pergamon 1971, p. 135.

7.28  R. Fuchs, K. L. Kliewer: Phys. Rev. **140** (1965) A2076.

7.29  H. Ibach: Phys. Rev. Lett. **24** (1970) 1416.

7.30  B. Hall, D. L. Mills, P. Zeppenfeld, K. Kern, U. Becher, G. Comsa: Phys. Rev. B **40** (1989) 6326.

7.31  H. Ibach, D. L. Mills, *Electron Energy Loss Spectroscopy and Surface Vibrations*, Academic Press, New York 1982.

7.32  J. C. Ariyasu, D. L. Mills: Phys. Rev. B **28** (1983) 2389.

7.33  F. M. Propst, T. C. Piper: J. Vac. Sci. Technol. **4** (1967) 53.

7.34  S. Lehwald, H. Ibach: unpublished.

7.35  M. Balden, S. Lehwald, H. Ibach: Phys. Rev. B **46** (1992) 4172.

7.36  M. Balden, S. Lehwald, H. Ibach, A. Ormeci, D. L. Mills: Journal of Electron Spectroscopy and Related Phenomena **64/65** (1993) 739.

7.37  S. Lehwald, H. Ibach, J. E. Demuth: Surf. Sci. **78** (1978) 577.

7.38  G. Brusdeylins, R. B. Doak, J. P. Toennies: Phys. Rev. Lett. **46** (1981) 437.

7.39  A. C. Levi, H. Suhl: Surf. Sci. **88** (1979) 221.

7.40  A. C. Levi: Nuovo Cim. **54** (1979) 357.

7.41  J. Harris, A. Liebsch: Phys. Rev. Lett. **49** (1982) 341.

7.42  J. L. Beeby: J. Phys. C **4** (1971) L359.

7.43  B. Gumhalter: Phys. Rep. **351** (2001) 1.

7.44  A. Siber, B. Gumhalter: Phys. Rev. B **59** (1999) 5898.

7.45  G. Armand: Surf. Sci. **63** (1977) 143.

7.46  H. Ibach, T. S. Rahman, in R. Vanselow, R. Hove (Eds.): *Chemistry and Physics of Solid Surfaces V*, Springer, Berlin, Heidelberg, New York 1984, p. 455.

7.47  R. B. Doak, U. Harten, J. P. Toennies: Phys. Rev. Lett. **51** (1983) 578.

7.48  H. Ibach: Journal of Electron Spectroscopy and Related Phenomena **64/65** (1993) 819.

7.49  R. E. Palmer, P. J. Rous: Rev. Mod. Phys. **64** (1992) 383.

7.50   J. Geiger, *Elektronen und Festkörper*, Fr. Vieweg u. Sohn, Braunschweig 1968.
7.51   H. Raether, in E. A. Niekisch (Ed.): *Springer Tracts in Modern Physics, Vol. 35*, Springer, Berlin, Heidelberg, New York 1965, p. 85.
7.52   J. Daniels, C. v. Festenberg, H. Raether, K. Zeppenfeld, in E. A. Niekisch (Ed.): *Springer Tracts in Modern Physics, Vol. 54*, Springer, Berlin, Heidelberg, New York 1970, p. 77.
7.53   A. A. Lucas, M. Sunjic: Phys. Rev. Lett. **26** (1971) 229.
7.54   E. Evans, D. L. Mills: Phys. Rev. B **5** (1972) 4126.
7.55   B. N. J. Persson, E. Zaremba: Phys. Rev. B **31** (1985) 1863.
7.56   H. Ibach, H. Lüth, *Solid-State Physics - An Introduction to Principles of Materials Science*, Springer-Verlag, Heidelberg 2003.
7.57   M.-L. Xu, B. M. Hall, S. Y. Tong, M. Rocca, H. Ibach, S. Lehwald, J. E. Black: Phys. Rev. Lett. **54** (1985) 1171.
7.58   S. Y. Tong, C. H. Li, D. L. Mills: Phys. Rev. Lett. **44** (1980) 407.
7.59   C. H. Li, S. Y. Tong, D. L. Mills: Phys. Rev. B **21** (1980) 3057.
7.60   R. G. Greenler: J. Chem. Phys. **44** (1966) 310.
7.61   J. D. E. McIntyre, D. E. Aspnes: Surf. Sci. **24** (1971) 417.
7.62   M. A. Chesters, J. Pritchard: Chem. Commun. **1970** 1454.
7.63   A. M. Bradshaw, J. Pritchard: Proc. Roy. Soc. (London) A **316** (1970) 169.
7.64   W. Erley: J. Electr. Spectr. Rel. Phen. **44** (1987) 65.
7.65   E. Kretschmann, H. Raether: Z. Naturforsch. A **23** (1968) 2135.
7.66   M. Osawa, K. Ataka, K. Yoshii, T. Yotsuyanagi: J. Electr. Spectr. Rel. Phen. **64/65** (1993) 371.
7.67   C. J. Hirschmugl, G. P. Williams, F. M. Hoffmann, Y. J. Chabal: Phys. Rev. Lett. **65** (1990) 480.
7.68   Y. J. Chabal: Phys. Rev. Lett. **55** (1985) 845.
7.69   B. N. J. Persson: Phys. Rev. B **44** (1991) 3277.
7.70   B. N. J. Persson, A. I. Volokitin: Surf. Sci. **310** (1994) 314.
7.71   D. Schumacher, *Springer Tracts in Modern Physics, Vol. 128*, Springer, Berlin, Heidelberg, New York 1992, p. 1.
7.72   R. K. Chang, T. E. Furtak, *Surface Enhanced Raman Scattering*, Plenum, New York 1982.
7.73   D. L. Mills, *Nonlinear Optics*, Springer, Berlin, Heidelberg, New York 1998.
7.74   Y. R. Shen: Pure&Applied Chem. **73** (2001) 1589.
7.75   Y. R. Shen, *Nonlinear Spectroscopy for Molecular Structure Determination*, Blackwell Science, Oxford 1998.
7.76   H. Ehrenreich, H. R. Philipp: Phys. Rev. **128** (1962) 1622.
7.77   H. Xu, J. Aizpurua, M. Käll, P. Apell: Phys. Rev. E **62** (2000) 4318.
7.78   D. H. Z. Jeong, Y. X.; Moskovits, M.: J. Phys. Chem. B **108** (2004) 12724.
7.79   I. Villegas, M. J. Weaver: J. Chem. Phys. **101** (1994) 1648.
7.80   M. Kuß, *Thesis Sz8483*, RWTH-Aachen, Aachen 2002.
7.81   P. K. Hansma, *Tunneling spectroscopy*, Plenum, New York 1982.
7.82   B. C. Stipe, M. A. Rezaei, W. Ho: Science **280** (1998) 1732.
7.83   B. C. Stipe, M. A. Rezaei, W. Ho: Phys. Rev. Lett. **82** (1999) 1724.
7.84   L. J. Lauhon, W. Ho: Phys. Rev. B **60** (1999) R8525.
7.85   N. Lorente, M. Persson: Phys. Rev. Lett. **85** (2000) 2997.
7.86   J. Tersoff, D. R. Hamann: Phys. Rev. Lett. **50** (1983) 1998.
7.87   N. Lorente, M. Persson, L. J. Lauhon, W. Ho: Phys. Rev. Lett. **86** (2001) 2593.

## Chapter 8

8.1   F. Sprosser-Prou, A. v. Felde, J. Fink: Phys. Rev. B **40** (1989) 5799.
8.2   P. T. Sprunger, G. M. Watson, E. W. Plummer: Surf. Sci. **269/270** (1992) 551.
8.3   C. Kunz: Z. Phys. **196** (1966) 311.
8.4   K.-D. Tsuei, E. W. Plummer, A. Liebsch, E. Pehlke, K. Kempa, P. Bakshi: Surf. Sci. **247** (1991) 302.
8.5   T. Kloos: Z. Phys. **265** (1973) 225.
8.6   H. Ehrenreich, H. R. Philipp: Phys. Rev. **128** (1962) 1622.
8.7   M. Rocca: Surf. Sci. Rep. **22** (1995) 1.
8.8   C. J. Powell, J. B. Swan: Phys. Rev. **115** (1959) 869.
8.9   J. Daniels, C. v. Festenberg, H. Raether, K. Zeppenfeld, in E. A. Niekisch (Ed.): *Springer Tracts in Modern Physics, Vol. 54*, Springer, Berlin, Heidelberg, New York 1970, p. 77.
8.10  K. D. Tsuei, E. W. Plummer, P. J. Feibelman: Phys. Rev. Lett. **63** (1989) 2256.
8.11  A. J. Bennett: Phys. Rev. B **1** (1970) 203.
8.12  K.-D. Tsuei, E. W. Plummer, A. Liebsch, E. Pehlke, K. Kempa, P. Bakshi: Surf. Sci. **247** (1991) 302.
8.13  A. Liebsch, *Electronic excitations at metal surfaces*, Plenum, New York 1997.
8.14  I. Tamm: Phys. Z. Soviet Union **I** (1932) 733.
8.15  W. Shockley: Phys. Rev. **56** (1939) 317.
8.16  R. Courths, S. Hüfner: Phys. Rep. **112** (1984) 53.
8.17  T. Warwick, P. Heimann, D. Mossessian, W. McKinney, H. Padmore: Rev. Sci. Instr. **66** (1995) 2037.
8.18  J. W. Cooper: Phys. Rev. Lett. **13** (1964) 762.
8.19  V. Dose: J. Appl. Phys. **14** (1977) 117.
8.20  A. Goldmann, M. Donath, W. Altmann, V. Dose: Phys. Rev. B **32** (1985) 837.
8.21  I. L. Shumay, U. Höfer, C. Reuß, U. Thomann, W. Wallauer, T. Fauster: Phys. Rev. B **58** (1998) 13974.
8.22  F. G. Allen, G. W. Gobeli: Phys. Rev. **127** (1962) 150.
8.23  G. Chiarotti, S. Nannarone, R. Pastore, P. Chiaradia: Phys. Rev. B **4** (1971) 3398.
8.24  H. Froitzheim, H. Ibach, D. L. Mills: Phys. Rev. B **11** (1975) 4980.
8.25  P. Krüger, J. Pollmann: Phys. Rev. Lett. **74** (1995) 1155.
8.26  R. I. G. Uhrberg, G. V. Hansson, J. M. Nicholls, S. A. Flodström: Phys. Rev. B **24** (1981) 4684.
8.27  L. S. O. Johansson, R. I. G. Uhrberg, P. Mårtensson, G. V. Hansson: Phys. Rev. B **42** (1990) 1305.
8.28  E. Landemark, C. J. Karlsson, L. S. O. Johansson, R. I. G. Uhrberg: Phys. Rev. B **49** (1994) 16523.
8.29  S. V. Dijken, H. J. W. Zandvliet, B. Poelsema: Surface Review and Letters **5** (1998) 15.
8.30  R. A. Wolkow: Phys. Rev. Lett. **68** (1992) 2636.
8.31  M. Sabisch, P. Krüger, J. Pollmann: Phys. Rev. **51** (1995) 13367.
8.32  H. Froitzheim, H. Ibach: Surf. Sci. **47** (1975) 713.
8.33  R. M. Feenstra: Phys. Rev. B **50** (1994) 4561.
8.34  R. M. Feenstra, J. A. Stroscio, J. Tersoff, A. P. Fein: Phys. Rev. Lett. **58** (1987) 1192.

8.35   P. Ebert, B. Engels, P. Richard, K. Schroeder, S. Blügel, C. Domke, M. Heinrich, K. Urban: Phys. Rev. Lett. **77** (1996) 2997.

8.36   W. Mönch, *Semiconductor Surfaces and Interfaces*, Springer, Berlin Heidelberg New York 1995.

8.37   J. E. Rowe, H. Ibach: Phys. Rev. Lett. **31** (1973) 102.

8.38   J. E. Rowe, H. Ibach: Phys. Rev. Lett. (1974) 421.

8.39   P. M. Echenique, J. B. Pendry: J. Phys. C **11** (1978) 2065.

8.40   N. Memmel: Surf. Sci. Rep. **32** (1998) 91.

8.41   P. Heimann, H. Neddermayer, H. F. Roloff: J. Phys. C **10** (1977) L17.

8.42   S. D. Kevan: Phys. Rev. Lett. **50** (1983) 526.

8.43   F. Reinert, G. Nicolay, S. Schmidt, D. Ehm, S. Hüfner: Phys. Rev. B **63** (2001) 115415.

8.44   R. Paniago, R. Matzdorf, G. Meister, A. Goldmann: Surf. Sci. **336** (1995) 113.

8.45   G. Neuhold, K. Horn: Phys. Rev. Lett. **78** (1997) 1327.

8.46   S. D. Kevan, R. H. Gaylord: Phys. Rev. B **36** (1987) 5809.

8.47   W. Chen, V. Madhavan, T. Jamneala, M. F. Crommie: Phys. Rev. Lett. **80** (1998) 1469.

8.48   A. Liebsch: Phys. Rev. B **17** (1978) 1653.

8.49   A. L. D. Kilcoyne, D. P. Woodruff, J. E. Rowe, R. H. Gaylord: Phys. Rev. **39** (1989) 12604.

8.50   S. Lizzit, A. Baraldi, A. Groso, K. Reuter, M. V. Ganduglia-Pirovano, C. Stampfl, M. Scheffler, M. Stichler, C. Keller, W. Wurth, D. Menzel: Phys. Rev. B **63** (2001) 205419.

8.51   D. P. Woodruff: Journal of Electron Spectroscopy and Related Phenomena **126** (2002) 55.

8.52   R. C. Jaklevic, J. Lambe, M. Mikkor, W. C. Vassell: Phys. Rev. Lett. **26** (1971) 88.

8.53   R. C. Jaklevic, J. Lambe: Phys. Rev. B **12** (1975) 4146.

8.54   B. T. Jonker, R. L. Park: Surf. Sci. **127** (1983) 183.

8.55   H. Iwasaki, B. T. Jonker, R. L. Park: Phys. Rev. B **32** (1985) 643.

8.56   P. D. Loly, J. B. Pendry: J. Phys. C **16** (1983) 423.

8.57   A. L. Wachs, A. P. Shapiro, T. C. Hsieh, T. C. Chiang: Phys. Rev. B **33** (1986) 1460.

8.58   R. Otero, A. L. VázquezdeParga, R. Miranda: Phys. Rev. B **66** (2002) 115401.

8.59   H. Okamoto, D. Chen, T. Yamada: Phys. Rev. Lett. **89** (2002) 256101.

8.60   P. Bruno: Phys. Rev. B **52** (1995) 411.

8.61   G. Neuhold, K. Horn: Phys. Rev. Lett. **78** (1997) 1327.

8.62   T.-C. Chiang: Surf. Sci. Rep. **39** (2000) 181.

8.63   D. A. Luh, T. Miller, J. J. Paggel, M. Y. Chou, T. C. Chiang: Science **292** (2000) 1131.

8.64   Y. Hasegawa, P. Avouris: Phys. Rev. Lett. **71** (1993) 1071.

8.65   M. F. Crommie, C. P. Lutz, D. M. Eigler: Science **262** (1993) 218.

8.66   J. Li, W.-D. Schneider, R. Berndt: Phys. Rev. B **56** (1997) 7656.

8.67   N. Garcia, P. A. Serena: Surf. Sci. **330** (1995) L665.

8.68   C. Steimer, M. Giesen, L. Verheij, H. Ibach: Phys. Rev. B **64** (2001) 085416.

8.69   J. M. G. O. Sánchez, P. Segovia, J. Alvarez, A. L. Vázquez de Parga, J. E. Ortega, M. Prietsch, and R. Miranda: Phys. Rev. B **52** (1995) 7894.

8.70   T. B. Grimley: Proc. Phys. Soc. **90** (1967) 751.

8.71   T. L. Einstein, J. R. Schrieffer: Phys. Rev. B **7** (1973) 3629.

8.72   K. H. Lau, W. Kohn: Surf. Sci. **75** (1978) 69.

8.73   P. Hyldgaard, M. Persson: J. Phys.: Cond. Mat. **12** (2000) L13.
8.74   J. Repp, F. Moresco, G. Meyer, K.-H. Rieder, P. Hyldgaard, M. Persson: Phys. Rev. Lett. **85** (2000) 2981.
8.75   G. Bergmann: Phys. Rep. **107** (1984) 1.
8.76   B. N. J. Persson: Phys. Rev. B **44** (1991) 3277.
8.77   B. N. J. Persson: Surf. Sci. **269-270** (1992) 103.
8.78   P. Wißmann, *Springer Tracts in Modern Physics, Vol. 77*, Springer, Berlin, Heidelberg, New York 1975, p. 1.
8.79   D.Schumacher, *Springer Tracts in Modern Physics, Vol. 128*, Springer, Berlin, Heidelberg, New York 1992, p. 1.
8.80   H. Winkes, D. Schumacher, A. Otto: Surf. Sci. **400** (1998) 44.
8.81   B. M. Ocko, J. X. Wang, T. Wandlowski: Phys. Rev. Lett. **79** (1997) 1511.
8.82   K. Fuchs: Proc. Camb. Phil. Soc. **34** (1938) 100.
8.83   E. H. Sondheimer: Adv. Phys. **1** (1952) 1.
8.84   R. B. Dingle: Proc. Roy. Soc. A **176** (1950) 545.
8.85   H. Ibach, H. Lüth, *Solid-State Physics - An Introduction to Principles of Materials Science*, Springer-Verlag, Heidelberg 2003.
8.86   J. R. Schrieffer, in R. H. Kingston (Ed.): *Semiconductor Surface Physics*, University of Pennsylvania Press, Philadelphia 1957, p. 68.
8.87   W. Mönch, *Semiconductor Surfaces and Interfaces*, Springer, Berlin Heidelberg New York 1995.
8.88   R. F. Greene: Crit. Rev. Sol. State Sci. **4** (1974) 477.
8.89   H. Lüth, *Surfaces and Interfaces of Solid Materials*, Springer, Berlin, Heidelberg, New York 2001.
8.90   T. C. Sah, T. H. Ning, L. L. Tschopp: Surf. Sci. **32** (1971) 561.
8.91   K. R. Roos, K. L. Roos, M. H.-v. Hoegen, F.-J. M. z. Heringdorf: J. Phys. Cond. Mat. **17** (2005) 1407.
8.92   M. Brand, M. Hartmann, C. Hassel, G.Dumpich: J. Phys. Cond. Mat**???** (2006).
8.93   N. Agrait, A. L. Yeyati, J. M. van Ruitenbeek: Phys. Rep. **377** (2003) 81.
8.94   K. Hansen, E. Laegsgaard, I. Stensgaard, F. Besenbacher: Phys. Rev. B **56** (1997) 2208.
8.95   L. Limot, J. Kröger, R. Berndt, A. Garcia-Lekue, W. A. Hofer: Phys. Rev. Lett. **94** (2005) 126102.

## Chapter 9

9.1    A. S. Arrot, in J. A. C. Bland, B. Heinrich (Eds.): *Ultrathin Magnetic Structures, Vol. I*, Springer, Berlin, Heidelberg, New York 1994, p. 7.
9.2    K. Dahmen, H. Ibach, D. Sander: J. Magn. Magn. Mat. **231** (2001) 74.
9.3    D. Sander: Rep. Prog. Phys. **62** (1999) 809.
9.4    K. Baberschke: Appl. Phys. **62** (1996) 417.
9.5    M. Farle, W. Platow, A. N. Anisimov, P. Poulopoulos, K. Baberschke: Phys. Rev. B **56** (1997) 5100.
9.6    R. Bergholz, U. Gradmann: J. Magn. Magn. Mater. **45** (1984) 389.
9.7    M. Grimsditch, P. Vavassori: J. Phys.Cond. Mat. **16** (2004) R275.
9.8    D. T. Pierce, F. Meier, P. Zurcher: Appl. Phys. Lett. **26** (1975) 670.
9.9    P. Drescher, e. al.: Appl. Phys. A **63** (1996) 203.

9.10  H. P. Oepen, H. Hopster, in H. Hopster, H. P. Oepen (Eds.): *Magnetic Microscopy of Nanostructures*, Springer, Berlin, Heidelberg, New York 2005, p. 137.

9.11  W. Kuch, in H. Hopster, H. P. Oepen (Eds.): *Magnetic Microscopy of Nanostructures*, Springer, Berlin, Heidelberg, New York 2005, p. 1.

9.12  A. Scholl, H. Ohldag, F. Nolting, S. Anders, J. Stöhr, in H. Hopster, H. P. Oepen (Eds.): *Magnetic Microscopy of Nanostructures*, Springer, Berlin, Heidelberg, New York 2005, p. 29.

9.13  M. Bode, R. Wiesendanger, in H. Hopster, H. P. Oepen (Eds.): *Magnetic Microscopy of Nanostructures*, Springer, Berlin, Heidelberg, New York 2005, p. 203.

9.14  W. Wulfhekel, in H. Hopster, H. P. Oepen (Eds.): *Magnetic Microscopy of Nanostructures*, Springer, Berlin, Heidelberg, New York 2005, p. 181.

9.15  L. Abelmann, A. v. d. Bos, C. Lodder, in H. Hopster, H. P. Oepen (Eds.): *Magnetic Microscopy of Nanostructures*, Springer, Berlin, Heidelberg, New York 2005, p. 254.

9.16  L. Néel: J. Phys. Rad. (1954) 225.

9.17  H. J. Elmers, J. Hauschild, U. Gradmann: Phys. Rev. B **59** (1199) 3688.

9.18  T. Duden, E. Bauer: Proc. Mat. Res. Soc. **475** (1997) 283.

9.19  H. P. Oepen, M. Speckmann, Y. Millev, J. Kirschner: Phys. Rev. B **55** (1997) 2752.

9.20  N. D. Mermin, H. Wagner: Phys. Rev. Lett. **17** (1966) 1133.

9.21  R. P. Erickson, D. L. Mills: Phys. Rev. B **43** (1991) 11527.

9.22  M. Bander, D. L. Mills: Phys. Rev. B **38** (1988) 12015.

9.23  H. Ibach, H. Lüth, *Solid-State Physics - An Introduction to Principles of Materials Science*, Springer-Verlag, Heidelberg 2003.

9.24  J. W. Lynn, H. A. Mook: Phys. Rev. B **23** (1981) 198.

9.25  H. Capellmann, V. Vieira: Solid State Commun. **43** (1982) 747.

9.26  H. Capellmann, V. Viera: Phys. Rev. B **25** (1982) 3333.

9.27  S. F. Alvarado, M. Campagna, F. Ciccacci, H. Hopster: J. Appl .Phys. **53** (1982) 7920.

9.28  D. L. Abraham, H. Hopster: Phys. Rev. Lett. **58** (1987) 1352.

9.29  A. Berger, U. Linke, H. P. Oepen: Phys. Rev. Lett. **68** (1992) 839.

9.30  A. Berger, H. P. Oepen: Phys. Rev. B **45** (1992) 12596.

9.31  W. H. Meiklejohn, C. P. Bean: Phys. Rev. **102** (1956) 1413.

9.32  M. N. Baibich, J. M. Broto, A. Fert, F. N. V. Dau, F. Petroff, P. Eitenne, G. Creuzet, A. Friederich, J. Chazelas: Phys. Rev. Lett. **61** (1988) 2475.

9.33  G. Binasch, P. Grünberg, F. Saurenbach, W. Zinn: Phys. Rev. B **39** (1989) 4828.

9.34  A. Paul, M. Buchmeier, D. E. Bürgler, P. Grünberg: J. Magn. Magn. Mater. **286** (2005) 258.

9.35  P. Grünberg, R. Schreiber, Y. Pang, M. B. Brodsky, H. Sowers: Phys. Rev. Lett. **57** (1986) 1986.

9.36  S. S. P. Parkin, N. More, K. P. Roche: Phys. Rev. Lett. **64** (1990) 2304.

9.37  S. S. P. Parkin: Phys. Rev. Lett. **67** (1991) 3598.

9.38  J. Unguris, R. J. Celotta, D. T. Pierce: Phys. Rev. Lett. **67** (1991) 140.

9.39  R. K. Kawakami, E. Rotenberg, E. J. Escorcia-Aparicio, H. J. Choi, T. R. Cummins, J. G. Tobin, N. V. Smith, Z. Q. Qiu: Phys. Rev. Lett. **80** (1998) 1754.

9.40  P. Bruno: Phys. Rev. B **52** (1995) 411.

9.41  H. A. Mook, D. M. Paul: Phys. Rev. Lett. **54** (1985) 227.

9.42  R. P. Erickson, D. L. Mills: Phys. Rev. B **46** (1992) 861.

9.43  R. Arias, D. L. Mills: Phys. Rev. B **60** (1999) 7395.

9.44  M. P. Gokhale, A. Ormeci, D. L. Mills: Phys. Rev. B **46** (1992) 8978.

9.45  H. Ibach, D. Bruchmann, R. Vollmer, M. Etzkorn, P. S. A. Kumar, J. Kirschner: Rev. Sci. Instrum. **74** (2003) 4089.

9.46   R. Vollmer, M. Etzkorn, P. S. A. Kumar, H. Ibach, J. Kirschner: Phys. Rev. Lett. **91** (2003) 147201.
9.47   D. L. Mills, in V. M. Aranovich, R. Loudon (Eds.): *Surface Excitations*, Elsevier, New York 1984, p. 379.
9.48   A. T. Costa, R. B. Muniz, D. L. Mills: Phys. Rev. B **69** (2004) 64413.
9.49   A. T. Costa, R. B. Muniz, D. L. Mills: Phys. Rev. B **70** (2004) 54406.

# Chapter 10

10.1   E. W. Müller: Z. Phys. **131** (1951) 136.
10.2   E. W. Müller: Z. Electrochem. **61** (1957) 43.
10.3   G. Ehrlich, F. G. Hudda: J. Chem. Phys. **44** (1966) 1039.
10.4   G. L. Kellogg: Surf. Sci. Rep. **21** (1994) 1.
10.5   G. L. Kellogg, A. F. Wright, M. S. Daw: J. Vac. Sci. Technol. A **9** (1991) 1757.
10.6   G. Ayrault, G. Ehrlich: J. Chem. Phys. **60** (1974) 281.
10.7   C. L. Chen, T. T. Tsong: Phys. Rev. Lett. **64** (1990) 3147.
10.8   G. L. Kellogg, P. J. Feibelman: Phys. Rev. Lett. **64** (1990) 3143.
10.9   C. Klünker, J. B. Hannon, M. Giesen, H. Ibach, G. Boisvert, L. J. Lewis: Phys. Rev. B **58** (1998) R7556.
10.10  U. Kürpick: Phys. Rev. B **64** (2001) 075418.
10.11  M. Giesen, G. Beltramo, J. Müller, H. Ibach, W. Schmickler: Surf. Sci. **595** (2005) 127.
10.12  M. Giesen: Prog. Surf. Sci. **68** (2001) 1.
10.13  D. C. Senft, G. Ehrlich: Phys. Rev. Lett. **74** (1995) 294.
10.14  S. Glasstone, K. J. Laidler, H. Eyring, *The Theory of Rate Processes*, McGraw-Hill, New York 1941.
10.15  Z. P. Shi, Z. Zhang, A. K. Swan, J. F. Wendelken: Phys. Rev. Lett. **76** (1996) 4927.
10.16  G. L. Kellogg, A. F. Voter: Phys. Rev. Lett. **67** (1991) 622.
10.17  J. C. Hamilton, M. S. Daw, S. M. Foiles: Phys. Rev. Lett. **74** (1995) 2760.
10.18  J. C. Hamilton: Phys. Rev. Lett. **77** (1996) 886.
10.19  K. Morgenstern, G. Rosenfeld, B. Poelsema, G. Comsa: Phys. Rev. Lett. **74** (1995) 2058.
10.20  S. V. Khare, T. L. Einstein: Phys. Rev. B **57** (1998) 4782.
10.21  D. C. Schlößer, K. Morgenstern, L. K. Verheij, G. Rosenfeld, F. Besenbacher, G. Comsa: Surf. Sci. **465** (2000) 19.
10.22  R. Gomer: Rep. Prog. Phys. **53** (1990) 917.
10.23  W. W. Mullins: J. Appl. Phys. **30** (1959) 77.
10.24  M. Giesen, G. Schulze Icking-Konert: Surf. Sci. **412/413** (1998) 645.
10.25  J. B. Ikonomov, *Mathematisch-Naturwissenschaftliche Fakultät*, Heirich-Heine-Universität, Düsseldorf 2003.
10.26  A. Pimpinelli, J. Villain, *Physics of Crystal Growth*, Cambridge University Press, Cambridge 1997.
10.27  R. J. Asaro, W. A. Tiller: Metall. Trans. **3** (1972) 1789.

10.28  M. A. Grinfeld: Sov. Phys. Dokl. **31** (1986) 831.

10.29  D. E. Jesson, S. J. Pennycook, J.-M. Baribeau, D. C. Houghton: Phys. Rev. Lett. **71** (1993) 1744.

10.30  W. H. Yang, D. J. Srolovitz: Phys. Rev. Lett. **71** (1993) 1593.

10.31  E. A. Brener, R. Spatschek: Phys. Rev. E **67** (2003) 16112.

10.32  R. L. Schwoebel, E. J. Shipsey: J. Appl. Phys. **37** (1966) 3682.

10.33  K. Mussawisade, T. Wichmann, K. W. Kehr: Surf. Sci. **412/413** (1998) 55.

10.34  M. Villarba, H. Jónsson: Surf. Sci. **317** (1994) 15.

10.35  P. J. Feibelman: Phys. Rev. Lett. **81** (1998) 168.

10.36  J. Merikoski, I. Vattulainen, J. Heinonen, T. Ala-Nissila: Surf. Sci. **387** (1997) 167.

10.37  Y. Li, A. E. DePristo: Surf. Sci. **319** (1994) 141.

10.38  U. Kürpick: Phys. Rev. B **69** (2004) 205410.

10.39  W. Ostwald: Z. Phys. Chem. (Leipzig) **34** (1900) 495.

10.40  I. M. Lifshitz, Y. V. Slyozov: J. Phys. Chem. Solids **19** (1961) 35.

10.41  M. Zinke-Allmang, L. C. Feldman, M. H. Grabow: Surf. Sci. Rep. **16** (1992) 377.

10.42  K. Morgenstern, G. Rosenfeld, G. Comsa: Phys. Rev. Lett. **76** (1996) 2113.

10.43  G. Schulze Icking-Konert: PhD Thesis, University of Aachen D82, Jül-Report 3588, Forschungszentrum Jülich (1998).

10.44  C. Steimer, M. Giesen, L. Verheij, H. Ibach: Phys. Rev. B **64** (2001) 085416.

10.45  G. Schulze Icking-Konert, M. Giesen, H. Ibach: Surf. Sci. **398** (1998) 37.

10.46  M. Giesen, H. Ibach: Surf. Sci. **431** (1999) 109.

10.47  M. Giesen, G. Schulze Icking-Konert, H. Ibach: Phys. Rev. Lett. **80** (1998) 552.

10.48  M. Giesen, H. Ibach: Surf. Sci. **464** (2000) L 697.

10.49  P. J. Feibelman: Surf. Sci. **478** (2001) L349.

10.50  K. Morgenstern, G. Rosenfeld, G. Comsa, M. R. Sørensen, B. Hammer, E. Lægsgaard, F. Besenbacher: Phys. Rev. B **63** (2001) 045412.

10.51  M. S. Zei, G. Ertl: Surf. Sci. **442** (1999) 19.

10.52  J. M. Doña, J. González-Velasco: Surf. Sci. **274** (1992) 205.

10.53  M. Hidalgo, M. L. Marcos, J. Gonzalez-Velasco: Appl. Phys. Lett. **67** (1995) 1486.

10.54  J. J. MartinezJubrias, M. Hidalgo, M. L. Marcos, J. Gonzalez-Velasco: Surf. Sci. **366** (1996) 239.

10.55  N. Hirai, H. Tanaka, S. Hara: Appl. Surf. Sci. **130-132** (1998) 505.

10.56  N. Hirai, K.-I. Watanabe, S. Hara: Surf. Sci. **493** (2001) 568.

10.57  K. Kubo, N. Hirai, T. Tanaka, S. Hara: Surf. Sci. **565** (2004) L271.

10.58  M. Giesen, M. Dietterle, D. Stapel, H. Ibach, D. M. Kolb: Proc. Mat. Res. Soc. **451** (1997) 9.

10.59  M. Giesen, M. Dietterle, D. Stapel, H. Ibach, D. M. Kolb: Surf. Sci. **384** (1997) 168.

10.60  M. Giesen, R. Randler, S. Baier, H. Ibach, D. M. Kolb: Electrochim. Acta **45** (1999) 527.

10.61  M. Giesen, D. M. Kolb: Surf. Sci. **468** (2000) 149.

10.62  M. Dietterle, T. Will, D. M. Kolb: Surf. Sci. **327** (1995) L495.

10.63  M. Giesen-Seibert, H. Ibach: Surf. Sci. **316** (1994) 205.

10.64  M. Giesen-Seibert, R. Jentjens, M. Poensgen, H. Ibach: Phys. Rev. Lett. **71** (1993) 3521.

10.65  B. Blagojevic, P. M. Duxbury, in P. M. Duxbury, T. J.-. Pence (Eds.): *Dynamics of Crystal Surfaces and Interfaces*, Plenum, New York 1997.
10.66  A. Pimpinelli, J. Villain, D. E. Wolf, J. J. Métois, J. C. Heyraud, I. Elkinani, G. Uimin: Surf. Sci. **295** (1993) 143.
10.67  M. Poensgen, J. F. Wolf, J. Frohn, M. Giesen, H. Ibach: Surf. Sci. **274** (1992) 430.
10.68  M. Giesen, S. Baier: J. Phys.: Condens. Matter **13** (2001) 5009.
10.69  J. M. Bermond, J. J. Metois, J. C. Heyraud, C. Alfonso: Surf. Sci. **331** (1995) 855.
10.70  M. Giesen-Seibert, F. Schmitz, R. Jentjens, H. Ibach: Surf. Sci. **329** (1995) 47.
10.71  F. Nita, A. Pimpinelli: Phys. Rev. Lett. **95** (2005) 106104.

# Chapter 11

11.1   D. Walton: J. Chem. Phys. **37** (1962) 53.
11.2   J. Wintterlin, R. Schuster, G. Ertl: Phys. Rev. Lett. **77** (1996) 123.
11.3   D. E. Sanders, A. E. DePristo: Surf. Sci. **254** (1991) 341.
11.4   M. Kalff, P. Smilauer, G. Comsa, T. Michely: Surf. Sci. **426** (1999) L447.
11.5   J. Krug: J. Stat. Phys. **87** (1997) 505.
11.6   B. Poelsema, G. Comsa, *Scattering of Thermal Energy Atoms*, Springer, Berlin, Heidelberg, New York 1989.
11.7   R. Kunkel, B. Poelsema, L. K. Verheij, G. Comsa: Phys. Rev. Lett. **65** (1990) 733.
11.8   M. C. Bartelt, J. W. Evans: Surf. Sci. **314** (1994) L835.
11.9   K. J. Caspersen, J. W. Evans: Phys. Rev. B **64** (2001) 075401.
11.10  G. M. White: J. Chem. Phys. **50** (1969) 4672.
11.11  J. E. McDonald: Am. J. Phys. **30** (1962) 870.
11.12  J. E. McDonald: J. Am. Phys. **31** (1963) 311.
11.13  Y. B. Zeldovich: J. Exp. Theoret. Phys. USSR **12** (1942) 525.
11.14  E. Budevski, G. Staikov, W. J. Lorenz, *Electrochemical Phase Formation and Growth*, VCH, Weinheim, New York 1996.
11.15  M. Giesen, M. Dietterle, D. Stapel, H. Ibach, D. M. Kolb: Proc. Mat. Res. Soc. **451** (1997) 9.
11.16  M. Giesen, M. Dietterle, D. Stapel, H. Ibach, D. M. Kolb: Surf. Sci. **384** (1997) 168.
11.17  K. Morgenstern, G. Rosenfeld, E. Laegsgaard, F. Besenbacher, G. Comsa: Phys. Rev. Lett. **80** (1998) 556.
11.18  C. Steimer, M. Giesen, L. Verheij, H. Ibach: Phys. Rev. B **64** (2001) 085416.
11.19  E. Budevski, V. Bostanov, G. Staikov: Ann. Rev. Mater. Sci. **10** (1980) 85.
11.20  E. Budevski, V. Bustanov, T. Vitanov, Z. Stoynov, A. Kotzeva, R. Kaischev: Electrochim. Acta **11** (1966) 1697.
11.21  W. Obretenov, V. Bostanov, V. Popov: J. Electroanal. Chem. **132** (1982) 273.
11.22  V. Bostanov, W. Obretenov, G. Staikov, D. Roe, E. Budevski: J. Cryst. Growth **52** (1981) 761.
11.23  H. Brune, K. Kern, in D. A. King, D. P. Woodruff (Eds.): *The Chemical Physics of Solid Surfaces and Heterogenous Catalysis, Vol. 8*, Elsevier 1997, p. 149.
11.24  J. A. Meyer, P. Schmid, R. J. Behm: Phys. Rev. Lett. **74** (1995) 3864.
11.25  H. Brune, K. Bromann, H. Röder, K. Kern: Phys. Rev. B **52** (1995) R14380.
11.26  C. Ratsch, A. P. Seitsonen, M. Scheffler: Phys. Rev. B **55** (1997) 6750.
11.27  H. Brune, K. Kern: Nature **369** (1994) 469.

11.28  M. Sulzberger, *Fakultät für Mathematik, Informatik und Naturwissenschaften*, RWTH-Aachen D82, Aachen 2004.

11.29  P. Raiteri, D. B. Migas, L. Miglio, A. Rastelli, H. v. Känel: Phys. Rev. Lett. **88** (2002) 256103.

11.30  M. Kästner, B. Voigtländer: Phys. Rev. Lett. **82** (1999) 2745.

11.31  B. Voigtländer: Surf. Sci. Rep. **43** (2001) 127.

11.32  I. Daruka, J. Tersoff, A.-L. Barabási: Phys. Rev. Lett. **82** (1999) 2753.

11.33  J. Tersoff, R. M. Tromp: Phys. Rev. Lett. **70** (1993) 2782.

11.34  F. C. Frank: Disc. Faraday Soc. **48** (1949) 67.

11.35  W. K. Burton, N. Cabrera, F. C. Frank: Phil. Trans. Roy. Soc. London Ser. A **243** (1951) 299.

11.36  F. C. Frank: Adv. Phys. **1** (1952) 91.

11.37  G. S. Bales, A. Zangwill: Phys. Rev. B **41** (1990) 5500.

11.38  L. Schwenger, R. L. Folkerts, H.-J. Ernst: Phys. Rev. B **55** (1997) R7406.

11.39  T. Maroutian, L. Douillard, H.-J. Ernst: Phys. Rev. Lett. **83** (1999) 4353.

11.40  T. Maroutian, L. Douillard, H.-J. Ernst: Phys. Rev. B **64** (2001) 165401.

11.41  O. Pierre-Louis, C. Misbah, Y. Saito, J. Krug, P. Politi: Phys. Rev. Lett. **80** (1998) 4221.

11.42  O. Pierre-Louis, M. R. D'Orsogna, T. L. Einstein: Phys. Rev. Lett. **82** (1999) 3661.

11.43  H. Emmerich: Phys. Rev. B **65** (2002) 233406.

11.44  J. Kallunki, J. Krug, M. Kotrla: Phys. Rev. B **65** (2002) 205411.

11.45  G. Schulze Icking-Konert: PhD Thesis, University of Aachen D82, Jül-Report 3588, Forschungszentrum Jülich (1998).

11.46  T. Michely, J. Krug, *Islands, Mounds and Atoms, Vol. 42*, Springer, Berlin Heidelberg 2004.

# Subject Index

# List of Common Acronyms

2PPE (*Two-Photon PhotoEmission*)

AES (*Auger Electron Spectroscopy*)
AFM (*Atomic Force Microscope*)
ASOS-model (*Absolute (height difference) Solid On Solid*)
ATG-instability (*Asaro-Tiller-Grinfeld*)
ATR (*Attenuated Total Reflection*)

CMA (*Cylindrical Mirror Analyzer*)
CVD (*Chemical Vapor Deposition*)

DAS-model (*Dimer Adatom Stacking fault*)
DGSOS-model (*Discrete Gaussian Solid On Solid*)

EAM (*Embedded Atom Model*)
EELS (*Electron Energy Loss Spectroscopy*)
EMT (*Effective Medium Theory*)
ESCA (*Electron Spectroscopy for Chemical Analysis* )

FIM (*Field Ion Microscope*)
FMR (*Ferro Magnetic Resonance*)
FWHM (*Full Width at Half Maximum*)
FTIR (*Fourier-Transform InfraRed*)

GIXRD (*Grazing Incidence X-Ray Diffraction*)
GMR (*Giant Magneto Resistance*)

HAS (*Helium Atom Scattering*)
HOMO (*Highest Occupied Molecular Orbital*)
HOPG (*Highly Ordered Pyrolytic Graphite*)
HREELS (*High Resolution Electron Energy Loss Spectroscopy*)

KESE (*Kink Ehrlich-Schwoebel Effect*)

LDA (*Local Density Approximation*)
LEED (*Low Energy Electron Diffraction*)
LEEM (*Low Energy Electron Microscopy*)
LPE (*Liquid Phase Epitaxy*)
LUMO (*Lowest Unoccupied Molecular Orbital*)

MBE (*Molecular Beam Epitaxy*)
MCXD (*Magnetic Circular X-ray Dichroism*)

MEED (*Medium Energy Electron Diffraction*)
MFM (*Tunnel Magneto Resistance*)
MIGS (*Metal Induced Gap States*)
MLXD (*Magnetic Linear X-ray Dichroism*)
MOKE (*Magneto-Optic Kerr Effect*)

PED (*Photo Electron Diffraction*)
PEEM (*Photo Emission Electron Microscope*)
pzc (*potential of zero charge*)

REM (*Reflection Electron Microscopy*)
RHEED (*Reflection High Energy Electron Diffraction*)
RHE (*UnderPotential Deposited*)
RIARS (*Reflection Absorption InfraRed Spectroscopy*)
RPA (*Random Phase Approximation*)

SACP (*Selected Area Channeling Patterns*)
SAW (*Surface Acoustic Waves*)
SBZ (*Surface Brillouin Zone*)
SCE (*Saturated Calomel Electrode*)
SEM (*Scanning Electron Microscopy*)
SEMPA (*Scanning Electron Microscopy with Polarization Analysis*)
SERS (*Surface Enhanced Raman Spectroscopy*)
SEXAFS (*Surface Extended X-ray Absorption Fine Structure*)
SFG (*Sum Frequency Generation*)
SPALEED (*Spot Profile Analysis Low Energy Electron Diffraction*)
SPLEED (*Spin Polarized Low Energy Electron Diffraction*)
SPLEEM (*Spin Polarized Low Energy Electron Microscopy*)
SPEELS (*Spin Polarized Electron Energy Loss Spectroscopy*)
SPSTM (*Spin Polarized Scanning Tunneling Microscope*)
STM (*Scanning Tunneling Microscopy*)
STS (*Scanning Tunneling Spectroscopy*)
SXRD (*Surface X-Ray Diffraction*)

TDS (*Thermal Desorption Spectroscopy*)
TMR (*Tunnel Magneto Resistance*)
TOM (*Torsion Oscillation Magnetometry*)
TPD (*Temperature Programmed Desorption*)
TSK-model (*Terrace Step Kink*)

UHV (*Ultra-High Vacuum*)
UPD (*UnderPotential Deposition*)
UPS (*Ultraviolet Photoemission Spectroscopy*)

VSFG (*Vibration Sum Frequency Spectroscopy*)

XHV (e*Xtreme High Vacuum*)
XPS (*X-ray Photoemission Spectroscopy* )